CONTEMPORARY ENDOCRINOLOGY

Series Editor:
P. Michael Conn, PhD
Oregon Health & Science University
Beaverton, OR, USA

For further volumes:
http://www.springer.com/series/7680

Evanthia Diamanti-Kandarakis • Andrea C. Gore
Editors

Endocrine Disruptors and Puberty

 Humana Press

Editors
Evanthia Diamanti-Kandarakis M.D., Ph.D.
Third Department of Medicine
Sotiria Hospital
Medical School University of Athens
Athens, 14578, Greece
akandara@otenet.gr

Andrea C. Gore Ph.D.
Gustavus and Louise Pfeiffer Professor
Pharmacology & Toxicology
College of Pharmacy
University of Texas at Austin
Austin, TX 78712, USA
andrea.gore@austin.utexas.edu

ISBN 978-1-60761-560-6   e-ISBN 978-1-60761-561-3
DOI 10.1007/978-1-60761-561-3
Springer New York Dordrecht Heidelberg London

Library of Congress Control Number: 2011941610

© Springer Science+Business Media, LLC 2012
All rights reserved. This work may not be translated or copied in whole or in part without the written
permission of the publisher (Humana Press, c/o Springer Science+Business Media, LLC, 233 Spring Street,
New York, NY 10013, USA), except for brief excerpts in connection with reviews or scholarly analysis.
Use in connection with any form of information storage and retrieval, electronic adaptation, computer
software, or by similar or dissimilar methodology now known or hereafter developed is forbidden.
The use in this publication of trade names, trademarks, service marks, and similar terms, even if they
are not identified as such, is not to be taken as an expression of opinion as to whether or not they are
subject to proprietary rights.
While the advice and information in this book are believed to be true and accurate at the date of going to
press, neither the authors nor the editors nor the publisher can accept any legal responsibility for any
errors or omissions that may be made. The publisher makes no warranty, express or implied, with respect
to the material contained herein.

Printed on acid-free paper

Humana Press is part of Springer Science+Business Media (www.springer.com)

*To Artemis, Eleni, Stelios and Anna,
who keep teaching me that love
is above all!*

*To my parents for their inspiration
and support and to all my patients
who trust in my care.*

EDK

*To David, Rachel, Sarah and Isaac
with all my love*

ACG

Preface

As the births of living creatures are ill-shapen, so are all innovations, which are the births of time.

Sir Francis Bacon (On Innovation, ' Essays: 1597)

Mounting evidence stemming from over 10 years of experimental , epidemiological, as well as clinical studies has transformed the once generally discounted subject of endocrine disruptors into, currently, an issue of great concern, not only within the scientific community but among society as a whole. Thus, from initially constituting the domain of purely experimental basic research, it has now emerged as a major medical challenge. Today, it is one of especial pertinence to health specialists in light of the increasing confirmation of environmental endocrine disruption in humans.

Equally significantly, it has been established that age at exposure represents an important parameter critically determining the clinical consequences of such exposure. In this respect, puberty, a crucial developmental stage, has been definitively identified as a key window of vulnerability with regard to EDs.

For these reasons, Dr. Michael Conn's proposal that I should undertake the editing of a volume titled *Endocrine Disruptors and Puberty*, comprising a component of a thematic series reflecting the foremost advances in present-day Endocrinology, was of immense interest to me. This is an issue largely absent from the field of scientific literature, an omission that certainly needed remedying. Exposure to endocrine disruptors at puberty could, after all, alter the trajectory of the rest of an individual's life cycle, and potentially those of their descendants. At this point I should like to express my warm appreciation to Dr. Conn both for initiating this most essential and worthy project and for his consideration of me as editor of this volume.

"In order to integrate the growing body of clinical evidence from data in humans with that of experimental scientific knowledge , I was delighted when Dr Andrea Gore, an authority on the basic research in this field, agreed to participate in this project as co-editor."

Working closely together with the authors furnishing the chapters of this volume, all well known in the field, we have made every effort to provide an up-to-date overview of this issue.

The main objective of this endeavor will have been achieved if the reader finds in the pages of this book valuable and stimulating facts and insights, as we indeed have done throughout its preparation.

Evanthia Diamanti-Kandarakis, M.D., Ph.D., Senior Editor

With special appreciation to Andrea Gore Ph.D., Co-editor

Acknowledgement

We gratefully acknowledge Richard Lansing for his continuous support from inception through production of this book. Kevin Wright provided expert editorial assistance and kept us on point.

Contents

1 Introduction to Endocrine Disruptors and Puberty 1
Evanthia Diamanti-Kandarakis and Andrea C. Gore

Part I EDCs, Endocrine Systems, and Puberty

**2 In Utero Exposure to Environmental Chemicals:
Lessons from Maternal Cigarette Smoking
and Its Effects on Gonad Development and Puberty** 11
Rebecca McKinlay, Peter O'Shaughnessey,
Richard M. Sharpe, and Paul A. Fowler

**3 Reproductive Neuroendocrine Targets of Developmental
Exposure to Endocrine Disruptors** 49
Sarah M. Dickerson, Stephanie L. Cunningham,
and Andrea C. Gore

**4 Endocrine Disruptors and Puberty Disorders
from Mice to Men (and Women)** 119
Alberto Mantovani

**5 Thyroid Hormone Regulation of Mammalian
Reproductive Development and the Potential Impact
of Endocrine-Disrupting Chemicals** 139
Kara Renee Thoemke, Thomas William Bastian,
and Grant Wesley Anderson

xii

Part II Developmental Exposure to Endocrine Disruptors and Adverse Reproductive Outcomes

6 **Developmental Exposure to Endocrine Disruptors and Ovarian Function**.. 177
Evanthia Diamanti-Kandarakis, Eleni Palioura,
Eleni A. Kandaraki

7 **Developmental Exposure to Environmental Endocrine Disruptors and Adverse Effects on Mammary Gland Development**.. 201
Suzanne E. Fenton, Lydia M. Beck, Aditi R. Borde,
and Jennifer L. Rayner

8 **Developmental Exposure to Endocrine Disruptors and Male Urogenital Tract Malformations**............................. 225
Mariana F. Fernandez and Nicolas Olea

9 **EDC Exposures and the Development of Reproductive and Nonreproductive Behaviors** 241
Craige C. Wrenn, Ashwini Mallappa, and Amy B. Wisniewski

Part III Developmental Exposure to Endocrine Disruptors and Metabolic Disorders

10 **Adipocytes as Target Cells for Endocrine Disruption** 255
Amanda Janesick and Bruce Blumberg

11 **Altered Glucose Homeostasis Resulting from Developmental Exposures to Endocrine Disruptors** 273
Alan Schneyer and Melissa Brown

12 **Developmental Exposure to Endocrine Disrupting Chemicals: Is There a Connection with Birth and Childhood Weights?** 283
Elizabeth E. Hatch, Jessica W. Nelson, Rebecca Troisi,
and Linda Titus

Part IV EDCs and Human Health

13 **The Impact of Endocrine Disruptors on Female Pubertal Timing** 325
Jean-Pierre Bourguignon and Anne-Simone Parent

Contents

14 The Influence of Endocrine Disruptors on Male Pubertal Timing 339
Xiufeng Wu, Ningning Zhang, and Mary M. Lee

15 Secular Trends in Pubertal Timing: A Role for Environmental Chemical Exposure? 357
Vincent F. Garry and Peter Truran

Index 373

Contributors

Grant Wesley Anderson Department of Pharmacy Practice and Pharmaceutical Sciences, College of Pharmacy Duluth, University of Minnesota, Duluth, USA

Thomas William Bastian Department of Pharmacy Practice and Pharmaceutical Sciences, College of Pharmacy Duluth, University of Minnesota, Duluth, USA

Lydia M. Beck Summers of Discovery Internship Program, National Institute of Environmental Health Sciences, Research Triangle Park, NC, USA

School of Medicine, University of Missouri, Columbia, MO, USA

Bruce Blumberg Department of Developmental and Cell Biology, 2011 Biological Sciences 3, University of California, Irvine, CA, USA

Department of Pharmaceutical Sciences , University of California, Irvine, CA, USA

Aditi R. Borde Summers of Discovery Internship Program, National Institute of Environmental Health Sciences, Research Triangle Park, NC, USA

Jean-Pierre Bourguignon Developmental Neuroendocrinology Unit, GIGA Neurosciences, University of Liège, Liège, Belgium

Department of Pediatrics, CHU ND des Bruyères, Chênée, Belgium

Melissa Brown Pioneer Valley Life Science Institute, University of Massachusetts Amherst, Springfield, MA, USA

Stephanie L. Cunningham Division of Pharmacology & Toxicology, College of Pharmacy, University of Texas at Austin, Austin, TX, USA

Evanthia Diamanti-Kandarakis Third Department of Medicine, Sotiria Hospital, Medical School University of Athens, Athens, Greece

Sarah M. Dickerson Division of Pharmacology & Toxicology, College of Pharmacy, University of Texas at Austin, Austin, TX, USA

Suzanne E. Fenton National Toxicology Program, National Institute of Environmental Health Sciences, Research Triangle Park, NC, USA

Mariana F. Fernandez Biomedical Research Centre, University of Granada. San Cecilio University Hospital, Granada, CIBERESP, Spain

Paul A. Fowler Centre for Reproductive Endocrinology and Medicine, Division of Applied Medicine, Institute of Medical Sciences, University of Aberdeen, Aberdeen, Scotland

Vincent F. Garry Laboratory Medicine and Pathology; Environmental Medicine Division, University of Minnesota Medical School, Minneapolis, MN, USA

Andrea C. Gore University of Texas at Austin, College of Pharmacy-Pharmacology, Austin, USA

Elizabeth E. Hatch Department of Epidemiology, Boston University School of Public Health, Boston, MA, USA

Amanda Janesick Department of Developmental and Cell Biology, 2011 Biological Sciences 3, University of California, Irvine, CA, USA

Eleni A. Kandaraki Department of Medicine, Huddersfield Royal Infirmary Hospital, Huddersfield, UK

Mary M. Lee Department of Pediatrics, University of Massachusetts Medical School, Worcester, MA, USA

Ashwini Mallappa Division of Pediatric Endocrinology, University of Oklahoma Health Sciences Center, Oklahoma City, OK, USA

Alberto Mantovani Food and Veterinary Toxicology Unit, Dept. Food Safety and Veterinary Public Health, Istituto Superiore di Sanità, Rome, Italy

Rebecca McKinlay Division of Applied Medicine, Centre for Reproductive Endocrinology and Medicine, Institute of Medical Sciences, University of Aberdeen, Aberdeen, Scotland

Jessica W. Nelson Department of Environmental Health, Boston University School of Public Health, Boston, MA, USA

Nicolas Olea School of Medicine, University of Granada. San Cecilio University Hospital, Granada, CIBERESP, Spain

Peter O'Shaughnessey Institute of Biodiversity, Animal Health and Comparative Medicine, School of Veterinary Medicine, University of Glasgow, Glasgow, Scotland

Eleni Palioura Third Department of Medicine, Sotiria Hospital, Medical School University of Athens, Athens, Greece

Contributors

Anne-Simone Parent Developmental Neuroendocrinology Unit, GIGA Neurosciences, University of Liège, Liège, Belgium

Department of Pediatrics, CHU ND des Bruyères, Chênée, Belgium

Jennifer L. Rayner SRC Inc., Arlington, VA, USA

Alan Schneyer Pioneer Valley Life Science Institute, University of Massachusetts Amherst, Springfield, MA, USA

Richard M. Sharpe Medical Research Council Human Reproductive Sciences Unit, Centre for Reproductive Biology, The Queens' Medical Research Institute, Edinburgh, Scotland

Kara Renee Thoemke Department of Biology, The College of St. Scholastica, Duluth, USA

Linda Titus Department of Community & Family Medicine and Pediatrics, Dartmouth Medical School and Dartmouth-Hitchcock Medical Center, Lebanon, NH, USA

Rebecca Troisi Epidemiology and Biostatatistics Program, Division of Cancer Epidemiology and Genetics, National Cancer Institute, National Institutes of Health, Department of Health and Human Services, Bethesda, MD, USA

Peter Truran Minnesota Center for the Philosophy of Science, University of Minnesota, Minneapolis, MN, USA

Amy B. Wisniewski Division of Pediatric Endocrinology, University of Oklahoma Health Sciences Center, Oklahoma City, OK, USA

Craige C. Wrenn Department of Pharmaceutical, Biomedical and Administrative Sciences, College of Pharmacy and Health Sciences, Drake University, Des Moines, IA, USA

Xiufeng Wu Department of Pediatrics, University of Massachusetts Medical School, Worcester, MA, USA

Ningning Zhang Department of Pediatrics, University of Massachusetts Medical School, Worcester, MA, USA

Chapter 1
Introduction to Endocrine Disruptors and Puberty

Evanthia Diamanti-Kandarakis and Andrea C. Gore

Abstract It is now widely accepted that chemical pollutants in the environment can interfere with the endocrine system. Indeed, all hormone-sensitive physiological systems are vulnerable to endocrine disruptors. However, the potential impact of endocrine-disrupting chemicals (EDCs) on human health still remains a medical issue of major concern among the scientific and general public. Evidence for associations between environmental chemicals and adverse health effects is traditionally derived from experimental studies and wildlife observations in contaminated ecosystems. There is currently a wealth of information from studies conducted on laboratory animals, most frequently rodents although more recently on sheep, nonhuman primates, and nonmammalian species, demonstrating a clear cause-and-effect relationship between EDC exposure and endocrine disease.

Keywords EDC • Endocrine disruptors • Pubertal development • Puberty

General Introduction to Endocrine Disruptors

It has been more than 10 years since the Weybridge European Workshop defined endocrine disruptors (EDs) as "an exogenous substance that causes adverse health effects in an intact organism, or its progeny, secondary to changes in endocrine function," [1] and the United States Environmental Protection Agency (EPA) defined

---

E. Diamanti-Kandarakis (✉)
Third Department of Medicine, Sotiria Hospital, Medical School University of Athens, Athens, Greece
e-mail: e.diamanti.kandarakis@gmail.com

A.C. Gore (✉)
University of Texas at Austin, College of Pharmacy-Pharmacology, Austin, USA
e-mail: andrea.gore@austin.utexas.edu

E. Diamanti-Kandarakis and A.C. Gore (eds.), *Endocrine Disruptors and Puberty*, Contemporary Endocrinology, DOI 10.1007/978-1-60761-561-3_1,
© Springer Science+Business Media, LLC 2012

them as "an exogenous agent that interferes with the production, release, transport, metabolism, binding, or elimination of natural hormones in the body responsible for the maintenance of homeostasis and the regulation of developmental processes" [2]. It is now widely accepted that chemical pollutants in the environment can interfere with the endocrine system. Indeed, all hormone-sensitive physiological systems are vulnerable to endocrine disruptors [3]. However, the potential impact of endocrine-disrupting chemicals (EDCs) on human health still remains a medical issue of major concern among the scientific and general public.

Evidence for associations between environmental chemicals and adverse health effects is traditionally derived from experimental studies and wildlife observations in contaminated ecosystems. There is currently a wealth of information from studies conducted on laboratory animals, most frequently rodents although more recently on sheep, nonhuman primates, and nonmammalian species, demonstrating a clear cause-and-effect relationship between EDC exposure and endocrine disease. Human data are relatively scant and mainly based on epidemiological and toxicological studies at the level of populations in a country or a region [4,5]. The assessment of potential environmental health risks in humans has several limitations to overcome, including the complexity of EDC in terms of their mechanisms, targets, composition, timing of exposure, and potential latency to the development of disease. Furthermore, the combined effects of mixture of chemicals, nonmonotonic dose–response curves, and sample inadequacy are some of the confounding factors involved in the investigation of human effects. Evidence for a causal link between environmental EDCs and health disorders, therefore, is limited to toxic accidents such as contamination of cooking oil with PCBs in Taiwan and Japan [6], the explosion of a factory in Seveso, Italy [7], that exposed the local population to high levels of dioxins, and a few other local incidents. Between these case studies and the conservation of human endocrine systems with other species with proven endocrine disruption, there is no doubt that the human endocrine system is a biologically plausible target for disruption by the large number of man-made chemicals released into the environment.

It is estimated that some of the developmental impairments reported in humans today have resulted from exposure to endocrine disruptors [8]. The paradigm of "DES daughters" is one of the most representative examples of such a detrimental interaction. The identification of variable adverse reproductive disorders and increased incidence of clear cell adenocarcinoma in female offspring born to mothers who received diethylstilbestrol (DES) during gestation clearly demonstrate the vulnerability of human health to early life exposure of estrogenic compounds [9,10]. This unfortunate incidence of DES exposure highlighted the causal link between prenatal exposure to a hormonally active substance and the latent development of an endocrine disease later in life. The "fetal basis of adult disease" [11] is an important part of research in endocrine disruption, as scientists and clinicians have begun to appreciate that the fetal environment – a combination of genes, epigenetic modifications of those genes, maternal hormones, the placenta, and compounds to which the mother is exposed – sets the stage for the predisposition to a disorder many years later.

To date, the list of chemicals documented or suspected to possess endocrine-disrupting properties is rather heterogeneous, though few seem to share common molecular structure and physicochemical characteristics. For instance, many EDCs remain highly persistent in the environment due to their lipophilic properties or their long half-lives. Other substances are extensively used in the industry and are ubiquitously detected in the atmosphere. Individuals can be exposed in a variety of sources (contaminated air, water, food, soil), and high-risk populations are considered those with occupational or accidental exposure to hormonally active compounds.

Among substances of concern are synthetic chemicals used in industry, agriculture, and pharmacy including pesticides, fungicides, herbicides, plastics, plasticizers, industrial solvents, and pharmaceutical agents. Due to the unprecedented increase in the production and use of industrial and agricultural chemicals, most of these substances are now widespread in the environment.

There are also natural potential endocrine disruptors. Phytoestrogens are substances in soy, alfalfa, and other plants that have estrogenic activity in estrogen-binding assays. There are questions about the benefits/risks of such substances [12]. While dietary soy may be healthy, particularly in relationship to a high-fat animal protein diet, there are questions about the use of soy or phytoestrogen supplements that are not subject to the same regulatory guidelines as food and drugs. The use of soy formula ought to be more carefully scrutinized, as developing infants with milk allergies or whose mothers choose not to breastfeed may take in virtually all of their nutrients from these products.

Historically, basic, clinical, and epidemiological research on EDCs began on reproductive systems of males and females. For example, the observation by Skakkebaek's group of a decline in sperm number and quality in men over the past 50 years [13] suggested that exposure to industrial contaminants since World War II has an inverse relationship to fertility. At present, scientific interest has expanded to other endocrine systems including thyroid and neuroendocrine systems and, more recently, metabolic dysregulation including obesity and diabetes mellitus, two major epidemics in the modern world. The potential association between developmental exposure to environmental contaminants and this wide spectrum of endocrine systems will be analyzed in the chapters of this book.

Exposure to Endocrine Disruptors During *Critical* and *Sensitive* Windows of Susceptibility

An important part of fully appreciating the consequences of EDCs' action is a consideration of the stage of development at which exposure occurs. Indeed, a fairly consistent and recurrent finding from many experimental studies is that timing of exposure is a decisive factor that ultimately affects the severity and reversibility (or lack thereof) of the outcomes observed. For instance, the developing fetuses or neonates are extremely sensitive to perturbation by chemicals with hormone-like activity [14]. This constitutes the basis of the developmental programming hypothesis, which proposes that exposure of the developing tissues/organs to an adverse stimulus or insult during critical or sensitive times of development can permanently reprogram normal physiological responses, leading to hormonal disorders later in life [11,15].

A number of potential exposure periods to environmental contaminants have been identified: preconceptional (maternal and paternal), preimplantation, postimplantation (organogenesis, first trimester), early and late fetal, premature infant, perinatal, neonatal (term), infant, toddler, preteen (prepubertal), adolescent, and adult [16]. Some of these developmental intervals have been recognized as windows of susceptibility to environmental insults though adverse effects can occur during any period of the life span.

More specifically, permanent and irreversible damage is believed to result from exposure during a *critical window of susceptibility* [17]. This term refers to a period characterized by marked cellular migration, proliferation, and development of specialized function during which interference with environmental chemicals could disrupt the cellular, tissue, and organ physiology, resulting in long-term diseases [16,18]. In contrast, the term *sensitive window of susceptibility* refers to a time-sensitive interval still characterized by adverse affects on development but with a reduced magnitude compared to exposure during a critical window. In other words, these latter stages appear to be somewhat less sensitive to environmental insults if exposures are limited to these periods of life [18].

It is evident that identification of critical and sensitive windows of chemical sensitivity is extremely important in the evaluation of the effects of all toxicant chemicals, including those suspected of causing endocrine disruption. In general, broad windows of susceptibility can be identified for many systems; however, even in the same developmental stage (e.g., prenatal life), not all organs/tissues show the same sensitivity to environmental insults [19]. For example, if an exposure occurs in early prenatal life (first 4 weeks of pregnancy), the central nervous system will be influenced but not the external genitalia, which develop later.

Exposure to Endocrine Disruptors and Effects on Pubertal Development

Puberty is a complex developmental process characterized by rapid physiological alterations that lead to the maturation of secondary sexual characteristics, acceleration of growth, and attainment of reproductive capacity. This sensitive developmental interval constitutes the transition from a nonreproductive to a reproductive state. Puberty is initiated during late childhood with the maturation of the hypothalamic-pituitary-gonadal axis and requires extensive interplay among a variety of hormones. There are a number of dramatic changes in reproductive hormones during this period, particularly estrogens in females and androgens in males. Considering that most endocrine disruptors act as estrogen mimics or may exhibit antagonistic effects to estrogen or androgen receptors, such exposures have been

1 Introduction to Endocrine Disruptors and Puberty

implicated as provoking pubertal abnormalities in humans by interrupting normal hormonal activity [20,21].

Recent interest has arisen about the earlier timing of puberty in girls today as compared to a century ago [22]. As first reported by Herman-Giddens in 1997 [23], and recently confirmed by two independent groups [24,25], girls are undergoing puberty several years earlier, and there are ethnic differences in the timing of the onset of puberty, although that ethnic gap appears to be closing [24]. Both thelarche (breast development) and menarche (first menses) are advanced in modern girls [26]. Some of these differences may be due to improvements in general health and nutrition; nevertheless, a potential impact of endocrine-disrupting chemicals seems likely. By contrast to this work in girls, no such observations have been made for boys. While this sex difference may seem puzzling, it could be explained at least in part by the far more important role of estrogens in female than male puberty and the panoply of estrogenic endocrine disruptors whose presence has increased dramatically over the past century. Exposures of prepubertal females to estrogens are known to cause early puberty in both laboratory animal models as well as in humans. By contrast, male puberty is much more dependent upon androgens, and there are relatively few androgenic endocrine disruptors – in fact, most endocrine disruptors that act upon androgen receptors are antiandrogenic and would therefore not stimulate male puberty. Although clearly a causal relationship between increased exposures to estrogenic endocrine disruptors and female puberty would need to be established, we speculate that these compounds may be contributing to this sexually dimorphic process.

The peripubertal/adolescent period constitutes a significant life stage characterized by many hormonal alterations that go well beyond estrogens and androgens. Other reproductive hormones such as progesterone, and other steroids such as adrenal hormones, increase in their levels during pubertal development. Nonsteroid hormones including thyroid hormones also undergo dramatic alterations during postnatal development. Profound modifications in body composition occur during this time, including alterations in the amount and regional distribution of body fat. Body fat can both produce and store hormones, making them depots of not just endogenous hormones but also of persistent endocrine disruptors that may be resistant to degradation or elimination. Furthermore, the onset of puberty is associated with modest physiological insulin resistance [27], making it a potentially sensitive period for disruption of insulin homeostasis. Beyond these physical changes, puberty is also accompanied with complex psychological and behavioral alterations that often cause self-conscious and aggressive behavior. Any disruptions of the timing or progress of puberty could be associated with perturbations of psychological well-being.

Although puberty as such is an important developmental interval, the impact of endocrine disruptors on all aspects of human puberty is not extensively represented in the current literature, and causal relationships are virtually impossible to establish. Nevertheless, there is considerable evidence in laboratory animal models that many estrogenic endocrine disruptors are associated with, or are indeed causal to, early puberty in female rodents, whereas males are less frequently affected [28].

These data in animals are consistent with the epidemiological observations of early female puberty in humans, strengthening the possibility for a role of endocrine disruptors in this latter process.

Goals of This Book

This introduction should make it clear that there are important gaps in knowledge about the links between endocrine disruptors and the timing of the onset of puberty. Not only must we acquire more information in humans, but stronger cause-and-effect links must be made when possible. In addition, the mechanisms for these processes need to be understood, and an appreciation of the multiplicity of mechanisms should be included in such analyses. For example, it is overly simplistic to attribute the epidemic of early puberty in girls to estrogenic endocrine disruptors alone. Not only are there other contributing factors such as genetics, body weight, eating habits, and social influences, there are other molecular pathways that need to be studied. Hormones and receptors involved in the mediation of adrenal function, adipogenesis, pancreatic insulin secretion, and other processes are also targets of environmental endocrine disruptors.

It is also important to go back to the point we have raised about critical and sensitive developmental windows. Puberty itself is a sensitive period, and exposures to endocrine disruptors at this life stage can alter the timing of puberty or other endocrine processes. Nevertheless, this point also needs to be put into the context of earlier life critical periods such as fetal and early postnatal development because life is a continuum of exposures. If there are exposures to endocrine disruptors at this time, this may set the stage for a perturbation of postnatal developmental processes, including puberty and leading to adult functionality or dysfunction. Even the lack of further exposure to endocrine disruptors at puberty may already result in an individual with a pubertal anomaly. What is important to consider is the interaction of these processes. An individual exposed early in life to a cocktail of endocrine disruptors may continue to be exposed postnatally. This is certainly the case for wildlife that does not or cannot migrate and for humans who are born in, and grow up in, a community with high exposures to endocrine disruptors. The combinatorial effect of early and later developmental exposures may result in a two-hit or multiple-hit situation that amplifies the problem of exposures across the life cycle.

The chapters in this book grapple with these and other issues about endocrine disruption and puberty, including the basic mechanisms, epidemiological studies, and clinical investigation. A comprehensive understanding of these processes requires information about more than just reproductive systems – it must include other endocrine and neurological processes. This book includes perspectives on reproductive systems of animal models and evidence from human studies, but it goes beyond this to include other physiological targets such as the thyroid, fat, neuroendocrine systems, and the pancreas. The elusive links between endocrine disruption and puberty are slowly beginning to reveal themselves.

References

1. European Commission. European workshop on the impact of endocrine disruptors on human health and wildlife. Paper presented at: Weybridge, UK, Report No. EUR 17549, Environment and Climate Research Programme, DG XXI.1996; Brussels.
2. Kavlock RJ, Daston GP, DeRosa C, et al. Research needs for the risk assessment of health and environmental effects of endocrine disruptors: a report of the U.S. EPA-sponsored workshop. Environ Health Perspect. 1996;104 Suppl 4:715–40.
3. Diamanti-Kandarakis E, Bourguignon J-P, Giudice LC, et al. Endocrine-disrupting chemicals: an endocrine society scientific statement. Endocr Rev. 2009;30:293–342.
4. Langer P, Kocan A, Tajtakova M, Trnovec T, Klimes I. What we learned from the study of exposed population to PCBs and pesticides. Open Environ Pollution Toxicol J. 2009;1: 54–65.
5. De Jager C, Farias P, Barraza-Villarreal A, et al. Reduced seminal parameters associated with environmental DDT exposure and p, p'-DDE concentrations in men in Chiapas, Mexico: a cross-sectional study. J Androl. 2006;27:16–27.
6. Yu ML, Guo YL, Hsu CC, Rogan WJ. Increased mortality from chronic liver disease and cirrhosis 13 years after the Taiwan "yucheng" ("oil disease") incident. Am J Ind Med. 1997; 31(2):172–5.
7. Bertazzi PA, Consonni D, Bachetti S, et al. Health effects of dioxin exposure: a 20-year mortality study. Am J Epidemiol. 2001;153:1031–44.
8. Colborn T, Clement C. Statement from the work session on chemically-induced alterations in sexual and functional development: the wildlife/human connection. In: Colborn T, Clement C, editors. Chemically-induced alterations in sexual and functional development: the wildlife/ human connection. Princeton: Princeton Scientific Publishing; 1992.
9. Herbst AL, Ulfelder H, Poskanzer DC. Adenocarcinoma of the vagina. Association of maternal stilbestrol therapy with tumor appearance in young women. N Engl J Med. 1971;284: 878–81.
10. Schrager S, Potter BE. Diethylstilbestrol exposure. Am Fam Physician. 2004;69(10): 2395–400.
11. Barker DJP. The developmental origins of adult disease. Eur J Epidemiol. 2003;18:733–6.
12. Patisaul HB, Jefferson W. The pros and cons of phytoestrogens. Front Neuroendocrinol. 2010;31:400. Epub.
13. Carlsen E, Giwercman A, Keiding N, Skakkebaek NE. Evidence for decreasing quality of semen during past 50 years. BMJ. 1992;305(6854):609–13.
14. Bern H. The fragile fetus. In: Colburn T, Clement C, editors. Chemically-induced alterations in sexual and functional development: the wildlife/human connection. Princeton: Princeton Scientific Publishing; 1992. p. 9–15.
15. Gluckman PD, Hanson MA. The fetal matrix: evolution, development and disease. Cambridge, UK: Cambridge University Press; 2005.
16. Morford LL, Henck JW, Bresline WJ, DeSesso JM. Hazard identification and predictability of children's health risk from animal data. Environ Health Perspec. 2004;112:266–71.
17. Ben-Shlomo Y, Kuh D. A life course approach to chronic disease epidemiology: conceptual models, empirical challenges and interdisciplinary perspectives. Int J Epidemiol. 2002;31: 285–93.
18. Louis RH. Periconception window: advising the pregnancy planning couple. Fertil Steril. 2008;89:119–21.
19. Selevan SG, Kimmel CA, Mendola P. Identifying critical windows of exposure for children's health. Environ Health Perspect. 2000;108 Suppl 3:451–5.
20. Schoeters G, Den Hond E, Dhooge W, van Larebeke N, Leijset M. Endocrine disruptors and abnormalities of pubertal development. Basic Clin Pharmacol Toxicol. 2008;102(2):168–75.
21. Roy JR, Chakraborty S, Chakraborty TR. Estrogen-like endocrine disrupting chemicals affecting puberty in humans – a review. Med Sci Monit. 2009;15:137–45.

22. Hughes SM, Gore AC. How the brain controls puberty, and implications for sex and ethnic differences. Fam Community Health. 2007;30(1 Suppl):S112–4.
23. Herman-Giddens ME, Slora EJ, Wasserman RC, et al. Secondary sexual characteristics and menses in young girls seen in office practice: a study from the pediatric research in office settings network. Pediatrics. 1997;99:505–12.
24. Maisonet M, Christensen KY, Rubin C, et al. Role of prenatal characteristics and early growth on pubertal attainment of British girls. Pediatrics. 2010;126:E591. Epub.
25. Biro FM, Galvez MP, Greenspan LC, et al. Pubertal assessment method and baseline characteristics in a mixed longitudinal study of girls. Pediatrics. 2010;126:e583.
26. Euling S, Herman-Giddens ME, Lee PA, et al. Examination of US puberty-timing data from 1940 to 1994 for secular trends: panel findings. Pediatrics. 2008;121(3):172–91.
27. Moran A, Jacobs DR, Steinberger J, et al. Insulin resistance during puberty. Results from clamp studies in 357 children. Diabetes. 1999;48:2039–44.
28. Dickerson SM, Gore AC. Estrogenic environmental endocrine-disrupting chemical effects on reproductive neuroendocrine function and dysfunction across the life cycle. Rev Endocr Metab Disord. 2007;8(2):143–59.

Part I
EDCs, Endocrine Systems, and Puberty

Chapter 2
In Utero Exposure to Environmental Chemicals: Lessons from Maternal Cigarette Smoking and Its Effects on Gonad Development and Puberty

Rebecca McKinlay, Peter O'Shaughnessey, Richard M. Sharpe, and Paul A. Fowler

Abstract Although tobacco use has declined in recent years, a significant minority of pregnant women still smoke and cessation during gestation is rare. Substantial passive exposure to tobacco smoke is an unavoidable hazard to many more. This is of great concern, since tobacco smoke contains a large number of toxins known to be hazardous to foetal development. These include carbon monoxide, nicotine and its breakdown products, polycyclic aromatic hydrocarbons, thiocyanate and many more. Some are known or suspected endocrine-disrupting chemicals (EDCs). In addition to providing further evidence to dissuade pregnant women and those around them from smoking, studying the effects of tobacco smoke on human development gives insight into effects of other EDCs and EDC mixtures which affect the same developmental processes and end points.

In utero development of the reproductive tract is a poorly understood process that is entirely dependent on the proper functioning of the maternal and foetal endocrine systems. It is vulnerable to disruption in both genders. Thus far, only the physiological effects of the disruption of oestrogen and androgen-mediated signalling and the hypothalamic-pituitary-gonadal (HPG) axis have been investigated in detail. Our knowledge of the process in both animals and humans is far from complete. Exposure of animal models to EDCs affecting sex steroid-mediated processes can cause a

---

R. McKinlay (✉) • P.A. Fowler
Division of Applied Medicine, Centre for Reproductive Endocrinology and Medicine,
Institute of Medical Sciences, University of Aberdeen, Aberdeen, Scotland
e-mail: McKinlay.1982@gmail.com

P. O'Shaughnessey
Institute of Biodiversity, Animal Health and Comparative Medicine,
School of Veterinary Medicine, University of Glasgow, Glasgow, Scotland

R.M. Sharpe
Medical Research Council Human Reproductive Sciences Unit,
Centre for Reproductive Biology, The Queens' Medical Research Institute,
Edinburgh, Scotland

E. Diamanti-Kandarakis and A.C. Gore (eds.), *Endocrine Disruptors and Puberty*,
Contemporary Endocrinology, DOI 10.1007/978-1-60761-561-3_2,
© Springer Science+Business Media, LLC 2012

wide range of adverse reproductive outcomes. These include impaired gametogenesis and later subfertility, early reproductive senescence, neuroendocrine dysregulation affecting fertility and sexual behaviour, genital deformities and increased susceptibility to certain cancers. Effects on humans are thus far uncertain, but many epidemiological studies have associated EDC exposure with decreasing fertility, reductions in the age of pubertal onset and increases in susceptibility to certain cancers and deformities of the reproductive tract. Studies of the physiological and cellular effects of individual EDCs and EDC mixtures in vitro and in vivo have added plausibility to these associations.

Prenatal human tobacco smoke exposure has multiple known adverse effects on reproductive development and may affect the age of puberty in females at least. Overall, maternal and foetal oestrogen and androgen synthesis are suppressed centrally by suppression of the HPG axis by the activation of the hypothalamic-pituitary-adrenal (HPA) axis and peripherally by the direct inhibition of their synthesis. The hypothalamic-pituitary-thyroid (HPT) axis and the peripheral transformation and utilization of thyroid hormones are also suppressed by activation of the HPA axis. Neuroendocrine programming of the HPA, HPT and HPG is perturbed. Decreased fertility and fecundity have been noted in both genders, and exposed females are more likely to experience early menopause. Some epidemiological studies have reported a connection with malformations of the male reproductive tract (testicular dysgenesis syndrome), the plausibility of which is supported by evidence that the production of desert hedgehog, a morphogen essential for male sexual development, is halved by maternal smoking. These outcomes are similar to those observed in epidemiological and animal studies of other EDCs and EDC mixtures, but a great deal of further research is required to determine their exact physiological effects and properly quantify the risks they pose. The effects of tobacco smoke and other EDC mixtures and EDCs on systems other than those directly mediated by the oestrogen receptors and androgen receptor have barely begun to be elucidated.

Keywords Endocrine disruption • Human • Pregnancy • Puberty • Reproductive development • Smoking • Tobacco

Introduction

Although smoking has become less popular in recent years, the direct and indirect exposure of pregnant women to tobacco smoke is still a major heath issue and is likely to remain so for some time. In 2005, 33% of pregnant women in the UK were active smokers immediately before or during their pregnancy, and one in six (17%) smoked throughout [1]. In some subpopulations, this percentage is higher. A recent survey of pregnant women in Scotland, for example, found that 25% smoke at maternity book in, but only 3.5% of these quit before delivery, despite the presence of interventions intended to help them stop [2]. There has been a downward trend in the number of pregnant women smoking, but it has been at best modest. The number of pregnant smokers in the UK decreased by only 2% between 2000 and 2005, and there was no change in Scotland [1]. In the USA, the downward trend in the number of smokers stopped in 2005 and has remained steady since then. Currently, 20% of adults in the USA smoke, and about half of children aged 3–11 are exposed regularly to second-hand smoke [3]. In developing nations, tobacco use is still rising, most notably in Africa, Southeast Asia and certain parts of Latin America [4].

In utero exposure to tobacco combustion products has a broad range of effects on foetal development. Thus far, much attention has been given to the effects of carbon monoxide-induced hypoxia [5, 6] and the effects of nicotine [7]. Other notable toxicants present include toxic trace elements such as cadmium, nickel and mercury [8] and polycyclic aromatic hydrocarbons (PAHs) [9]. The consequences faced by the children of smokers are diverse and worrying. Babies are lighter at birth [10], and the incidences of pre-, peri- and postnatal mortality [11] are increased. Asthma is more common amongst the children of smokers [12], as are cognitive deficits [13], and in both males and females, impaired fertility and fecundity [14–17].

Endocrine-disrupting compounds (EDCs) have been studied extensively in recent years, and there are great concerns about their possible effects on human development, including alterations to neuroendocrinological function; disrupted breast and prostate development and increased risks of breast and prostate cancer; disruptions in the regulation of metabolism and an increasing predisposition to obesity; and the disruption of cardiovascular and reproductive endocrinology and development in both genders [18]. Both natural and man-made EDCs have been found to affect male and female reproductive development in animal experimental studies, including the processes involved in puberty [19–21]. Some of the chemicals found in tobacco smoke such as PAHs, cadmium, arsenic and nicotine are EDCs, but their contribution, if any, to the described effects of smoke exposure in utero is unknown. Environmental EDC exposures typically result from contact with mixtures of compounds, which although often present individually at concentrations lower than would be expected to cause adverse effects, act together cumulatively under most circumstances [22]. Unlike most exposures, however, prenatal tobacco smoke exposure can be studied in humans as well as in animal models. The impacts on foetal development are similar to those thought to be at least partly caused by exposure to other EDCs and EDC mixtures. Therefore, by discovering how developmental processes are affected, we can learn more about EDC actions at the molecular level. Tobacco use is voluntary, but concomitant exposure to environmental tobacco smoke may not be if an individual lives with or spends time with smokers. In the UK, roughly 80% of young women drink (40% heavily) and up to 20% use recreational drugs [23]. Cessation upon falling pregnant is not marked, with <5% of women quitting smoking while pregnant. While drug-using women tend to stop taking ecstasy and reduce their alcohol consumption when they become pregnant, 50% continue to use tobacco and cannabis [24]. By better quantifying the effects of these drugs, the information available to try to convince expectant mothers and those around them to give up can hopefully be made more persuasive.

Endocrine-Disrupting Chemicals and Reproductive Development

Development of the Reproductive Tract

Between the third and eighth weeks of gestation, the human embryo undergoes a remarkable and rapid transformation, beginning as a flat, relatively undifferentiated plate of cells roughly 0.4 mm in diameter and by the end is recognizably human (Fig. 2.1), bearing recognizable structures both internally and externally. The foundations of the human reproductive tract are laid during this time, and its development carries on throughout gestation, childhood and puberty. A timeline of the events involved is shown in Fig. 2.2. Development is the same in both sexes until the seventh week. The primordial germ cells differentiate and migrate into the developing gonads. The urogenital folds and tubercle, which will become the external genitalia, appear. After this point, the sexes diverge dramatically, and three distinct phases of sexual development occur. The first occurs in utero as the basic structures of the reproductive organs develop and assume their position in the body. The second occurs perinatally as surges of gonadotropins trigger gamete maturation in both sexes and the proliferation of the cell types supporting spermatogenesis in males. The third and final stage is the process of puberty, which marks the attainment of reproductive capacity and sexually dimorphic secondary adult sexual characteristics.

Male

During the fifth week of human gestation, a population of cells from the yolk sac migrates through the gut mesentery and into the body of the embryo. They invade an area of coelomic epithelium, medial and ventral to the developing kidney. This area is stimulated to proliferate and thicken, forming a pair of genital ridges which develop in a bipotential manner until the end of the sixth week. In normal male foetuses, a subtle cellular change during the sixth week triggers the pattern of gene expression which ultimately leads to the male phenotype [26]. Several comprehensive reviews of this can be found in the literature [27–30]. The *SRY* gene is the key genetic switch, and its expression triggers the earliest morphological signs of sexual differentiation: the beginning of testis cord formation. SRY expression is Sertoli cell specific, and the Sertoli cells are central to the process of testicular morphogenesis. During the seventh week, the pre-Sertoli cells mature, polarizing and aggregating around the germ cells to form the seminiferous cords. This process divides the gonad into tubular testis cords, which are composed of Sertoli and germ cells, and the interstitial spaces between the cords. These spaces are not empty, but contain several important cell types. This process is dependent on the migration of mesonephric endothelial cells into the gonad during vascularization [29]. As vascularization commences, peritubular myoid cells, which are resident in the testis from the

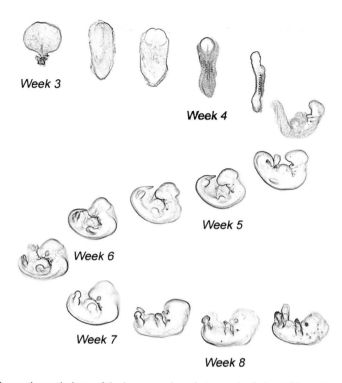

Fig. 2.1 External morphology of the human embryo between the 3rd and 8th weeks of gestation. The embryo begins as a disc of cells. In the 3rd week, this folds in on itself to form the notochord, the precursor of the vertebral column, and the neural plate, which will become the brain and spinal cord begins to form on the embryos surface. A portion of the yolk sac is incorporated into the embryo, which will become the lining of the gastrointestinal tract. The neural plate, which will become the brain and spinal cord also begins to form at this time, and the notochord begins to form somites, projections in the middle of the embryo which eventually form vertebrae and ribs. They continue to appear until midway through the 5th week. The neural plate folds in on itself, closing in the 4th week. The optic and lens placodes, which will form the eyes, form during week 4, as does the heart prominence and the pharyngeal arches which will form the pharynx. During the 5th week, the nasal pit, otic placode and otocysts appear, and the limb buds begin to develop. Retinal pigments appear during the 6th week, and at the same time, the nasal pits migrate to the ventral side of the embryo, and the limb buds form hand plates as the arms begin to develop. Digital rays appear on the hand plates during week 6, and footplates appear as the legs begin to form. The liver prominence begins to grow during this week, and the eyes and acoustic meatus become visible. The skeleton begins to ossify during the 7th week, the eyelids form and the body of the embryo begins to straighten, which together with its developing fingers and rounding head begins to give it a distinctly human appearance. This continues during the 8th week, and as the embryo uncurls, the limbs lengthen, the hands and feet turn inwards and the nose, ears and mouth become distinct[25]

second week of gestation, form a single layer of flattened cells around the Sertoli cells. These smooth muscle-like cells cooperate with the Sertoli cells to deposit the basal lamina, enclosing the seminiferous cords. Undifferentiated fibroblast-like cells then form the connective tissue in the interstitial spaces, which are also penetrated by cells for the gonadal vasculature migrating in from the mesonephros.

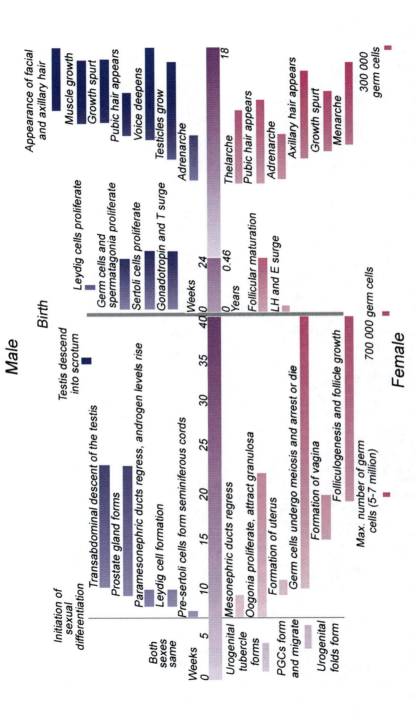

Fig. 2.2 Timeline of the development of the human reproductive tract from conception until the end of puberty in the male and female. PGCs primordial germ cells, LH luteinizing hormone, E oestrogen, T testosterone

Between the eighth and the tenth week, another important cell type follows this path and differentiates in these spaces. Those remaining in the mesonephros form the steroid-secreting cells in the adrenal cortex, and those that enter the gonad form the Leydig cells in males. These are the cells that are stimulated to secrete androgens and drive the further masculinization of the embryo. Initially, Leydig cell development occurs independently of the influence of gonadotropins in rodents, and possibly in humans [31]. This is a relatively short phase, and in humans, early androgen secretion rapidly comes under the control of human chorionic gonadotropin (hCG), levels of which peak during the tenth week [32]. Later in gestation, the foetal HPG axis plays a more dominating role [31–34]. The development of the male gonads can occur without the presence of the germ cells, although there is some evidence that they may support Sertoli cell differentiation [35].

In addition to the gonads, several internal and external structures in the sexually undifferentiated embryo must successfully develop and differentiate [29]. Two pairs of genital ducts, the mesonephric (Wolffian) and paramesonephric (Müllerian) ducts, exist in the undifferentiated embryo. In the male, the secretion of anti-Müllerian hormone (AMH) by the Sertoli cells causes the paramesonephric ducts to actively degenerate in the eighth–tenth weeks of gestation. They leave behind the appendix testis, a vestigial remnant usually present on the top of the testis attached to its covering membrane (the tunica vaginalis), and a small pouch in the area of the prostatic urethra which eventually forms the prostatic utricle. The mesonephric duct differentiates into the epididymis, vas deferens and seminal vesicles, which together form a ductal system connecting the testis to the urethra. The development of this system is stimulated by testosterone and is dependent on region-specific interactions between the mesenchyme and the mesonephric duct epithelium.

Early in development, an anterior-posterior androgen gradient forms, and the mesenchymal cells alone express androgen receptors. The anterior portion of the mesonephric duct begins to elongate and convolute to form the structure which will become the epididymis, the posterior forms the vas deferens and outgrowths from the posterior end of the vas deferens form the seminal vesicles [36, 37]. Rising androgen levels between the eighth and tenth weeks of gestation also stimulate prostate gland formation. Although some tissues within the prostate remain oestrogen responsive, embryonic, foetal and postnatal prostate growth and development is dependent on the presence of testosterone. In the embryo, it triggers the epithelium of the urogenital sinus to form an outgrowth into its mesenchyme, where it forms numerous solid buds which go on to become smooth muscle-lined prostatic ducts. During the regression of the paramesonephric duct, some of its epithelial cells undergo epithelial-mesenchymal transition, and a portion of these cells make up part of the mesenchyme that participates in the formation of the prostate [38]. This tissue remains responsive to oestrogens and starts to proliferate under the influence of maternal oestrogen during week 12 [38, 39]. Maternal oestrogens drive the proliferation of the prostatic stroma until week 23.

The genital tubercle is a small outgrowth which forms during the first month of gestation from tissue associated with the urogenital sinus. Its early development is guided by similar processes to limb development and is sex hormone independent.

Early in the fifth week, it is joined by the urogenital folds. These consist of an inner pair of folds which arise as an enlargement of the underlying ectoderm and extend along the edge of the developing cloaca, and an outer pair arising from the mesoderm form shortly after. These are the labioscrotal swellings. Although these structures are fully formed by the seventh week of gestation, the external genitalia of both genders remain relatively undifferentiated until about the 12th. Testosterone secretion masculinizes the external genitalia of the male, a process that becomes obvious at around weeks 14–16. However, recent data from the rat suggest that differentiation of a masculine reproductive system is programmed by the action of Leydig cell-derived androgens prior to morphologically apparent differentiation, and this time window in humans is estimated to be in the period 8–14 weeks [31, 40]. The anogenital distance increases, and the genital tubercle enlarges and elongates to form the phallus. This is accompanied by the closure of the urogenital folds, extending the urethra to the tip of the fully formed penis, and the fusion of the labioscrotal swellings to form the scrotum [41, 42]. This is an exceptionally delicate androgen-mediated developmental process. Its disruption results in the birth of boys with a spectrum of genital deformities, ranging from mild hypospadias to full pseudohermaphroditism.

Hypospadias and pseudohermaphroditism are the result of the incomplete closure of the urogenital folds. In mild cases of hypospadias, the end of the urethra falls a little short of the tip of the penis, resulting in an enlarged urethral opening or small slit in the glans. In more severe cases, the urethra opens further down the shaft or even at the base of the penis, and reconstructive surgery is required to ensure the penis will be functional later in life. The non-closure of both the urogenital and the labioscrotal folds results in pseudohermaphroditic genitalia, making it difficult to ascertain the baby's gender at birth [43]. Hypospadias is the most common congenital human deformity, affecting 1 in 125 live male births, and both genetic and environmental factors are implicated in its aetiology [43, 44]. Pseudohermaphroditism is less frequent and usually results from genetic mutations that affect either the androgen receptor [45–47] or testosterone/dihydrotestosterone synthesis [47]. Associations made between increasing incidences of hypospadias, cryptorchidism, and later poor semen quality and testicular cancer have led to the hypothesis that these conditions may be part of a single underlying condition, testicular dysgenesis syndrome, which arises from disruption of the androgen-dependant processes involved in the development of the male reproductive tract [48].

The descent of the testis from the abdominal cavity to the scrotum is one of the final gestational events for the human male foetus. It represents the culmination of a much longer process involving two distinct stages, transabdominal and inguinoscrotal descent [41]. These stages occur separately and are mediated by different sets of factors. Transabdominal descent starts at about weeks 10–14 and finishes at about weeks 20–23. Two pairs of gubernacula form from the cranial suspensory ligament of the testis in an androgen-dependant process. One pair is located above the testis, extending from the upper gonadal tips to the posterior abdominal wall, and degenerate to allow testicular descent. The others, which are much more robust, form between the lower poles of the testis and an area of the perineum which develops during this process into the inner ring of the inguinal canal. During their development, the lower gubernacular segments connect with a larger extra abdominal process extending through the inguinal canal and attaching to the scrotum. Gubernacular development and transabdominal testicular descent are dependent on the production of INSL3 in rodents, an insulin-like peptide hormone structurally similar to relaxin. Its action is slightly enhanced by androgens, but androgens are not required for gubernacular formation. The inguinoscrotal stage is mediated by androgens alone and usually occurs by week 35 of gestation. Both processes can be disrupted in animal experimental studies by anti-androgens, oestrogens and oestrogen-like compounds. During the inguinoscrotal stage, the testes pass through the inguinoscrotal canal quickly and then settle within the scrotum. The gubernacula contract and their collagen content increases as they undergo involution to form the small band of ligamentous tissue that attaches the testis to the scrotal floor.

Female

In the absence of SRY, activation of the Wnt4 pathway, and possibly other factors, blocks the expression of the male gonadal phenotype in a poorly understood process that enables ovarian development [28, 29]. Gonadal development commences slightly later in the female than in the male. Upon entering the ovary, the primordial germ cells are surrounded by support cells derived from the ovarian mesenchyme. After a few divisions, the germ cells differentiate into oogonia, which unlike male germ cells continue to divide, forming 'nests' of clustered oocytes. This proliferation continues from the seventh to the end of the 22nd week by which time most of the oocytes will be in mitotic arrest. Shortly after their formation, the oocyte nests begin to break down as support cells invade the nest and differentiate into pre-granulosa cells which surround the developing oocytes [49].

Granulosa cells are essential for oocyte survival. As well as physically surrounding the oocytes, they are responsive to gonadotropins and a variety of other factors. In response, they secrete a number of paracrine and endocrine factors that both sustain the oocyte and developing follicle and participate in endocrine signalling. Oocytes that attract a sufficient number of granulosa cells will go on to develop into primordial follicles, whilst those that do not will undergo apoptosis. These primordial follicles undergo several stages in development, and at each stage large numbers die. The processes governing the selection of oocytes and follicles are poorly understood, but in humans, the initial population of 5–7 million germ cells will produce an ovarian reserve at birth of about 700,000 follicles containing oocytes held in a state of mitotic arrest. This does not represent a static population, however, since follicles will leave this state of quiescence, grow and die during both prenatal and postnatal life. Many of the follicles that grow prenatally are abnormal, indicating that this process could at least in part be a way to eliminate damaged or suboptimal oocytes. From birth to puberty, the ovarian reserve diminishes from around 700,000 to 300,000, and post-pubertally numbers continue to decline, either through ovulation or atresia, until the ovarian reserve is depleted and menopause occurs. During follicle development, theca cells surround the follicular granulosa cells and are analogous to the Leydig cells in the male. One of their primary functions is to secrete androgens and progestins under gonadotropin control. Luteinizing hormone (LH) stimulates production of androgens which are then converted into oestrogens by aromatase, which is produced by growing follicles in response to follicle-stimulating hormone (FSH). They differentiate from the stroma, appearing around follicles after the formation of two or more layers of granulosa cells [49, 50].

In the absence of androgens, the female genital phenotype manifests [51, 52]. There is no proportional increase in anogenital distance or fusion of the labioscrotal or urogenital folds. The labioscrotal folds instead become the labia majora, and the urogenital fold becomes the labia minora. Rather than extending forwards and growing into a phallus, the genital tubercle enlarges and folds inwards slightly to form the clitoris. The urogenital sinus enters a period of rapid growth in week 11, forming a series of evaginations known collectively as the sinovaginal bulbs. These fuse with the uterovaginal canal at 15 weeks to form the vaginal plate. This does not acquire a lumen until around week 20. Its caudal end slides down the urethra and acquires an independent perineal opening. The oviducts, uterus and the upper third of the vagina differentiate from the paramesonephric duct. The mesonephric duct degenerates in the absence of androgens and can leave three remnants. These are the epoöphoron and paroöphoron in the ovarian mesentery, and Gartner's cysts, a scattering of small inclusion cysts located near the vagina. Where the two paramesonephric ducts touch at the posterior wall of the pelvic urethra, they fuse to form a short tube called the uterovaginal or genital canal. This gives rise to the uterus and upper portion of the vagina, and the unfused portions become the oviducts. The differentiation of these three structures is largely controlled by cues from underlying mesenchymal cells which trigger the expression of different HOX protein–controlled signalling pathways [53]. Gubernacula similar to those found in males form. The upper set attached to the ovaries and the lower extend down through rudimentary inguinal canals on either side of the vagina and attach to the labioscrotal folds. Unlike the male versions, these gubernacula do not shorten or thicken but attach to the paramesonephric ducts during the seventh week of gestation. As these are drawn together to form the uterus, the gubernacula pull the ovaries into their final position in the broad ligament of the uterus, a fold in the peritoneum. The gubernacula persist throughout development and adulthood, the upper pair becoming the round ligament of the ovary and the lower becoming the round ligament of the uterus [26].

EDC Effects on Reproductive Development

The reproductive tract, particularly that of the male, is the most studied developmental target of endocrine-disrupting chemicals (EDCs). Exposure to exogenous oestrogens, xeno-oestrogens or anti-androgens has been consistently shown to hinder testicular development via multiple pathways [54–57]. Similarly, development of the ovary is adversely affected by anti-oestrogens, xeno-oestrogens, over- or underexposure to endogenous oestrogens and exposure to androgens [58–61]. EDCs can affect gonadal development in three ways: (1) direct alteration of gene expression by mimicking the action of an endogenous hormone, or interfering with its synthesis breakdown or transport; (2) changing the neuroendocrine mediation of the later stages of gonadal development; and (3) altering the epigenetic regulation of gene transcription [62]. Many EDCs interfere with endocrine-mediated signalling pathways other than those controlled by androgens and oestrogens, but little research on other types in the context of reproductive development has been carried out to date [18]. For this reason, this review is mostly restricted to compounds affecting the oestrogen receptors (ERs), androgen receptor (AR) or other components of the HPG system. Elucidating the effects of individual endocrine-disrupting chemicals on this process can be difficult, since they tend to be present in the environment together with many other EDCs, and many form numerous different metabolites which can have very different effects to their parent compounds. Methoxychlor, for example, is an organochlorine insecticide with a weak affinity for ERα and a slightly greater affinity for ERβ [63]. Its major metabolites, however, include a compound which is an ERα agonist and an ERβ and AR antagonist and shows oestrogenic activity which is approximately 100 times greater than that of the parent methoxychlor compound [64].

The effects of certain xeno-oestrogens on folliculogenesis are a good example of the consequences of the potential for direct actions of endocrine disruptors. In the normally developing ovary, inhibin and activin act as mutually antagonistic paracrine factors which regulate follicular development [65]. In mice, rats and baboons, oocytes and pre-granulosa cells express activin receptors abundantly whilst inhibin receptor expression is minimal. This is regulated by intraovarian oestrogen, and if insufficient oestrogen is present, activin receptor expression decreases whilst inhibin receptor expression increases and folliculogenesis is disrupted [66, 67]. An increased number of multi-oocyte follicles is an indication that this pathway has been perturbed [68]. Overexposure to oestrogens or exposure to synthetic oestrogens has been shown in a mouse model to impede follicular formation by disrupting this balance, inducing the production of multi-oocytic follicles, decreasing the number of small antral follicles present and decreasing activin synthesis [69]. These phenomena and reductions in follicular reserve have been found in rats injected with high doses (20 mg/kg/day) of bisphenol A during follicle formation [70] and in the ovaries of alligators from lakes heavily contaminated with EDCs [71], although there is no evidence that humans are similarly affected.

The chromophils-producing FSH and LH appear in the foetal adenohypophysis at around 22–28 weeks of gestation [72], enabling the development of the hypothalamus-pituitary-gonadal neuroendocrine axis. A functional foetal hypothalamus-pituitary-gonadal axis during the third trimester of gestation appears to be essential for development of the normally formed gonads. The gonads of male anencephalic foetuses are generally reported to be small, as are their penises, although both are normally formed [73]. Such females have normal numbers of oocytes but lack the small and growing antral follicles seen in healthy foetuses, and both genders have reduced populations of gonadal interstitial cells [73]. In order to correctly regulate reproductive function and development, the HPG axis must adopt either a male or female phenotype post-pubertally. In rodent models, sexually dimorphic areas include the anteroventral periventricular nucleus (AVPV), the sexually dimorphic nucleus of the pre-optic area (SDN-POA), areas which are rich in the oestrogen receptors (ERα, ERβ), androgen receptor (AR) and progesterone receptor (PGR) and are central to the regulation of the HPG axis. They are subject to extensive steroid-mediated sexually dimorphic prenatal programming which is exquisitely vulnerable to the actions of certain EDCs [74]. In the absence of testosterone, the HPG adopts a female phenotype. In adulthood, the pulsatile release of gonadotropin-releasing hormone (GnRH) varies over the course of the ovulatory cycle, stimulating the production of follicle-stimulating hormone (FSH) to drive follicle maturation and oestrogen production, and a surge of luteinizing hormone (LH), triggering ovulation. If testosterone is present, then a male phenotype is adopted and the GnRH pulses are released at a constant rate, stimulating FSH, LH and testosterone production and enabling constant spermatogenesis [75].

Sexual differentiation of the rodent brain, including the hypothalamic-pituitary unit, is determined around the time of birth. The male brain is masculinized primarily by the actions of oestrogen, although androgens also play a role, and exposure to both natural and man-made oestrogenic compounds can masculinize the female brain if they are present in sufficient quantities. In humans, however, masculinization is solely androgen mediated [76]. Therefore, only EDCs with anti-androgenic properties are expected to possibly impair human brain masculinization.

To date, most research in this arena has been done in rodents. In experimental studies, different classes of EDC affect sexually dimorphic brain areas such as the rodent AVPV and SDN-POA in different ways causing different effects. Some phyto-oestrogens and their metabolites, for example, act as weak oestrogen receptor agonists or antagonists [77], only targeting the oestrogen receptors, whilst the effects of PCBs are less predictable. Their modes of action are poorly understood, involving the perturbation of multiple hormonal targets including androgens and the thyroid hormones, and affecting the dopamine, serotonin and acetylcholine neurotransmitter systems as well as binding to oestrogen receptors [78]. Prenatal exposure to phyto-oestrogens and other oestrogenic and anti-androgenic EDCs such as bisphenol A and methoxychlor have been shown to decrease GnRH activation in adulthood [79, 80]. In rabbits, the anti-androgenic fungicide vinclozolin decreases the number of GnRH neurons in the mediobasal hypothalamus of both males and females but increases GnRH neuron number in the terminals in the region of the median eminence [81]. Since GnRH neurons do not express androgen receptors themselves, the mechanism by which vinclozolin causes this decline and the contribution, if any, that this makes to vinclozolins' effects on reproduction are still to be determined. Foetal vinclozolin exposure causes reproductive malformations identical to the androgen antagonist flutamide in rats [82]. What effects these chemicals may have, if any, on the sexual differentiation of the human brain remains to be elucidated.

Experimental models of the effects of foetal EDC exposure usually use high doses of single compounds rather than mixtures of low concentrations of these compounds, their metabolites and natural EDCs as usually occurs outside the laboratory.

Compounds have frequently been found to act together additively, even when compounds have been present at levels below those conventionally thought as toxic [83]. For example, foetal sheep from ewes grazed on pastures treated with sewage sludge, which contains a variety of EDCs, have been found to have significant disruption of the hypothalamic pituitary axis even although none of the EDCs present reach the animals in concentrations that are deemed hazardous by the 'no observed effect levels' reported by studies assessing their toxicity and endocrine-disrupting potential individually. Effects include reduced hypothalamic GnRH mRNA expression and reduced hypothalamic and pituitary GnRH receptor, galanin receptor and KISS1 expression [84]. Ovarian and testicular development [85, 86], maternal bone density [87] and the behaviour of adult offspring [88] have also been shown to be adversely affected by foetal exposure using this model.

As they migrate down the genital ridge and into the developing gonad, primordial germ cells undergo a unique process. Their genes are completely demethylated, erasing previous epigenetic programming [89]. During sexual differentiation, the genes are remethylated, receiving sex-specific programming appropriate for the type of gamete they will later become [90]. Altered DNA methylation within the somatic cells resulting from EDC exposures in previous generations is postulated as the means by which epigenetic changes are passed down the germ cell line [91]. Although true transgenerational effects resulting from exposure to EDCs have been documented in animal models, they have yet to be demonstrated in humans. This is unsurprising given the limited number of studies available, the length of human lifespan and reproductive cycles, their genetic and epigenetic heterogeneity, and the diverse chemical exposures to which they are subject.

Effects of Prenatal Endocrine-Disrupting Chemical Exposure on Puberty

In Western Europe and North America, puberty typically occurs between the ages of 12–13 and 17–18 in boys and 10–11 and 16–17 in girls. To give a means of objectively measuring the onset and progress of puberty, the external features of children and adolescents can be classified according to five stages defined by James Tanner in 1969 and henceforth known as Tanner stages [92]. In girls (Fig. 2.3) and boys (Fig. 2.4), stage 1 represents the prepubescent child. Individuals lack pubic hair; the breasts of females lack glandular tissue and have areolae that lie flush with the surrounding skin. Boys have testis with volumes of less than 1.5 ml and a penis of 3 cm or less. As puberty progresses in females, the breast buds and associated glandular tissue begin to form; the areolae enlarge slightly and a small amount of long, slightly pigmented but fine hair appears along the labia majora during stage 2. In males, the scrotum thins and reddens, the testicles begin to enlarge, reaching 1.6–6 ml and downy pubic hair begins to emerge around the base of the penis and scrotum. Stage 3 is defined by the beginning of penile growth in boys, as the penis elongates to around 6 cm. The scrotum enlarges as the testicles reach between 6 and 12 ml in

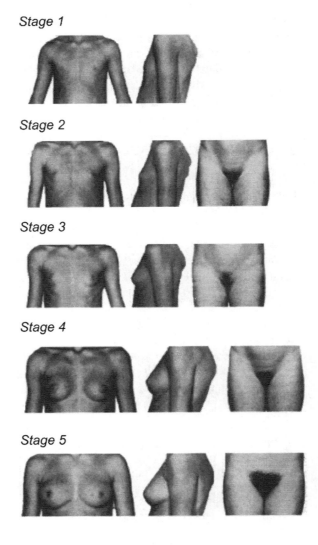

Fig. 2.3 Development of the breasts and pubic hair at the five Tanner stages

volume, and the breasts of girls begin to become elevated compared to the surrounding skin and begin to extend beyond the areola, which continue to enlarge. In both sexes, the pubic hair extends laterally and begins to darken, coarsen and curl. In stage 4, the pubic hair becomes fully adult in texture, covering the pubis but not the inner thigh. The quantity of pubic hair present and the rate at which it is acquired can vary between individuals. Testicular volume at this stage should lie between 12 and 20 ml, and the penis should be around 10 cm in length. In girls, the breasts continue to grow, and the areolae and papillae begin to stand proud of the breast, forming a secondary mound. By stage 5, this secondary mound disappears, and the areolae once again lie flush with the surrounding breast. The breasts reach their adult size. In boys, the penis reaches a flaccid length of around 15 cm, although this

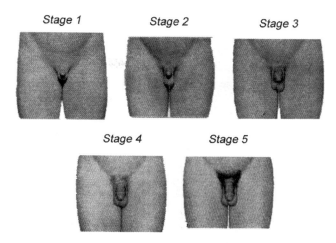

Fig. 2.4 Development of the penis, scrotum and pubic hair at the five Tanner stages

can differ widely between individuals and the testicular volume exceeds 20 ml. The pubic hair of both sexes attains its adult distribution, extending to the medial surface of the thighs and up the central line of the abdomen on many individuals.

Known effects of EDCs on the prenatal programming of the central neuroendocrine systems controlling puberty have recently been reviewed [93, 94]. Many of these effects involve changes in the central regulation of the HPG and hypothalamic-pituitary-thyroid (HPT) axes. Although the hypothalamic-pituitary-adrenal (HPA) axis is likely to be similarly vulnerable, it remains relatively unstudied [95], probably because its role and importance is radically different in primates and evolutionarily distinct animals such as rodents. Effects on puberty are gender specific, which at least partly reflects sex differences in the development of the brain regions involved in central control of the neuroendocrine axes, namely the hypothalamus and limbic systems. In crucial sexually dimorphic areas of the rodent brain such as the sexually dimorphic nucleus of the pre-optic area of the hypothalamus (SDN-POA), the mediobasal hypothalamus and the anteroventral periventricular nucleus (AVPV), exposure to EDCs can change both the number and phenotype of cells. Prenatal exposure to high doses of bisphenol A has been shown to benignly advance pubertal maturation of spermatogenesis slightly in male rats, probably by increasing FSH levels [96]. Similar exposure to the phyto-oestrogen genistein changes pubertal timing in female rats, with some studies showing an acceleration and others showing a delay [97, 98]. At puberty, male rats receiving the same compound have lower body weights, smaller testis and are less likely to achieve preputial separation than unexposed males. They also have lower circulating testosterone levels and impaired masculine sexual behaviour later in life [99, 100].

The peripheral control of puberty by endocrine organs and tissues can also be affected by exposure to EDCs in utero. The gonads are possibly the best-researched peripheral target of EDC action. As discussed above, exposure of the developing testes or ovaries to exogenous oestrogens and xeno-oestrogens, or the incorrect levels of endogenous sex steroids, and exposure of the developing ovaries to anti-oestrogens has been consistently shown to hinder their development via multiple pathways [54–61]. Every step of thyroid hormone production and metabolism is sensitive to disruption by pollutants, and many have been shown to disrupt foetal thyroid development [101]. The mammary gland is also affected (see Chap. 7). Prenatal and perinatal exposure to environmentally relevant concentrations of bisphenol A disrupts mammary gland development and sensitizes it to the action of oestrogen later in life. Overall development in utero is hastened. Absolute ductal area and ductal extension are increased, as is epithelial proliferation and penetration into the fat pad [102, 103]. The sensitivity of the tissue to estradiol and its expression of progesterone receptors at puberty are increased dramatically. Fewer of the terminal end bud cells undergo apoptosis, which, together with the increased progesterone expression, may explain why bisphenol A-treated animals tend to have mammary glands that have more branched ducts and are more densely packed with alveolar buds later in life [104]. Other EDCs have been shown to hasten or delay mammary development in animal models [105–107], and there is concern that such prenatal changes may contribute towards early or late thelarche in humans [108–110], although there is little direct evidence for this.

Finally, since the initiation of puberty depends to a large extent on energy availability, its disruption by EDCs may at least partly be due to the alteration of neuroendocrine set points regulating homeostasis or by direct effects on adipocytes [97] (see Chap. 10). Adipogenesis and lipid accumulation have been demonstrated in vitro in response to EDCs such as organotins, 4-nonylphenol and bisphenol A [111, 112]. At a neuroendocrine level, EDCs can change the sensitivity of the brain to peripheral signalling agents such as leptin, adiponectin and IGF-1, changing the expression of hypothalamic factors involved in homeostasis such as Agouti-related protein (AgRP) and GnRH [18, 97, 113] (See Chap 10).

Tobacco and Prenatal Development

General Effects

Over 4,800 chemicals are found in tobacco smoke, including a wide range of reproductive toxicants such as carbon monoxide, nicotine, arsenic, numerous heavy metals and polycyclic aromatic hydrocarbons (PAHs), and many of the adverse affects of exposure to it are well documented [114–116]. One of the most marked differences between babies of smokers and non-smokers is size. The babies of smokers suffer restricted growth in the womb [117, 118], and this effect is more pronounced in males than in female foetuses [119]. Like other causes of intrauterine growth restriction, this causes changes in the neuroendocrine programming affecting growth and homeostasis, which although intended to be of benefit in a hostile, hungry environment, can lead to the development of illnesses later in life if the postnatal environment does not match that anticipated by the prenatal conditions [120, 121]. Smoking-exposed babies tend to be smaller and thinner than average at birth, but display accelerated 'catch up' growth and greater adiposity during childhood, and have a higher incidence of obesity, cardiovascular disease and type 2 diabetes later in life [122–124], changes that are common to many babies whose foetal growth was subnormal irrespective of the primary cause. As will be explored in the remainder of this section, maternal smoking also has adverse effects on the development of the foetal reproductive tract and later fertility and fecundity, the neuroendocrine axes regulating growth and reproduction, and is thought to influence pubertal timing. In addition, many of the agents in tobacco smoke, including its principle addictive component nicotine, target the nervous and respiratory systems. Associations between maternal smoking and neurodevelopmental defects such as attention deficit disorder, impaired visual motor coordination, reduced IQ and autism are becoming well established [13, 125, 126]. Disorders of the respiratory tract and the regulation of breathing are more common in the infants and children of smokers. Maternal smoking and an infant age of <9 months are the only two consistent background characteristics of infants who succumb to sudden infant death syndrome (SIDS). Prenatal exposure impairs lung development, and the response of the respiratory system to hypoxia causes neurotransmitter abnormalities in multiple areas in the infant brainstem, midbrain and forebrain and can increase pro-inflammatory responses to later tobacco smoke in some infants, factors which are thought to contribute to the fourfold increase in the risk of SIDS in prenatally exposed babies [127]. It is interesting to note that lung development, just like Sertoli cell proliferation, is modulated by locally produced androgens and therefore could be vulnerable to the effects of anti-androgens [128]. Wheezing and asthma are also more prevalent in the children of smokers [129–131]. A fuller review of the medical conditions prenatal tobacco exposure contributes to is available in the recent literature [114].

The Reproductive Tract

Adverse effects of maternal smoking on gonadogenesis and later fertility and fecundity have been observed in both genders. In females, reductions are seen in ovarian follicle numbers [132] and the uteri of exposed girls tend to be smaller [133]. Fertility and fecundity in adulthood is reduced, particularly if the women exposed in utero themselves smoke [16, 134–136]. There is also evidence that they are less likely to carry to term babies conceived using assisted reproductive technologies [137] and that they go through the menopause earlier than the daughters of non-smokers [138]. A large part of the adverse effects on female fertility and fecundity are thought to arise from depletion of the foetal ovarian reserve. A decrease of 41% in germ cell numbers and a 29% reduction in somatic cell numbers in females exposed during the first trimester versus controls was found in a recent study of human foetuses [139]. By suppressing the action of aromatase, nicotine and cotinine inhibit oestrogen-mediated oocyte development, reducing the number of

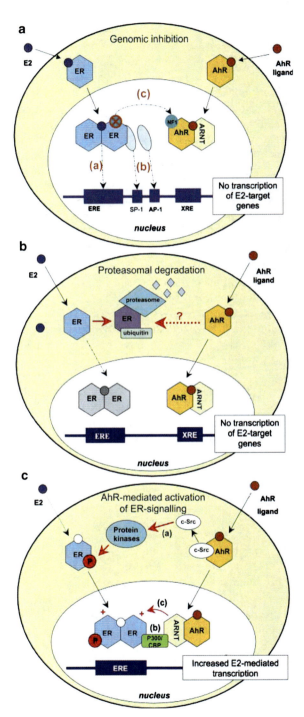

Fig. 2.5 (**a**) Mechanisms of genomic inhibition of steroid-mediated transcription by ligand-activated AhR. Transcriptional inhibition depends on iXREs within promoter regions of oestrogen-regulated

oocytes reaching metaphase II, and may reduce oocyte fertilization rates in adulthood [140]. The aryl hydrocarbon receptor (AhR) plays a key role in the death of oocytes during ovarian development [141]. The interaction of polycyclic aromatic hydrocarbons (PAHs) in tobacco smoke with the AhR is thought to be at least partly responsible for the reduced oocyte numbers and viability observed in the daughters of smokers, as emphasized by a recent study of differentiated human embryonic stem cells as an analogue of germ cells [142]. By increasing oocyte expression of Bax, a pro-apoptotic cell death regulator from the Bcl-2 family, PAHs can induce oocyte apoptosis by inducing expression of the Bcl-2 protein Harakiri [143]. In addition, nicotine, but not cotinine, has a direct toxic effect on oocytes [144] and inhibits meiotic maturation in foetal oocytes [145, 146].

In addition to depleting the reserve of oocytes formed prenatally, maternal smoking may change hormone production and responsiveness later in life, increasing the incidences of certain reproductive disorders. Prenatal nicotine exposure increases plasma testosterone in female adolescent mice and sheep [147]. In humans, increased serum testosterone levels have been reported in women who smoke during pregnancy and then later in their adult daughters [148]. Elevated serum testosterone levels are associated with polycystic ovary syndrome (PCOS), a reproductive and metabolic disorder thought to arise due to disturbances in GnRH secretion and that may be caused at least in part by prenatal exposure to androgens [149–152]. Any link existing between maternal smoking and PCOS, however, remains to be confirmed.

Steroid hormone receptors are highly vulnerable to disruption by PAHs since, by binding to the AhR receptor, PAHs can inhibit or increase oestrogen receptor-mediated transcription, as shown in Fig. 2.5. In addition to the effects on steroid hormone synthesis, the action of PAHs on the AhR in reproductive organs other than the ovary may be responsible for some of the abnormalities affecting the foetuses of smokers. The AhR is expressed abundantly throughout the oviduct, uterus and vagina. Studies of AhR-knockout (AhRKO) mice and the effects of the toxic PAH TCDD have demonstrated that the AhR plays numerous roles contributing to the successful initiation and maintenance of pregnancy and that these can be compromised by exogenous ligands and can be found reviewed in depth by [153]. Mice exposed in utero to the PAHs in tobacco smoke have severely reduced ovarian

---

Fig. 2.5 (continued) genes which prevent ER transcriptional activity by (a) being located close by to neighbouring EREs, (b) disrupting the formation of ER–SP-1 or ER–AP-1 DNA complexes and (c) competing for limited cofactors (e.g. NF-1). These mechanisms apply to interactions of the AhR with all sex steroid hormone receptors; interactions are shown here for the oestrogen receptor (ER). (**b**) Activation of AhR by exogenous ligands can rapidly reduce the levels of ER protein by proteasomal degradation. The exact mechanism by which liganded AhR initiates that degradation is presently unknown. (**c**) AhR-mediated mechanisms leading to the activation of sex steroid hormone signalling in the absence of oestrogen. (a) Upon binding of the ligand, a conformational change causes the release of active c-Src from the cytoplasmic AhR–protein complex. c-Src activates several protein kinases leading to phosphorylation and activation of ER. (b) A direct association of AhR–ARNT complexes with nuclear ER and the cofactor p300/CBP in the absence of E2. (c) Activation of steroid hormone receptor by direct protein interaction with the AhR–ARNT complex in the absence of hormone has so far only been observed for the ER[160] ©Society for Reproduction and Fertility 2005, reproduced by permission

reserves due to the activation of the AhR-mediated initiation of the transcription of Harakiri, one of the genes which regulate apoptosis [154].

Heavy metal exposure via maternal smoking is also likely to affect gonadogenesis in both genders. In an experimental mouse model, cadmium reduces primordial germ cell numbers and motility, leading to a smaller genital ridge size. Primordial germ cell degeneration is increased, and gonad size is reduced. In adulthood, males but not females exposed in utero showed impaired fertility, and their sperm had less capacity to fertilize eggs in vitro [155]. Arsenic and lead suppress estradiol and progesterone production in adult mice exposed in utero [156], and irregularities in the oestrus cycle and corpus luteum failure are both seen in the lead-exposed animals [157, 158]. In males, lead and arsenic exposure in adulthood damage spermatocyte DNA, reducing sperm viability [159].

The male reproductive tract is very sensitive to the effects of maternal smoking. A recent study of human foetuses exposed during the first trimester showed that germ cell numbers are reduced by 55% [138]. Men exposed in utero have smaller testes, their sperm count is reduced by up to ~40% and they are less likely to father children [134, 160, 161]. As babies, they have increased incidences of the reproductive tract defects associated with testicular dysgenesis syndrome (TDS) such as undescended testicles, cryptorchidism [141, 162, 163] and hypospadias [164]. The risk of cryptorchidism may also be increased by the use of nicotine patches as a tobacco replacement [165] although not all data are supportive [166, 167].

Foetal development of Sertoli and Leydig cells is a key process which is both vulnerable to disruption by EDCs and strongly associated with TDS [40, 48]. Decreased Leydig cell function has been reported in adult smokers and animal models [168, 169]. Whether the number of Sertoli or Leydig cells present is reduced by maternal smoking is unknown, but a 37% decrease in the number of somatic cells present in the testis of human fetuses exposed in utero was found in a recent study [139], and the fall in sperm counts in adult men whose mothers smoke is thought to be explained by reduced Sertoli cell number [170]. There is also compelling evidence of developmental impairment at the molecular level. The morphogen desert hedgehog (Dhh) is expressed specifically in the foetal Sertoli cells [171, 172] and is essential for Leydig cell development and subsequent normal development of the male reproductive tract [173]. Maternal smoking halves Dhh expression during the second trimester, which might be expected to affect foetal Leydig cell development [174].

The Neuroendocrine Axes

Cigarette smoking affects the neuroendocrine control of peripheral endocrine tissues in both the mother and foetus and alters the programming of set points in the foetal neuroendocrine axis that guide later reproductive development and function. The hypothalamic-pituitary-adrenal axis affects reproductive development because its activation suppresses that of the hypothalamic-pituitary-gonadal axis [175, 176]. Nicotine strongly activates the HPA axis [177] and has been shown to inhibit normal testosterone synthesis in foetal and neonatal rats, an effect which can be reversed by treatment with the corticosterone-blocking drug metyrapone [178]. Intrauterine hypoxia has also been shown in animal models to activate the HPA axis and make it hyperactive later in life [179]. In the foetuses of smokers, carbon monoxide levels are higher than in the mother [180]. Carboxyhaemoglobin, which is formed by the binding of carbon monoxide to haemoglobin, is removed slightly less efficiently from the foetal blood stream than from the adult, [181] reducing the foetus's ability to transport oxygen more than the mothers. Chronic exposure to nicotine contributes further to the hypoxic intrauterine environment by causing the arteries supplying blood to the uterus to contract [182, 183], both via its direct action on the arterial walls and by activating the maternal HPA axis. In exposed newborn human infants, adrenocorticotrophic hormone (ACTH) levels have been found to be significantly elevated without a concurrent rise in cortisol, indicating that sensitivity to or production of ACTH is changed [184].

In rodent models, maternal smoking affects the HPG axis in a sex-dependent manner, demasculinizing the sexually dimorphic areas of the brain involved in its control in males and masculinizing them in females. Nicotine and cotinine exert a demasculinizing effect on these areas of the male brain without impacting upon the sexual dimorphism of the female brain by inhibiting aromatase activity, with cotinine being approximately twice as potent as nicotine [182, 185]. In females, testosterone synthesis is increased both pre- and postnatally [147, 185], which has the potential in rodents to masculinize their GnRH production and SPN-POA and AVPV volumes [60, 186]. In female animals experimentally exposed to androgens in utero, oocyte numbers are reduced, and oestrus cycles are impaired in adulthood. This occurs via increased GnRH neuron activity, inhibiting the sensitivity of the HPG axis to the progesterone-mediated negative feedback necessary to promote follicle-stimulating hormone (FSH) synthesis and ovulation [187, 188].

AhR ligands can have a lasting effect on the sexually dimorphic neuroendocrine programming of the HPG axis. The distribution of AhR gene expression in the brain follows the expression of the genes encoding for glutamic acid decarboxylase 65 and 67 [189, 190], which are essential for GABA synthesis. GABAergic neurons in the AVPV and mediobasal hypothalamus and the AhR and ER expressing GABA/glutamatergic neurons unique to the AVPV [191] are essential for transmission, amongst other things, of the estradiol-mediated signals needed to trigger the GnRH-stimulated LH surge in the female [192]. The gender-specific apoptosis of these neurons in the AVPV, mediobasal hypothalamus and SDN-POA, which allows their sexual differentiation, is mediated by the AhR via Bax. Exposure to AhR ligands, such as 2,3,7,8-Tetrachlorodibenzo-p-dioxin (TCDD), during periods of gender-specific brain differentiation in rodents produces lifelong changes in sex-specific behaviour and reproductive competence [193]. Male rats exposed to TCDD have demasculinized patterns of GnRH release and sexual behaviour [193, 194]. In females, the SDN-POA is masculinized and masculine behaviour results [195].

The effect of nicotine on maternal sex hormone production and metabolism may have a significant effect on the reproductive development of both genders. In addition to suppressing oestrogen production at all levels via the activation of the HPA axis, nicotine and cotinine actively inhibit their production by ovarian granulosa cells and have a synergistic effect when present together [196]. In female foetuses, reduced levels of the maternal sex steroids cause premature nest breakdown and a reduction in oocyte number, although follicle maturation can still occur [197, 198]. In adult female rats prenatally exposed to nicotine, steroidogenesis is altered (decreased oestrogen, increased progesterone), and the time to pregnancy is increased [199]. In the human, maternal chorionic gonadotropin (hCG) production is lowered, which, in turn, may affect LH and subsequent testosterone production in male foetuses. The ratio of the active isoform of the LH receptor is increased relative to the inactive form in exposed foetuses, which may be part of an adaptive strategy [33].

Exposure to the heavy metals present in tobacco, such as cadmium and lead, can alter maternal HPG function. Gonadal and placental oestrogen and progesterone secretion are suppressed by cadmium [200–203]. Hypothalamic serotonin, dopamine and noradrenaline are reduced by both cadmium and lead, diminishing gonadotropin secretion [204].

Some of the adverse effects of maternal smoking on reproductive development may be mediated by the hypothalamic-pituitary-thyroid (HPT) axis. Its function and the conversion of T_4 to T_3 in peripheral tissues are suppressed by activation of the HPA axis [205–207]. As with the HPA and HPG axes, the set points controlling the sensitivity of the HPT axis are organized before birth and are sensitive to change by exogenous pollutants [208]. During adolescence, HPT underactivity is usually associated with delayed puberty [209] and overactivity with precocious puberty [210]. Cigarette smoke contains numerous chemicals capable of interfering with thyroid activity, although the effects of some may be moderated by iodine intake and existing thyroid status [211]. Smoke may exhibit a weakly pro-thyroid action in subjects with normally functioning thyroid glands [212] and an anti-thyroid action in those with existing hypothyroidism [213]. Thiocyanate was one of the first thyroid-disrupting components of tobacco smoke identified. It competitively inhibits iodine uptake [214], inhibits its organification and reduces the synthesis of thyroid hormones [215]. Benzopyrene and other PAHs in tobacco have been found to affect the action and excretion of thyroid hormones in a number of different ways. Some interact directly with T_3 receptors, acting as agonists [216], others can act as antagonist by dissociating the thyroid hormone response element from the thyroid receptor [217]. T_4 levels can be reduced by the induction of glucuronyl transferase activity, increasing the formation and subsequent excretion of T_4 glucuronides [218]. The structure of the thyroid-stimulating hormone receptor may be altered by maternal smoking in such a way that it makes the thyroid more immunogenic and therefore increases susceptibility to autoimmune thyroid diseases [219]. In addition, maternal smoking reduces TSH levels, increases thyroglobulin levels and decreases the effect of thyroid hormones on peripheral tissues [219]. Thyroid volume is also increased in pregnant smokers [220] even though TSH production is reduced [221], while neonates have elevated thyroglobulin levels which persist into infancy [222, 223].

Puberty

There are few studies of the effects of maternal smoking on the timing and progression of human puberty, and the results are scant and conflicting. Two small early studies in Poland found that the daughters of smokers tended to have earlier menarche than those from non-smoking families, that maternal smoking was a greater factor than paternal smoking and that girls from smoking families were also at greater risk of precocious puberty [224]. One later large study linked heavy maternal smoking during pregnancy with a higher ponderal index and earlier voice change and initiation of shaving in adolescent boys after adjustment for other factors. The ponderal index is calculated by dividing an individual's body weight, in kilograms, by their height in metres cubed. It gives a measure of body composition, with larger values indicating greater fatness. No association was found between smoking and the onset of menarche or self-reported breast development in girls, but the likelihood of late menarche was reduced [225]. Two further large longitudinal studies have found conflicting associations between the onset of menarche and maternal smoking or smoke exposure. The first involved 994 female adolescents and amongst those whose mothers were exposed to high levels of ETS during pregnancy, the adjusted mean age was 4 months earlier than in unexposed girls. Those whose mothers smoked over 20 cigarettes a day and who were also heavily exposed to ETS were half as likely to start menarche late but showed no reduction in age of menarche after adjustment for other factors known to affect menarcheal age such as body weight [226]. The second enrolled a cohort of 1556 girls and found a mean delay of 4 months amongst the daughters of heavy smokers after adjustment [227]. Results of a smaller longitudinal study of 262 daughters from a US cohort were published recently. It was found that girls exposed to tobacco prenatally, to environmental tobacco smoke (ETS) as children or both were more likely to start menarche later than the median age of 12 [228].

Puberty in humans can be hard to measure experimentally, especially in males since markers that can be used to gauge the onset of puberty in females, such as breast growth and menarche, are more visible than testicular growth. Therefore, more data are available for girls than boys [94]. Different events during puberty are controlled by different processes – development of the reproductive tract, for example, is regulated by the HPG axis in females, whilst pubic hair development is stimulated by adrenal androgens [229, 230]. Markers used to measure pubertal processes must therefore be chosen carefully since, for instance, the appearance of pubic hair reveals little about the maturation or functional status of the ovaries. Environmental insults affecting puberty, particularly those, such as maternal smoking, which result in exposure to many different compounds with different modes of action, are therefore likely to have multiple measurable end points. Studies using markers that cover a very limited range of end points, such as age of menarche, or markers which result from numerous physiological processes, such as the age of voice change in boys, could easily miss effects on the process of puberty as a whole. The combinations of varying environmental chemical exposures and the genetic variations between individuals in a population make it difficult to distinguish the effects of individual chemicals [231] especially since phenotypic variation often increases with exposure to chemicals and other environmental insults [232].

Maternal Smoking and Reproductive Development: Lessons

It is hard to imagine a more noxious combination of prenatal chemical insults than those delivered by maternal cigarette smoking. In addition to the direct toxicity of nicotine, heavy metals and tobacco combustion products that cross the placenta, and the perturbation of vital maternal endocrine signals, the foetus is subject to an environment that is oxygen deprived. It is no surprise that women who smoke suffer more adverse pregnancy outcomes and reduced fecundity compared to those who do not and that their children are at greater risk of numerous developmental effects [114]. The adverse effects resulting from prenatal exposure to environmental tobacco smoke demonstrate that the chemical mixture plays a significant role in this, independently of hypoxia [233]. Unlike studies involving laboratory animals or wildlife-based observations, human studies of the effects of prenatal tobacco exposure give indications of how a mixture of EDCs and other reproductive toxicants affects human physiology. It affects many of the same targets, such as the AhR, as do other environmental EDCs and EDC mixtures. Measuring the effects of EDCs in a human population, however, can be very challenging. As with any outbred animal population, there is considerable genetic variation amongst humans that can affect how they respond to these toxins. Many of the processes affected, such as puberty, are very complex involving multiple tissues and signalling pathways that are vulnerable to perturbation by EDCs and other toxicants. Identifying measurable end points that are reliable indicators of the effects of prenatal tobacco exposure on these processes is essential, but can be very difficult, especially in human studies. Multiple processes can affect individual end points, and the use of some end points is limited by expense, time and ethical considerations. This is especially true for those requiring repeated measurements, such as testicular growth, or which can only be measured using invasive procedures and/or tissue samples, such as amniocentesis or the collection of human foetal tissues from aborted or miscarried foetuses. Further longitudinal studies of environmental factors affecting prenatal, postnatal infant, child and adolescent sexual development that measure multiple developmental end points are certainly required.

Many of the reproductive end points identified in studies of prenatal tobacco exposure have been observed to result from prenatal exposure to other EDCs, some of which have been studied extensively. These include oocyte and male gonocyte loss, reproductive tract malformations and reduced fertility and fecundity in adulthood. For these disorders, a number of common aetiologies have been identified. Loss of germ cells, for example, can occur due to direct toxicity to the developing cells, the activation of AhR-mediated apoptosis or the alteration of foetal or maternal hormone levels. The activation of the maternal and foetal HPA axis and subsequent

HPG and HPT axis suppression, however, has not been studied to any great extent in any EDC exposure models. Maternal and foetal HPA axis activation and the subsequent programming of the foetal HPA are implicated extensively as a factor in adverse birth outcomes such as intrauterine growth retardation (IUGR), and developmental and neurodevelopmental disorders such as diabetes and mental illnesses which manifest later in life [234–236]. The HPA axis is susceptible to many toxins, including EDCs, but little is known about the extent to which these affect humans [18, 236]. At least two components present in tobacco smoke, carbon monoxide and nicotine, strongly activate the HPA axis and other toxins present such as cadmium are highly toxic to the adrenal gland [237]. One key lesson which must be learned from the effects of prenatal tobacco exposure is that perturbations of the foetal and maternal HPA axis function and foetal HPA axis development can occur as the result of exposure to environmental toxins and play an important role in the effects of the exposure on development. Further research into the effects of EDCs on adrenal gland and HPA axis is therefore vital.

Quantification of the roles epigenetic changes play in the effects of maternal smoking is still in its early stages. Two studies of children have found changes both in global and gene-specific methylation induced by maternal smoking. The first found an overall increase in DNA methylation in mononuclear cells isolated from blood [238]. The second, which used buccal cells and a more sensitive technique, found a global decrease which was linked to the maternal genotype of GSTM1, a gene involved in the metabolism of tobacco smoke [239]. What these changes mean in terms of reproductive development is yet to be elucidated, but a recent study found changes in the methylation of brain-derived neurotrophic factor-6, which plays a role in neuron and axon survival and is perturbed in neurodegenerative disease [240]. So far, transgenerational effects on human reproduction arising as a consequence of grandparental smoking have not been documented. However, there is evidence that other adverse effects are transmissible for more than one generation. Grandmaternal smoking is associated with asthma in humans [241], and elevated insulin levels have been found in the F2 generation of offspring from nicotine-exposed rats [242]. Transgenerational reproductive effects should be considered as a possible outcome in future studies of maternal smoking and environmental tobacco smoke exposure and could have much to teach us.

Like virtually all exposures to EDCs and other toxins that occur outside laboratory studies, prenatal tobacco exposure via maternal smoking involves a complex mixture of compounds, many of which have not yet had their toxicities or effects on human reproduction quantified. Some of these compounds or compound classes, such as benzopyrene and other PAHs, can reach humans via a variety of other sources, and the study of the effects of maternal smoking can be used to inform investigations of the consequences of exposure from them, and vice versa. Others such as nicotine are unlikely to reach humans via other pathways, but knowledge of their effects on development at the molecular level can still give insight into the effects of other chemicals which target the same processes as nicotine within the body. Studies of the interaction between maternal smoking, genetic factors and developmental outcomes in their children can pinpoint genetic polymorphisms vulnerable to environmental perturbations. So far, little work has been done looking at how these affect foetal reproductive development, but gene polymorphisms have been identified which are associated with reduced infant birth weights, orofacial clefts and neurodevelopmental abnormalities in the children of smoking mothers [243–245]. Being exposed to tobacco combustion products in utero, the children of smokers are subject to a genetic roulette wheel of developmental risk. The consequences may not be apparent immediately at birth but can manifest later in life, and future generations may be at risk. It is therefore imperative that anti-smoking measures continue to be implemented, that prospective parents be strongly encouraged to give up smoking and supported in their efforts to do so.

References

1. Bolling K. Infant feeding survey 2005: early results. 2006. http://www.ic.nhs.uk/pubs/breast-feed2005. Accessed 23 Dec 2010.
2. Tappin D, MacAskill S, Bauld L, Eadie D, Shipton D, Galbraith L. Smoking prevalence and smoking cessation services for pregnant women in Scotland. Subst Abuse Treat Prev Policy. 2010;5(1):1. http://www.substanceabusepolicy.com/content/5/1/1.
3. US Department of Health and Human Resources Centres for Disease Control and Prevention. Decline in US adult smoking rate stalled: half of children still exposed to secondhand smoke. 2010. http://www.cdc.gov/media/pressrel/2010/r100907.htm. Accessed 23 Dec 2010.
4. World Health Organisation. Tobacco control country profiles. 2003. www.wpro.who.int/internet/resources.ashx/TFI/TCCP2.pdf. Accessed 23 Dec 2020.
5. Longo L. Carbon monoxide: effects on oxygenation of the fetus in utero. Science. 1976;194(4264):523–5.
6. Haustein KO. Cigarette smoking, nicotine and pregnancy. Int J Clin Pharmacol Ther. 1999;37(9):417–27.
7. Bruin JE, Gerstein HC, Holloway AC. Long-term consequences of fetal and neonatal nicotine exposure: a critical review. Toxicol Sci. 2010;116:364.
8. Chiba M, Masironi R. Toxic and trace elements in tobacco and tobacco smoke. Bull World Health Organ. 1992;70(2):269–75.
9. Detmar J, Rennie M, Whiteley K, Qu D, Taniuchi Y, Shang X, Casper RF, Adamson SL, Sled JG, Jurisicova A. Fetal growth restriction triggered by polycyclic aromatic hydrocarbons is associated with altered placental vasculature and AhR-dependent changes in cell death. Am J Phisol-Endoc M. 2008;295(2):E519–30.
10. Leonardi-Bee J, Smyth A, Britton J, Coleman T. Environmental tobacco smoke and fetal health: systematic review and meta-analysis. Arch Dis Child. Fetal Neonatal Edn. 2008;93(5):351–61.
11. Salihu H, Wilson R. Epidemiology of prenatal smoking and perinatal outcomes. Early Hum Dev. 2007;83(11):713–20.
12. Prescott S, Clifton V. Asthma and pregnancy: emerging evidence of epigenetic interactions in utero. Curr Opin Allergy Clin Immun. 2009;9(5):417–26.
13. Cornelius MD, Day NL. Developmental consequences of prenatal tobacco exposure. Curr Opin Neurol. 2009;22(2):121–5.
14. Jensen TK, Joffe M, Scheike T, Skytthe A, Gaist D, Petersen I, Christensen K. Early exposure to smoking and future fecundity among Danish twins. Int J Androl. 2007;29(6):603–13.

15. Jensen T, Jørgensen N, Punab M, Haugen TB, Suominen J, Zilaitiene B, Horte A, Andersen AG, Carlsen E, Magnus Ø, Matulevicius V, Nermoen I, Vierula M, Keiding N, Toppari J, Skakkebaek NE. Association of in utero exposure to maternal smoking with reduced semen quality and testis size in adulthood: a cross-sectional study of 1,770 young men from the general population in five European countries. Am J Epidemiol. 2004;159(1):49.
16. Bolumar F, Olsen J, Boldsen J, The European Study Group on Infertility and Subfecundity. Smoking reduces fecundity: a European multicenter study on infertility and subfecundity. Am J Epidemiol. 1996;143(6):578–87.
17. Ye X, Skjaerven R, Basso O, Baird DD, Eggesbo M, Uicab LA, Haug K, Longnecker MP. In utero exposure to tobacco smoke and subsequent reduced fertility in females. Hum Reprod. 2010;25(11):2901–6.
18. Diamanti-Kandarakis E, Bourguignon J, Giudice LC, Hauser R, Prins GS, Soto AM, Zoeller RT, Gore AC. Endocrine-disrupting chemicals: an endocrine society scientific statement. Endocr Rev. 2009;30(4):293–342.
19. Toppari J, Juul A. Trends in puberty timing in humans and environmental modifiers. Mol Cell Endocrinol. 2010;324(1–2):39–44.
20. Zama AM, Uzumcu M. Epigenetic effects of endocrine-disrupting chemicals on female reproduction: an ovarian perspective. Front Neuroendocrinol. 2010;31(4):420–39.
21. Main KM, Skakkebæk NE, Virtanen HE, Toppari J. Genital anomalies in boys and the environment. Best Pract Res Clin Endocrinol Metab. 2010;24(2):279–89.
22. Kortenkamp A. Ten years of mixing cocktails: a review of combination effects of endocrine-disrupting chemicals. Environ Health Perspect. 2007;115(S1):98–105.
23. Crome IB, Kumar MT. Epidemiology of drug and alcohol use in young women. Semin Fet Neonat Med. 2007;12(2):98–105.
24. Moore DG, Turner JD, Parrott AC, Goodwin JE, Fulton SE, Min MO, Fox HC, Braddick FM, Axelsson EL, Lynch S, Ribeiro H, Frostick CJ, Singer LT. During pregnancy, recreational drug-using women stop taking ecstasy (3,4-methylenedioxy-N-methylamphetamine) and reduce alcohol consumption, but continue to smoke tobacco and cannabis: initial findings from the Development and Infancy Study. J Psychopharmacol. 2010;24(9):1403–10.
25. Hill, M. Carnegie stages. University of New South Wales embryology. 2009. http://php.med.unsw.edu.au/embryology/index.php?title=Main_Page. Accessed 14 April 2011.
26. Schoenwolf G, Bleyl S, Brauer P, Francis-West P. Larsen's human embryology. Philadelphia: Churchill Livingston; 2008. 712.
27. Brennan J, Capel B. One tissue, two fates: molecular genetic events that underlie testis versus ovary development. Nat Rev Genet. 2004;5(7):509–21.
28. DeFalco T, Capel B. Gonad morphogenesis in vertebrates: divergent means to a convergent end. Annu Rev Cell Dev Biol. 2009;25:457–82.
29. Wilhelm D, Palmer S, Koopman P. Sex determination and gonadal development in mammals. Physiol Rev. 2007;87(1):1–28.
30. Cederroth CR, Pitetti J, Papaioannou MD, Nef S. Genetic programs that regulate testicular and ovarian development. Mol Cell Endocrinol. 2007;265–266:3–9.
31. Scott HM, Mason JI, Sharpe RM. Steroidogenesis in the fetal testis and its susceptibility to disruption by exogenous compounds. Endocr Rev. 2009;30(7):883–925.
32. Fowler PA, Evans LW, Groome NP, Templeton A, Knight PG. A longitudinal study of maternal serum inhibin-A, inhibin-B, activin-A, activin-AB, pro-alphaC and follistatin during pregnancy. Hum Reprod. 1998;13(12):3530–6.
33. Fowler P, Bhattacharya S, Gromoll J, Monteiro A, O'Shaughnessy P. Maternal smoking and developmental changes in luteinizing hormone (LH) and the LH receptor in the fetal testis. J Clin Endocrinol Metab. 2009;94(12):4688–95.
34. O'Shaughnessey P, Fowler P. Endocrinology of the mammalian fetal testis. Reproduction. 2011;141(1):37–46.
35. Adams IR, McLaren A. Sexually dimorphic development of mouse primordial germ cells: switching from oogenesis to spermatogenesis. Development. 2002;129(5):1155–64.

36. Hannema SE, Hughes IA. Regulation of Wolffian duct development. Horm Res. 2006;67(3): 142–51.
37. Archambeault DR, Tomaszewski J, Joseph A, Hinton BT, Yao HH. Epithelial-mesenchymal crosstalk in Wolffian duct and fetal testis cord development. Genesis. 2009;47(1):40.
38. Cai Y. Participation of caudal müllerian mesenchyma in prostate development. J Urol. 2008;180(5):1898–903.
39. Bierhoff E, Walljasper U, Hofmann D, Vogel J, Wernert N, Pfeifer U. Morphological analogies of fetal prostate stroma and stromal nodules in BPH. Prostate. 1997;31(4):234–40.
40. Welsh M, Saunders PTK, Fisken M, Scott HM, Hutchison GR, Smith LB, Sharpe RM. Identification in rats of a programming window for reproductive tract masculinization, disruption of which leads to hypospadias and cryptorchidism. J Clin Invest. 2008;118(4):1479–90.
41. Klonisch TF, Fowler PA, Hombach-Klonisch S. Molecular and genetic regulation of testis descent and external genitalia development. Dev Biol. 2004;270(1):18.
42. Wilhelm D, Koopman P. The makings of maleness: towards an integrated view of male sexual development. Nature Rev Gene. 2006;7(8):620–31.
43. Baskin LS, Ebbers MB. Hypospadias: anatomy, etiology, and technique. J Pediatr Surg. 2006;41(3):463–72.
44. Willingham E, Baskin LS. Candidate genes and their response to environmental agents in the etiology of hypospadias. Nat Clin Pract Urol. 2007;4:270–9.
45. Shah R. Testicular feminization: the androgen insensitivity syndrome. J Pediatr Surg. 1992; 27(6):757–60.
46. Wu X, Zhou Q, Mao J, Lu S, Nie M. Mutational analysis of androgen receptor gene in four Chinese patients with male pseudohermaphroditism. Fertil Steril. 2010;93(6):2076.e1–4.
47. Al-Attia HM. Male pseudohermaphroditism due to 5 alpha-reductase-2 deficiency in an Arab kindred. Postgrad Med J. 1997;73(866):802–7.
48. Skakkebak NE, Rajpert-De Meyts E, Main KM. Testicular dysgenesis syndrome: an increasingly common developmental disorder with environmental aspects: opinion. Hum Reprod. 2001;16(5):972–8.
49. Hartshorne GM, Lyrakou S, Hamoda H, Oloto E, Ghafari F. Oogenesis and cell death in human prenatal ovaries: what are the criteria for oocyte selection? Mol Hum Reprod. 2009; 15(12):805–19.
50. Young J, McNeilly AS. Theca – the forgotten cell of the ovarian follicle. Reproduction. 2010;140:489–504.
51. Wünsch L, Schober J. Imaging and examination strategies of normal male and female sex development and anatomy. Best Pract Res Cl En Met. 2007;21(3):367–79.
52. Rey R, Picard J. Embryology and endocrinology of genital development. Baillière Clin Endoc. 1998;12(1):17–33.
53. Yin Y, Ma L. Development of the mammalian female reproductive tract. J Biochem. 2005; 137(6):677–83.
54. Viguier-Martinez MC, de Reviers MTH, Barenton B, Perreau C. Effect of a non-steroidal antiandrogen, flutamide, on the hypothalamo-pituitary axis, genital tract and testis in growing male rats: endocrinological and histological data. Acta Endocrinol. 1983;102(2):299–306.
55. Yasuda Y, Kihara T, Tanimura T. Effect of ethinyl estradiol on the differentiation of mouse fetal testis. Teratology. 1985;32(1):113–8.
56. Sharpe RM. Pathways of endocrine disruption during male sexual differentiation and masculinisation. Best Pract Res Clin Endocrinol Metab. 2006;20(1):91–110.
57. Habert R, Muczynski V, Lehraiki A, Lambrot R, Lécureuil C, Levacher C, Coffigny H, Pairault C, Moison D, Frydman R, Rouiller-Fabre V. Adverse effects of endocrine disruptors on the foetal testis development: focus on the phthalates. Folia Histochem Cytobiol. 2009;47(5):S67–74.
58. Newbold RR, Bullock BC, Mc Lachlan JA. Exposure to diethylstilbestrol during pregnancy permanently alters the ovary and oviduct. Biol Reprod. 1983;28(3):735–44.
59. Witcher JA, Clemens LG. A prenatal source for defeminization of female rats is the maternal ovary. Horm Behav. 1987;21(1):36–43.

60. Abbott DH, Padmanabhan V, Dumesic DA. Contributions of androgen and estrogen to fetal programming of ovarian dysfunction. Reprod Biol and Endoc. 2006;4:17.
61. Adewale HB, Jefferson WN, Newbold RR, Patisaul HB. Neonatal bisphenol-A exposure alters rat reproductive development and ovarian morphology without impairing activation of gonadotropin-releasing hormone neurons. Biol Reprod. 2009;81(4):690–9.
62. Crain A, Janssen SJ, Edwards TM, Heindel J, Ho SM, Hunt P, Iguchi T, Juul A, McLachlan JA, Schwartz J, Skakkebaek N, Soto AM, Swan S, Walker C, Woodruff TK, Woodruff TJ, Giudice LC, Guillette Jr LJ. Female reproductive disorders: the roles of endocrine-disrupting compounds and developmental timing. Fertil Steril. 2008;90(4):911–40.
63. Kuiper GGJM, Carlsson B, Grandien K, Enmark E, Häggblad J, Nilsson S, Gustafsson JA. Comparison of the ligand binding specificity and transcript tissue distribution of estrogen receptors α and β. Endocrinology. 1997;138(3):863–70.
64. Gaido KW, Maness SC, McDonnell DP, Dehal SS, Kupfer D, Safe S. Interaction of methoxychlor and related compounds with estrogen receptor α and β, and androgen receptor: structure-activity studies. Mol Pharmacol. 2000;58(4):852–8.
65. Lerch TF, Xu M, Jardetzky TS, Radhakrishnan I, Kazer R, Shea LD, Woodruff TK. The structures that underlie normal reproductive function. Mol Cell Endocrinol. 2007;267 (1–2):1–5.
66. Findlay JK, Drummond AE, Britt KL, Dyson M, Wreford NG, Robertson DM, Groome NP, Jones ME, Simpson ER. The roles of activins, inhibins and estrogen in early committed follicles. Mol Cell Endocrinol. 2000;163(1–2):81–7.
67. Billiar RB, Zachos NC, Burch MG, Albrecht ED, Pepe GJ. Up-regulation of α-inhibin expression in the fetal ovary of estrogen-suppressed baboons is associated with impaired fetal ovarian folliculogenesis. Biol Reprod. 2003;68(6):1989–96.
68. Pepe GJ, Billiar RB, Albrecht ED. Regulation of baboon fetal ovarian folliculogenesis by estrogen. Mol Cell Endocrinol. 2006;247(1–2):41–6.
69. Kipp JL, Kilen SM, Bristol-Gould S, Woodruff TK, Mayo KE. Neonatal exposure to estrogens suppresses activin expression and signaling in the mouse ovary. Endocrinology. 2007; 148(5):1968–76.
70. Rodríguez HA, Santambrosio N, Santamaría CG, Muñoz-de-Toro M, Luque EH. Neonatal exposure to bisphenol A reduces the pool of primordial follicles in the rat ovary. Reprod Toxicol. 2010;30(4):550–7. Epub 2010.
71. Moore BC, Kohno S, Cook RW, Alvers AL, Hamlin HJ, Woodruff TK, Guillette LJ. Altered sex hormone concentrations and gonadal mRNA expression levels of activin signaling factors in hatchling alligators from a contaminated Florida lake. J Exp Zoo Part A: Ecological Genetics Physiol. 2010;313A(4):218–30.
72. Conklin JL. The development of the human fetal adenohypophysis. Anat Rec. 1968;160(1): 79–91.
73. Baker TG, Scrimgeour JB. Development of the gonad in normal and anencephalic human fetuses. J Reprod Fertil. 1980;60(1):193–9.
74. Gore AC. Developmental programming and endocrine disruptor effects on reproductive neuroendocrine systems. Front Neuroendocrinol. 2008;29(3):358–74.
75. Bliss SP, Navratil AM, Xie J, Roberson MS. GnRH signaling, the gonadotrope and endocrine control of fertility. Front Neuroendocrinol. 2010;31(3):322–40.
76. McCarthy MM. How it's made: organisational effects of hormones on the developing brain. J Neuroendocrinol. 2010;22(7):736–42.
77. Mueller SO, Simon S, Chae K, Metzler M, Korach KS. Phytoestrogens and their human metabolites show distinct agonistic and antagonistic properties on estrogen receptor α (ERα) and ERβ in human cells. Toxicol Sci. 2004;80(1):14–25.
78. Dickerson SM, Gore AC. Estrogenic environmental endocrine-disrupting chemical effects on reproductive neuroendocrine function and dysfunction across the life cycle. Rev Endoc Metlic Dis. 2007;8(2):143–59.
79. Bateman HL, Patisaul HB. Disrupted female reproductive physiology following neonatal exposure to phytoestrogens or estrogen specific ligands is associated with decreased GnRH

activation and kisspeptin fiber density in the hypothalamus. Neurotoxicology. 2008;29(6): 988–97.

80. Mahoney MM, Padmanabhan V. Developmental programming: Impact of fetal exposure to endocrine-disrupting chemicals on gonadotropin-releasing hormone and estrogen receptor mRNA in sheep hypothalamus. Toxicol Appl Pharmacol. 2010;247(2):98–104.

81. Wadas B, Hartshorn C, Aurand E, Palmer JS, Roselli CE, Noel ML, Gore AC, Veeramachaneni DN, Tobet SA. Prenatal exposure to vinclozolin disrupts selective aspects of the gonadotrophin-releasing hormone neuronal system of the rabbit. J Neuroendocrinol. 2010;22(6): 518–26.

82. Gray LE, Ostby JS, Kelce WR. Developmental effects of an environmental antiandrogen: the fungicide vinclozolin alters sex differentiation of the male rat. Toxicol Appl Pharmacol. 1994;129(1):46–52.

83. Bellingham M, Fowler PA, Amezaga MR, Whitelaw CM, Rhind SM, Cotinot C, Mandon-Pepin B, Sharpe RM, Evans NP. Foetal hypothalamic and pituitary expression of gonadotrophin-releasing hormone and galanin systems is disturbed by exposure to sewage sludge chemicals via maternal ingestion. J Neuroendocrinol. 2010;22(6):527–33.

84. Christiansen S, Scholze M, Axelstad M, Boberg J, Kortenkamp A, Hass U. Combined exposure to anti-androgens causes markedly increased frequencies of hypospadias in the rat. Int J Androl. 2008;31(2).

85. Paul C, Rhind SM, Kyle CE, Hayley Scott H, Chris McKinnell C, Richard M, Sharpe RM. Cellular and hormonal disruption of fetal testis development in sheep reared on pasture treated with sewage sludge. Environ Health Perspect. 2005;113(11):87–8.

86. Fowler PA, Dora NJ, McFerran H, Amezaga MR, Miller DW, Lea RG, Cash P, McNeilly AS, Evans NP, Cotinot C, Sharpe RM, Rhind SM. In utero exposure to low doses of environmental pollutants disrupts fetal ovarian development in sheep. Mol Hum Reprod. 2008;14(5): 269–80.

87. Lind PM, Gustafsson M, Hermsen SA, Larsson S, Kyle CE, Orberg J, Rhind SM. Exposure to pastures fertilised with sewage sludge disrupts bone tissue homeostasis in sheep. Sci Total Environ. 2009;407(7):2200.

88. Erhard HW, Rhind SM. Prenatal and postnatal exposure to environmental pollutants in sewage sludge alters emotional reactivity and exploratory behaviour in sheep. Sci Total Environ. 2004;332(1–3):101–8.

89. Hajkova P, Erhardt S, Lane N, Haaf T, El-Maarri O, Reik W, Walter J, Surani MA. Epigenetic reprogramming in mouse primordial germ cells. Mech Dev. 2002;117(1–2):15–23.

90. Li E. Chromatin modification and epigenetic reprogramming in mammalian development. Nature Rev Genet. 2002;3:662–73.

91. Anway MD, Skinner MK. Epigenetic transgenerational actions of endocrine disruptors. Endocrinology. 2006;147(6):s43.

92. Marshall WA, Tanner JM. Growth and endocrinology of the adolescent. In: Gardner L, editor. Endocrine and genetic diseases of childhood. 1969. 19.

93. Euling SY, Selevan SG, Pescovitz OH, Skakkebaek NE. Role of environmental factors in the timing of puberty. Pediatrics. 2008;121:S167–71.

94. Gajdos ZKZ, Henderson KD, Hirschhorn JN, Palmert MR. Genetic determinants of pubertal timing in the general population. Mol Cell Endocrinol. 2010;324(1–2):21–9.

95. Bourguignon J, Parent A. Early homeostatic disturbances of human growth and maturation by endocrine disrupters. Curr Opin Pediatr. 2010;22(4):470–7.

96. Mouritsen A, Aksglaede L, Sørensen K, Mogensen SS, Leffers H, Main KM, Frederiksen H, Andersson AM, Skakkebaek NE, Juul A. Hypothesis: exposure to endocrine-disrupting chemicals may interfere with timing of puberty. Int J Androl. 2010;33(2):346–59.

97. Gore A. Neuroendocrine targets of endocrine disruptors. Hormones. 2010;9(1):16–27.

98. Casanova M, You L, Gaido K, Archibeque-Engle S, Janszen D, Heck H. Developmental effects of dietary phytoestrogens in Sprague-Dawley rats and interactions of genistein and daidzein with rat estrogen receptors alpha and beta in vitro. Toxicol Sci. 1999;51(2): 236–44.

99. Atanassova N, McKinnell C, Turner KJ, Walker M, Fisher JS, Morley M, Millar MR, Groome NP, Sharpe RM. Comparative effects of neonatal exposure of male rats to potent and weak (environmental) estrogens on spermatogenesis at puberty and the relationship to adult testis size and fertility: evidence for stimulatory effects of low estrogen levels. Endocrinology. 2000;141(10):3898–907.

100. Levy JR, Faber KA, Ayyash L, Hughes CL. The effect of prenatal exposure to the phytoestrogen genistein on sexual differentiation in rats. Proc Soc Exp Biol Med. 1995;208(1):60–6.

101. Wisniewski AB, Klein SL, Lakshmanan Y, Gearhart JP. Exposure to genistein during gestation and lactation demasculinizes the reproductive system in rats. J Urol. 2003;169(4):1582–6.

102. Howdeshell K. A model of the development of the brain as a construct of the thyroid system. Environ Health Perspect. 2002;110(Suppl 3):337–48.

103. Vandenberg LN, Maffini MV, Wadia PR, Sonnenschein C, Rubin BS, Soto AM. Exposure to environmentally relevant doses of the xenoestrogen bisphenol-A alters development of the fetal mouse mammary gland. Endocrinology. 2007;148(1):116–27.

104. Markey CM, Luque EH, Munoz de Toro M, Sonnenschein C, Soto AM. In utero exposure to bisphenol A alters the development and tissue organization of the mouse mammary gland. Biol Reprod. 2001;65(4):1215–23.

105. Munoz-de-Toro M, Markey CM, Wadia PR, Luque EH, Rubin BS, Sonnenschein C, Soto AM. Perinatal exposure to bisphenol-A alters peripubertal mammary gland development in mice. Endocrinology. 2005;146(9):4138–47.

106. Rayner JL, Wood C, Fenton SE. Exposure parameters necessary for delayed puberty and mammary gland development in long–evans rats exposed in utero to atrazine. Toxicol Appl Pharmacol. 2004;195(1):23–34.

107. Hilakivi-Clarke L, Cho E, Onojafe I, Liao DJ, Clarke R. Maternal exposure to tamoxifen during pregnancy increases carcinogen-induced mammary tumorigenesis among female rat offspring. Clin Cancer Res. 2000;6(1):305–8.

108. Moral R, Wang R, Russo I, Mailo D, Lamartiniere C, Russo J. The plasticizer butyl benzyl phthalate induces genomic changes in rat mammary gland after neonatal/prepubertal exposure. BMC Genomics. 2007;8(1):453–65.

109. Wolff MS, Teitelbaum SL, Pinney SM, Windham G, Liao L, Biro F, Kushi LH, Erdmann C, Hiatt RA, Rybak ME, Calafat AM, Breast Cancer and Environment Research Centers. Investigation of relationships between urinary biomarkers of phytoestrogens, phthalates, and phenols and pubertal stages in girls. Environ Health Perspect. 2010;118(7):1039–46.

110. Colón I, Caro D, Bourdony CJ, Rosario O. Identification of phthalate esters in the serum of young Puerto Rican girls with premature breast development. Environ Health Perspect. 2000;108(9):895–900.

111. Selevan SG, Rice D, Hogan K, Euling S, Pfahles-Hutchens A, Bethel J. Blood lead concentration and delayed puberty in girls. New Engl J Med. 2003;348(16):1527–36.

112. Grun F, Watanabe H, Zamanian Z, Maeda L, Arima K, Cubacha R, Gardiner DM, Kanno J, Iguchi T, Blumberg B. Endocrine-disrupting organotin compounds are potent inducers of adipogenesis in vertebrates. Mol Endocrinol. 2006;20(9):2141–55.

113. Wada K, Sakamoto H, Nishikawa K, Sakuma S, Nakajima A, Fujimoto Y, Kamisaki Y. Life style-related diseases of the digestive system: endocrine disruptors stimulate lipid accumulation in target cells related to metabolic syndrome. J Pharmacol Sci. 2007;105(2):133–7.

114. Bourguignon J, Rasier G, Lebrethon M, Gérard A, Naveau E, Parent A. Neuroendocrine disruption of pubertal timing and interactions between homeostasis of reproduction and energy balance. Mol Cell Endocrinol. 2010;324(1–2):110–20.

115. Rogers JM. Tobacco and pregnancy. Reprod Toxic. 2009;28(2):152–60.

116. Rogers JM. Tobacco and pregancy: overview of exposures and effects. Birth Defects Res C Embryo Today. 2008;84:1–15.

117. Adam T, McAughey J, McGrath C, Mocker C, Zimmermann R. Simultaneous on-line size and chemical analysis of gas phase and particulate phase of cigarette mainstream smoke. Anal Bioanal Chem. 2009;394(4):1193–203.

118. Reeves S, Bernstein I. Effects of maternal tobacco-smoke exposure on fetal growth and neonatal size. Expert Rev Obst Gyn. 2008;3(6):719–30.
119. Fried PA, O'Connell CM. A comparison of the effects of prenatal exposure to tobacco, alcohol, cannabis and caffeine on birth size and subsequent growth. Neurotoxicol Teratol. 1987;9(2):79–85.
120. Zarén B, Lindmark G, Bakketeig L. Maternal smoking affects fetal growth more in the male fetus. Paediatr Perinat Epidemiol. 2000;14(2):118–26.
121. Langley-Evans SC. Developmental programming of health and disease. Proc Nutr Soc. 2006;65(1):97–105.
122. Ross MG, Beall MH. Adult sequelae of intrauterine growth restriction. Semin Perinatol. 2008;32(3):213–8.
123. von Kries R, Toschke AM, Koletzko B, Slikker Jr W. Maternal smoking during pregnancy and childhood obesity. Am J Epidemiol. 2002;156(10):954–61.
124. Sharma AJ, Cogswell ME, Li R. Dose-response associations between maternal smoking during pregnancy and subsequent childhood obesity: effect modification by maternal race/ethnicity in a low-income US cohort. Am J Epidemiol. 2008;168(9):995–1007.
125. Somm E, Schwitzgebel VM, Vauthay DM, Aubert ML, Hüppi PS. Prenatal nicotine exposure and the programming of metabolic and cardiovascular disorders. Mol Cell Endocrinol. 2009;304(1–2):69.
126. Willford JA, Chandler LS, Goldschmidt L, Day NL. Effects of prenatal tobacco, alcohol and marijuana exposure on processing speed, visual–motor coordination, and interhemispheric transfer. Neurotoxicol Teratol. 2010;32:580. In Press, Corrected Proof.
127. Hultman CM, Sparén P, Cnattingius S. Perinatal risk factors for infantile autism. Epidemiology. 2002;13(4):417–23.
128. Fleming P, Blair PS. Sudden infant death syndrome and parental smoking. Early Hum Dev. 2007;83(11):721–5.
129. Carey MA, Card JW, Voltz JW, Arbes Jr SJ, Germolec DR, Korach KS, Zeldin DC. It's all about sex: gender, lung development and lung disease. Trends Endoc Met. 2007;18(8):308–13.
130. Lux AL, Henderson AJ, Pocock SJ, The ALSPAC Study Team. Wheeze associated with prenatal tobacco smoke exposure: a prospective, longitudinal study. Arch Dis Child. 2000; 83(4):307–12.
131. Magnusson LL, Olesen AB, Wennborg H, Olsen J. Wheezing, asthma, hayfever, and atopic eczema in childhood following exposure to tobacco smoke in fetal life. Clin Exp Allergy. 2005;35(12):1550–6.
132. Best D, Committee on Environmental Health, Committee on Native American Child Health, Committee on Adolescence. Secondhand and prenatal tobacco smoke exposure. Pediatrics. 2009;124(5):e1017–44.
133. Lutterodt MC, Sorensen KP, Larsen KB, Skouby SO, Andersen CY, Byskov AG. The number of oogonia and somatic cells in the human female embryo and fetus in relation to whether or not exposed to maternal cigarette smoking. Hum Reprod. 2009;24(10):2558–66.
134. Hart R, Sloboda D, Doherty D, Norman RJ, Atkinson HC, Newnham JP, Dickinson JE, Hickey M. Prenatal determinants of uterine volume and ovarian reserve in adolescence. J Clin Endocrinol Metab. 2009;94(12):4931.
135. Jensen TK, Henriksen TB, Hjollund NHI, Scheike T, Kolstad H, Giwercman A, Ernst E, Bonde JP, Skakkebaek NE, Olsen J. Adult and prenatal exposures to tobacco smoke as risk indicators of fertility among 430 danish couples. Am J Epidemiol. 1998;148(10):992–7.
136. Weinberg CR, Wilcox AJ, Baird DD. Reduced fecundability in women with prenatal exposure to cigarette smoking. Am J Epidemiol. 1989;129(5):1072–8.
137. Thorup J, Cortes D, Petersen BL. The incidence of bilateral cryptorchidism is increased and the fertility potential is reduced in sons born to mothers who have smoked during pregnancy. J Urol. 2006;176(2):734–7.
138. Meeker JD, Missmer SA, Cramer DW, Hauser R. Maternal exposure to second-hand tobacco smoke and pregnancy outcome among couples undergoing assisted reproduction. Hum Reprod. 2007;22(2):337–45.

139. Strohsnitter WC, Hatch EE, Hyer M, Troisi R, Kaufman RH, Robboy SJ, Palmer JR, Titus-Ernstoff L, Anderson D, Hoover RN, Noller KL. The association between in utero cigarette smoke exposure and age at menopause. Am J Epidemiol. 2008;167(6):727–33.
140. Mamsen LS, Lutterodt MC, Andersen EW, Skouby SO, Sørensen KP, Andersen CY, Byskov AG. Cigarette smoking during early pregnancy reduces the number of embryonic germ and somatic cells. Hum Reprod. 2010;25(11):2755–61.
141. Danforth DR. Endocrine and paracrine control of oocyte development. Obstet Gynecol. 1995;172(2, Part 2):747–52.
142. Karman BN, Flaws JA. The aryl hydrocarbon receptor may regulate the number of germ cell nests during neonatal life in the mouse. Biol Reprod. 2008;78(223):715.
143. Kee K, Flores M, Cedars MI, Pera RAR. Human primordial germ cell formation is diminished by exposure to environmental toxicants acting through the AHR signaling pathway. Toxicol Sci. 2010;117(1):218–24.
144. Jurisicova A, Taniuchi A, Li H, Shang Y, Antenos M, Detmar J, Xu J, Matikainen T, Benito Hernández A, Nunez G, Casper RF. Maternal exposure to polycyclic aromatic hydrocarbons diminishes murine ovarian reserve via induction of Harakiri. J Clin Invest. 2007;117(12):3971–8.
145. Blackburn CW, Peterson CA, Hales HA, Carrell DT, Jones KP, Urry RL, Peterson CM. Nicotine, but not cotinine, has a direct toxic effect on ovarian function in the immature gonadotropin-stimulated rat. Reprod Toxicol. 1994;8(4):325–31.
146. Racowsky C, Hendricks RC, Baldwin KV. Direct effects of nicotine on the meiotic maturation of hamster oocytes. Reprod Toxicol. 1989;3(1):13–21.
147. Liu Y, Li G, Rickords LF, White KL, Sessions BR, Aston KI, Bunch TD. Effect of nicotine on in vitro maturation of bovine oocytes. Anim Reprod Sci. 2008;103(1–2):13–24.
148. Smith LM, Cloak CC, Poland RE, Torday J, Ross MG. Prenatal nicotine increases testosterone levels in the fetus and female offspring. Nicotine Tob Res. 2003;5(3):369–74.
149. Kandel DB, Udry JR. Prenatal effects of maternal smoking on daughters' smoking: nicotine or testosterone exposure? Am J Public Health. 1999;89(9):1377–83.
150. Ehrmann DA. Polycystic ovary syndrome. N Engl J Med. 2005;352(12):1223–36.
151. de Zegher F, Ibanez L. Early origins of polycystic ovary syndrome: hypotheses may change without notice. J Clin Endocrinol Metab. 2009;94(10):3682–5.
152. Steckler TL, Herkimer C, Dumesic DA, Padmanabhan V. Developmental programming: excess weight gain amplifies the effects of prenatal testosterone excess on reproductive cyclicity–implication for polycystic ovary syndrome. Endocrinology. 2009;150(3):1456–65.
153. Homburg R. Androgen circle of polycystic ovary syndrome. Hum Reprod. 2009;24(7):1548–55.
154. Hernández-Ochoa I, Karman BN, Flaws JA. The role of the aryl hydrocarbon receptor in the female reproductive system. Biochem Pharmacol. 2009;77(4):547–59.
155. Pocar P, Fischer B, Klonisch T, Hombach-Klonisch S. Molecular interactions of the aryl hydrocarbon receptor and its biological and toxicological relevance for reproduction. Reproduction. 2005;129(4):379–89.
156. Tam PPL, Liu WK. Gonadal development and fertility of mice treated prenatally with cadmium during the early organogenesis stages. Teratology. 1985;32(3):453–62.
157. Dávila-Esqueda ME, Jiménez-Capdeville ME, Delgado JM, De la Cruz E, Aradillas-García C, Jiménez-Suárez V, Escobedo RF, Llerenas JR. Effects of arsenic exposure during the pre- and postnatal development on the puberty of female offspring. Exp Toxicol Path. 2010.Jun 25. [Epub ahead of print].
158. McGivern RF, Sokol RZ, Berman NG. Prenatal lead exposure in the rat during the third week of gestation: long-term behavioral, physiological, and anatomical effects associated with reproduction. Toxicol Appl Pharmacol. 1991;110(2):206–15.
159. Ronis MJ, Badger TM, Shema SJ, Roberson PK, Shaikh F. Reproductive toxicity and growth effects in rats exposed to lead at different periods during development. Toxicol Appl Pharmacol. 1996;136(2):361–71.

160. Nava-Hernández MP, Hauad-Marroquín LA, Bassol-Mayagoitia S, García-Arenas G, Mercado-Hernández R, Echávarri-Guzmán MA, Cerda-Flores RM. Lead-, cadmium-, and arsenic-induced DNA damage in rat germinal cells. DNA Cell Biol. 2009;28(5):241–8.
161. Ramlau-Hansen CH, Thulstrup AM, Storgaard L, Toft G, Olsen J, Bonde JP. Is prenatal exposure to tobacco smoking a cause of poor semen quality? A follow-up study. Am J Epidemiol. 2007;165(12):1372–9.
162. Swan SH, Elkin EP, Fenster L. The question of declining sperm density revisited: an analysis of 101 studies published 1934–1996. Environ Health Perspect. 2000;108(10):961–6.
163. Jensen MS, Toft G, Thulstrup AM, Bonde JP, Olsen J. Cryptorchidism according to maternal gestational smoking. Epidemiology. 2007;18(2):220–5.
164. Biggs ML, Baer A, Critchlow CW. Maternal, delivery, and perinatal characteristics associated with cryptorchidism: a population-based case-control study among births in Washington State. Epidemiology. 2002;13(2):197–204.
165. Brouwers MM, Feitz WF, Roelofs LA, Kiemeney LA, de Gier RP, Roeleveld N. Risk factors for hypospadias. Eur J Pediatr. 2007;166(7):671–8.
166. Morales-Suárez-Varela MM, Bille C, Christensen K, Olsen J. Smoking habits, nicotine use, and congenital malformations. Obstet Gynecol. 2006;107(1):51–7.
167. Pierik FH, Burdorf A, Deddens J, Juttmann R, Weber R. Maternal and paternal risk factors for cryptorchidism and hypospadias: a case–control study in newborn boys. Environ Health Perspect. 2004;112(15):1570–6.
168. McBride ML, Van den Steen N, Lamb CW, Gallagher RP. Maternal and gestational factors in cryptorchidism. Int J Epidemiol. 1991;20(4):964–70.
169. Yamamoto Y, Isoyama E, Sofikitis N, Miyagawa I. Effects of smoking on testicular function and fertilizing potential in rats. Urol Res. 1998;26(1):45–8.
170. Sofikitis N, Miyagawa I, Dimitriadis D, Zavos P, Sikka S, Hellstrom W. Effects of smoking on testicular function, semen quality and sperm fertilizing capacity. J Urol. 1995;154(3):1030–4.
171. Sharpe RM. Environmental/lifestyle effects on spermatogenesis. Philos T Roy Soc B. 2010;365(1546):1697–712.
172. Clark AM, Garland KK, Russell LD. Desert hedgehog (Dhh) gene is required in the mouse testis for formation of adult-type Leydig cells and normal development of peritubular cells and seminiferous tubules. Biol Reprod. 2000;63(6):1825–38.
173. Canto P, Vilchis F, Soderlund D, Reyes E, Mendez JP. A heterozygous mutation in the desert hedgehog gene in patients with mixed gonadal dysgenesis. Mol Hum Reprod. 2005;11(11):833–6.
174. Svechnikov K, Landreh L, Weisser J, Izzo G, Colón E, Svechnikova I, Söder O. Origin, development and regulation of human leydig cells. Horm Res Paed. 2010;73(2):93–101.
175. Fowler PA, Cassie S, Rhind SM, Brewer MJ, Collinson JM, Lea RG, Baker PJ, Bhattacharya S, O'Shaughnessy PJ. Maternal smoking during pregnancy specifically reduces human fetal desert hedgehog gene expression during testis development. J Clin Endocrinol Metab. 2008;93(2):619–26.
176. Rabin D, Gold PW, Margioris A, Chrousos GP. Stress and reproduction: interactions between the stress and reproductive axis. In: Chrousos GP, Loriaux DL, Gold PW, editors. Mech phys emot stress. New York: Prentice-Hall; 1988. p. 377–87.
177. Kalantaridou SN, Zoumakis E, Makrigiannakis A, Lavasidis LG, Vrekoussis T, Chrousos GP. Corticotropin-releasing hormone, stress and human reproduction: an update. J Reprod Immune. 2010;85(1):33–9.
178. Balfour DJK. Chapter 16 Influence of nicotine on the release of monoamines in the brain. In: Nordberg A, Fuxe K, Holmstedt B, Sundwall A, editors. Prog Brain Res. Vol. 79. New York: Elsevier; 1989. pp. 165–172. doi: 10.1016/S0079–6123(08)62476–0.
179. Sarasin A, Schlumpf M, Müller M, Fleischmann I, Lauber ME, Lichtensteiger W. Adrenal-mediated rather than direct effects of nicotine as a basis of altered sex steroid synthesis in fetal and neonatal rat. Reprod Toxicol. 2003;17(2):153–62.

180. Fan JM, Chen XQ, Jin H, Du JZ. Gestational hypoxia alone or combined with restraint sensitizes the hypothalamic–pituitary–adrenal axis and induces anxiety-like behavior in adult male rat offspring. Neuroscience. 2009;159(4):1363–73.
181. Bureau MA, Shapcott D, Berthiaume Y, Monette J, Blouin D, Blanchard P, Begin R. Maternal cigarette smoking and fetal oxygen transport: a study of P50, 2, 3-diphosphoglycerate, total hemoglobin, hematocrit, and type F hemoglobin in fetal blood. Pediatrics. 1983;72(1):22.
182. Longo LD. Carbon monoxide in the pregnant mother and fetus and its exchange across the placenta. Ann N Y Acad Sci. 1970;174:313–41. Biological Effects of Carbon Monoxide.
183. Xiao D, Huang X, Yang S, Zhang L. Direct effects of nicotine on contractility of the uterine artery in pregnancy. J Pharmacol Exp Ther. 2007;322(1):180–5.
184. Milart P, Kauffels W, Schneider J. Vasoactive effects of nicotine in human umbilical arteries. Zbl Gynäkol. 1994;116(4):217–9.
185. McDonald S, Walker M, Perkins S, Beyene J, Murphy K, Gibb W, Ohlsson A. The effect of tobacco exposure on the fetal hypothalamic-pituitary-adrenal axis. BJOG. 2007;113(11):1289–95.
186. Von Zigler NI, Schlumpf M, Lichtensteiger W. Prenatal nicotine exposure selectively affects perinatal forebrain aromatase activity and fetal adrenal function in male rats. Dev Brain Res. 1992;62(1):23–31.
187. Huang M, Wei N, Hu J, Xu Y. Effect of exogenous androgen on structures of sexually dimorphism nucleus in pre-optic area and anteroventral periventricular nucleus before sexual differentiation in female rats. Zhejiang Daxue Xuebao. 2008;37(5):483–6.
188. Sullivan SD, Moenter SM. Prenatal androgens alter GABAergic drive to gonadotropin-releasing hormone neurons: Implications for a common fertility disorder. P Natl Acad Sci USA. 2004;101(18):7129–34.
189. Dumesic DA, Patankar MS, Barnett DK, Lesnick TG, Hutcherson BA, Abbott DH. Early prenatal androgenization results in diminished ovarian reserve in adult female rhesus monkeys. Hum Reprod. 2009;24(12):3188–95.
190. Petersen SL, Curran M, Marconi S, Carpenter C, Lubbers L, McAbee M. Distribution of mRNAs encoding the arylhydrocarbon receptor, arylhydrocarbon receptor nuclear translocator, and arylhydrocarbon receptor nuclear translocator-2 in the rat brain and brainstem. J Comp Neurol. 2000;427(3):428–39.
191. Erlander MG, Tobin AJ. The structural and functional heterogeneity of glutamic acid decarboxylase: a review. Neurochem Res. 1991;16(3):215–26.
192. Ottem EN, Godwin JG, Krishnan S, Petersen SL. Dual-phenotype GABA/glutamate neurons in adult preoptic area: sexual dimorphism and function. J Neurosci. 2004;24(37):8097–105.
193. Petersen SL, Ottem EN, Carpenter CD. Direct and indirect regulation of gonadotropin-releasing hormone neurons by estradiol. Biol Reprod. 2003;69(6):1771–8.
194. Mably TA, Bjerke DL, Moore RW, Gendron-Fitzpatrick A, Peterson RE. In utero and lactational exposure of male rats to 2, 3, 7, 8-tetrachlorodibenzo-p-dioxin 3. Effects on spermatogenesis and reproductive capability. Toxicol Appl Pharmacol. 1992;114(1):118–26.
195. Bjerke DL, Brown TJ, MacLusky NJ, Hochberg RB, Peterson RE. Partial demasculinization and feminization of sex behavior in male rats by in utero and lactational exposure to 2,3,7,8-tetrachlorodibenzo-p-dioxin is not associated with alterations in estrogen receptor binding or volumes of sexually differentiated brain nuclei. Toxicol Appl Pharmacol. 1994;127(2):258–67.
196. Weiss B. Sexually dimorphic nonreproductive behaviors as indicators of endocrine disruption. Environ Health Perspect. 2002;110(Suppl 3):387–91.
197. Gocze PM, Szabo I, Freeman DA. Influence of nicotine, cotinine, anabasine and cigarette smoke extract on human granulosa cell progesterone and estradiol synthesis. Gynecol Endocrinol. 1999;13(4):266–72.
198. Chen Y, Jefferson W, Newbold R, Padilla-Banks E, Pepling M. Estradiol, progesterone, and genistein inhibit oocyte nest breakdown and primordial follicle assembly in the neonatal mouse ovary in vitro and in vivo. Endocrinology. 2007;148(8):3580–90.

199. Guo Y, Guo K, Huang L, Tong X, Li X. Effect of estrogen deprivation on follicle/oocyte maturation and embryo development in mice. Chin Med J. 2004;117(4):498–502.
200. Holloway A, Kellenberger L, Petrik J. Fetal and neonatal exposure to nicotine disrupts ovarian function and fertility in adult female rats. Endocrine. 2006;30(2):213–6.
201. Zhang W, Pang F, Huang Y, Yan P, Lin W. Cadmium exerts toxic effects on ovarian steroid hormone release in rats. Toxicol Lett. 2008;182(1–3):18–23.
202. Zhang W, Wu Z, Li H, Tatsuo E, Takehiko K. Effects of cadmium as a possible endocrine disruptor upon the serum level of sex steroids and the secretion of gonadotropins from pituitary in adult rats. Acta Med Nagasaki. 2002;47(1–2):53–6.
203. Henson MC, Chedrese PJ. Endocrine disruption by cadmium, a common environmental toxicant with paradoxical effects on reproduction. Exp Biol Med. 2004;229(5):383–92.
204. Lafuente A, Cano P, Esquifino A. Are cadmium effects on plasma gonadotropins, prolactin, ACTH, GH and TSH levels, dose-dependent? Biometals. 2003;16(2):243–50.
205. Pillai S, Priya L, Gupta S. Effects of combined exposure to lead and cadmium on the hypothalamic–pituitary axis function in proestrous rats. Food Chem Toxicol. 2003;41(3):379–84.
206. Bianco AC, Nunes MT, Hell NS, Maciel RM. The role of glucocorticoids in the stress-induced reduction of extrathyroidal 3, 5, 3′′-triiodothyronine generation in rats. Endocrinology. 1987;120(3):1033–8.
207. Kakucska I, Qi Y, Lechan RM. Changes in adrenal status affect hypothalamic thyrotropin-releasing hormone gene expression in parallel with corticotropin-releasing hormone. Endocrinology. 1995;136(7):2793–4.
208. Helmreich DL, Parfitta DB, Lub X, Akilb H, Watson SJ. Relation between the hypothalamic-pituitary-thyroid (HPT) axis and the hypothalamic-pituitary-adrenal (HPA) axis during repeated stress. Neuroendocrinology. 2005;81(3):183–92.
209. Decherf S, Seugnet I, Fini J, Clerget-Froidevaux M, Demeneix BA. Disruption of thyroid hormone-dependent hypothalamic set-points by environmental contaminants. Mol Cell Endocrinol. 2010;323(2):172–82.
210. Pringle PJ, Stanhope R, Hindmarsh P, Brook CGD. Abnormal pubertal development in primary hypothyroidism. Clin Endocrinol (Oxf). 1988;28(5):479–86.
211. Pesce L, Zimmerman D. Early puberty and hyperthyroidism. In: Early puberty and hyperthyroidism, A Case-based guide to clinical endocrinology, contemporary endocrinology. Totowa: Humana Press; 2008. p. 437–43. doi:10.1007/978–1–60327–103–5_48.
212. Tziomalos K, Charsoulis F. Endocrine effects of tobacco smoking. Clin Endocrinol (Oxf). 2004;61(6):664–74.
213. Utiger R. Effects of smoking on thyroid function. Eur J Endocrinol. 1998;138(4):368–9.
214. Muller B, Zulewski H, Huber P, Ratcliffe JG, Staub JJ. Impaired action of thyroid hormone associated with smoking in women with hypothyroidism. N Eng J Med. 1995;333(15):964–9.
215. Vanderlaan JE, Vanderlaan WP. The iodide concentrating mechanism of the rat thyroid and its inhibition by thiocyanate. Endocrinology. 1947;40(6):403–16.
216. Fukayama H, Nasu M, Murakami S, Sugawara M. Examination of antithyroid effects of smoking products in cultured thyroid follicles: only thiocyanate is a potent antithyroid agent. Acta Endocrinol. 1992;127(6):520–5.
217. Fritsche E, Cline J, Nguyen N, Scanlan T, Abel J. Polychlorinated biphenyls disturb differentiation of normal human neural progenitor cells: clue for involvement of thyroid hormone receptors. Environ Health Perspect. 2005;113(7):871.
218. Miyazaki W, Iwasaki T, Takeshita A, Tohyama C, Koibuchi N. Identification of the functional domain of thyroid hormone receptor responsible for polychlorinated biphenyl–mediated suppression of its action in vitro. Environ Health Perspect. 2008;116(9):1231–6.
219. Wu C, Zukerberg L, Ngwu C, Harlow E, Lees J. In vivo association of E2F and DP family proteins. Mol Cell Biol. 1995;15(5):2536–46.
220. Pontikides N, Krassas G. Influence of cigarette smoking on thyroid function, goiter formation and autoimmune thyroid disorders. Hormones. 2007;1(2):91–8.

221. Knudsen N, Bulow I, Laurberg P, Ovesen L, Perrild H, Jorgensen T. Parity is associated with increased thyroid volume solely among smokers in an area with moderate to mild iodine deficiency. Eur J Endocrinol. 2002;146(1):39–43.
222. McDonald SD, Walker MC, Ohlsson A, Murphy KE, Beyene J, Perkins SL. The effect of tobacco exposure on maternal and fetal thyroid function. Eur J Obstet Gyn R B. 2008;140(1): 38–42.
223. Ericsson UB, Ivarsson SA, Persson PH. Thyroglobulin in cord blood. The influence of the mode of delivery and the smoking habits of the mother. Eur J Ped. 1987;146(1):44–7.
224. Gasparoni A, Autelli M, Ravagni-Probizer M, Bartoli A, Regazzi-Bonora M, Chirico G, Rondini G. Effect of passive smoking on thyroid function in infants. Eur J Endocrinol. 1998;138(4):379–82.
225. Kolasa E, Hulanicka B, Waliszko A. Does exposure to cigarette smoke influence girls maturation? Przegląd epidemiologiczny. 1998;52(3):339–50.
226. Fried PA, James DS, Watkinson B. Growth and pubertal milestones during adolescence in offspring prenatally exposed to cigarettes and marihuana. Neurotoxicol Teratol. 2001;23(5): 431–6.
227. Windham GC, Bottomley C, Birner C, Fenster L. Age at menarche in relation to maternal use of tobacco, alcohol, coffee, and tea during pregnancy. Am J Epidemiol. 2004;159(9): 862–71.
228. Windham GC, Zhang L, Longnecker MP, Klebanoff M. Maternal smoking, demographic and lifestyle factors in relation to daughter's age at menarche. Paediatr Perinat Epidemiol. 2008;22(6):551–61.
229. Ferris JS, Flom JD, Tehranifar P, Mayne ST, Terry MB. Prenatal and childhood environmental tobacco smoke exposure and age at menarche. Paediatr Perinat Epidemiol. 2010;24(6): 515–23.
230. Palmert MR, Hayden DL, Mansfield MJ, Crigler Jr JF, Crowley Jr WF, Chandler DW, Boepple PA. The longitudinal study of adrenal maturation during gonadal suppression: evidence that adrenarche is a gradual process. J Clin Endocrinol Metab. 2001;86(9):4536–42.
231. Sklar CA, Kaplan SL, Grumbach MM. Evidence for dissociation between adrenarche and gonadarche: studies in patients with idiopathic precocious puberty, gonadal dysgenesis, isolated gonadotropin deficiency, and constitutionally delayed growth and adolescence. J Clin Endocrinol Metab. 1980;51(3):548–56.
232. Den Hond E, Schoeters G. Endocrine disrupters and human puberty. Int J Androl. 2006;29(1): 264–90.
233. Orlando EF, Guillette Jr LJ. A re-examination of variation associated with environmentally stressed organisms. Hum Reprod Update. 2001;7(3):265–72.
234. DiFranza JR, Aligne CA, Weitzman M. Prenatal and postnatal environmental tobacco smoke exposure and children's health. Pediatrics. 2004;113(4):1007–15.
235. Van den Bergh BRH, Van Calster B, Smits T, Van Huffel S, Lagae L. Antenatal maternal anxiety is related to HPA-Axis dysregulation and self-reported depressive symptoms in adolescence: a prospective study on the fetal origins of depressed mood. Neuropsychopharmacology. 2008;33:536–45.
236. Ward AMV, Syddall HE, Wood PJ, Chrousos GP, Phillips DIW. Fetal programming of the hypothalamic-pituitary-adrenal (HPA) axis: low birth weight and central HPA regulation. J Clin Endocrinol Metab. 2004;89(3):1227–33.
237. Kapoor A, Dunn E, Kostaki A, Andrews MH, Matthews SG. Fetal programming of hypothalamo-pituitary-adrenal function: prenatal stress and glucocorticoids. J Physiol. 2006; 572(1):31–44.
238. Hinson JP, Raven PW. Effects of endocrine-disrupting chemicals on adrenal function. J Clin End Metab. 2006;20(1):111–20.
239. Terry MB, Ferris JS, Pilsner R, Flom JD, Tehranifar P, Santella RM, Gamble MV, Susser E. Genomic DNA methylation among women in a multiethnic New York city birth cohort. Cancer Epidem Biomar. 2008;17(9):2306–10.

240. Breton CV, Byun H, Wenten M, Pan F, Yang A, Gilliland FD. Prenatal tobacco smoke exposure affects global and gene-specific DNA methylation. Am J Respir Crit Care Med. 2009; 180(5):462–7.
241. Toledo-Rodriguez M, Lotfipour S, Leonard G, Perron M, Richer L, Veillette S, Pausova Z, Paus T. Maternal smoking during pregnancy is associated with epigenetic modifications of the brain-derived neurotrophic factor-6 exon in adolescent offspring. Am J Med Genet B Neuropsychiatr Genet. 2010;153(7):1350.
242. Hylkema MN, Blacquiere MJ. Intrauterine effects of maternal smoking on sensitization, asthma, and chronic obstructive pulmonary disease. Proc Am Thorac Soc. 2009;6(8):660–2.
243. Holloway AC, Cuu DQ, Morrison KM, Gerstein HC, Tarnopolsky MA. Transgenerational effects of fetal and neonatal exposure to nicotine. Endocrine. 2007;31(3):254–9.
244. Sasaki S, Kondo T, Sata F, Saijo Y, Katoh S, Nakajima S, Ishizuka M, Fujita S, Kishi R. Maternal smoking during pregnancy and genetic polymorphisms in the Ah receptor, CYP1A1 and GSTM1 affect infant birth size in Japanese subjects. Mol Hum Reprod. 2006;12(2): 77–83.
245. Lammer EJ, Shaw GM, Iovannisci DM, Van Waes J, Finnell RH. Maternal smoking and the risk of orofacial clefts: susceptibility with NAT1 and NAT2 polymorphisms. Epidemiology. 2004;15(2):150–6.

Chapter 3
Reproductive Neuroendocrine Targets of Developmental Exposure to Endocrine Disruptors

Sarah M. Dickerson, Stephanie L. Cunningham, and Andrea C. Gore

Abstract Recent scientific evidence has advanced our understanding of how exogenous environmental chemicals influence developing organisms. The documented effects of industrial compounds introduced into the environment by humans include actions on hormonal systems, including the reproductive, growth, thyroid, and lactotrophic axes. The central focus of this chapter is the impact of environmental endocrine-disrupting chemicals (EDCs) on the developing reproductive neuroendocrine axis. Because the developing brain and reproductive organs are sculpted via processes that are largely influenced by steroid hormones, organisms are particularly vulnerable to the actions of EDCs during the period of life spanning late gestation and early postnatal life. A review of the literature regarding the effects of five classes of EDCs on the developing mammal is provided, with specific regard to sexual differentiation of the reproductive hypothalamic-pituitary-gonadal axis. Although the majority of the studies summarized herein focus on laboratory animals, epidemiological information regarding the known effects of EDCs on humans will be presented when available. We conclude that the available experimental animal and epidemiological data are supportive of a potential role of EDCs on alterations of neuroendocrine development, and on pubertal onset.

Keywords 5-alpha reductase • AGD – anogenital distance • AhR – aryl hydrocarbon receptor • Alfalfa • Alpha-fetoprotein • Apoptosis • AR – androgen receptor • ARC – arcuate nucleus • Aroclor • Aromatase • AVPV – anteroventral periventricular nucleus • BPA – bisphenol A • Calbindin • Chlorpyrifos • CNS – central nervous system • Coumestrol • Critical period • Daidzein • DDT – dichlorodiphenyltrichloroethane • Dioxin • EDCs – Endocrine-disrupting chemicals • ER – estrogen receptor • Fenitrothion • FSH – follicle-stimulating hormone • Genistein

---

S.M. Dickerson • S.L. Cunningham • A.C. Gore (✉)
Division of Pharmacology & Toxicology, College of Pharmacy,
University of Texas at Austin, Austin, TX, USA
e-mail: smdickerson2010@gmail.com; slcunningham@utexas.edu;
andrea.gore@austin.utexas.edu

E. Diamanti-Kandarakis and A.C. Gore (eds.), *Endocrine Disruptors and Puberty*,
Contemporary Endocrinology, DOI 10.1007/978-1-60761-561-3_3,
© Springer Science+Business Media, LLC 2012

• Glucocorticoid • GnRH – gonadotropin-releasing hormone • Gonadotropin • Growth • GT1 • HPG – hypothalamic-pituitary-gonadal • Hypothalamus • Isoflavone • Kisspeptin • LH – luteinizing hormone • Methoxychlor • Neuroendocrine (neuroendocrinology) • Organochlorine • Organohalogen • PBB – polybrominated biphenyl • PBDE – polybrominated diphenyl ether • PCB – polychlorinated biphenyl • Pesticide • Phthalate • Phytoestrogen • Pituitary • POA – preoptic area • POP – persistent organic pollutant • PR – progesterone receptor • Puberty • Pyrethroid • Resveratrol • SDN-POA – sexually dimorphic nucleus of the preoptic area • Sexual behavior • Sexual differentiation • Soy • Tefluthrin • Thyroid • Tyrosine hydroxylase

Introduction

Hypothalamic Control of Reproduction

Endocrine-disrupting chemicals (EDCs) are chemicals in our environment that interfere with the normal functioning of an organism's endocrine systems, including those involved in reproduction. In vertebrate species, reproductive function is regulated by the hypothalamic-pituitary-gonadal (HPG) axis, which includes the gonadotropin-releasing hormone (GnRH), pituitary gonadotropins [luteinizing hormone (LH) and follicle-stimulating hormone (FSH)], and gonadal steroid hormones. The hypothalamus, located at the base of the brain, maintains homeostasis of a variety of physiologic functions by acting as a central processor, integrating converging inputs from sensory and autonomic systems. Importantly, the hypothalamus coordinates cues related to the organism's internal and external environment with the timing of reproductive capacity, in order to ensure that the energetically costly process of reproduction occurs only during the most favorable conditions. Although the hypothalamic GnRH neurons provide the primary driving force onto other HPG levels, all three levels must function properly for the appropriate control of reproduction.

The GnRH neuroendocrine cells controlling reproduction have their neural somata (cell bodies) in the hypothalamus, and extend an axon to the base of the hypothalamus, the median eminence. For the control of HPG processes, GnRH-releasing cells respond to changing environmental conditions by rapidly adjusting their release of the GnRH peptide into the portal capillary vasculature that transports the peptide to the anterior pituitary gland. From the pituitary, LH and FSH are released into the general circulatory system, and target the gonads to drive steroidogenesis and gametogenesis. Gonadal steroids then act upon steroid hormone receptors, primarily estrogen receptors (ERs), androgen receptors (ARs), and progesterone receptors (PRs) that are widely and heterogeneously expressed in target tissues, both reproductive and nonreproductive, including the reproductive tract and genitalia, breast, fat, bone, muscle, kidney, liver, and many other organ systems. Of relevance to the regulation of HPG function, steroid hormone receptors are also expressed in the central nervous system (CNS) and pituitary gland.

These receptors enable steroid feedback to be exerted on the hypothalamic-pituitary levels of the HPG axis.

While GnRH neurons in the hypothalamus control pituitary and gonadal hormone release through feedforward regulatory mechanisms, and in turn are regulated by gonadal steroid hormones through feedback regulatory mechanisms, it is important to note that GnRH neurons themselves do not express most of the nuclear sex steroid hormone receptors. This raises the question: how are GnRH neurons regulated by steroid feedback? The answer is provided by observations that the CNS, and particularly the hypothalamus, is abundant in ERs, ARs, and PRs [1]. A network of these steroid-sensitive neurons in the hypothalamus converges upon the GnRH cells. This neural circuitry is quite heterogeneous, as the neurons can arise from a variety of locations, and they can release a diversity of neurotransmitters. This circuit may seem unduly complex, but it makes sense in considering that reproduction must be properly coordinated with the internal and external environment. Many CNS pathways are needed to convey all of the specific types of information to the GnRH system, which is the final common pathway to the pituitary gland.

Sex Differences in HPG Function

There are important sex differences in the control of reproductive physiology and behavior. Although both male and female mammals release GnRH in a pulsatile manner, at intervals of about 30–120 min depending upon species, the GnRH neurosecretory system differs between males and females in a fundamental way. In females, the amplitude and frequency of GnRH release varies profoundly across the reproductive cycle, whereas in males, GnRH is fairly consistently released in pulses. Moreover, feedback of gonadal steroid hormones onto GnRH cells is always negative in males, whereas it shifts from negative to positive during the preovulatory surge in females. CNS inputs to GnRH neurons are also sexually dimorphic in their regulation of GnRH release and in their neuroanatomical properties. For example, important inputs to GnRH cells arise from the anteroventral periventricular nucleus (AVPV), which regulates GnRH release via expression of ERs [2, 3], and is approximately two times larger in volume and cell number/density in adult females than in males [4, 5]. The preovulatory GnRH/gonadotropin surge in females is regulated by inputs from the AVPV to GnRH neurons, a process that does not occur in males. Conversely, the sexually dimorphic nucleus of the preoptic area (SDN-POA), the size of which is correlated with regulation of sexual behavior and levels of gonadotropins, has a volume roughly two to four times higher in male rats than in females [6, 7]. Normal reproductive function depends on proper development of these and other sexually dimorphic neural circuits that take shape during the process of brain sexual differentiation, which in rodents occurs during late embryonic/early postnatal development [8].

The concept of a critical period of brain sexual differentiation has been proposed and studied by neuroendocrinologists for decades. However, its importance to environmental endocrine disruption is only just recently becoming clear. A growing body of scientific evidence has illuminated the vulnerability of the developing neonate to exposure to environmental EDCs that may disrupt sexual differentiation, but effects of which may not be manifested until much later in life. The importance of developmental exposures is a focus of the current article, and will be elaborated upon below.

Endocrine Disruption of Neuroendocrine Systems

Environmental endocrine-disrupting chemicals (EDCs) were originally defined by the United States Environmental Protection Agency (EPA) as "exogenous agents that interfere with the synthesis, secretion, transport, binding action, or elimination of natural hormones in the body that are responsible for the maintenance of homeostasis, reproduction, development, and/or, behavior" [9]. The EPA has gone on to expand its definition to: "An exogenous substance that changes endocrine function and causes effects at the level of the organism, its progeny, and/or (sub)populations of organisms." Regardless of which definition one uses, it is clear that EDCs have a diversity of target and act through a number of pathways related to endocrine control of homeostasis. In addition to their effects on the HPG axis, EDCs may also have actions on other neuroendocrine hormonal systems, including the thyroid (hypothalamic-pituitary-thyroid axis), stress (hypothalamic-pituitary-adrenal axis), growth (somatotropic axis), lactotrophic axis, neuropeptides, and other metabolic hormones. This chapter will primarily focus on the effects of developmental EDC exposure on reproductive neuroendocrine targets in the perinatal, pubertal, and adult mammal. Regarding the effects of developmental EDC exposure on other hormonal systems, refer to the following reviews on thyroid [10–12]; glucocorticoids [13, 14]; growth [15]; and general metabolism [16]. Endocrine disruption is also a significant concern for nonmammalian species. Invertebrates and aquatic species may be particularly vulnerable, as they have a large surface-to-volume ratio and lack a barrier such as skin to protect against environmental toxicants, and they are swimming in the very substance (water) in which many EDCs are dissolved or in solution. Egg-laying animals also have vulnerability through exposure of the eggshell to the environment. Although this chapter focuses on effects of EDCs on the HPG axis of mammals, we refer readers to the following articles and reviews on other species/ classes: nematodes [17], Drosophila [18], aquatic invertebrates [19–21], fish [22], reptiles [23, 24], and birds [25].

As will be discussed in greater detail, the mediation of the endocrine-disrupting effects of EDCs on neuroendocrine reproductive function often occurs through (but is not limited to) steroid hormone receptors, particularly ERs and ARs, which are expressed abundantly in the hypothalamus, but are differentially affected by EDCs. Some effects may also occur directly upon GnRH neurons, as illustrated by the next section on GnRH cell lines. Later in this chapter, we will highlight the evidence for neuroendocrine effects of five of the most extensively studied classes of EDCs: phytoestrogens, industrial organohalogen compounds, pesticides, bisphenol A, and phthalates.

EDC Effects on GnRH Cell Lines

GnRH neurons in the hypothalamus are notoriously difficult to study. Their location at the base of the brain makes them inaccessible to traditional electrophysiological measures, as does the fact that they are scattered across a continuum in the preoptic area-hypothalamus. These have been significant obstacles to research in the hypothalamic control of HPG function. The GnRH GT1 cell lines, created in the late 1980s [26] have proven to be valuable resources for understanding GnRH properties. They have many neural properties, and they synthesize and release the GnRH peptide in a pulsatile manner [27]. To date, this cell line has been used to study effects of the EDCs, PCBs, dioxin (TCDD) and organochlorine (methoxychlor, chlorpyrifos) and pyrethroid (tefluthrin) pesticides, and the phytoestrogen coumestrol. A brief summary of those results is provided, with evidence that most of these compounds directly affect these immortalized GnRH cells (Table 3.1).

The first studies to investigate actions of EDs on GT1-7 cells were those from our laboratory, investigating effects of PCBs and organochlorine pesticides. Dose-response analysis of the PCB mixtures, Aroclor 1221 and Aroclor 1254, for 24 h demonstrated differential effects on cell morphology, GnRH release, and GnRH gene expression [28]. Specifically, GnRH peptide release was stimulated by Aroclor 1221 (but not Aroclor 1254), an effect that was blocked by the estrogen receptor antagonist, ICI 182780. GnRH mRNA levels were elevated at the intermediate concentration of 1 μM but were unchanged at higher/lower concentrations. By contrast, Aroclor 1254 stimulated GnRH mRNA levels only at the lowest concentration (0.01 μM). Finally, GT1-7 cell morphology was altered by PCB treatment, with Aroclor 1221 increasing cell confluence and promoting neurite outgrowth. Aroclor 1254 tended to cause some cell death and for GT1-7 cells to retract their neural processes. As a whole, these data support the effects of PCB mixtures on this immortalized cell line.

We have since gone on to investigate the mechanisms by which PCBs may cause neurotoxicity in GT1-7 cells [29]. In that study, we tested three different individual PCB compounds representing different classes: PCB74 (coplanar), PCB118 (dioxin-like coplanar), and PCB153 (noncoplanar), or their combination. Those results, summarized in Table 3.1, demonstrate effects of the individual PCBs on loss of GT1-7 cell viability, attributable to both increased necrosis and apoptosis. In general, the individual PCBs caused necrosis at the highest dosages and times of incubation, whereas apoptosis occurred at lower doses and shorter time points. Surprisingly, we did not find any additivity or synergism of the PCBs in combination, although this may be attributable to our only studying one combination that did not fully explore the range of dosages needed for such phenomena to be observed. GnRH peptide levels were measured in this same study [29] demonstrating generally stimulatory effects at lower dosages/shorter time points, and inhibitory effects at longer time points and higher dosages.

Effects of the pesticides, chlorpyrifos and methoxychlor, have been studied on GnRH peptide and mRNA levels in GT1-7 cells [28]. Neither pesticide affected

Table 3.1 Effects of environmental endocrine disruptors on GnRH GT1 cell lines

ED treatment	GT1 cell morphology/ viability	GnRH peptide concentrations	GnRH gene expression	Cell death mechanism	Reference
PCB (Aroclor 1221) for 24 hrs @ 0.01–100 µM	Stimulate neurite outgrowth and proliferation	Stimulate at 10 µM dose	Stimulate at 1 µM (no effect higher/lower doses)	N.S.	Gore et al. [28]
PCB (Aroclor 1254) for 24 hrs @ 0.01–100 µM	Cause some cell death; retraction of neural processes	No effect	Stimulate at low (0.01 µM) doses (no effect higher doses)	N.S.	Gore et al. [28]
PCB74 for 1, 4, 8, or 24 hrs @ 0.1–100 µM	Decrease viability beginning 1 hr	Stimulate at 1 and 10 µM (4 hr), inhibit at 0.1 and 100 µM (8 hr), inhibit at all doses (24 hr)	N.S.	Increase necrosis at 24 hr (100 µM). Increase apoptosis at 4 hr (all doses), 8 and 24 hr (0.1–10 µM) and activation of cleaved caspase 3/7 (0.1–10 µM)	Dickerson et al. [29]
PCB118 for 1, 4, 8, or 24 hrs @ 0.1–100 µM	Decrease viability beginning 1 hr, especially 0.1–10 µM	Stimulate at 0.1–10 µM (4 hr), inhibit at 1 µM (8 hr), inhibit at 1–100 µM (24 hr)	N.S.	Increase necrosis at 8 hr (100 µM) and 24 hr (all doses). Increase apoptosis at 1–24 hr (especially 0.1–10 µM) and activation of cleaved caspase 3/7 (0.1, 1 µM)	Dickerson et al. [29]
PCB153 for 1, 4, 8, or 24 hrs @ 0.1–100 µM	Decrease viability beginning 4 hr, especially 100 µM	Stimulate at 0.1 and 100 µM (1 hr), stimulate at 100 µM (4 hr), inhibit at 1–100 µM (24 hr)	N.S.	Increase necrosis at 8 hr (100 µM) and 24 hr (all doses). Increase apoptosis at 8 hr (0.1 and 100 µM) and 24 hr (0.1 µM). No effect on cleaved caspase 3/7 activation	Dickerson et al. [29]

Chlorpyrifos for 24 hrs @ 0.01–100 μM	Neurite extension and increased confluency	No effect	Stimulate at 0.01–1 μM; inhibit at 1–100 μM	N.S.	Gore [28]
Methoxychlor for 24 hrs @ 0.01–100 μM	Mild neurotoxicity	No effect	Stimulate at 0.01–1 μM; inhibit at 1–100 μM	N.S.	Gore [28]
TCDD (1–100 nM) for 1–72 hrs	No effect on GT1-7 cell number	No effect on GnRH peptide concentrations	No effect on GnRH promoter-reporter expression in transfected cells	N.S.	Petroff et al. [30]
Coumestrol (1 nM–10 μM) for 6–48 hr	N.S.	N.S.	Coumestrol (1 μM) inhibits GnRH mRNA; greatest effect of 1 μM @ 6 hr	N.S.	Bowe et al. [31]

N.S. = not studied.

GnRH peptide concentrations. Chlorpyrifos increased confluency of the cells and stimulated neurite extension, whereas methoxychlor was mildly neurotoxic. Effects of the pesticides on GnRH gene expression were similar: both stimulated GnRH mRNA levels at lower concentrations, and inhibited levels at the highest concentrations.

Other laboratories have investigated actions of other EDs on more limited endpoints in GT1 cells. Petroff et al. ([30]; Table 3.1) showed that the dioxin TCDD had no effect on GnRH secretion, on promoter-reporter expression in transfected GT1-7 cells, or on GT1-7 cell number [30]. They also demonstrated that their GT1-7 cells did not express the aryl hydrocarbon receptor, possibly explaining the lack of effect of dioxin on this orphan receptor.

The effects of the pyrethroid insecticide tefluthrin were tested on ion currents in GT1-7 cells [32]. Cells were treated with this compound at 10 μM, and currents were measured. Current amplitude and firing frequency were both increased by tefluthrin application. It will be interesting to examine other effects of this pyrethroid, including GnRH peptide release and gene expression.

Finally, coumestrol has been studied for its effects on GnRH gene expression in GT1-7 cells, with a focus on the mediation of these actions by the ERβ [31]. In general, coumestrol had inhibitory effects on GnRH mRNA levels, with the greatest effect seen at the 6 h time point and the 1 μM concentration (Table 3.1). A role for ERβ was established through use of the specific antagonist, R,R-diethyl tetrahydrochrysene (R,R-THC), which blocked the inhibitory effect of coumestrol on GnRH mRNA levels [31]. These data support the actions of this phytoestrogen directly on GnRH immortalized neurons through this ER subtype.

As a whole, these studies on the GT1-7 cell line support the likelihood that environmental EDCs can directly target GnRH neurons. Although it is critical to extend these studies to in vivo models, they provide a potential mechanism for effects of these compounds on HPG systems.

Properties of Five Classes of EDCs

Regardless of whether they are naturally occurring or man-made, EDCs are chemicals in our environment that act through numerous mechanisms to block, imitate, or otherwise modify normal hormonal function. These alterations to endogenous endocrine activity may involve the direct or indirect interaction of EDCs with steroid hormone or neurotransmitter receptors, alterations in levels of bioavailable endogenous hormones or their receptors by EDCs, or actions on enzymes involved in the biosynthesis or degradation of endogenous hormones, among other potential mechanisms. From a historical perspective, the first endocrine disruptors to be identified were estrogenic in nature, although it is now known that EDCs may also be antiestrogenic or antiandrogenic in nature. Hundreds of chemicals have already been identified as known EDCs, and systematic screening by the Endocrine Disruptor Screening Program (EDSP) by the EPA of pesticides and other environmental contaminants is

3 Reproductive Neuroendocrine Targets of Developmental Exposure...

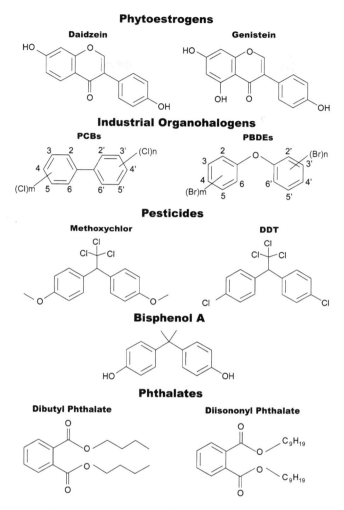

Fig. 3.1 Structures of representative endocrine disruptors

currently underway. This section will highlight five classes of EDCs known to target reproductive neuroendocrine systems: phytoestrogens, industrial organohalogens, pesticides, bisphenol A, and phthalates, structures of which are shown in Fig. 3.1. In this section, we provide background information about these EDCs, and in subsequent sections, we review evidence for their actions on neuroendocrine systems, particularly when exposure occurs during critical developmental windows such as prenatal, early postnatal, and pubertal.

Phytoestrogens: While many consider endocrine disruption to be a relatively recent discipline, the field has its early roots in studies published more than 60 years ago, reporting that plant chemicals could have estrogenic actions in livestock.

For example, breeding difficulties were observed in sheep feeding on red clover pastures in Australia [33], and later similar effects were reported in other species (California quail [34] and deer mice [35]). Exposure to these naturally occurring nonsteroidal plant compounds collectively referred to as phytoestrogens is mainly through the diet, and, due to similarities between their chemical structure and endogenous estradiol, they may have actions as weak estrogens or antiestrogens in an organism. For example, phytoestrogens such as the isoflavones genistein and daidzein bind both estrogen receptors (ERα and ERβ) and initiate ER-dependent transcription [36, 37], although their binding affinity is several orders of magnitude lower than endogenous estradiol [38]. Competition binding assays of resveratrol with recombinant ERα and ERβ show that it has weak binding activity relative to estradiol to both ERs, with about tenfold greater affinity for the ERβ than the ERα [39]. In addition to their interaction with the ER, phytoestrogens have other actions on endocrine systems, and have been shown to increase circulating sex-hormone-binding globulins [40], modulate enzymatic activities of steroidogenic enzymes [41–43], act as antioxidants [44, 45], and alter cellular signal transduction [46]. Not surprisingly, phytoestrogens have an effect on neuroendocrine reproductive physiology across the life cycle, and their effects are more pronounced when exposure occurs during development.

Though diverse in nature, phytoestrogens are categorized into several groups, including the flavones (e.g., luteolin), isoflavones (e.g., genistein, daidzein; Fig. 3.1), flavanols (e.g., catechin), lignans (e.g., pinoresinol), and coumestans (e.g., coumestrol). Dietary sources are the main route of exposure, as phytoestrogens are found in a number of plants, including soy, alfalfa, herbs, fruits, vegetables, nuts, and flax seed, as well as in nutritional supplements, wine (resveratrol), soy-based infant formula, and soy-based baby food. Interestingly, although the public widely regards synthetic EDCs with growing concern, the general perception of phytoestrogens is that they must be beneficial to health, by virtue of their natural origins. Indeed, a diet rich in fruit and vegetables is generally a healthy one relative to one with high meat/fat consumption. However, plant phytoestrogens share a number of cellular and molecular targets in common with man-made EDCs, and increasing evidence suggests that when consumed in excess, particularly through dietary supplementation (supplements are not regulated by the FDA and contain variable and often excessive amounts of a compound), phytoestrogens may also share detrimental effects on reproductive physiological in common with EDCs of industrial origins.

Industrial Organohalogens (PCBs, PBDEs): Industrial organohalogens encompass an array of compounds used for or produced as a by-product of industrial applications. They include a diverse group of chemicals such as polychlorinated biphenyls (PCBs), polybrominated biphenyls (PBBs), polybrominated diphenyl ether flame-retardants (PBDEs), chlorinated paraffins, furans, and dioxins. As a whole, members of this family of persistent organic pollutants (POPs) are generally structurally stable and resistant to environmental or biodegradation, and are routinely detected in the tissues of humans and wildlife.

Much of the information regarding the effects of industrial organohalogens on the reproductive axis comes from studies involving PCBs (Fig. 3.1), a focus of this chapter.

PCBs are a group of related compounds that consist of chlorine atoms arranged around a biphenyl core, a feature that contributes to their ability to interact with various hormone and neurotransmitter receptors as agonists, antagonists, or with mixed activity. Endogenous estradiol has a phenolic moiety in its A ring, and the interaction of this moiety with ERs in particular is mimicked by environmental EDCs such as PCBs. Although widely used between the 1930s and the 1970s, manufacture of PCBs was prohibited in 1977 due to their toxicity. Ironically, the same qualities that made use of PCBs desirable for such industrial applications as capacitors, transformer oil, sealants, paints, etc., contributed to their persistence in the environment and their toxicity. For instance, PCBs resist degradation, and because they bioaccumulate and biomagnify, they are still prevalent as complex mixtures throughout the food chain and in environmental samples. Remarkably, they contaminate even remote areas, such as the Arctic and Antarctic regions [47, 48], due to movement of marine biota associated with fluctuating climate, as well as through air and water currents. Exposure of humans and wildlife to PCBs occurs mainly through consumption of contaminated food, and has been linked to reproductive deficits. For example, farmed salmon has relatively high concentrations of PCBs (among other EDCs) [49] and pregnant women have been advised to limit consumption of this otherwise healthy food choice to once monthly.

Based on their chemical structure, PCBs are grouped into one of the following three categories: coplanar, dioxin-like coplanar, or noncoplanar [29]. In addition, the structural properties of each PCB class determine whether and how (i.e., agonist, antagonist) they interact with hormone and neurotransmitter receptors. Along with actions on ERs, PCBs target a broad range of hormone receptors (thyroid hormone receptor, androgen receptor), neurotransmitter receptors (e.g., acetylcholine, dopamine, GABA, and serotonin), hormone-binding proteins (e.g., thyroid hormone binding protein), as well as orphan receptors (e.g., aryl hydrocarbon receptor) and intracellular signaling receptors (e.g., ryanodine receptor). In vitro studies have shown that the hydroxylated metabolites of PCBs inhibit estradiol sulfotransferase, which may prolong the bioavailability of endogenous estradiol [50]. Furthermore, depending on their structure, individual PCBs may be estrogenic, antiestrogenic, or antiandrogenic. To complicate matters, because PCBs are present in human and wildlife samples as mixtures of the most persistent congeners and their hydroxylated metabolites, additive, synergistic, or antagonistic effects between PCBs may occur [51].

Pesticides (DDT, methoxychlor): Pesticides have a long history of global use, and have a broad spectrum of applications in agricultural and commercial pest control. Insecticides are one of the most widely used classes of pesticides, are categorized as either organochlorine or organophosphorus, and share in common the inactivation of acetylcholine esterase as a mechanism of action. Other classes of pesticides include the herbicides and fungicides, which target unwanted plants and fungi. While these agents have been useful for eradicating undesirable pests and disease vectors, pesticides also have unintended actions on nontargeted species including humans, and recent studies have shown that even extremely low levels of exposure can cause damage to the developing reproductive axis.

Like PCBs, organochlorine pesticides are lipophilic, building up in the fatty tissues and remaining in the body for years. Organochlorine insecticides (Fig. 3.1) include such agents as dichlorodiphenyltrichloroethane (DDT), methoxychlor, dicofol, heptachlor, chlordane, endosulfan, aldrin, dieldrin, endrin, and mirex. Humans are predominantly exposed to organochlorine pesticides as residues on produce or through pest control. Individuals working directly with insecticides come into direct contact through inhalation and skin. Starting in the late 1970s, use of many organochlorine pesticides, such as the insecticide DDT, was banned in response to growing public concern and increasing scientific evidence linking organochlorine pesticide exposure to adverse health effects in both humans and wildlife. In the context of endocrine disruption, most organochlorine pesticides such as kepone, o,p'-DDT, dieldrin, and methoxychlor are weakly estrogenic, although some have also been found to possess antiestrogenic and/or antiandrogenic activities. In addition, because they have actions on neurotransmitter synthesis/degradation and/or receptors, some of which have inputs to the hypothalamus (e.g., cholinergic, GABAergic, dopamine), they may exert indirect effects on neuroendocrine function.

Organochlorine pesticides were replaced by the less persistent, albeit more acutely toxic, organophosphorus pesticides. Although these compounds are not lipophilic and degrade more readily than organochlorine pesticides, because they are applied ubiquitously, humans are routinely exposed to organophosphorus pesticides through contact with contaminated surfaces and consumption of contaminated produce, air, and water. Organophosphorus pesticides include such agents as chlorpyrifos (now banned), parathion, malathion, diazinon, and dichlorvos. Although their effects on the thyroid axis have been reported, the impact of organophosphorus pesticides on reproductive neuroendocrine endpoints has not been rigorously studied, and merits further investigation.

Bisphenol A: Bisphenol A (4,4'-isopropylidenediphenol; BPA; Fig. 3.1), an organic compound with two hydroxyl-substituted phenol groups, was first synthesized in 1891 [52] and is a basic building block of polycarbonate plastic. Originally synthesized for pharmaceutical use as a synthetic estrogen, BPA is formed as an intermediate compound in the manufacturing process for polymers, epoxy resins, polysulfone and polyester resins, dyes, flame-retardants, and rubber chemicals. Yearly global use of BPA is estimated at over 2 million metric tons, and it is found in a multitude of consumer products including water and baby bottles, medical and dental devices, compact discs, optical and automotive lenses, household electronics, carbonless copy paper, color developer, and as coatings on the inside of food and beverage cans.

Traditionally, BPA has been thought to differ from organophosphate compounds in that it does not accumulate in tissues, has a short half-life, and is rapidly excreted by the body. However, BPA can be detected in human fetuses [53, 54], suggesting maternal-fetal transfer. Furthermore, its ubiquitous presence in the environment poses a constant source of exposure to both humans and wildlife alike, and epidemiological studies suggest a link between BPA concentrations and cardiovascular disease, diabetes and abnormal liver enzyme levels [55]. Though it was known since 1936 that BPA was an estrogenic compound [56], it was not discovered that BPA could be leached from polycarbonate plastics in sufficiently

3 Reproductive Neuroendocrine Targets of Developmental Exposure...

high concentrations to affect reproductive processes until the past decade [57]. Of relevance to neuroendocrinologists, exposure during the pre- or postnatal period causes deleterious effects on female reproductive physiology, including pubertal timing, reproductive cycling, and induction of early reproductive senescence [58–61]. In laboratory animals, BPA can disrupt sexual differentiation of the brain, alter expression of sex steroid receptors in the hypothalamus and pituitary, and regulation of gonadotropin levels. As a result of public concern, many retailers have removed BPA-containing products off their shelves, and many vendors have reformulated their products to eliminate BPA as a component.

Phthalates (DEHP, DBP, DIDP, DINP): Phthalates (Fig. 3.1) are used as plasticizers to increase flexibility and durability of plastics. They are alkyl or dialkyl esters of phthalic acid and are used in a wide variety of applications, ranging from gelling agents, adhesives, stabilizers, lubricants, binders, emulsifying agents, personal care products, medical devices, children's toys, paints, food wrappers and containers, pharmaceuticals, and detergents. Some widely used phthalates include benzyl butyl phthalate (BBP), dibutyl phthalate (DBP), dioctyl phthalate (DOP), diisodecyl phthalate (DIDP), di-isononyl phthalate (DINP), di(2-ethylhexyl) phthalate (DEHP), bis(2-ethylhexyl) adipate (DEHA), 4-tert-octylphenol (tOP), 4-chloro-3-methylphenol (CMP), 2,4-dichlorophenol (2,4-DCP), 2-phenylphenol (2-PP), and resorcinol.

In contrast to the other EDCs previously summarized in this chapter, phthalates are not estrogenic in nature. Rather, they exert antiandrogenic actions on endocrine systems, and are thought to impact primarily male development. However, phthalates do not interact with the androgen receptor, unlike other antiandrogens such as flutamide. Rather, by causing a reduction in testosterone synthesis in the fetal testis, exposure to phthalates during early development is associated with a host of disorders to the male reproductive tract (reduced anogenital distance, nipple retention, hypospadias, undescended testes, etc.) [62]. Although most of the research involving phthalates focuses on the male gonad, a few recent studies have also shown that this group of chemicals may also affect neuroendocrine development in laboratory animals, as will be discussed below.

Neuroendocrine Action of EDs During Perinatal Development

Brain development is a highly complex and coordinated sequence of events that begins early on during embryogenesis and continues through puberty and into early adulthood. The perinatal brain is particularly vulnerable to disruption by EDCs, as it is during this time that proper development of the central nervous system and neuroendocrine reproductive axis is fine-tuned by factors that are influenced by endocrine hormones, including genes, neurotransmitters, and growth/survival factors [63, 64]. Importantly, through the developmental process of brain sexual differentiation, the organism acquires sex-appropriate brain morphology and function. As will be discussed in greater detail below, disruption of any of these development processes may permanently alter brain morphology and function, causing disruptions in adult reproductive physiology and behavior later in life. We will begin this section by briefly summarizing the process of brain sexual differentiation. Later sections will highlight evidence that early exposures to EDCs may disrupt this crucial process.

Sexual Differentiation of the Hypothalamus: Hormones and Apoptosis

In vertebrate species, sex differences in reproductive physiology and behavior are attributed to phenotypic differences in the neuroanatomy and neurochemistry of specific brain regions. These differences become organized during critical developmental windows, particularly the late embryonic and early postnatal periods, and are permanent. At these times, sex differences in levels of gonadal hormones are large, and contribute to morphological and functional differences in the size, cell number, phenotype, and neurochemistry of hypothalamic brain regions [5, 7, 65–67]. A key component in the shaping of the brain into a male or female typical pattern is the presence of the testicular hormone testosterone, and its aromatization to estradiol. Neonatal female rodents have relatively low levels of gonadal hormones because the perinatal ovary is quiescent, and the steroid-binding protein α-fetoprotein forms a complex with any circulating endogenous estrogen, which prevents it from entering the brain. Conversely, around the time of birth the testis releases a surge of testosterone in neonatal males, which is not bound by α-fetoprotein [68]. Once it enters the brain, testosterone may act directly upon androgen receptors (ARs), or may it may be converted to estradiol, the endogenous ligand for estrogen receptors (ERs), by the enzyme aromatase. Consequently, while the female rodent brain develops under relatively low levels of steroid hormones, the male brain is exposed to high levels of both testosterone and estradiol. Thus, testosterone and estradiol act directly upon their cognate receptors in a sexually dimorphic manner to organize the brain into a sex-appropriate pattern. Not surprisingly, alterations in the normal hormonal milieu during this phase of development can have deleterious effects, which may manifest as deficits in reproductive function and behavior during adulthood.

Apoptosis, an essential component of normal neuronal development, is one of the mechanisms through which hormones sculpt the brain during sexual differentiation [69]. Depending upon the hypothalamic nucleus, estradiol may stimulate or inhibit apoptosis through the interaction of ligand-bound ER dimers with nuclear response elements that promote cell survival or death. For instance, the number of apoptotic cells in the medial preoptic nucleus, spinal nucleus of the bulbocavernosis (SNB) [70], bed nucleus of the stria terminalis (BNST) [71], and SDN-POA [72] is higher in female rodents than in males during early postnatal development. In males, these regions are involved in penile erection, and in masculine sexual physiology and behavior. On the other hand, the number of apoptotic cells in the anteroventral periventricular nucleus (AVPV) is greater in male rodents compared to females during a similar time period [5, 73], and this region is accordingly larger in females than in males.

There are discrete hormone-sensitive time periods during which the volumes of these particular regions are vulnerable to substances that act on steroid hormone receptors [74, 75]. For example, exposure of developing female rats to testosterone or excessive levels of estradiol (more than can be bound by circulating alpha-fetoprotein) results in a larger, masculinized SDN-POA in adulthood, while castration of neonatal males results in a smaller, feminized SDN-POA [74]. Similarly, the rat AVPV is sensitive to hormone influence during a period spanning late prenatal and early postnatal development [76]. Orchidectomy of neonatal males and treatment of neonatal females with testosterone or estradiol abolishes the sex difference in AVPV volume and cell number [5, 66, 72, 77]. As will be discussed in greater detail below, perinatal exposure to environmental EDCs, which do not bind to the protective alpha-fetoprotein, also causes similar changes in the volume of sexually dimorphic nuclei [78–80].

Sexual dimorphism is not limited solely to anatomical differences or number of cells in these brain nuclei; there are also marked differences in the phenotype of cells in sexually dimorphic brain regions. For instance, the AVPV of female rodents expresses a greater number of cells that are immunopositive for tyrosine hydroxylase [69, 81], kisspeptin [82–86], and estrogen receptor [81, 87]. In the end, adult sex-specific reproductive behaviors are the manifestation of proper organization of the brain by sex steroid hormones during perinatal development. Thus, late prenatal and early postnatal development comprises a critical window during which the functional morphology and neurochemistry of these sexually dimorphic regions are established. Exposure to endocrine active exogenous substances or even to maternal or endogenous hormones at the inappropriate time could cause permanent alterations in adult reproductive function. To follow is a summary of the literature investigating the effects of phytoestrogens, industrial organohalogens, pesticides, bisphenol A, and phthalates on neonatal development of reproductive neuroendocrine function. For more detailed information regarding compounds, age and duration of treatment, dosage, exposure route, and species, refer to Tables 3.2, 3.3, 3.4, 3.5, 3.6.

Effects of Five EDC Classes on Brain Sexual Differentiation

Phytoestrogens

While the majority of EDCs are synthetic in origin and have been the focus of much scrutiny in the public press, phytoestrogens have mainly been touted as having positive health effects. However, a fair comparison should also consider that, as outlined below, phytoestrogens share some cellular and molecular targets in common with man-made EDCs, and have higher relative binding affinities for the estrogen receptor, particularly ERβ relative to many industrial and chemical compounds (see Table 3.2). Studies with laboratory animals have shown that the phytoestrogens

Table 3.2 Effects of developmental phytoestrogen exposure on reproductive neuroendocrine function

Effect	Sex	Compound	Age at treatment	Method of exposure (Dose)	Age at testing	Organism	Reference
Neuroendocrine gene/protein Expression							
↓ ERβ mRNA- hypothalamus	M	Daidzein	E85–E114	Maternal diet (8 mg/kg)	P1	Pig	Ren et al. [91]
No Δ: GnRH mRNA- hypothalamus	M/F	Daidzein	E1–P1	Maternal/personal diet (5 mg/kg)	P1	Duck	Zhao et al. [92]
↑ GnRH mRNA- hypothalamus			E1–P21		P21		
↑ tyrosine hydroxylase protein – AVPV	M	Genistein	P1–P2	SC inj (1000 μg)	P19	Rat	Patisaul et al. [93]
↓ tyrosine hydroxylase + ERα dual immunore- activity – AVPV	F						
No Δ: ERα, ERβ, PR, and SRC1 gene expression – MPOA	M/F	Genistein	E15–P10	Maternal diet (1000 ppm)	P10	Rat	Takagi et al. [94]
↓ Calbindin-D28K protein	F	Isoflavones	E1–E20	Maternal diet (200 ppm)	E20	Rat	Taylor et al. [95]
CREB phosphorylation in MPOA/AVPV	F	Genistein	P14	SC inj (1, 10, 100 μg/30 g BW)	P14	Rat	Sarvari et al. [96]
Sexual differentiation of the brain							
No Δ: SDN-POA volume	M	Genistein	P1–P21	SC inj (4 mg/kg) P1-P6; daily gavage (40 mg/kg) P7-P21	P22	Rat	Lewis et al. [97]
↑ SDN-POA volume	F						
Hormones and anogenital distance							
↓ AGD	F/M	Genistein	E16–E20	SC inj (5 mg)	P1	Rat	Levy et al. [98]
No Δ: AGD	F/M	Isoflavones	E1–P21	Maternal diet (5000 ppm)	P7, P21	Mice	Takashima-Sasaki et al. [99]
No Δ: AGD	F/M	Genistein	E5–P13	Maternal diet (0.5 g/kg diet)	P1	Rat	Tousen et al. [100]

No Δ: AGD	M	Isoflavones	E0–P21	Maternal diet (5 ppm) Maternal diet (500, 1000 ppm)	P5	Rat	Akingbemi et al. [101]
↓ AGD: F1, F2, F3 generations	F	Genistein	E0–P2	Maternal diet (500 ppm)	P2	Rat	Delclos et al. [102]
No Δ: AGD	M/F	Genistein alone	E1–P1	Maternal and personal diet (0.1% by weight)	P1	Rat	Casanova et al. [103]
↑ AGD	F	Genistein + Daidzein					
↓ AGD	M	Genistein	E1–P21	Maternal diet (300 mg/kg)	P21	Mice	Wisniewski [104]
No Δ: AGD	M	Genistein	Gestation	Maternal diet (5 ppm)	P14, P21	Rat	Ball et al. [105]
↓ AGD			Lactation Gestation and lactation				
No Δ: estradiol, progesterone	F	Genistein	P1–P5	SC inj (0.5, 5, 50 mg/kg)	P19	Mice	Jefferson et al. [106]
↑ Serum testosterone	F	Coumestrol	P1–P10	Maternal diet (100 µg/g)	P10	Rat	Whitten et al. [107]
↓ serum estradiol, progesterone	F	Genistein	E0–P21	Maternal diet (5 µg/day)	P21	Rat	Awoniyi et al. [108]
↑ Serum testosterone	M	Isoflavones	E0–P21	Maternal diet (5 ppm)	P21	Rat	Akingbemi et al. [101]
↓ Serum testosterone, ↑ Androsterone				Maternal diet (500, 1000 ppm)			

↑ increased, ↓ decreased, *AGD* anogenital distance, *AVPV* anteroventral periventricular nucleus of the hypothalamus, *bw* Body weight, *CREB* camp response element binding, *E* embryonic day, *ERα* estrogen receptor alpha, *ERβ* estrogen receptor beta, *F* female, *GnRH* gonadotropin-releasing hormone, *Inj* injection, *M* male, *MPOA* medial preoptic area, *mRNA* messenger RNA, *No Δ* unchanged, *P* postnatal day, *ppm* parts per million, *PR* progesterone receptor, *SC* subcutaneous, *SDN-POA* sexually dimorphic nucleus of the preoptic area, *SRC1* steroid receptor coactivator one

Table 3.3 Effects of developmental organohalogen exposure on reproductive neuroendocrine function

Effect	Sex	Compound	Age at Treatment	Method of Exposure	Age at Testing	Organism	Reference
Neuroendocrine gene/protein expression							
↓ Androgen Receptor protein expression – hypothalamus	F	A1254	E15–E19	Maternal gavage (25 mg/kg)	E20	Rat	Colciago et al. [127]
↓ Aromatase activity	M	PCBs (RM)	E0–P0	Maternal diet (4 mg/kg bw)	P0	Rat	Hany et al. [128]
↑ aromatase expression	M	PCBs (RM)	E15–P21	SC inj (10 mg/kg/day from E15–E19, then twice weekly thereafter)	P21	Rat	Colciago et al. [129]
No Δ: aromatase expression, ↓ 5 α reductase-1	F						
↑ Aryl hydrocarbon receptor gene expression – hypothalamus	M	A1254	E15–E19	Maternal gavage (25 mg/kg)	E20	Rat	Pravettoni et al. [130]
↑ ERα, No Δ ERβ gene expression- VMN,↓ Progesterone receptor gene expression- VMN	F	A1254	E10–E18	SC inj (10 mg/kg bw/day)	E18	Rat	Lichtensteiger et al. [131]
↑ α-fetoprotein gene expression in brain	M	A1254	E5–P1	Maternal gavage (18 mg/kg every third day)	P1	Mice	Shimada et al. [132]
Sexual differentiation of the brain							
↑ ERα immunoreactivity MPN	F	A1221, PCBs (RM)	E16–P21	Maternal exposure (1 mg/kg by IP inj)	P1	Rat	Dickerson and Gore, [259]
Hormones and anogenital distance							
↓ Serum estradiol and ↓ testosterone	F	PCBs (RM)	E0-P21	Maternal diet (2, 4 mg/kg bw)	P21	Rat	Kaya et al. [133]
↓ Serum testosterone	M						

S.M. Dickerson et al.

Effect	Sex	Compound	Exposure window	Dose/route	Age	Species	Reference
↓ Serum testosterone	M	PCB126 PCB169	E7-E21	Maternal gavage (3, 30 µg/kg/day)	P21	Rat	Yamamoto et al. [134]
↓ Serum estradiol ↓ Anogenital distance	M F	A1254	E10-E18	SC inj (30 mg/kg bw)	P21	Rat	Lilienthal et al. [135]
↑ Anogenital distance	M	PCB118	E6	Maternal gavage (375 µg/kg)	P3, P15, P21	Rat	Kuriyama et al. [136]
↑ Anogenital distance	F	A1254	E10-E18	SC inj (10 mg/kg/bw)	P2	Rat	Ceccatelli et al. [137]
↑ Anogenital distance	F	PCB47 PCB77	E7-E18	IP inj (20 mg/kg bw); (0.25 mg, 1 mg/kg bw)	P1	Rat	Wang et al. [138]
↓ AGD, ↓ serum P4 ↑ AGD, ↑ serum P4, ↓ serum T	F M	A1221 PCBs (RM)	E16 – P21	Maternal IP inj (1 mg/kg)	P1	Rat	Dickerson and Gore, [259]
	M F	PCBs (RM)	E15-P21	SC inj (10 mg/kg/day from E15 to E19, then twice weekly thereafter)	P21	Rat	Cocchi et al. [139]
↓ AGD ↓ serum estradiol	M/F M	PBDE-99	E10-E18	SC inj (10 mg/kg bw)	P21	Rat	Lilienthal et al. [135]

↑ increased, ↓ decreased, *A1221* Aroclor 1221, *A1254* Aroclor 1254, *AGD* anogenital distance, *bw* Body weight, *E* embryonic day, *ERα* estrogen receptor alpha, *ERβ* estrogen receptor beta, *F* female, *GnRH* gonadotropin-releasing hormone, *Inj* injection, *IP* Intraperitoneal, *M* male, *MPN* medial preoptic nucleus, *mRNA* messenger RNA, *No Δ* unchanged, *P* postnatal day, *P4* progesterone, *PCB* polychlorinated biphenyl, *ppm* parts per million, *PR* progesterone receptor, *RM* reconstituted mixture, *SC* subcutaneous, *VMN* ventromedial nucleus of the hypothalamus

Table 3.4 Effects of developmental pesticide exposure on reproductive neuroendocrine function

Effect	Sex	Compound	Age at treatment	Method of exposure	Age at testing	Organism	Reference
Neuroendocrine Gene/Protein Expression							
↓ PR gene expression in MPOA	M	Methoxychlor	E15-P10	Maternal diet (240 ppm)	P10	Rat	Takagi et al. [94]
↑ PR gene expression in MPOA	F						
↓ ERβ gene expression in MPOA	F			(1200 ppm)			
↓ SRC1 gene expression in MPOA	F			(240 ppm)			
Hormones and Anogenital Distance							
Inhibited testosterone surge	M	Linuron	E13-E18	Maternal gavage (50 or 75 mg/kg/day)	E18	Rat	Wilson et al. [141]
No Δ: fetal testis steroidogenesis	M	DDE	E13-E17	Maternal gavage (50 and 100 mg/kg/day)	E19.5	Rat	Adamsson et al. [142]
No Δ: serum LH	F	Methoxychlor	E30 – E90	SC inj (5 mg/kg/day)	P1-P30	Sheep	Savabieasfahani et al. [143]
No Δ: AGD	M				P1		
↓ AGD	M	Methoxychlor	E2-E4	SC inj (5 mg/kg/day)	P21	Mice	Amstislavsky et al. [144]
No Δ in AGD	F						
↓ AGD (0.2 and 20 mg/kg/day dose)	M	Methoxychlor	E11-E17	Maternal gavage (0.2–100 mg/kg/day)	P1	Mice	Palanza et al. [145]
↑ AGD (100 mg/kg/day dose)							
↓ AGD (20 mg/kg/day dose)	F	Methoxychlor	E11-E17	Maternal gavage (20 mg/kg)	P1	Mice	Palanza et al. [145]
↑ AGD (100 mg/kg/day dose)				100 mg/kg/day			

No Δ in AGD	M/F	Methoxychlor	E15-P1	Maternal diet (24–1200 ppm)	P1	Rat	Masutomi et al. [79]
↑ AGD	M	o,p'-DDT	E11-E17	Maternal gavage (100 mg/kg/day)	P1	Mice	Palanza et al. [145]
↑ AGD	F	o,p'-DDT	E11-E17	Maternal gavage (0.2 and 100 mg/day)	P1	Mice	Palanza et al. [145]
↓ AGD	M	Fenitrothion	E12-E21	Maternal gavage (20 or 25 mg/kg/day)	P1	Rat	Struve et al. [146]
No Δ in AGD	M/F	fenitrothion	E0 – P21	Maternal diet (10, 20, 60 ppm)	P1, P21	Rat	Okahashi et al. [147]
↓ AGD No Δ in AGD	M F	fenitrothion	E12 – E21	Maternal gavage (20, 25 mg/kg)	P1	Rat	Turner et al. [148]

↑ increased, ↓ decreased, *AGD* anogenital distance, *bw* Body weight, *DDE* dichlorodiphenyldichloroethylene, *DDT* dichlorodiphenyltrichloroethane, *E* embryonic day, *ERβ* estrogen receptor beta, *F* female, *FSH* follicle-stimulating hormone, *GnRH* gonadotropin-releasing hormone, *Inj* injection, *IP* Intraperitoneal, *LH* luteinizing hormone, *M* male, *MPOA* medial preoptic area, *No Δ* unchanged, *o* ortho, *p'* para, *P* postnatal day, *ppm* parts per million, *PR* progesterone receptor, *SC* subcutaneous, *SRC1* steroid receptor coactivator one, *T* testosterone

Table 3.5 Effects of developmental bisphenol a exposure on reproductive neuroendocrine function

Effect	Sex	Compound	Age at treatment	Method of exposure (Dose)	Age at testing	Organism	Reference
Neuroendocrine gene/protein expression							
↑ ERα mRNA, protein in POA	F	BPA	P1-P7	SC inj (2 mg/kg bw total) (80 mg/ kg bw total)	P8, P21	Rat	Monje et al. [153]
↓ ERα mRNA, protein (P8) ↑ (P21) in POA							
↑ ERα mRNA in MBH	F	BPA	P1-P5	SC inj (100 μg/day)	P30	Rat	Khurana et al. [154]
↑ ERβ gene expression in POA	M	BPA	E8-E23	SC osmotic pump (25 or 250 μg/kg)	P30	Rat	Ramos et al. [155]
Sexual differentiation of the brain							
↓ TH-ir in AVPV; no Δ ARC volume	F	BPA	E8-P16	SC osmotic pump (25 ng, 250 ng/kg bw)	P22-P24	Mice	Rubin et al. [156]
No Δ volume SDN-POA	F	BPA	E11-P10	Maternal gavage (320 mg/kg)	P10	Rat	Kwon et al. [157]
↑ TH-ir cells in AVPV (demasculinized)	M	BPA	P1-P2	SC inj (1 mg total)	P19	Rat	Patisaul et al. [93]
No Δ: ERα-ir in AVPV	M/F						
↓ TH/ERα-ir (double labeled) in AVPV	M/F						
Hormones and anogenital distance							
↑ AGD	M	BPA	E16-E18	Maternal feed (50 μg/kg bw)	P3	Mice	Gupta [158]
No Δ: AGD	M	BPA	E7-P2	Maternal gavage (2–200 μg/kg/day)	P2	Rat	Howdeshell et al. [159]
No Δ: AGD	F	BPA	E7-P2	Maternal gavage (2–200 μg/kg/day)	P2	Rat	Ryan et al. [160]

3 Reproductive Neuroendocrine Targets of Developmental Exposure…

Effect	Sex	Compound	Exposure	Dose/route	Age	Species	Reference
No Δ: AGD	M/F	BPA	E15-P2	Maternal diet (60–3000 ppm)	P2	Rat	Takagi et al. [161]
No Δ: AGD	M/F	BPA	E6-E21	Maternal gavage (20 μg/kg–50 mg/kg/day)	P1	Rat	Tinwell et al. [162]
No Δ: AGD	M/F	BPA	E6-P20	Maternal gavage (4–40 mg/kg)	P7, P14, P21	Rat	Kobayashi et al. [163]
↑ AGD in F2 generation	F	BPA	NA	Born to prenatally exposed (via maternal diet F1 females)	P1	Rat	Tyl et al. [164]
No Δ: AGD	M						
↓ anoscrotal/anonavel ratio	M	BPA	E30-E90	SC inj (5 mg/kg bw)	P1	Sheep	Savabieasfahani et al. [143]
Suppressed postnatal T peak	M	BPA	E1-P0	Maternal water (0.2, 2, 20, 200 μg/ml)	P0	Rat	Tanaka et al. [165]
No Δ: serum estradiol	F	BPA	P1-P7	SC inj (2 mg/kg bw total) (80 mg/kg bw total)	P21	Rat	Monje et al. [153]
↓basal serum LH, ↓ GnRH-induced LH release, ↑ GnRH pulsatility	F	BPA	P1-P10	SC inj (5 mg total)	P13	Rat	Ferna ndez et al. [166]

↑ increased, ↓ decreased, AGD anogenital distance, ARC arcuate nucleus, AVPV anteroventral periventricular nucleus of the hypothalamus, BPA bisphenol A, bw Body weight, E embryonic day, ERα estrogen receptor alpha, ERβ estrogen receptor beta, F female, GnRH gonadotropin-releasing hormone, Inj injection, IP Intraperitoneal, M male, MBH mediobasal hypothalamus, mRNA messenger RNA, No Δ unchanged, P postnatal day, ppm parts per million, POA preoptic area, PR progesterone receptor, SDN-POA sexually dimorphic nucleus of the preoptic area, T testosterone, TH tyrosine hydroxylase

Table 3.6 Effects of developmental phthalate exposure on reproductive neuroendocrine function

Effect	Sex	Compound	Age at treatment	Method of exposure (Dose)	Age at testing	Organism	Reference
Neuroendocrine Gene/Protein Expression							
↓ PR mRNA in MPOA	F	DINP	E3-P10	Maternal diet (20,000 ppm)	P10	Rat	Takagi et al. [94]
↓ POA aromatase	M	DEHP	E6-P21	Maternal gavage (0.135,0.405 mg/ kg) (.015–1.215 mg/kg)	P1, P22	Rat	Andrade et al. [177]
↑ POA aromatase	F				P22		
Sexual differentiation of the brain							
No Δ volume SDN-POA	M/F	DINP	E15-P10	Maternal diet (24–1200 ppm)	P27	Rat	Masutomi et al. [79]
Hormones and anogenital distance							
↓ AGD	M	BBP	E0-P21	Maternal gavage (20–500 mg/kg)	P1	Rat	Nagao et al. [178]
↓ serum FSH					P22		
No change fetal testis steroidogenesis	M	DINP	E13-E17	Maternal gavage (250 or 750 mg/kg/day)	E19.5	Rat	Adamsson et al. [142]
↓ AGD	M	DBP	E15-16,18-19	Maternal gavage (500 mg/kg)	P1, P13	Rat	Carruthers and Foster [179]
↓ AGD, ↓ testicular testosterone	M	DBP	E12-E19	Maternal diet (112 or 582 mg/kg/ day)	E19, E20	Rat	Struve et al. [180]
↓ AGD, ↓ testicular testosterone	M	DBP	E15.5–E18.5	Maternal gavage (100 or 500 mg/kg/day)	E17.5, E21.5	Rat	Macleod et al. [181]
↓ AGD					P25		
↓ AGD	M	DnHP DEHP	E12–E21	Maternal gavage (125–500 mg/kg)	P1	Rat	Saillenfait et al. [182]
↓ AGD	M	BBP	E14–P1	Maternal diet (3750, 11250 ppm)	P0	Rat	Tyl et al. [183]
↓ AGD	M/F	DBP DINP	E15–P21	Maternal diet (20–10000 ppm) Maternal diet (40–20000 ppm)	P1	Rat	Lee et al. [184]

↓ E2 (DBP 200 ppm dose and DINP 40 ppm dose)	F	DBP DINP	E15–P21	Maternal diet (20–10000 ppm) Maternal diet (40–20000 ppm)	P7	Rat	Lee et al. [184]
↑ AGD	F	DEHP	E6–E21	Maternal gavage (37.5–300 mg/kg/day)	P7, P21	Rat	Piepenbrink et al. [185]
↓ AGD, ↓ fetal testicular T production	M	DEHP	E14–P0	Maternal gavage (234, 469, 750, 938 mg/kg bw)	P21, E20	Rat	Culty et al. [186]
↓ AGD, ↓ serum T ↑ AGD	M F	DiBP	E7–E21	Maternal gavage (600 mg/kg)	E19, E21 E21	Rat	Borsch et al. [187]
↓ Fetal testicular T production ↓ AGD	M	DEHP	E14–P3	Maternal gavage (750 mg/kg/day)	E17–E20 P2	Rat	Parks et al. [188]
↓ Fetal testicular T production	M	DBP	E12–E21	Maternal gavage (500 mg/kg/day)	E18, E21	Rat	Mylchreest et al. [189]
↓ Fetal testicular T production	M	DBP	E13–E21	Maternal gavage (500 mg/kg/day)	E19	Rat	Fisher et al. [190]
↓ AGD	M	DEHP	E6–P20	Maternal gavage (500 mg/kg)	P4	Rat	Yamasaki et al. [191]
↓ AGD	M	DBP	E14–E18	Maternal gastric intubation (750 mg/kg bw)	P7	Rat	Zhu et al. [192]
↓ AGD	M	DBP	E15–E21	Maternal diet (10,000 ppm)	P2, P14	Rat	Lee et al. [193]
No Δ: AGD	M/F	DINP	E15–P10	Maternal diet (24–1200 ppm)	P2	Rat	Masutomi et al. [79]

↑ increased, ↓ decreased, *BBP* benzyl butyl phthalate, *bw* Body weight, *DBP* dibutyl phthalate, *DEHP* di(2-ethylhexyl) phthalate, *DINP* di-isononylphthalate, *DnHP* di-n-butyl phthalate, *E* embryonic day, *E2* estradiol, *F* female, *FSH* follicle-stimulating hormone, *GnRH* gonadotropin-releasing hormone, *M* male, *MPOA* medial preoptic area, *mRNA* messenger RNA, *No Δ* unchanged, *P* postnatal day, *POA* preoptic area, *ppm* parts per million, *PR* progesterone receptor, *SDN-POA* sexually dimorphic nucleus of the preoptic area, *T* testosterone

from maternal dietary exposure are transferred to pups during gestation. Following birth, the developing neonate may also be exposed to phytoestrogens via the diet, either through mother's milk [88], or in the case of humans, through soy-based infant formula or soy-based baby food. Indeed, phytoestrogens have been detected in the serum and amniotic fluid of pregnant mothers, as well as in the serum and placental tissues of newborns. Levels of phytoestrogens in infants fed soy-based formulas may reach as high as 2.5 μM [89, 90].

Sexual Differentiation of the Brain: There is increasing evidence to suggest that developmental exposure to phytoestrogens may have sex-dependent effects on sexually dimorphic systems in the brain by mimicking the action of endogenous estradiol. Two regions of the brain have been most studied in this regard: the AVPV and the SDN-POA. In the AVPV, a sex difference in the volume and a population of neurons co-expressing ERα and tyrosine hydroxylase, the rate-limiting enzyme for dopamine synthesis, is evident in postnatal rats by the time of weaning. More specifically, the female AVPV is larger in volume and has a higher number of tyrosine-hydroxylase-expressing cells compared to males, and has approximately three times as many cells that co-express both ERα and tyrosine hydroxylase [81]. A recent study by Patisaul et al. [93] reported that brief postnatal exposure to genistein had lasting sex-dependent effects on expression of ERα and tyrosine hydroxylase at postnatal day 19 [93]. Specifically, genistein treatment demasculinized tyrosine hydroxylase immunoreactivity in the male AVPV, while in females the number of cells co-expressing ERα and tyrosine hydroxylase was defeminized compared to untreated female controls.

The AVPV is not the only sexually dimorphic hypothalamic nucleus targeted by phytoestrogens. For instance, Lewis et al. [97] found that the SDN-POA was masculinized in postnatal female rats with long-term postnatal dietary exposure to genistein, while males were unaffected. Charles River rat pups were exposed to genistein (100 or 1,000 μg) from postnatal days 1 to 10 and castrated on day 21 [97]. As young adults, the SDN-POA volume was measured [109]. The females exposed to the higher dose of genistein had significantly larger SDN-POA volumes than control or low-dose genistein females. Effects in the males were not detected.

Recent studies have characterized a sexually dimorphic population of neurons expressing calbindin-D28k (CALB) located within the SDN-POA [110–113]. During late gestation, CALB expression is greater in males than females, and this calcium-binding protein is thought to contribute to sexual differentiation of the SDN by protecting neurons against apoptotic cell death through modulation of intracellular calcium [114, 115]. Interestingly, dietary isoflavones in standard laboratory chow may play a role in the establishment of the sexually dimorphic expression pattern of CALB, at least in rodents, as the sex difference is abolished in fetuses of dams fed a phytoestrogen-free diet [113]. In the study by Scallet et al. [113], embryonic females in the phytoestrogen-free group had CALB levels that were indistinguishable from their male siblings or control males. Although there is a paucity of information regarding the effects of developmental phytoestrogen exposure on the brain as it undergoes the process of sexual differentiation, there are a number of studies that have evaluated the results of this type of exposure on the adult, differentiated brain (See section 4.5.1 for a summary).

Resveratrol was studied for its effects on brain sexual differentiation in Sprague-Dawley rats [116]. Pregnant dams were exposed to resveratrol through drinking water during lactation (postnatal days 1–22), and their offspring evaluated for volumetric measurements of the AVPV and SDN-POA in adulthood. Males exposed to resveratrol had reduced SDN-POA volume, and increased AVPV volume, in concert with a diminution of mounting behavior, compared to vehicle males. Female AVPV and SDN-POA volumes were unaffected, and their sexual behavior (lordosis behavior) and estrous cyclicity was similarly unaffected.

GnRH Gene Expression and Activation: GnRH gene expression following developmental exposure to daidzein via maternal diet has been measured in the duckling. Daidzein had no effect on GnRH gene expression in the duckling preoptic area (POA) at hatching, but when this endpoint was evaluated in exposed offspring 4 weeks after birth, a 21% decrease in hypothalamic GnRH mRNA was observed [92]. In treated 415-day-old duck breeders, hypothalamic GnRH was upregulated, while ERβ mRNA was downregulated following 9 weeks of dietary daidzein exposure [92]. These interesting results merit follow-up to clarify whether the downregulation of hypothalamic GnRH mRNA observed in exposed offspring affects sexual maturation or reproductive performances in adult life.

In contrast to results in birds, a study conducted by Takagi et al. [94] found that GnRH gene expression in postnatal rat pups is not affected by maternal exposure to genistein through the diet, and assayed at postnatal day 10 [94]. In another study, Bateman and Patisaul [82] treated newborn pups daily (for 4 days) with genistein. In adulthood, the co-expression of fos (an immediate early gene product indicating transcriptional activation) in GnRH neurons, using a regimen of ovariectomy followed by estradiol + progesterone induction of the GnRH/LH surge, was monitored in the female rats. Early life treatment with genistein significantly reduced co-expression of fos in GnRH neurons [82]. Differences between these results may be due to the endpoint (GnRH gene expression vs. fos co-expression) and the age of analysis (postnatal day 10 vs. adulthood).

Nuclear Steroid Hormone Receptors: The majority of information regarding phytoestrogen actions on neuroendocrine gene and protein expression comes from studies performed in laboratory animals, primarily rodents. Nuclear steroid hormone receptor gene expression seems to be a sensitive target of perinatal exposure to phytoestrogens, and may be related to the effects of phytoestrogens on developmental apoptosis described earlier. For instance, gene expression of the two nuclear ERs, ERα and ERβ, is influenced by phytoestrogen exposure. In a study investigating the effects of maternal dietary exposure to daidzein on the reproductive development of exposed offspring, researchers found that male piglets gestationally exposed to this phytoestrogen had reduced hypothalamic ERβ gene expression at birth [91]. The vulnerability of ER gene expression to disruption may depend on the structure of individual phytoestrogens, and there may also be species differences.

Takagi et al. [94] reported that maternal dietary exposure to genistein did not affect gene expression of ERα or ERβ in the medial preoptic area of postnatal male or female rat pups at P10 [94]. Whether or not these effects of phytoestrogens at the level of gene expression are reflected by changes at the protein level under these experimental conditions has not yet been evaluated. Furthermore, to our knowledge, such studies are limited to ERs, and other target receptors merit investigation. Considering the role that sex steroid hormone receptors play in the process of brain sexual differentiation, this type of data would help establish the mechanism(s) through which phytoestrogens affect adult volumes of sexually dimorphic regions.

A recent study has taken a first step at understanding the nonclassical estrogen-like effects of phytoestrogens in the neonatal hypothalamus. Sarvari et al. [96] found that acute exposure to doses between 1 and 100 μg genistein rapidly induces phophorylation of cAMP response element-binding (CREB) protein in the female postnatal hypothalamus [96]. These nongenomic changes were observed in the medial preoptic area and AVPV 4 h after treatment. These data have relevance for human exposure, as the highest doses used in this study were slightly lower than the typical isoflavone content of human soy-based infant formula.

Serum Hormones and Anogenital Distance: There is a dearth of information regarding the effects of neonatal phytoestrogen exposure on early postnatal serum hormones, although several studies have shown that this endpoint may be affected. The study described earlier by Zhao et al. [92], found that, in addition to downregulated GnRH mRNA, ducklings gestationally exposed to daidzein also had reductions in serum thyroid hormone levels [92]. Jefferson et al. [106] found that genistein exposure during the five days following birth did not affect prepubertal levels of estradiol or progesterone in female mice [106]. Using a rat model, Awoniyi et al. [108] found that genistein exposure via maternal diet during the period of gestation and lactation reduced serum estradiol and progesterone at weaning in female rat offspring [108]. Similarly, a recent study by Akingbemi et al. [101] found that prepubertal male rats exposed to low levels of isoflavones in the maternal diet (5 ppm) had increased serum testosterone levels at weaning, presumably due to observed increases in Leydig cell testosterone production [101]. This effect was dose dependent, as prepubertal males exposed to higher levels of isoflavones in the maternal diet (500 and 1,000 ppm) had reduced serum testosterone that coincided with increased serum androsterone. Further studies investigating the effects of developmental exposure to phytoestrogens on sex steroid hormone levels in the early postnatal period are warranted, given the crucial role of gonadal hormones in the sculpting of sexually dimorphic brain regions.

Numerous studies have investigated postnatal anogenital distance (AGD), a developmental marker that is sensitive to the perinatal hormonal status [117]. Some (but not all) studies suggest that AGD may not be particularly sensitive to prenatal phytoestrogen exposure. For instance, the AGD of postnatal and weanling mice exposed to relatively high levels of isoflavones (5000 ppm) through maternal diet was not affected in either male or female pups at P7 or P21 [99]. Similarly, Tousen et al. [100]

3 Reproductive Neuroendocrine Targets of Developmental Exposure... 77

found no effect of maternal dietary exposure to genistein on early postnatal AGD at a dose of 0.5 g/kg from pregnancy day 5 to postnatal day 13 in male or female rats [100]. Akingbemi et al. [101] also found no effect of maternal dietary exposure to genistein and daidzein at levels between 5 and 1,000 ppm on AGD in male Long-Evans rat pups at P5 [101].

By contrast, Delclos et al. [102] found a reduction in early postnatal anogenital distance of female rat pups gestationally exposed via maternal diet to 500 ppm genistein [102]. This trend continued to the subsequent F2 and F3 generations of nonexposed female offspring. Interestingly, Becker et al. [118] found that male and female offspring of dams fed a phytoestrogen-free diet had larger AGD than control counterparts fed standard lab chow, or those fed a diet high in daidzein and genistein [118]. Differences between these studies may possibly be explained by the composition of phytoestrogens used (single, mixture), and suggest that the presence or absence of phytoestrogens in standard laboratory chow is an important consideration for endocrine disruption studies. A study by Casanova et al. [103] highlights the idea that phytoestrogen mixtures may have distinct effects on AGD compared to individual compounds [103]. They found that although fetal exposure of rats to genistein via maternal diet had no effect on AGD at birth in males or females, genistein plus daidzein resulted in an increased AGD in females compared to control.

Wisniewski et al. [104] found that, at weaning, anogenital distance was reduced in postnatal male offspring of dams fed a diet supplemented with 300 ppm genistein [104]. However, follow-up studies revealed that the effects of genistein on AGD may be highly dependent upon the timing of exposure (i.e., during gestation only, during lactation only, during both gestation and lactation). When exposure was limited to gestation only or lactation only, Ball et al. [105] found that maternal exposure to genistein (at 5 ppm) did not affect AGD in male offspring at P14 and P21 [105]. However, when exposure was extended throughout the period of gestation and lactation, the AGD was reduced in male offspring at these same time points.

Taken together, the studies mentioned in this section suggest that exposure of the developing neonate to phytoestrogens through maternal diet affects early neuroendocrine development of the HPG axis, including brain sexual differentiation. While the effects observed in the studies mentioned herein are dependent upon the chemical studied, as well as exposure route and timing, and duration, many of these treatment paradigms resulted in circulating serum levels of phytoestrogens comparable to the levels measured in human infants on a soy-based diet [119], suggesting their results are of particular relevance for human health. Anogenital distance, while not a neuroendocrine marker per se, may reflect hormonal changes during early development and may serve as a biomarker for early life exposure.

Organohalogens (PCBs, PBDEs)

As mentioned earlier, neonatal exposure of developing humans and animals to organohalogen compounds is a global public health and environmental concern, as humans and wildlife even from remote areas have appreciable levels of this class of

EDC in their body burden. Early studies with pregnant laboratory animals showed that PCBs readily cross the placenta, and are also passed to the suckling young [120–122]. Transplacental transfer of PCBs has been confirmed in humans, as levels in the serum of newborns corresponds with maternal levels [123–125]. Moreover, since they are small, lipophilic compounds, PCBs also cross the blood-brain barrier, and have been detected in human fetal brain tissue at an average concentration of 50 ppb [126]. Age at parity may be an important consideration, as presumably the body burden of organohalogens increases with age, and many women are delaying childbirth in industrialized nations. Indeed, epidemiological evidence suggests that children born to older mothers may have increased exposure to compounds such as PCBs [123]. This type of exposure may have relevance for human health, as a growing body of evidence shows that developmental exposure to organohalogens such as PCBs have effects on the developing neuroendocrine system of laboratory animals (Table 3.3).

Brain Sexual Differentiation and Gene/Protein Expression: The effects of neonatal exposure to organohalogens on brain sexual differentiation have been studied in several animal models, and collectively these data suggest that this endpoint is sensitive to disruption by these environmental pollutants. The bulk of the literature has focused on the effects of perinatal PCB exposure. Several studies have reported that such exposure to PCBs alters expression of nuclear hormone receptors and aromatase activity, an effect that seems to be dependent upon the specific PCB molecule(s). For instance, Hany et al. [128] found that newborn male pups gestationally exposed to a reconstituted mixture of PCBs formulated after the congener pattern found in human milk exhibited significantly reduced aromatase activity in the hypothalamus/preoptic area, while no effect was observed in pups similarly exposed to Aroclor 1254 (A1254) [128]. A study by Colciago et al. [127] found that prenatal Aroclor 1254 exposure during late gestation had no effect on hypothalamic aromatase expression or activity [127]. Hypothalamic androgen receptor protein expression is sensitive to even short-term prenatal PCB exposure, as acute A1254 exposure during embryonic development reduced hypothalamic AR protein expression in embryonic female rats at embryonic day 20 in rats, which is late in gestation. In a follow-up study by the same group, Colciago et al. [129] found that perinatal exposure to a reconstituted PCB mixture caused an increase in hypothalamic aromatase expression in male rats at weaning, while females were unaffected [129]. In contrast, only female rats had decreased levels of hypothalamic 5-alpha reductase, the enzyme that converts testosterone into the more potent androgen dihydrotestosterone, at the time of weaning. Whether these reductions in aromatase and 5-alpha reductase activities result in reduced androgen action in the developing female brain or impairments on adult female sexual behavior are not clear.

Estrogen receptor and estrogen-dependent gene expression are also affected by developmental exposure to PCBs. For instance, Lichtensteiger et al. [131] found that gestational exposure to A1254 increased ERα gene expression in the ventromedial nucleus of the hypothalamus (VMN), a region important for feminine sexual behavior, in female rat embryos [131]. In contrast, progesterone receptor gene

3 Reproductive Neuroendocrine Targets of Developmental Exposure... 79

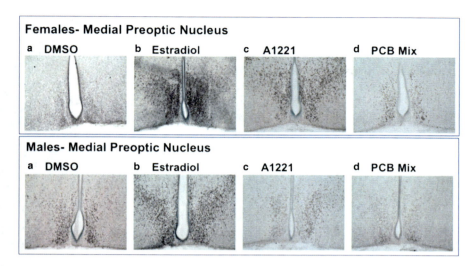

Fig. 3.2 Perinatal exposure to PCBs disrupts estrogen receptor alpha (ERα) protein expression in the medial preoptic nucleus of Sprague-Dawley rats the day following birth. In females, EDC treatment masculinizes the ERα expression pattern

expression in the female VMN was decreased by treatment in the same study. Our laboratory has performed an investigation of effects of prenatal exposure to PCBs on embryonic days 16 and 18 on expression of ERα immunoreactive cell numbers in the hypothalamus [259]. Results demonstrate that prenatal exposure to estradiol increased expression of ERα immunoreactive cells (Fig. 3.2). In a recent study by Shimada et al. [132] investigating the toxicogenomics of prenatal exposure to A1254 in male mice at birth, the gene expression of alpha-fetoprotein in whole brain extracts was increased several fold (~6.8-fold change) compared to control [132]. Since PCBs do not bind alpha-fetoprotein, this could have implications for the bioavailability of endogenous estradiol aromatized from fetal testis. As a whole, these studies suggest acute disruption of hormone receptors and metabolism/biosynthesis of hormones by organohalogen EDCs.

Serum Hormones and Anogenital Distance: In addition to disruptions in the normal pattern of gene and protein expression in the developing brain, early exposure to PCBs also alters serum hormone levels and anogenital distance (AGD) in exposed offspring. For example, Kaya et al. [133] reported decreased serum estradiol and testosterone in female rats exposed to a reconstituted mixture of PCBs throughout the period of gestation and lactation, while only serum testosterone was reduced in exposed males [133]. The effect of PCBs on serum hormones may be dependent upon the composition of the congeners and the dose used. For instance, gestational exposure to low doses (3–30 ug/kg/day) of the individual PCB congeners PCB126 or PCB169 resulted in reduced levels of serum testosterone in weanling males [134]. Similarly, a study in our laboratory found that perinatal exposure to a low-dose (2 mg/kg) reconstituted mixture of PCB138, PCB153, and PCB180 reduced serum testosterone the day after birth in male rats [259]. In addition, we found that serum progesterone levels were reduced in female offspring. In contrast, another group showed that when rats were exposed to a higher dose (4 mg/kg/day) throughout the period of late gestation until weaning, PCB126 combined with PCB138, PCB153, and PCB180 had no effect on testosterone in exposed males [139]. Finally, Lilienthal et al. [135] found that gestational exposure to A1254 caused reductions in serum estradiol in weanling male rats, while female estradiol levels were not affected. As a whole, these data support effects of PCBs on hormone levels in early life [135].

PCB effects on neonatal AGD are dependent upon compound and sex. For instance, a study investigating the effects of gestational and lactational exposure to the dioxin-like coplanar congener PCB118 found that male rats had hypermasculinized AGD throughout early postnatal development, although the effects on female AGD were not reported [136]. Interestingly, A1254 exposure hyperfeminized (shortened) the AGD in female pups, while male AGDs were unchanged compared to control [135]. The effect of A1254 on female AGD seems to be dose dependent, as gestational exposure to A1254 over the same time period with a lower dose caused an increase in the AGD of treated females, compared to control females [137]. However, the same study found that similar doses of PBDE-99, a major congener found in human breast milk samples, had no effect on AGD in treated females. Similarly, exposure to the noncoplanar congener PCB47 and the dioxin-like congener PCB77 during gestation defeminized AGD in female neonates, while males were unaffected [138]. Studies from our laboratory have observed a sex-dependent effect of PCBs on AGD at birth. In neonatally exposed females, 2 mg/kg of a reconstituted PCB mixture (congeners 138, 153, and 180) caused a decrease in AGD at birth, while exposed males had an increased AGD compared to control animals [259].

Collectively, the studies mentioned herein demonstrate the sensitivity of sex steroid hormone-dependent gene expression in the brain, anogenital distance, and serum steroid hormone levels in the neonate to developmental PCB exposure. More studies need to be conducted, in order to understand the impact of developmental PCB exposure on sexually dimorphic brain regions in exposed neonates.

Organochlorine Pesticides (DDT, Methoxychlor)

Due to the widespread agricultural use of pesticides, most humans are exposed to low levels of these compounds through consumption of trace residues on produce. Studies of the endocrine-disrupting effects of organochlorine pesticides began in 1960, when it was discovered that DDT, perhaps the best-known pesticide, is weakly estrogenic in birds [140]. Although most insecticides share the common action of inhibition of acetylcholine esterase, not all pesticides have direct estrogenic or anti-androgenic activities. Rather, because some of these agents have the capacity to inhibit cytochrome p450 enzymes that metabolize steroid hormones such as estradiol, testosterone, and estrone, it is possible that they interfere with steroid hormone metabolism, thus disturbing normal hormone balance. Since the earliest studies focusing on DDT, other organochlorine and organophosphorus pesticides have been extensively studies with regard to their potential to disrupt central nervous system function and development. The bulk of the literature focuses on the effects of organochlorine pesticides on neurotransmitter and thyroid function, or on adult exposures. However, a small number of studies have investigated the effects of perinatal exposure on neuroendocrine HPG development. While details of individual experiments vary (see Table 3.4), the literature supports the probability that exposure to pesticides during the critical developmental window of brain sexual differentiation can permanently affect reproductive physiology.

Brain Sexual Differentiation and Gene/Protein Expression: Although studies investigating the effects of neonatal pesticide exposure on neuroendocrine development are scarce, they do suggest that like other estrogenic EDs, organochlorine insecticides can disrupt sexual differentiation of brain regions that are important for reproduction. For instance, several studies have reported that estrogen-responsive genes are sensitive to early exposure to methoxychlor. Progesterone receptor (PR) expression is estrogen responsive and sexually dimorphic in the medial preoptic area, with male rats having higher levels than females, beginning at P10 [149]. Takagi et al. [94] found that maternal dietary exposure to high levels of methoxychlor caused sex-dependent changes in PR expression in the medial preoptic area at P10 [94]. Compared to their control counterparts, exposed females had increased expression, while exposed males had reduced PR expression. These results suggest that at this high dose, methoxychlor may masculinize the female MPOA and demasculinize the male MPOA. The same study found no effect of methoxychlor on PR expression at lower doses, although gene expression of estrogen receptor β and steroid receptor coactivator 1 was reduced in females. In addition, studies with the organophosphorus pesticide fenitrothion have shown that the adult volumes of sexually dimorphic nuclei are sensitive to perinatal exposure to this compound. Struve et al. [146] found a dose-related increase in the SDN-POA volume in males and a dose-related decrease in SDN-POA volume in females exposed to fenitrothion during late gestation [146].

Serum Gonadotropins, Hormones, and Anogenital Distance: Although prenatal maternal exposure to estrogenic pesticides such as DDE, a metabolite of DDT, does not affect steroidogenesis in fetal rat testis [150], other pesticides with antiandrogenic activity have been shown to reduce fetal testosterone production [141]. For example, Wilson et al. [141] found that maternal exposure to the herbicide linuron during late gestation inhibited the perinatal testosterone surge production in male rat fetuses. Gonadotropins may not be sensitive to gestational exposure to pesticides; female sheep born to mothers injected with methoxychlor had serum concentrations of LH that were not significantly different from control females [143].

Male mice born to mothers injected subcutaneously with methoxychlor with 5 mg/kg on embryonic days 2–4, had shorter AGD at postnatal day 21 compared to controls, while females were unaffected [144]. Similarly, Palanza et al. [145] found that gestational exposure to methoxychlor from embryonic days 11–17 resulted in shorter AGD for males at birth [145]. Effects of methoxychlor on female AGD depended upon dose, with lower doses (200 and 2000 ug/kg/day) resulting in decreased AGD, and a higher dose (20,000 ug/kg/day) resulting in increased AGD. In addition, this study evaluated the effect of DDT on AGD, and found that both males and females gestationally exposed to DDT had longer AGD compared to control animals. Masutomi et al. [79] found that perinatal exposure to methoxychlor via maternal diet did not affect either male or female AGD the day after birth, at the dose range used in the study (24–1,200 ppm) [79].

Studies with organophosphorus pesticide fenitrothion have shown that the male AGD is feminized by perinatal treatment. Turner et al. [148] found that maternal exposure during a period spanning late gestation to 25 mg/kg/day fenitrothion reduced the male AGD at birth, while females were not affected [148]. However, this effect was transient, as the AGD length measured in these same animals during adulthood was not different from control animals. A study by a different group found that maternal dietary exposure to lower doses of fenitrothion (10–60 ppm) resulted in no effects on anogenital distance or areolae retention [147].

Bisphenol A (BPA)

The ubiquitous presence of BPA in consumer and food products provides abundant potential for developmental exposure of the neonate, indirectly through gestational transfer, and directly, through ingestion of foods stored in lined cans, infant formulas in baby bottles, or maternal milk. Indeed, the United States Centers for Disease Control estimates that nearly all Americans have detectable BPA in their bodies, and recent studies have demonstrated that children have higher body burdens of BPA compared to adults. In addition to readily crossing the placenta [151], BPA also apparently crosses the blood-brain barrier, at least in laboratory animals, where it has been shown to partition into the brain, including the hypothalamus, as well as into the pituitary and reproductive organs [152]. A summary of studies on BPA is provided in Table 3.5.

Sexual Differentiation of the Brain: As BPA has been measured in fetal plasma, umbilical cord blood, placental tissues, and in mother's milk, it is clear that developing neonates are being exposed through both gestational and lactational transfer. Because the binding affinity for alpha-fetoprotein is several orders of magnitude lower than estradiol, it has enhanced bioavailability compared to estradiol. Therefore, it is not surprising that this estrogenic compound could impact the organization of sexually dimorphic brain regions. Indeed, the effect of developmental exposure to BPA on the differentiation of several sexually dimorphic hypothalamic nuclei has been reported, including the AVPV, the SDN-POA, and the arcuate nucleus (ARC).

As mentioned earlier, the AVPV is sexually dimorphic in rodents, with females having a larger volume and a higher number of ERα, ERβ, and TH-expressing cells. Patisaul et al. [93] found that acute postnatal treatment of rat pups with 250 ug BPA demasculinized TH-ir in the postnatal male AVPV around the time of weaning [93]. In addition to having more TH-ir cells, female rats also have three times greater the number of cells that co-express both TH and ERα compared to males, but BPA treatment defeminized this expression pattern. In another study, early postnatal exposure to a higher dose of BPA (0.05 mg/kg bw) resulted in increased AVPV ERα expression in females, while yet a higher dose (50 mg/kg/day) at 20 mg/kg resulted in receptor expression decrease [153]. Such results lend to the knowledge of an inverted-U dose-response curve for bisphenol A [167]. Current literature suggests significant morphological and functional deficits in mammals at levels established by the US EPA as "safe" for humans (BPA human reference dose 0.05 mg/kg/day) [168]. For example, studies using very low doses of BPA found that the sex difference in tyrosine hydroxylase-immunoreactive neurons within the AVPV was abolished following gestational and lactational exposure [156]. Studies within both the arcuate nucleus (ARC) and the preoptic area (SDN-POA) showed no detectable changes in volume changes in the neonate after developmental exposure [156, 157]. However, changes to regional volume are probably less sensitive than changes to subpopulations of specific neural phenotypes, and it is possible that future work investigating neurotransmitters and receptors in these brain regions will demonstrate effects of perinatal BPA exposure.

Neuroendocrine Gene/Protein Expression: As with other estrogenic EDCs, BPA exposure during the neonatal period can alter neuroendocrine gene and protein expression during early perinatal development, as has been shown for gene expression of the nuclear estrogen receptors (ERα and ERβ). This is important because activation of ERs plays a critical role in the sculpting of the postnatal brain. The sensitivity of the developing female preoptic area to developmental BPA exposure has been explored by a series of experiments performed in the laboratory of Ramos. This group found that postnatal exposure to BPA spanning the seven days following birth altered preoptic area (POA) gene and protein expression of ERα in a dose-dependent manner. A high dose (total exposure 80 mg/kg) resulted in decreased POA ERα gene and protein expression the day following the final dose (P8), while a low dose (total exposure 2 mg/kg) resulted in an increase in ERα gene and protein expression [153]. By the time of weaning (postnatal day 21, which is 2 weeks following final dose), both the low and high doses of BPA were associated with increased gene expression of ERα in the female POA. Interestingly, serum estradiol was not affected in females at weaning, suggesting that the changes in POA gene and protein expression observed 2 weeks following treatment might be attributed to an organizational disruption in the transcriptional control of estrogen-sensitive genes rather than alterations in endogenous estradiol. Follow-up studies by this same group have shown that the effects of developmental BPA exposure on hypothalamic ERα gene expression in female rats persist into adulthood [169]. Neuroendocrine protein expression may also constitute a target for BPA exposure. Patisaul et al. [93] found that early postnatal exposure to 1-mg BPA during the two days following birth resulted in a demasculinized (feminized-increased) expression pattern for tyrosine hydroxylase-ir in the male rat AVPV around the time of weaning [93]. Interestingly, the number of cells co-labeled for tyrosine hydroxylase and ERα was defeminized in both females and males.

A recent study by Fernandez et al. [166] investigated the effects of early postnatal BPA exposure on hypothalamic and pituitary function in female rats injected daily the first ten days of life with either a low (0.5 mg total) or high (5 mg total) dose. In females treated with the highest dose of BPA, basal gonadotropin levels and GnRH-induced LH release were reduced compared to control females, while GnRH pulsatility was increased, an effect that persisted into adulthood [166]. In an effort to get at the mechanism underlying this effect, the group investigated the in vitro responsiveness of primary pituitary cultures obtained from treated animals to exogenous GnRH. They found a reduction in the GnRH-induced inositol triphosphate formation, and LH release, suggesting that, in female rats, the pituitary response to GnRH may be impaired by neonatal BPA exposure.

Serum Hormones and Anogenital Distance: In addition to causing disruptions in the normal pattern of postnatal neuroendocrine gene and protein expression, serum hormone levels and anogenital distances of animals perinatally exposed to BPA may also be altered. For example, male rats born to mothers exposed to BPA in drinking water throughout the period of gestation had serum testosterone levels comparable to control animals during late gestation, but had reduced testosterone surges following birth [165]. This is important because, as discussed earlier, the perinatal testosterone surge is necessary for proper development of a male-typical brain.

The postnatal AGD does not appear to be particularly sensitive to neonatal BPA. Although a few studies have found effects of BPA on postnatal AGD [158, 170], most other studies have not been able to confirm these findings. For instance, the postnatal AGD of Long-Evans rats gestationally exposed to 2–200 ug/kg/day BPA was not affected the day following birth in males [159] or females [160]. Similarly, no effect of BPA on postnatal AGD was found in Sprague-Dawley rats exposed to 60–3,000 ppm during a similar time period via maternal diet [161], via maternal gavage to doses ranging from 20 ug/kg/day to 50 mg/kg/day [162], or to even higher doses ranging from 4 to 40 mg/kg/day [163]. Other BPA derivatives have shown no effect on postnatal AGD; for example, Hyoung et al. [171] found no effect of gestational and lactational exposure to bisphenol A diglycidyl ether on the AGD in male offspring throughout postnatal development [171]. Although perinatal BPA exposure may not have effects on the F1 generation, the AGD of the F2 generation offspring may be affected. In a three-generation reproductive toxicity study performed by Tyl et al. [164], the group found that female rats born to mothers who were themselves prenatally exposed to BPA via maternal diet (in other words, the F2 generation) had increased AGD at birth, while males of the F2 generation were unaffected [164].

Studies in sheep have shown that early postnatal gonadotropins may be sensitive to gestational BPA exposure. For instance, Savabieasfahani et al. (2006) found that female sheep prenatally exposed to BPA had higher levels of LH compared to control animals, although FSH levels were unchanged [143]. Follow-up studies showed that this treatment paradigm resulted in delayed and severely dampened or absent preovulatory LH surges [172].

Phthalates (DINP, DIDP, DBP, DEHP)

Phthalates are a family of compounds commonly found in consumer products including plastics, food packaging, and cosmetics. Studies in laboratory animals have shown that phthalates are transferred through the placenta to the developing fetus [173, 174]; therefore, the offspring of exposed pregnant mothers may be susceptible to the endocrine-disrupting effects of phthalates. The vast majority of studies investigating the effects of phthalates on developing mammals have focused primarily on the male reproductive tract, with a particular focus on the gonad, which is especially vulnerable target of early phthalate exposure [175, 176]. However, a few recent studies have also investigated the neuroendocrine impact of developmental phthalate exposure, and their results are summarized here (Table 3.6).

Developmental Effects: To our knowledge, at the time of this writing only one study has evaluated the effects of neonatal phthalate exposure on sexual differentiation of volume of sexually dimorphic brain regions in rats [79]. In the study, developing rats were exposed via maternal diet to relatively low levels of di-isononyl phthalate (24–1,200 ppm), and no effects on early postnatal AGD, pubertal onset, or prepubertal SDN-POA volume were observed in males or females. Another study by the same laboratory looked at gene expression of ERα, ERβ, SRC-1, SRC-2, and progesterone receptor in the medial preoptic area of P10 rat pups exposed via maternal dietary DINP exposure [94]. Gene expression was not affected in males, while in females only PR expression was reduced compared to control.

The majority of studies evaluating the effect of neonatal phthalate exposure on AGD have found that phthalates reduce (demasculinize) the male AGD in late gestation [180], early postnatal [181, 194] or throughout both periods [183, 193]. One study found a dose- and sex-dependent effect of maternal dietaryl exposure to di-isononyl phthalate (DINP), with males of all doses having a reduced AGD at birth, while the highest dose caused an increase in female AGD [184]. Other studies have also found that female AGD is masculinized by phthalate exposure [185]. Serum testosterone levels at P7 were not affected by treatment in either males or females, but estradiol levels were reduced in females from the lowest dose (40 ppm). At adulthood, exposed males and females had reductions in sex-typical reproductive behaviors (mounts, ejaculations in males and lordosis in females) without changes in serum hormone levels [185].

As mentioned earlier, the developing testis secretes testosterone, which peaks around the time of birth and then decreases until the peripubertal period. This perinatal testosterone surge is crucial for male reproductive development. Several studies have shown that prenatal phthalate exposure reduces fetal testosterone during critical developmental windows [180, 188–190, 195], an effect that could cause latent deficits in adult reproductive function.

One study investigated the effect of phthalates on preoptic area aromatase activity in the postnatal brain, and found that, at birth, while females were not affected at any dose used in the study, males had reduced POA aromatase activity at low doses and increase aromatase activity at higher doses. In contrast, when littermates of these animals were assayed at weaning, female POA aromatase was increased by all doses, whereas male aromatase activity was only affected by a higher dose [177].

Effects of EDC Exposure on the Pubertal Transition

Puberty is the developmental process whereby the juvenile organism reaches adulthood, and is initiated by the activation of the HPG axis. This protracted transitional period is marked by acceleration of growth, development of secondary sexual characteristics, and culminates with the attainment of adult reproductive function. Following organization of the brain throughout early postnatal development, during which the HPG axis is active, a quiescent phase characterized by low levels of gonadotropins and circulating steroid hormones maintains inhibitory control over HPG activity. During the pubertal transition to adult reproductive capacity, reactivation of the GnRH neurosecretory system is manifested as increased pulse frequency and amplitude of hypothalamic GnRH release. This in turn, promotes pulsatile release of the gonadotropins LH and FSH from the anterior pituitary and subsequent activation of the gonads. The increase in levels of circulating gonadal sex steroid hormones results in development of secondary sex characteristics. Furthermore, the sex-specific brain morphology and neurochemical circuits organized during early postnatal development become activated to coordinate reproductive capacity with sexual behaviors.

Recent studies have provided much insight into the central mechanisms governing pubertal initiation in mammals. The use of laboratory animal models as sentinels for the potential impact of EDCs on human puberty should be considered with the caveat that inhibitory control over GnRH secretion during the quiescent phase differs in some ways between rodents and primates. While steroid inhibitory control represses GnRH neurosecretory activity in rodents, this is much less potent in primates. Central mechanisms that operate independently of gonadal hormones operate to suppress GnRH in primates, as well as in rodents. Because the bulk of research on endocrine disruption is performed in rodents, most of this discussion will focus on those species. In rodents, the onset of puberty is traditionally reflected by the external markers of vaginal opening (an estrogen-mediated event) in females, and preputial separation (an androgen-mediated event) in males. In addition, in female rodents the final endpoint for pubertal onset is the age at first estrus, as it follows the first preovulatory gonadotropin surge.

A large body of evidence supports the idea that developmental exposure to environmental EDCs influences the onset and progression of puberty. Indeed, precocious (early) puberty, delayed puberty, or no change in timing of puberty, can occur in response to developmental EDC exposure, depending upon the nature of the endocrine disruptor, the dose, the sex, and the timing of exposure. These studies on laboratory rodents are likely to reflect processes in humans, as epidemiological and clinical studies have shown a correlation between environmental chemical exposures and pubertal timing [196–198].

Phytoestrogens

The onset of puberty may be advanced, delayed, or unaffected by early phytoestrogen exposure, with females generally more sensitive than males, and results dependent upon numerous factors, including composition (single, mixture), dose, timing, and duration of exposure (Table 3.7). Not all phytoestrogens impair pubertal onset. For instance, Nikaido et al. [199] found that maternal exposure to 2 or 40 mg/kg daily doses of the phytoestrogens genistein or zearalenone, during the period spanning late gestation, advanced vaginal opening in female mice [199]. In contrast, developmental exposure to equivalent doses of other phytoestrogens, such as daidzein [97, 103] or resveratrol [199] have not altered pubertal onset in rats. Timing of exposure is also important. A single dose of coumestrol in early postnatal life hastened the onset of puberty in female rats [200]. On the contrary, gestational exposure to genistein delayed the onset of puberty [98], while postnatal exposure to genistein hastened the onset of puberty in female rats, no matter whether the exposure was acute and low dose [201] or long term and high dose [97]. Likewise, while postnatal exposure to resveratrol alone had no effect on the timing of puberty in female rats, when exposure was expanded to include both gestational and early postnatal periods, pubertal onset was delayed [116, 202]. The observed effects of phytoestrogens on puberty may be sex specific though, as gestational and lactational exposure to genistein [203] or an isoflavone mixure [99], or short-term postnatal exposure [106] to genistein had no effect on timing of puberty in male mice.

There is a scarcity of information regarding the effect of neonatal phytoestrogen exposure on serum hormone levels prior to or during the pubertal transition. A recent study by Gunnarsson et al. [204] found that postnatal dietary exposure to a mixture of genistein, daidzein, biochanin A, and formononetin had a stimulatory effect on the establishment of testosterone production during puberty in male goat kids, which was preceded by an increase in thyroid hormone levels [204]. On the other hand, a study investigating the hormonal effects of soy phytoestrogens in humans found no effect of long-term soy formula on serum estradiol in males or females ranging in age from 7 to 96 months. None of the children enrolled in the retrospective study showed signs of precocious puberty [205]. It is not clear whether the timing of puberty was normal or delayed, because the children were not followed until puberty. Similarly, a prospective epidemiological study evaluating reproductive outcomes in a larger cohort of individuals fed soy formula during infancy [206] found no effect of short-term exposure (0–16 weeks of age) on pubertal onset in males or females.

To our knowledge, there has been a single study investigating the effect of pubertal exposure to phytoestrogens on the attainment of adult reproductive capacity [207]. The group found that dietary exposure of male mice to genistein starting at weaning throughout pubertal development did not significantly disrupt male reproductive function, including sperm count and quality. However, the investigators did not determine whether exposed males displayed normal mating behavior or sired

Table 3.7 ED effects on peripubertal neuroendocrine endpoints

Effect	Sex	Compound	Age at treatment	Method of exposure (Dose)	Age at testing	Organism	Reference
A. Phytoestrogens (Genistein, Daidzein, Isoflavones)							
Neuroendocrine gene/protein expression							
↓ AVPV Kisspeptin fiber density	F	Genistein	P1–P4	SC inj (10 mg/kg BW)	Puberty – 27 weeks	Rat	Bateman and Patisaul [82]
Sexual differentiation of the brain							
↑ SDN-POA volume	F	Genistein	P1–P10	SC inj- (1000 μg; 500, 1000 μg)	P49	Rat	Faber et al. [109, 246]
No Δ: SDN-POA volume	M				P42		
↓ SDN-POA volume	M	Genistein	E7–P50	Maternal and personal diet	P50	Rat	Slikker et al. [247]
No Δ: SDN-POA volume	F						
↓ SDN-POA,↑ AVPV volume	M	Resveratrol	P1–P22	Maternal drinking water (5 and 50 μM)	P25 – puberty	Rat	Henry and Witt [116]
No Δ: SDN-POA, AVPV volume	F						
Neuroendocrine reproductive function and behavior							
↓ Estradiol-stimulated induction of PR in POA and VMN	F	Coumestrol	P45–P50 12 days duration	Diet- 0.02%	P57–P62	Mice	Jacob et al. [248]
↓ Lordosis quotient, persistent estrus	F	Coumestrol	P5	SC inj (3 mg)	P76-P82	Rat	Kouki et al. [200]
Irregular estrous cycles	F	Genistein	P1–P21	SC inj (4 mg/kg) P1–P6; daily gavage (40 mg/kg) P7–P21	Puberty – 13 weeks	Rat	Lewis et al. [97]
↓ Lordosis response Irregular estrous cycles	F	Genistein	P1–P5	SC inj (1 mg)	P60	Rat	Kouki et al. [201]
↑ GnRH-induced LH	F	Genistein	P1–P10	SC inj- (100 μg; 1000 μg)	P42	Rat	Faber et al. [109]
↓ pituitary response to GnRH	M						

	Sex	Compound	Exposure	Dose/route	Age	Species	Reference
↑ pituitary response to GnRH; ↓ GnRH-induced LH	F	Genistein	P1–P10	SC inj- (10 μg; 100, 500, and 1000 μg)	P42	Rat	Faber et al. [246]
↓ GnRH activation in response to E2 and P4 stimulation, Irregular estrous cycles	F	Genistein	P1–P4	SC inj (10 mg/kg BW)	Puberty - 27 weeks	Rat	Bateman and Patisaul [82]
↓ Lordosis response, ↓ Sexual receptivity, Irregular estrous cycles	F	Resveratrol	E1–P21	Maternal drinking water (1500 μg/kg bw/day)	P77–P84	Rat	Kubo et al. [202]
↓ Intromissions, No Δ: mounting	M						
Pubertal onset							
Early pubertal onset	F	Coumestrol	P5	SC inj (1, 3 mg)	P20–P38	Rat	Kouki et al. [200]
Delayed pubertal onset	F	Genistein	E16–E20	SC inj (5 mg)	Puberty	Rat	Levy et al. [98]
Delayed pubertal onset	F	Genistein	P1–P5	Gavage (37.5 mg/kg)	Puberty	Mice	Jefferson et al. [249]
Early onset puberty	F	Genistein Zearalenone	E15–E18	SC inj (2 or 40 mg/kg/day)	Puberty	Mice	Nikaido et al. [199]
Early pubertal onset	F	Genistein	P1–P21	SC inj (4 mg/kg) P1-P6; daily gavage (40 mg/kg) P7- P21	P29 – puberty	Rat	Lewis et al. [97]
Early pubertal onset	F	Genistein	E1–P32	Maternal and personal diet (0.1% by weight)	Puberty	Rat	Casanova et al. [103]
Early pubertal onset	F	Genistein	P1–P5	SC inj (1 mg)	P25-P35	Mice	Kouki et al. [201]
Early pubertal onset	F	Genistein	P1–P4	SC inj (10 mg/kg BW)	Puberty	Rat	Bateman Patisaul [82]
No change in pubertal onset	M	Genistein	E1–P21	Maternal diet (5 or 300 mg/kg)	Puberty	Mice	Wisniewski [104]

(continued)

Table 3.7 (continued)

Effect	Sex	Compound	Age at treatment	Method of exposure (Dose)	Age at testing	Organism	Reference
Early pubertal onset	F	Isoflavone mixture	E1–P56	Maternal and personal diet (0.05% crude isoflavone)	Puberty	Mice	Takashima-Sasaki et al. [99]
No Δ in time to pubertal onset	M						
Delayed pubertal onset	F	Resveratrol	E1–P21	Maternal drinking water (1500 μg/kg bw/day)	P29- puberty	Rat	Kubo et al. [202]
Advanced puberty	F	Resveratrol	E15–E18	SC inj (40 mg/kg)	Puberty	Mice	Nikaido et al. [199]
No Δ in time to pubertal onset	F	Resveratrol	P1–P22	Maternal drinking water (5 and 50 μM)	P25 – puberty	Rat	Henry and Witt [116]
Advanced puberty	F	Zearalenone	E15–E18	SC inj (2 mg/kg)	Puberty	Mice	Nikaido et al. [199]

B. Industrial organohalogens (PCBs, PBBs, PBDEs, Dioxins)

Neuroendocrine gene/protein expression

Effect	Sex	Compound	Age at treatment	Method of exposure (Dose)	Age at testing	Organism	Reference
↓ ERβ protein in AVPV	F	A1221	E16, P1, P4	IP inj (0.34 mg/kg)	P42	Rat	Salama et al. [250]

Sexual differentiation of the brain

Effect	Sex	Compound	Age at treatment	Method of exposure (Dose)	Age at testing	Organism	Reference
↓ KISS immunoreactivity AVPV	F	A1221	E16–P21	Maternal exposure (1 mg/kg by IP inj)	P60	Rat	Dickerson and Gore [260]
↓ GnRH activation on proestrus							

Neuroendocrine reproductive function and behavior

Effect	Sex	Compound	Age at treatment	Method of exposure (Dose)	Age at testing	Organism	Reference
↓ Sexual Receptivity and Lordosis	F	A1254	P1–P7	IP inj (2.5, 5 mg)	P60	Rat	Chung et al. [251]
Impaired paced mating behaviors	F	A1221	E16, E18	IP inj (0.1, 1, 10 mg/kg)	P50	Rat	Steinberg et al. [252]
Irregular estrous cycles	F	A1254	P1, 3, 5, 7, 9	Maternal gavage (32, 64 ppm)	Puberty	Rat	Sager and Girard [215]

Effect	Sex	PCB	Exposure window	Dose (route)	Age/endpoint	Species	Reference
↓ Lordosis; ↓ Lordosis, ↓ Sexual receptivity	F	PCB47 PCB77	E7–E18	IP inj (20 mg/kg bw; 0.25 mg, 1 mg/kg bw)	P71	Rat	Wang et al. [138]
Pubertal onset							
Delayed pubertal onset	F	PCB126	E0–P20	Maternal gavage (3 µg/kg/day)	Puberty	Rat	Shirota et al. [210]
Delayed pubertal onset; ↓ Anogenital distance	F M	PCB126	E15	Maternal gavage (10 µg/kg)	Puberty	Rat	Faqi et al. [211]
Delayed pubertal onset, delayed start of regular estrous cycle, ↓ Serum estradiol and ↓ progesterone	F	PCB126	E13–E19	Maternal gavage (250 ng and 7.5 µg/kg bw)	Puberty P50	Rat	Muto et al. [212]
↑ Serum testosterone- prepubertal peak, ↓ Serum LH (PCB153 only)	M	PCB126 PCB153	E60–E150	Maternal gavage (49 ng/kg bw; 98 ng/kg bw)	12–40 weeks	Goat	Oskam et al. [214]
↓ Serum testosterone and LH	M	PCB126 PCB169	E7–E21	Maternal gavage (3, 30 µg/kg/day)	P42	Rat	Yamamoto et al. [134]
↓ Prepubertal serum LH, early pubertal onset, ↑ Serum progesterone during ovulation	F	PCB153	E60–P91	Maternal gavage (98 µg/kg bw)	9 months	Goat	Lyche et al. [213]
Advanced puberty	F	PCB126, 138, 153, 180	E15–P21	Maternal Sc inj (110 mg/kg total)	Puberty	Rat	Colciago et al. [129]
Delayed testicular descent	M						
↓ 5 alpha reductase mRNA-hypothalamus	F	PCB126, 138, 153, 180	E15–P21	Maternal Sc inj (110 mg/kg total)	P21	Rat	Colciago et al. [129]
Delayed pubertal onset	M/F	A1254	E10–E18	SC inj (30 mg/kg bw)	Puberty	Rat	Lilienthal et al. [135]

(continued)

Table 3.7 (continued)

Effect	Sex	Compound	Age at treatment	Method of exposure (Dose)	Age at testing	Organism	Reference
Delayed pubertal onset	F	A1254	E8–P21	Maternal gavage (10–50 mg/kg daily)	Puberty	Rat	Lee et al. [216]
Delayed pubertal onset Delayed first estrus	F	A1254	P1, 3, 5, 7, 9	Maternal gavage (32, 64 ppm)	Puberty	Rat	Sager and Girard [215]
↓ Tryptophan hydroxylase activity- hypothalamus	M	A1254	~P35	Gavage (0.33 mg/g bw)	~P42	Rat	Khan and Thomas [253]
Slightly earlier time to pubertal onset	F	A1221	E16, P1, P4	IP inj (0.34 mg/kg)	P42	Rat	Salama et al. [250]
Early pubertal onset	F	A1221	P2, P3	SC inj (1 mg/kg bw)	Puberty	Rat	Gellert [217]
Delayed pubertal onset	F	PBDE-99	E10–E18	SC inj (10 mg/kg bw)	Puberty	Rat	Lilienthal et al. [135]
No change in pubertal onset	F	PBDE-99	E10–E18	SC inj (9 or 90 mg/kg total dose)	Puberty	Rat	Ceccatelli et al. [137]
Delayed pubertal onset	F M	PBDE-71	P22–P41 P23-P53	Gavage (60 mg/kg) Gavage (30, 60 mg/kg)	Puberty	Rat	Stoker et al. [218]
C. Pesticides							
Neuroendocrine gene/protein expression							
↑ Hypothalamic oxytocin protein expression, No Δ: prolactin	M/F	Chlorpyrifos	E15–E18	Maternal gavage (3 or 6 mg/kg/day) and SC inj (1 or 3 mg/kg/day)	5 months	Mice	Tait et al. [229]
↓ vasopressin expression	M		P11–P14				
↓ Hypothalamic Gal, Sst, Penk1 gene expression	M	DDT	P21–P49	Gavage (0.06 mg/kg/day)	P49	Mice	Shuhtoh et al. [230]
Sexual differentiation of the brain							
↑ SDN-POA volume	M	Fenitrothion	E12–E21	Maternal gavage (20 or 25 mg/kg/day)	P96	Rat	Struve et al. [146]
↓ SDN-POA volume	F				P60		
Neuroendocrine reproductive function and behavior							

3 Reproductive Neuroendocrine Targets of Developmental Exposure...

Effect	Sex	Compound	Exposure	Dose (route)	Age	Species	Reference
↓ Basal and GnRH-induced release of LH	M	o,p'-DDT	P1–P10	SC inj (0.5 mg/day)	P42	Rat	Faber et al. [109]
No Δ: LH	F						
No Δ: basal LH, FSH, T ↑ GnRH stimulated T secretion	M	o,p'-DDT	E15–P28	Maternal gavage (8.85 or 88.5 mg/kg/day)	6 weeks	Rabbit	Veeramachaneni et al. [225]
↓ GnRH interpulse interval	F	o,p'-DDT	P6–P10	SC inj (10 mg/kg/day)	P15	Rat	Rasier et al. [228]
↓ GnRH-induced LH response					P21		
Pubertal onset							
Advanced pubertal onset	F	o,p'-DDT	P1–P27	SC inj (500 µg/kg/day)	Puberty	Rat	Gellert et al. [227]
Advanced pubertal onset	F	Methoxychlor	E15–P21	Maternal diet (1200 ppm)	Puberty	Rat	Masutomi et al. [79]
Delayed pubertal onset	M						
Advanced pubertal onset	F	Methoxychlor	P1–P10	IP inj (0.5 and 1 mg/kg/day)	Puberty	Mice	Eroschenko and Cooke [254]
Advanced pubertal onset	F	Methoxychlor	E2–E4	SC inj (5 mg/kg/day)	Puberty	Mice	Amstislavsky et al. [144]
No Δ: pubertal onset	F	Methoxychlor	E30–E90	SC inj (5 mg/kg/day)	Puberty	Sheep	Savabieasfahani et al. [143]
Advanced pubertal onset	F	o,p'-DDT	P6–P10	SC inj (10 mg/kg/day)	Puberty	Rat	Rasier et al. [228]
D. Bisphenol A							
Neuroendocrine gene/protein expression							
↑ ERβ mRNA in POA	M	BPA	E8–birth	SC osmotic pump (25 µg, 250 µg/kg bw)	P120	Rat	Ramos et al. [155]
↓ hypothalamic KiSS-1 mRNA	M	BPA	P1–P5	SC inj (100 µg, 500 µg/day)	P75	Rat	Navarro et al. [255]
↑ ERα protein in VMH	F	BPA	P23–P30	Gavage (40 µg/kg/day)	P37	Rat	Ceccarelli et al. [234]

(continued)

Table 3.7 (continued)

Effect	Sex	Compound	Age at treatment	Method of exposure (Dose)	Age at testing	Organism	Reference
↑ SRC-1 protein in MPN ↓ SRC-1 protein in VMH ↑ REA protein in VMH	F	BPA	P1–P7	SC inj every 48 h (.05 mg/kg bw) (.05 mg/kg bw) (.05, 20 mg/kg bw)	P102	Rat	Monje et al. [169]
↓ KiSS-ir fiber density in ARC	F	BPA	P0–P3	SC inj (50 mg/kg bw)	6 months	Rat	Patisaul et al. [61]
↓ ERα protein in MPN, VMH ↓ PR protein in VMH	F	BPA	P1–P7	SC inj every 48 h (20 mg/kg bw) (.05 mg/kg bw)	P102	Rat	Monje et al. [169]
Sexual differentiation of the brain							
No Δ volume SDN-POA	M/F	BPA	E0–P21	Maternal water (.1,1 ppm)	14 weeks	Rat	Kubo et al. [202]
No Δ volume SDN-POA	M	BPA	P1–P5	SC inj (300 µg/g bw)	14 weeks	Rat	Nagao et al. [256]
No Δ volume SDN-POA, AVPV ↑ total Calbindin-ir in SDN, No GnRH activation – unable to induce fos-expression	M	BPA	P1–P2	SC inj (250 µg every 12 hrs – 4 inj total)	P98	Rat	Patisaul et al. [76]
Neuroendocrine reproductive function and behavior							
↓ Serum testosterone	M	BPA	P23–P30	Gavage (40 µg/kg/day)	P37	Rat	Ceccarelli et al. [234]
Dampened preovulatory LH surge	F	BPA	E30–E90	SC inj (5 mg/kg bw)	Puberty	Sheep	Savabieasfahani et al. [143]
↑ Serum T, ↓ serum E2 ↓ serum LH	M	BPA	E12–P21	Maternal gavage (1.2, 2.4 µg/kg/day) (2.4 µg/kg bw)	P125	Rat	Salian et al. [257]

3 Reproductive Neuroendocrine Targets of Developmental Exposure…

Effect	Sex	Compound	Exposure	Route (dose)	Age	Species	Reference
No Δ: serum T	M	BPA	E11–E17	Maternal gavage (2, 20 µg/kg)	5, 8, 12, 13 weeks	Mouse	Kawai et al. [237, 258]
Prolonged diestrus	F	BPA	E15–E18	SC inj (40 mg/kg)	Postpuberty	Mice	Nikaido et al. [199]
↓ serum T	M	BPA	P23–P30	Gavage (40 µg/kg)	P37, P105	Rat	Della Seta et al. [236]
Advanced acyclicity	F	BPA	P0–P3	SC inj (50 mg/kg bw)	Adult	Rat	Adewale et al. [232]
Advanced acyclicity ↓ serum LH	F	BPA	E6–P21	Maternal water (1.2 mg/kg bw)	4 months, 9 months	Rat	Rubin et al. [58]
↑ AGD	M	BPA	E16–E18	Maternal diet (50 µg/kg bw)	P60	Mouse	Gupta [158]
↑ AGD	M	BPA	E11 E17	SC inj (2 µg, 20 µg/kg bw)	P60	Mice	Honma et al. [170]
↓ Serum T, ↓ Leydig-produced T ↓ serum T, ↓ testis weight/bw	M	BPA	P21–P90 E12–P21	Oral gavage (2.4 µg/kg bw) Maternal gavage (2.4 µg/kg bw)	P90	Rat	Akingbemi et al. [235]
Altered mating behavior (↓ hops, darts)	F	BPA	P1–P7	SC inj (.05,20 mg/kg bw)	P100–P102	Rat	Monje et al. [169]
Pubertal onset							
Advanced pubertal onset	F	BPA	E15–E18	SC inj (40 mg/kg)	Puberty	Mice	Nikaido et al. [199]
Advanced pubertal onset	F	BPA	E11–E17	SC inj (2 µg, 20 µg/kg bw)	Puberty	Mice	Honma et al. [170]
Advanced pubertal onset	F	BPA	P1–P10	SC inj (5 mg total)	Puberty	Rat	Fernandez et al. [166]

(continued)

Table 3.7 (continued)

Effect	Sex	Compound	Age at treatment	Method of exposure (Dose)	Age at testing	Organism	Reference
Advanced pubertal onset	F	BPA	P0–P3	SC inj (50 mg/kg bw)	Puberty	Rat	Adewale et al. [232]
No Δ: pubertal onset	F	BPA	E7–P2	Maternal gavage (2–200 µg/kg/day)	Puberty	Rat	Ryan et al. [160]
No Δ: pubertal onset	F	BPA	E30–E90	SC inj (5 mg/kg bw)	Puberty	Sheep	Savabieasfahani et al. [143]

E. Phthalates

Neuroendocrine reproductive function and behavior

Effect	Sex	Compound	Age at treatment	Method of exposure (Dose)	Age at testing	Organism	Reference
Reduced sex-typical behavior No Δ: serum hormones	M/F	DEHP	E6–E21	Maternal gavage (37.5–300 mg/kg/day)	11 weeks	Rat	Piepenbrink et al. [185]
↑ Serum T	M	DEHP	E6–P21	Maternal gavage (.045, .135, .405, 1.2, 15 mg/kg)	~P144	Rat	Andrade et al. [177]
↓ Serum T, ↓ FSH	M	BBP	E0–P21	Maternal gavage (500 mg/kg)	23 weeks	Rat	Nagao et al. [178]
↓ AGD	M	DBP	E15–16,18–19	Maternal gavage (500 mg/kg)	P90	Rat	Carruthers and Foster [179]
↓ Serum T	M	DEHP	E15–E29	Maternal gavage (1250 mg/kg)	P60	Rat	Culty et al. [186]
↓ Lordosis quotient	F	DBP	E15–P21	Maternal diet (20–10000 ppm)	20 weeks	Rat	Lee et al. [184]
		DINP		Maternal diet (40–20000 ppm)			

Pubertal onset

Effect	Sex	Compound	Age at treatment	Method of exposure (Dose)	Age at testing	Organism	Reference
No Δ: pubertal onset	M/F	DINP	E15–P10	Maternal diet (24–1200 ppm)	Puberty	Rat	Masutomi et al. [79]

No Δ: pubertal onset	M	DEHP	E1–P21	Maternal gavage (20–500 mg/kg/day)	Puberty	Rat	Dalsenter et al. [243]
Delayed pubertal onset	M	DEHP	E12–E21	Maternal gavage (125–500 mg/kg)	Puberty	Rat	Saillenfait et al. [240]
Delayed pubertal onset	M	DEHP	P22–Puberty	Gavage (300, 900 mg/kg/day)	Puberty	Rat	Noriega et al. [241]
Delayed pubertal onset	M/F	BBP	E14–P1	Maternal diet (11250 ppm)	Puberty	Rat	Tyl et al. [183]
Advanced pubertal onset Delayed pubertal onset	M	DEHP	P21–P49	Gavage 10 mg/kg Gavage 750 mg/kg	Puberty	Rat	Ge et al. [242]

↑ increased, ↓ decreased, *A1221* Aroclor 1221, *A1254* Aroclor 1254, *AGD* anogenital distance, *ARC* arcuate nucleus, *AVPV* anteroventral periventricular nucleus of the hypothalamus, *BBP* benzyl butyl phthalate, *BPA* bisphenol A, *bw* Body weight, *CREB* camp response element binding, *DBP* dibutyl phthalate, *DDE* dichlorodiphenyldichloroethylene, *DDT* dichlorodiphenyltrichloroethane, *DEHP* di(2-ethylhexyl) phthalate, *DINP* di-isononylphthalate, *DnHP* di-n-butyl phthalate, *E* embryonic day, *E2* estradiol, *ERα* estrogen receptor alpha, *ERβ* estrogen receptor beta, *F* female, *FSH* follicle-stimulating hormone, *Gal* galanin, *GnRH* gonadotropin-releasing hormone, *Inj* injection, *IP* Intraperitoneal, *KISS-1* kisspeptin, *M* male, *MBH* mediobasal hypothalamus, *MPN* medial preoptic nucleus, *MPOA* medial preoptic area, *mRNA* messenger RNA, *No Δ* unchanged, *o,p'* ortho, para, *P* postnatal day, *P4* progesterone, *PCB* polychlorinated biphenyl, *Penk1* preproenkephalin, *POA* preoptic area, *ppm* parts per million, *PR* progesterone receptor, *RM* reconstituted mixture, *SC* subcutaneous, *SDN-POA* sexually dimorphic nucleus of the preoptic area, *SRC1* steroid receptor coactivator one, *Sst* somatostatin, *T* testosterone, *VMN* ventromedial nucleus of the hypothalamus

offspring. Interestingly, pubertal exposure to phytoestrogens may affect adult serum hormone levels, as long-term exposure to phytoestrogens starting during early puberty (P28) until adulthood increased serum testosterone levels in postpubertal adult male Syrian hamsters [208].

Recent studies have highlighted the sensitivity of kisspeptin as a target of endocrine disruption [209], and this system of neurons appears to be sensitive to phytoestrogen exposure. This is important because the product of the Kiss-1 gene, kisspeptin, has been shown to be of crucial importance for the onset of puberty and maintenance of reproductive cycles in females. Studies from the laboratory of Heather Patisaul have discovered that the effects of phytoestrogens on puberty may be related to impairments in AVPV kisspeptin signaling or GnRH cell activation following developmental exposure to the phytoestrogen genistein. For example, that group found that neonatal exposure to genistein advanced pubertal onset in female rats, an effect that was followed by irregular estrous cyclicity by 10 weeks postpuberty. Interestingly, kisspeptin fiber density in the AVPV was reduced [61], and GnRH activation following stimulation by estradiol and progesterone was absent.

Organohalogens (PCBs, PBDEs)

Developmental exposure to organohalogen compounds such as PCBs or PBDEs may affect the timing of puberty, as well as serum gonadotropins and hormones, with effects varying by compound and sex (summarized in Table 3.7). In contrast to phytoestrogens, both males and females seem to be susceptible to pubertal disruption caused by organohalogens. The bulk of the information regarding the effects of organohalogens on pubertal timing comes from studies utilizing PCBs as a prototypical compound, and thus much of the literature summarized in this section focuses on PCBs.

As mentioned earlier, PCBs are categorized into three classes based upon arrangement of chlorine atoms around the biphenyl core. The available literature suggests that their three-dimensional structure greatly impacts their potential for disruption of pubertal onset, which is presumably related to their ability to impact interactions of hormones with their cognate steroid hormone receptors. In addition, the effects of individual PCBs may be sex specific. Studies with the dioxin-like coplanar congener PCB126 have consistently indicated a delay in female puberty following gestational exposure over a range of doses in rats [210–212], while male pubertal onset is unaffected. The effects of PCB126 may be species specific though, as pubertal onset in female goats was unaffected by gestational and lactational exposure [213]. Although the onset of male puberty is not affected by PCB126, a demasculinization of the AGD was observed in male rats [211], and an increase in the prepubertal testosterone peak in male goats was reported [214].

The effects of PCB mixtures on puberty have also been reported, including for the once commercially available Aroclors, which are named according to the average percentage chlorine of the congeners that comprise the mixture. The effects of

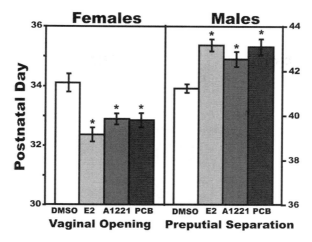

Fig. 3.3 Sex-dependent effect of perinatal PCB exposure on pubertal onset in Sprague-Dawley rats. In females, neonatal exposure to estradiol benzoate, Aroclor 1221, or a reconstituted mixture of PCBs (PCB138, PCB153, PCB180) advanced pubertal onset compared to vehicle counterparts ($*p<0.05$), while a delay was observed in EDC-treated males ($*p<0.05$). Modified from [260]

mixtures also seem to be sex dependent. For instance, the PCB mixture A1254 (54% chlorine content) delays the onset of puberty in gestationally but not early postnatally exposed male rats, while in female rats a delay in pubertal onset was noted regardless of whether exposure is restricted to gestation [135] or early postnatal development [215]. Delayed female puberty in response to A1254 exposure is consistently observed, even at high doses [216]. Conversely, the more lightly chlorinated commercial PCB mixture, A1221 (21% chlorine content), has been shown to advanced onset of puberty in female rats whether exposure occurs throughout gestation and lactation ([260]; Fig. 3.3), or acutely exposed after birth [217]. One potential explanation for the opposite effects of A1221 and A1254 on pubertal timing in females is the difference in percentage chlorine composition of each technical Aroclor mixture. For instance, because A1254 has a greater proportion of more highly chlorinated biphenyls than A1221, these structural disparities may manifest as differences in binding affinity to hormone and neurotransmitter receptors, and their ability to act as an estrogen agonist or antagonist. While studies investigating the effects of Aroclors on puberty are informative, the effects of environmentally relevant mixtures of individual PCB congeners have also been studied. For example, our laboratory and others have observed a sex-dependent effect of puberty in rats exposed throughout the period of gestation and lactation to a reconstituted mixture of PCBs comprised of the most prevalent congeners measured in human samples. An advancement of vaginal opening and first estrus was seen in females, while a delay of preputial separation was observed in males [129] ([260]; Fig. 3.3). It should be noted that prenatal exposure to estradiol and to A1221 also delayed puberty in the male rats in that preliminary study, a result that we are following up on in the laboratory.

In addition to their effects on pubertal timing, PCBs have also been reported to affect serum gonadotropin and hormone levels during the peripubertal transition. For example, in goats gestationally exposed to PCB153, a delay in pubertal onset, reduced prepubertal serum LH, and reduced serum progesterone during ovulation were observed in females [213], while a decrease in serum LH was noted in males. Similar results were obtained with male goats by a separate lab. Oskam et al. [214] found that PCB126 and PCB153 exposure during embryonic development caused a reduction in prepubertal LH, an effect that persisted into the postpubertal period [214]. Furthermore, that group also observed an increase in the prepubertal testosterone peak followed by a lasting reduction in testosterone for the 5 weeks following puberty. Effects of PCBs on serum hormones are not limited to goats; both serum estradiol and progesterone were decreased at late puberty/early adulthood (P50) in female rats gestationally exposed to PCB126 [212].

To our knowledge, the known effects on puberty of other classes of organohalogens such as PBDEs is limited to mixtures, but suggests that pubertal onset in rodents may be less susceptible to disruption by these compounds, compared to PCBs. For instance, Ceccatelli et al. [137] found that the mixture PBDE-99 did not alter vaginal opening in female rats prenatally exposed to either low (9 mg/kg total) or high (90 mg/kg total) doses of PBDE-99 [137]. In contrast, Stoker et al. [218] found that the mixture PBDE-71 delayed puberty in both males and females, and suppressed androgen-dependent growth of reproductive tissues in male rats in a dose-dependent manner following a peripubertal exposure [218]. In addition, serum thyroid hormones were reduced and prolactin levels were increased in males, although serum testosterone levels were not affected. Follow-up in vitro studies from this laboratory found that PBDE-71 acts as a competitive inhibitor of the androgen receptor, a mechanism that may underlie their observed antiandrogenic effects in vivo [219].

The question of whether early exposure to PCBs causes disruptions in human puberty remains largely unanswered [220]. For example, epidemiologic studies have reported no association between maternal exposure to PCBs and breast development [221–223] or age at menarche in developing girls [197, 224].

Pesticides (Atrazine, DDT, Endosulfan, Vinclozolin)

Developmental exposure to pesticides may also interfere with the process of puberty in a sex-dependent manner (Table 3.7). For example, Masutomi et al. [73] found that methoxychlor via maternal diet advanced female puberty in rats at the highest dose (1,200 ppm), while a delay in puberty was observed in males [79]. Similar results were obtained by Amstislavsky et al. [144], who found that early gestational exposure to methoxychlor advanced pubertal onset in females (males were not evaluated because the study ended before preputial separation) [144]. In addition, the pituitary and gonadal response to neuroendocrine inputs from the hypothalamus may be affected by developmental pesticide exposure. Faber et al. [109] found that neonatal exposure to DDT significantly suppressed basal LH levels and blunted GnRH-induced release in postpubertal male rats, while females were not similarly affected [109]. Veeramachaneni et al. [225] found that developmental exposure to DDT in male rabbits did not affect basal LH, FSH, or testosterone at 6 weeks of age, although they found increased secretion of testosterone in response to stimulation by administered GnRH [225].

Initial studies with DDT found that developmental exposure caused an advancement of puberty in laboratory animals [226], an observation that proved relevant for human public health, as DDT derivatives have been measured in the serum of human females with precocious puberty following migration from countries where its use is still prevalent. Following the report that DDT had inhibitory effects on pituitary gonadotropin secretion [226], researchers in this field hypothesized that DDT causes early activation of the hypothalamus [227]. Recent ex vivo and in vitro studies in the laboratory of Jean-Pierre Bourguignon have aimed for a better understanding of the potential mechanisms underlying DDT-induced precocious puberty in females. For instance, they found a reduction in the interpulse interval (reflective of an increase in pulse frequency) of GnRH release from hypothalamic explants obtained from postnatal female rats neonatally exposed to DDT, an effect that was prevented by blockade of estrogen receptors [228]. Furthermore, they observed a reduction in the serum LH response to stimulation by administered GnRH in weanling females, and early vaginal opening in littermates from the individuals used for the ex vivo experiment.

Recent studies have shown that other neuroendocrine systems are also affected by postnatal exposure to organochlorine pesticides. For instance, Tait et al. [229] found that neonatal exposure to chlorpyrifos in mice altered hypothalamic levels of oxytocin and vasopressin [229]. Similarly, a study by Shutoh et al. [230] investigated the effects of postnatal DDT exposure from the time of weaning until 7 weeks of age on the methylation of hypothalamic genes, including gene expression of several neuropeptides [230]. The group found that although global methylation was not affected by treatment, gene expression of galanin (Gal), somatostatin (Sst), and preproenkephalin (Penk1) were all reduced, likely a result of the observed hypomethylation of CpG islands located in the transcription start site for these genes. With respect to sexual differentiation of the brain, Sst is crucial for proper development of the SDN-POA [231]. However, the study did not evaluate whether DDT exposure affected SDN-POA volume or sexual differentiation, something that merits future research.

Bisphenol A

Recently, much public concern has been raised over the exposure of developing humans to components in plastics, including BPA, and the issue of whether this compound can affect human reproductive development and function is a topic of much controversy. While little direct evidence links developmental BPA exposure to pubertal effects in humans, a number of studies in laboratory animals, which reach puberty relatively quickly, have provided some insight into this question (Table 3.7). Perinatal exposure has been associated with advancement of pubertal onset in female rats [166, 232] and mice [170, 199], though some studies have found no treatment effect [143, 160]. Discrepancies between the observed effects (or lack thereof) of BPA on female puberty may be dependent upon timing, dose, and route of exposure. For example, the studies performed by Adewale et al. [232] and Fernandez et al. [166] found an advancement of vaginal opening in female rats acutely exposed (via subcutaneous injection) during early postnatal life to a low dose of BPA (200 ug/kg–5 mg total), while a higher dose (200 mg/kg total) had no effect. In contrast, Ryan et al. [160] and Savabieasfahani et al. [143] found that maternal exposure via gavage to a range of doses had no effect on puberty of female offspring. Researchers have also been discussing the importance of the strain of animals and relative sensitivity to hormones. For example, the rats in the study of Ryan et al. [160] did not respond to the estrogen control (ethinyl estradiol) except at the highest doses, raising the question of whether their rat strain may be relatively estrogen insensitive.

Developmental exposure to BPA may also modulate gonadotropin and serum hormone levels. For example, following the onset of puberty, young female rats typically cycle every 4–5 days, though BPA exposure has resulted in prolonged diestrus [199]. Cycle disruptions are not limited to rats; BPA exposure has also been shown to dampen the preovulatory LH surge in young sheep [143]. Effects of BPA on gonadotropins are also seen in males. When male rats were postnatally exposed to high levels of BPA (500 ug/rat), this resulted in increased levels of FSH during the peripubertal period (P25) [233]. Similarly, although male pubertal onset may not be sensitive to BPA exposure as this process in females, a decrease in both peripubertal estradiol and testosterone levels following prepubertal exposure has been observed in male rats [234–236]. The effects of BPA exposure on peripubertal testosterone levels may also be dependent upon timing and route of exposure, as Kawai et al. [237] found that maternal exposure to very low levels of BPA (2 ng/g) during a period spanning late gestation did not change serum testosterone in male mice offspring at 5 (peripubertal) or 13 weeks (postpubertal) of age [237].

BPA exposure may also have direct central effects on the neuroendocrine HPG axis. Importantly, Kim et al. [238] found that following oral dosing, BPA was found in hypothalamus, indicating that it crosses the blood-brain barrier and thus may act directly on CNS tissues to effect neuroendocrine function [238]. As mentioned earlier, kisspeptin signaling may constitute a sensitive target of estrogenic endocrine-disrupting chemicals [209]. Like phytoestrogens, the effects of BPA on puberty may be related to impairment of kisspeptin signaling. Manuel Tena-Sempere's laboratory found that neonatal BPA exposure decreased hypothalamic gene expression of kisspeptin in the peripubertal (P30) male and female rats. This effect may be limited to the peripubertal period, or depend on basal hormone status, as studies by the Patisaul lab using ovariectomized adult female found no impairment of signaling from kisspeptin neurons in the AVPV to GnRH neurons. In this series of experiments, Patisaul's group found that although puberty is advanced and estrous cyclicity is impaired in female rats following developmental BPA exposure, postpubertal GnRH activation is not impaired [232], nor is AVPV or ARC kisspeptin fiber density changed in postpubertal rats following developmental BPA exposure [61].

While kisspeptin signaling may not be responsible for the effects of BPA on puberty, disrupted signaling through estrogen receptors may be a mechanism by which BPA exerts neuroendocrine effects. For example, one study evaluated the effect of BPA administered during the peripubertal period on ERα expression in the hypothalamus of male and female rats, and found that in pubertal females, developmental BPA exposure increased the number of cells immunoreactive for ERα in the ventromedial hypothalamus (but not arcuate nucleus or POA), while males were unaffected in any of these regions [234]. Likewise, a study by Ramos et al. [155] found that prenatal BPA exposure caused an increase in male POA gene expression of ERβ, but no effect on ERα, during the peripubertal period (P30), an effect that persisted to adulthood (P120) [155].

Phthalates

Compared to other endocrine-disrupting chemicals, less is known about the effects of phthalates on pubertal timing (Table 3.7). Human studies have failed to show a correlation between serum phthalate concentrations and precocious puberty in females [239], although there may be a correlation of phthalate exposure with premature thelarche [196]. However, studies with laboratory animals have found that prenatal [240] and pubertal [241] exposure to diethylhexyl phthalate (DEHP) delays pubertal onset in male Sprague-Dawley and Long-Evans rats. This effect appears to be dose related though as another group found that low doses of DEHP advances preputial separation while higher doses delay puberty in Long-Evans male rats [242]. These observed effects may be strain dependent though, as pubertal onset in male Wistar rats was not affected by gestational and lactational exposure to a broad range of DEHP doses [243]. In addition to the timing of puberty, serum gonadotropins and hormones may also be affected in both males and females. Prepubertal female rats exposed to inhaled DEHP had advanced pubertal onset, and increased serum LH and estradiol [244]. In prepubertal males, inhalation exposure to DEHP caused increased plasma testosterone just after puberty compared to control, while LH and FSH were unaffected [245].

Overall Conclusions

Taken together, the data summarized in this chapter support a role for these five classes of EDCs on development of the reproductive neuroendocrine axis. The effects of EDCs on the developing HPG axis range from disruptions of sexual differentiation of dimorphic brain regions, to alterations in gene and protein expression, enzymatic activity, serum gonadotropin and hormone levels, to interference with feedback responses between the different levels of the HPG axis. While there is limited human data, and there are inconsistencies in the available animal data, it can be concluded that the developing reproductive neuroendocrine system constitutes a sensitive target of developmental EDC exposure. There appears to be increased interest in research on neuroendocrine endpoints, and we expect that the future will bring novel information, and potentially insights into mechanisms by which environmental EDCs cause neuroendocrine dysfunctions.

References

1. Chakraborty TR, Gore AC. Aging-related changes in ovarian hormones, their receptors, and neuroendocrine function. Exp Biol Med (Maywood). 2004;229(10):977–87.
2. Petersen SL, Barraclough CA. Suppression of spontaneous LH surges in estrogen-treated ovariectomized rats by microimplants of antiestrogens into the preoptic brain. Brain Res. 1989;484(1–2):279–89.
3. Wintermantel TM, Campbell RE, Porteous R, et al. Definition of estrogen receptor pathway critical for estrogen positive feedback to gonadotropin-releasing hormone neurons and fertility. Neuron. 2006;52(2):271–80.
4. Bleier R, Byne W, Siggelkow I. Cytoarchitectonic sexual dimorphisms of the medial preoptic and anterior hypothalamic areas in guinea pig, rat, hamster, and mouse. J Comp Neurol. 1982;212(2):118–30.
5. Sumida H, Nishizuka M, Kano Y, Arai Y. Sex differences in the anteroventral periventricular nucleus of the preoptic area and in the related effects of androgen in prenatal rats. Neurosci Lett. 1993;151(1):41–4.
6. Gorski RA, Gordon JH, Shryne JE, Southam AM. Evidence for a morphological sex difference within the medial preoptic area of the rat brain. Brain Res. 1978;148(2):333–46.
7. Gorski RA, Harlan RE, Jacobson CD, Shryne JE, Southam AM. Evidence for the existence of a sexually dimorphic nucleus in the preoptic area of the rat. J Comp Neurol. 1980;193(2): 529–39.
8. Barraclough CA. Production of anovulatory, sterile rats by single injections of testosterone propionate. Endocrinology. 1961;68:62–7.
9. Kavlock RJ, Daston GP, DeRosa C, et al. Research needs for the risk assessment of health and environmental effects of endocrine disruptors: a report of the U.S. EPA-sponsored workshop. Environ Health Perspect. 1996;104 Suppl 4:715–40.
10. Zoeller RT. Environmental chemicals as thyroid hormone analogues: new studies indicate that thyroid hormone receptors are targets of industrial chemicals? Mol Cell Endocrinol. 2005;242(1–2):10–5.
11. Zoeller RT. Environmental chemicals impacting the thyroid: targets and consequences. Thyroid. 2007;17(9):811–7.
12. Anderson G. Environmental chemicals and thyroid function in childhood. In: Gore AC, editors. Endocrine disruptors and puberty: Humana Press Totowa, NJ; 2010.
13. Desvergne B, Feige JN, Casals-Casas C. PPAR-mediated activity of phthalates: a link to the obesity epidemic? Mol Cell Endocrinol. 2009;304(1–2):43–8.
14. Schantz SL, Widholm JJ. Cognitive effects of endocrine-disrupting chemicals in animals. Environ Health Perspect. 2001;109(12):1197–206.
15. Scarth JP. Modulation of the growth hormone-insulin-like growth factor (GH-IGF) axis by pharmaceutical, nutraceutical and environmental xenobiotics: an emerging role for xenobiotic-metabolizing enzymes and the transcription factors regulating their expression. A review. Xenobiotica. 2006;36(2–3):119–218.

3 Reproductive Neuroendocrine Targets of Developmental Exposure...

16. Swedenborg E, Ruegg J, Makela S, Pongratz I. Endocrine disruptive chemicals: mechanisms of action and involvement in metabolic disorders. J Mol Endocrinol. 2009;43(1):1–10.
17. Hoss S, Weltje L. Endocrine disruption in nematodes: effects and mechanisms. Ecotoxicology. 2007;16(1):15–28.
18. Hirsch HV, Possidente D, Possidente B. Pb2+: an endocrine disruptor in drosophila? Physiol Behav. 2010;99(2):254–9.
19. Hotchkiss AK, Rider CV, Blystone CR, et al. Fifteen years after "Wingspread"–environmental endocrine disrupters and human and wildlife health: where we are today and where we need to go. Toxicol Sci. 2008;105(2):235–59.
20. Kloas W, Urbatzka R, Opitz R, et al. Endocrine disruption in aquatic vertebrates. Ann N Y Acad Sci. 2009;1163:187–200.
21. McLachlan JA, Newbold RR, Burow ME, Li SF. From malformations to molecular mechanisms in the male: three decades of research on endocrine disrupters. APMIS. 2001;109(4): 263–72.
22. Rempel MA, Schlenk D. Effects of environmental estrogens and antiandrogens on endocrine function, gene regulation, and health in fish. Int Rev Cell Mol Biol. 2008;267:207–52.
23. Sheehan DM, Willingham E, Gaylor D, Bergeron JM, Crews D. No threshold dose for estradiol-induced sex reversal of turtle embryos: how little is too much? Environ Health Perspect. 1999;107(2):155–9.
24. Guillette Jr LJ, Gunderson MP. Alterations in development of reproductive and endocrine systems of wildlife populations exposed to endocrine-disrupting contaminants. Reproduction. 2001;122(6):857–64.
25. Panzica GC, Viglietti-Panzica C, Mura E, et al. Effects of xenoestrogens on the differentiation of behaviorally-relevant neural circuits. Front Neuroendocrinol. 2007;28(4):179–200.
26. Mellon PL, Windle JJ, Goldsmith PC, Padula CA, Roberts JL, Weiner RI. Immortalization of hypothalamic GnRH neurons by genetically targeted tumorigenesis. Neuron. 1990;5(1): 1–10.
27. Martinez de la Escalera G, Clapp C. Regulation of gonadotropin-releasing hormone secretion: insights from GT1 immortal GnRH neurons. Arch Med Res.2001;32(6):486–498.
28. Gore AC. Organochlorine pesticides directly regulate gonadotropin-releasing hormone gene expression and biosynthesis in the GT1-7 hypothalamic cell line. Mol Cell Endocrinol. 2002;192(1–2):157–70.
29. Dickerson SM, Guevara E, Woller MJ, Gore AC. Cell death mechanisms in GT1–7 GnRH cells exposed to polychlorinated biphenyls PCB74, PCB118, and PCB153. Toxicol Appl Pharmacol. 2009;237(2):237–45.
30. Petroff BK, Croutch CR, Hunter DM, Wierman ME, Gao X. 2,3,7,8-Tetrachlorodibenzo-p-dioxin (TCDD) stimulates gonadotropin secretion in the immature female Sprague-Dawley rat through a pentobarbital- and estradiol-sensitive mechanism but does not alter gonadotropin-releasing hormone (GnRH) secretion by immortalized GnRH neurons in vitro. Biol Reprod. 2003;68(6):2100–6.
31. Bowe J, Li XF, Sugden D, Katzenellenbogen JA, Katzenellenbogen BS, O'Byrne KT. The effects of the phytoestrogen, coumestrol, on gonadotropin-releasing hormone (GnRH) mRNA expression in GT1-7 GnRH neurones. J Neuroendocrinol. 2003;15(2):105–8.
32. Wu SN, Wu YH, Chen BS, Lo YC, Liu YC. Underlying mechanism of actions of tefluthrin, a pyrethroid insecticide, on voltage-gated ion currents and on action currents in pituitary tumor (GH3) cells and GnRH-secreting (GT1–7) neurons. Toxicology. 2009;258(1):70–7.
33. Bennetts HW, Underwood EJ, Shier FL. A specific breeding problem of sheep on subterranean clover pastures in Western Australia. Aust Vet J. 1946;22:2–12.
34. Leopold AS, Erwin M, Oh J, Browning B. Phytoestrogens: adverse effects on reproduction in California quail. Science. 1976;191(4222):98–100.
35. Berger PJ, Sanders EH, Gardner PD, Negus NC. Phenolic plant compounds functioning as reproductive inhibitors in *Microtus montanus*. Science. 1977;195(4278):575–7.
36. Cos P, De Bruyne T, Apers S, Vanden Berghe D, Pieters L, Vlietinck AJ. Phytoestrogens: recent developments. Planta Med. 2003;69(7):589–99.

37. Barkhem T, Carlsson B, Nilsson Y, Enmark E, Gustafsson J, Nilsson S. Differential response of estrogen receptor alpha and estrogen receptor beta to partial estrogen agonists/antagonists. Mol Pharmacol. 1998;54(1):105–12.
38. Kuiper GG, Lemmen JG, Carlsson B, et al. Interaction of estrogenic chemicals and phytoestrogens with estrogen receptor beta. Endocrinology. 1998;139(10):4252–63.
39. Bottner M, Christoffel J, Jarry H, Wuttke W. Effects of long-term treatment with resveratrol and subcutaneous and oral estradiol administration on pituitary function in rats. J Endocrinol. 2006;189(1):77–88.
40. Mousavi Y, Adlercreutz H. Genistein is an effective stimulator of sex hormone-binding globulin production in hepatocarcinoma human liver cancer cells and suppresses proliferation of these cells in culture. Steroids. 1993;58(7):301–4.
41. Weber KS, Jacobson NA, Setchell KD, Lephart ED. Brain aromatase and 5alpha-reductase, regulatory behaviors and testosterone levels in adult rats on phytoestrogen diets. Proc Soc Exp Biol Med. 1999;221(2):131–5.
42. Harris RM, Wood DM, Bottomley L, et al. Phytoestrogens are potent inhibitors of estrogen sulfation: implications for breast cancer risk and treatment. J Clin Endocrinol Metab. 2004;89(4):1779–87.
43. Whitehead SARS. Endocrine-disrupting chemicals as modulators of sex steroid synthesis. Best Pract Res Clin Endocrinol Metab. 2006;20(1):45–61.
44. Wilson T, March H, Ban WJ, et al. Antioxidant effects of phyto-and synthetic-estrogens on cupric ion-induced oxidation of human low-density lipoproteins in vitro. Life Sci. 2002;70(19):2287–97.
45. Siow RC, Li FY, Rowlands DJ, de Winter P, Mann GE. Cardiovascular targets for estrogens and phytoestrogens: transcriptional regulation of nitric oxide synthase and antioxidant defense genes. Free Radic Biol Med. 2007;42(7):909–25.
46. Agarwal R. Cell signaling and regulators of cell cycle as molecular targets for prostate cancer prevention by dietary agents. Biochem Pharmacol. 2000;60(8):1051–9.
47. Jenssen BM. Endocrine-disrupting chemicals and climate change: a worst-case combination for arctic marine mammals and seabirds? Environ Health Perspect. 2006;114 Suppl 1:76–80.
48. Letcher RJ, Bustnes JO, Dietz R, et al. Exposure and effects assessment of persistent organohalogen contaminants in arctic wildlife and fish. Sci Total Environ. 2010;408(15): 2995–3043.
49. Hites RA, Foran JA, Carpenter DO, Hamilton MC, Knuth BA, Schwager SJ. Global assessment of organic contaminants in farmed salmon. Science. 2004;303(5655):226–9.
50. Kester MH, Bulduk S, Tibboel D, et al. Potent inhibition of estrogen sulfotransferase by hydroxylated PCB metabolites: a novel pathway explaining the estrogenic activity of PCBs. Endocrinology. 2000;141(5):1897–900.
51. Bergeron JM, Crews D, McLachlan JA. PCBs as environmental estrogens: turtle sex determination as a biomarker of environmental contamination. Environ Health Perspect. 1994;102(9):780–1.
52. Dianin A. Condensation of phenol with unsaturated ketones. Zhurnal Russkogo Fiziko-Khimicheskogo Obshchestva (Journal of the Russian Physicochemical Society). 1891;23:492.
53. Ikezuki Y, Tsutsumi O, Takai Y, Kamei Y, Taketani Y. Determination of bisphenol A concentrations in human biological fluids reveals significant early prenatal exposure. Hum Reprod. 2002;17(11):2839–41.
54. Schonfelder G, Wittfoht W, Hopp H, Talsness CE, Paul M, Chahoud I. Parent bisphenol A accumulation in the human maternal-fetal-placental unit. Environ Health Perspect. 2002;110(11):A703–7.
55. Lang IA, Galloway TS, Scarlett A, et al. Association of urinary bisphenol A concentration with medical disorders and laboratory abnormalities in adults. J Am Med Assoc. 2008; 300(11):1303–10.

56. Dodds EC, Lawson W. Synthetic oestrogenic agents without the phenanthrene nucleus. Nature. 1936;137(3476):996.
57. Hunt PA, Koehler KE, Susiarjo M, et al. Bisphenol a exposure causes meiotic aneuploidy in the female mouse. Curr Biol. 2003;13(7):546–53.
58. Rubin BS, Murray MK, Damassa DA, King JC, Soto AM. Perinatal exposure to low doses of bisphenol A affects body weight, patterns of estrous cyclicity, and plasma LH levels. Environ Health Perspect. 2001;109(7):675–80.
59. Saal FS, Hughes C. An extensive new literature concerning low-dose effects of bisphenol A shows the need for a new risk assessment. Environ Health Perspect. 2005;113(8):926–33.
60. Welshons WV, Nagel SC, vom Saal FS. Large effects from small exposures. III. Endocrine mechanisms mediating effects of bisphenol A at levels of human exposure. Endocrinology. 2006;147(6 Suppl):S56–69.
61. Patisaul HB, Todd KL, Mickens JA, Adewale HB. Impact of neonatal exposure to the ERalpha agonist PPT, bisphenol-A or phytoestrogens on hypothalamic kisspeptin fiber density in male and female rats. Neurotoxicology. 2009;30(3):350–7.
62. Howdeshell KL, Rider CV, Wilson VS, Gray Jr LE. Mechanisms of action of phthalate esters, individually and in combination, to induce abnormal reproductive development in male laboratory rats. Environ Res. 2008;108(2):168–76.
63. Dickerson SM, Gore AC. Estrogenic environmental endocrine-disrupting chemical effects on reproductive neuroendocrine function and dysfunction across the life cycle. Rev Endocr Metab Disord. 2007;8(2):143–59.
64. Gore AC. Neuroendocrine systems as targets for environmental endocrine-disrupting chemicals. Fertil Steril. 2008;89(2 Suppl):e101–2.
65. Simerly RB, Swanson LW, Gorski RA. The cells of origin of a sexually dimorphic serotonergic input to the medial preoptic nucleus of the rat. Brain Res. 1984;324(1):185–9.
66. Simerly RB, Swanson LW, Handa RJ, Gorski RA. Influence of perinatal androgen on the sexually dimorphic distribution of tyrosine hydroxylase-immunoreactive cells and fibers in the anteroventral periventricular nucleus of the rat. Neuroendocrinology. 1985;40(6):501–10.
67. De Vries GJ, al-Shamma HA. Sex differences in hormonal responses of vasopressin pathways in the rat brain. J Neurobiol. 1990;21(5):686–93.
68. Bakker J, Baum MJ. Role for estradiol in female-typical brain and behavioral sexual differentiation. Front Neuroendocrinol. 2008;29(1):1–16.
69. Forger NG, Rosen GJ, Waters EM, Jacob D, Simerly RB, de Vries GJ. Deletion of Bax eliminates sex differences in the mouse forebrain. Proc Natl Acad Sci USA. 2004;101(37):13666–71.
70. Nordeen EJ, Nordeen KW, Sengelaub DR, Arnold AP. Androgens prevent normally occurring cell death in a sexually dimorphic spinal nucleus. Science. 1985;229(4714):671–3.
71. Chung WC, Swaab DF, De Vries GJ. Apoptosis during sexual differentiation of the bed nucleus of the stria terminalis in the rat brain. J Neurobiol. 2000;43(3):234–43.
72. Davis EC, Shryne JE, Gorski RA. Structural sexual dimorphisms in the anteroventral periventricular nucleus of the rat hypothalamus are sensitive to gonadal steroids perinatally, but develop peripubertally. Neuroendocrinology. 1996;63(2):142–8.
73. Yoshida M, Yuri K, Kizaki Z, Sawada T, Kawata M. The distributions of apoptotic cells in the medial preoptic areas of male and female neonatal rats. Neurosci Res. 2000;36(1):1–7.
74. Rhees RW, Shryne JE, Gorski RA. Onset of the hormone-sensitive perinatal period for sexual differentiation of the sexually dimorphic nucleus of the preoptic area in female rats. J Neurobiol. 1990;21(5):781–6.
75. Rhees RW, Shryne JE, Gorski RA. Termination of the hormone-sensitive period for differentiation of the sexually dimorphic nucleus of the preoptic area in male and female rats. Brain Res Dev Brain Res. 1990;52(1–2):17–23.
76. Patisaul HB, Fortino AE, Polston EK. Differential disruption of nuclear volume and neuronal phenotype in the preoptic area by neonatal exposure to genistein and bisphenol-A. Neurotoxicology. 2007;28(1):1–12.

77. Simerly RB. Hormonal control of the development and regulation of tyrosine hydroxylase expression within a sexually dimorphic population of dopaminergic cells in the hypothalamus. Brain Res Mol Brain Res. 1989;6(4):297–310.
78. Ikeda M, Mitsui T, Setani K, et al. In utero and lactational exposure to 2,3,7,8-tetrachlorod-ibenzo-p-dioxin in rats disrupts brain sexual differentiation. Toxicol Appl Pharmacol. 2005;205(1):98–105.
79. Masutomi N, Shibutani M, Takagi H, Uneyama C, Takahashi N, Hirose M. Impact of dietary exposure to methoxychlor, genistein, or diisononyl phthalate during the perinatal period on the development of the rat endocrine/reproductive systems in later life. Toxicology. 2003;192(2–3):149–70.
80. Yamamoto M, Shirai M, Tamura A, et al. Effects of maternal exposure to a low dose of diethylstilbestrol on sexual dimorphic nucleus volume and male reproductive system in rat offspring. J Toxicol Sci. 2005;30(1):7–18.
81. Simerly RB, Zee MC, Pendleton JW, Lubahn DB, Korach KS. Estrogen receptor-dependent sexual differentiation of dopaminergic neurons in the preoptic region of the mouse. Proc Natl Acad Sci USA. 1997;94(25):14077–82.
82. Bateman HL, Patisaul HB. Disrupted female reproductive physiology following neonatal exposure to phytoestrogens or estrogen specific ligands is associated with decreased GnRH activation and kisspeptin fiber density in the hypothalamus. Neurotoxicology. 2008;29(6): 988–97.
83. Clarkson J, Herbison AE. Postnatal development of kisspeptin neurons in mouse hypothalamus; sexual dimorphism and projections to gonadotropin-releasing hormone neurons. Endocrinology. 2006;147(12):5817–25.
84. Gonzalez-Martinez D, De Mees C, Douhard Q, Szpirer C, Bakker J. Absence of gonadotropin-releasing hormone 1 and kiss1 activation in alpha-fetoprotein knockout mice: prenatal estrogens defeminize the potential to show preovulatory luteinizing hormone surges. Endocrinology. 2008;149(5):2333–40.
85. Kauffman AS, Gottsch ML, Roa J, et al. Sexual differentiation of kiss1 gene expression in the brain of the rat. Endocrinology. 2007;148(4):1774–83.
86. Simerly RB. Wired for reproduction: organization and development of sexually dimorphic circuits in the mammalian forebrain. Annu Rev Neurosci. 2002;25:507–36.
87. Orikasa C, Kondo Y, Hayashi S, McEwen BS, Sakuma Y. Sexually dimorphic expression of estrogen receptor beta in the anteroventral periventricular nucleus of the rat preoptic area: implication in luteinizing hormone surge. Proc Natl Acad Sci USA. 2002;99(5):3306–11.
88. Doerge DR, Twaddle NC, Churchwell MI, Newbold RR, Delclos KB. Lactational transfer of the soy isoflavone, genistein, in Sprague-Dawley rats consuming dietary genistein. Reprod Toxicol. 2006;21(3):307–12.
89. Setchell KD, Zimmer-Nechemias L, Cai J, Heubi JE. Exposure of infants to phyto-oestrogens from soy-based infant formula. Lancet. 1997;350(9070):23–7.
90. Setchell KD, Zimmer-Nechemias L, Cai J, Heubi JE. Isoflavone content of infant formulas and the metabolic fate of these phytoestrogens in early life. Am J Clin Nutr. 1998;68 (6 Suppl):1453S–61.
91. Ren MQ, Kuhn G, Wegner J, Nurnberg G, Chen J, Ender K. Feeding daidzein to late pregnant sows influences the estrogen receptor beta and type 1 insulin-like growth factor receptor mRNA expression in newborn piglets. J Endocrinol. 2001;170(1):129–35.
92. Zhao R, Wang Y, Zhou Y, et al. Dietary daidzein influences laying performance of ducks (Anas platyrhynchos) and early post-hatch growth of their hatchlings by modulating gene expression. Comp Biochem Physiol A Mol Integr Physiol. 2004;138(4):459–66.
93. Patisaul HB, Fortino AE, Polston EK. Neonatal genistein or bisphenol-A exposure alters sexual differentiation of the AVPV. Neurotoxicol Teratol. 2006;28(1):111–8.
94. Takagi H, Shibutani M, Lee KY, et al. Impact of maternal dietary exposure to endocrine-acting chemicals on progesterone receptor expression in microdissected hypothalamic medial preoptic areas of rat offspring. Toxicol Appl Pharmacol. 2005;208(2):127–36.

95. Taylor H, Quintero EM, Iacopino AM, Lephart ED. Phytoestrogens alter hypothalamic calbindin-D28k levels during prenatal development. Brain Res Dev Brain Res. 1999; 114(2):277–81.
96. Sarvari M, Szego EM, Barabas K, et al. Genistein induces phosphorylation of cAMP response element-binding protein in neonatal hypothalamus in vivo. J Neuroendocrinol. 2009;21(12): 1024–8.
97. Lewis RW, Brooks N, Milburn GM, et al. The effects of the phytoestrogen genistein on the postnatal development of the rat. Toxicol Sci. 2003;71(1):74–83.
98. Levy JR, Faber KA, Ayyash L, Hughes Jr CL. The effect of prenatal exposure to the phytoestrogen genistein on sexual differentiation in rats. Proc Soc Exp Biol Med. 1995;208(1): 60–6.
99. Takashima-Sasaki K, Komiyama M, Adachi T, et al. Effect of exposure to high isoflavone-containing diets on prenatal and postnatal offspring mice. Biosci Biotechnol Biochem. 2006;70(12):2874–82.
100. Tousen Y, Umeki M, Nakashima Y, Ishimi Y, Ikegami S. Effects of genistein, an isoflavone, on pregnancy outcome and organ weights of pregnant and lactating rats and development of their suckling pups. J Nutr Sci Vitaminol (Tokyo). 2006;52(3):174–82.
101. Akingbemi BT, Braden TD, Kemppainen BW, et al. Exposure to phytoestrogens in the perinatal period affects androgen secretion by testicular Leydig cells in the adult rat. Endocrinology. 2007;148(9):4475–88.
102. Delclos KB, Weis CC, Bucci TJ, et al. Overlapping but distinct effects of genistein and ethinyl estradiol (EE(2)) in female Sprague-Dawley rats in multigenerational reproductive and chronic toxicity studies. Reprod Toxicol. 2009;27(2):117–32.
103. Casanova M, You L, Gaido KW, Archibeque-Engle S, Janszen DB, Heck HA. Developmental effects of dietary phytoestrogens in Sprague-Dawley rats and interactions of genistein and daidzein with rat estrogen receptors alpha and beta in vitro. Toxicol Sci. 1999;51(2):236–44.
104. Wisniewski AB, Klein SL, Lakshmanan Y, Gearhart JP. Exposure to genistein during gestation and lactation demasculinizes the reproductive system in rats. J Urol. 2003;169(4): 1582–6.
105. Ball ER, Caniglia MK, Wilcox JL, et al. Effects of genistein in the maternal diet on reproductive development and spatial learning in male rats. Horm Behav. 2010;57(3):313–22.
106. Jefferson WN, Padilla-Banks E, Newbold RR. Adverse effects on female development and reproduction in CD-1 mice following neonatal exposure to the phytoestrogen genistein at environmentally relevant doses. Biol Reprod. 2005;73(4):798–806.
107. Whitten PL, Lewis C, Russell E, Naftolin F. Phytoestrogen influences on the development of behavior and gonadotropin function. Proc Soc Exp Biol Med. 1995;208(1):82–6.
108. Awoniyi CA, Roberts D, Veeramachaneni DN, Hurst BS, Tucker KE, Schlaff WD. Reproductive sequelae in female rats after in utero and neonatal exposure to the phytoestrogen genistein. Fertil Steril. 1998;70(3):440–7.
109. Faber KA, Hughes Jr CL. The effect of neonatal exposure to diethylstilbestrol, genistein, and zearalenone on pituitary responsiveness and sexually dimorphic nucleus volume in the castrated adult rat. Biol Reprod. 1991;45(4):649–53.
110. Lephart ED, Taylor H, Jacobson NA, Watson MA. Calretinin and calbindin-D28K in male rats during postnatal development. Neurobiol Aging. 1998;19(3):253–7.
111. Brager DH, Sickel MJ, McCarthy MM. Developmental sex differences in calbindin-D(28 K) and calretinin immunoreactivity in the neonatal rat hypothalamus. J Neurobiol. 2000;42(3): 315–22.
112. Sickel MJ, McCarthy MM. Calbindin-D28k immunoreactivity is a marker for a subdivision of the sexually dimorphic nucleus of the preoptic area of the rat: developmental profile and gonadal steroid modulation. J Neuroendocrinol. 2000;12(5):397–402.
113. Scallet AC, Divine RL, Newbold RR, Delclos KB. Increased volume of the calbindin D28k-labeled sexually dimorphic hypothalamus in genistein and nonylphenol-treated male rats. Toxicol Sci. 2004;82(2):570–6.

114. Dowd DR, MacDonald PN, Komm BS, Haussler MR, Miesfeld RL. Stable expression of the calbindin-D28K complementary DNA interferes with the apoptotic pathway in lymphocytes. Mol Endocrinol. 1992;6(11):1843–8.
115. McMahon A, Wong BS, Iacopino AM, Ng MC, Chi S, German DC. Calbindin-D28k buffers intracellular calcium and promotes resistance to degeneration in PC12 cells. Brain Res Mol Brain Res. 1998;54(1):56–63.
116. Henry LA, Witt DM. Effects of neonatal resveratrol exposure on adult male and female reproductive physiology and behavior. Dev Neurosci. 2006;28(3):186–95.
117. Marois G. Action of progesterone, testosterone and estradiol on the anogenital distance and somatic sexual differentiation in rats. Biol Med (Paris). 1968;57(1):44–90.
118. Becker LA, Kunkel AJ, Brown MR, Ball EE, Williams MT. Effects of dietary phytoestrogen exposure during perinatal period. Neurotoxicol Teratol. 2005;27(6):825–34.
119. Cao Y, Calafat AM, Doerge DR, et al. Isoflavones in urine, saliva, and blood of infants: data from a pilot study on the estrogenic activity of soy formula. J Expo Sci Environ Epidemiol. 2009;19(2):223–34.
120. Masuda Y, Kagawa R, Kuroki H, et al. Transfer of polychlorinated biphenyls from mothers to foetuses and infants. Food Cosmet Toxicol. 1978;16(6):543–6.
121. Masuda Y, Kagawa R, Tokudome S, Kuratsune M. Transfer of polychlorinated biphenyls to the foetuses and offspring of mice. Food Cosmet Toxicol. 1978;16(1):33–7.
122. Ando M, Saito H, Wakisaka I. Transfer of polychlorinated biphenyls (PCBs) to newborn infants through the placenta and mothers' milk. Arch Environ Contam Toxicol. 1985;14(1): 51–7.
123. Lackmann GM, Angerer J, Salzberger U, Tollner U. Influence of maternal age and duration of pregnancy on serum concentrations of polychlorinated biphenyls and hexachlorobenzene in full-term neonates. Biol Neonate. 1999;76(4):214–9.
124. Covaci A, Jorens P, Jacquemyn Y, Schepens P. Distribution of PCBs and organochlorine pesticides in umbilical cord and maternal serum. Sci Total Environ. 2002;298(1–3):45–53.
125. Park JS, Bergman A, Linderholm L, et al. Placental transfer of polychlorinated biphenyls, their hydroxylated metabolites and pentachlorophenol in pregnant women from eastern Slovakia. Chemosphere. 2008;70(9):1676–84.
126. Lanting CI, Huisman M, Muskiet FA, van der Paauw CG, Essed CE, Boersma ER. Polychlorinated biphenyls in adipose tissue, liver, and brain from nine stillborns of varying gestational ages. Pediatr Res. 1998;44(2):222–5.
127. Colciago A, Negri-Cesi P, Pravettoni A, Mornati O, Casati L, Celotti F. Prenatal Aroclor 1254 exposure and brain sexual differentiation: effect on the expression of testosterone metabolizing enzymes and androgen receptors in the hypothalamus of male and female rats. Reprod Toxicol. 2006;22(4):738–45.
128. Hany J, Lilienthal H, Sarasin A, et al. Developmental exposure of rats to a reconstituted PCB mixture or aroclor 1254: effects on organ weights, aromatase activity, sex hormone levels, and sweet preference behavior. Toxicol Appl Pharmacol. 1999;158(3):231–43.
129. Colciago A, Casati L, Mornati O, et al. Chronic treatment with polychlorinated biphenyls (PCB) during pregnancy and lactation in the rat Part 2: Effects on reproductive parameters, on sex behavior, on memory retention and on hypothalamic expression of aromatase and 5alpha-reductases in the offspring. Toxicol Appl Pharmacol. 2009;239(1):46–54.
130. Pravettoni A, Colciago A, Negri-Cesi P, Villa S, Celotti F. Ontogenetic development, sexual differentiation, and effects of Aroclor 1254 exposure on expression of the arylhydrocarbon receptor and of the arylhydrocarbon receptor nuclear translocator in the rat hypothalamus. Reprod Toxicol. 2005;20(4):521–30.
131. Lichtensteiger WCR, Faass O, Ma R, Schlumpf M. Effect of polybrominated diphenylether and PCB on the development of the brain–gonadal axis and gene expression in rats. Organohalog Compd. 2003;61:84–7.
132. Shimada M, Kameo S, Sugawara N, et al. Gene expression profiles in the brain of the neonate mouse perinatally exposed to methylmercury and/or polychlorinated biphenyls. Arch Toxicol. 2010;84(4):271–86.

133. Kaya H, Hany J, Fastabend A, Roth-Harer A, Winneke G, Lilienthal H. Effects of maternal exposure to a reconstituted mixture of polychlorinated biphenyls on sex-dependent behaviors and steroid hormone concentrations in rats: dose-response relationship. Toxicol Appl Pharmacol. 2002;178(2):71–81.

134. Yamamoto M, Narita A, Kagohata M, Shirai M, Akahori F, Arishima K. Effects of maternal exposure to 3,3',4,4',5-pentachlorobiphenyl (PCB126) or 3,3',4,4',5,5'-hexachlorobiphenyl (PCB169) on testicular steroidogenesis and spermatogenesis in male offspring rats. J Androl. 2005;26(2):205–14.

135. Lilienthal H, Hack A, Roth-Harer A, Grande SW, Talsness CE. Effects of developmental exposure to 2,2,4,4,5-pentabromodiphenyl ether (PBDE-99) on sex steroids, sexual development, and sexually dimorphic behavior in rats. Environ Health Perspect. 2006;114(2): 194–201.

136. Kuriyama SN, Chahoud I. In utero exposure to low-dose 2,3',4,4',5-pentachlorobiphenyl (PCB 118) impairs male fertility and alters neurobehavior in rat offspring. Toxicology. 2004;202(3):185–97.

137. Ceccatelli R, Faass O, Schlumpf M, Lichtensteiger W. Gene expression and estrogen sensitivity in rat uterus after developmental exposure to the polybrominated diphenylether PBDE 99 and PCB. Toxicology. 2006;220(2–3):104–16.

138. Wang XQ, Fang J, Nunez AA, Clemens LG. Developmental exposure to polychlorinated biphenyls affects sexual behavior of rats. Physiol Behav. 2002;75(5):689–96.

139. Cocchi D, Tulipano G, Colciago A, et al. Chronic treatment with polychlorinated biphenyls (PCB) during pregnancy and lactation in the rat: Part 1: effects on somatic growth, growth hormone-axis activity and bone mass in the offspring. Toxicol Appl Pharmacol. 2009;237(2):127–36.

140. Colborn T, Dumanoski D, Peterson Myers J. Our stolen future. New York: Dutton, Penguin Books; 1996.

141. Wilson VS, Lambright CR, Furr JR, Howdeshell KL, Earl Gray Jr L. The herbicide linuron reduces testosterone production from the fetal rat testis during both in utero and in vitro exposures. Toxicol Lett. 2009;186(2):73–7.

142. Adamsson A, Salonen V, Paranko J, Toppari J. Effects of maternal exposure to di-isononylphthalate (DINP) and 1,1-dichloro-2,2-bis(p-chlorophenyl)ethylene (p, p'-DDE) on steroidogenesis in the fetal rat testis and adrenal gland. Reprod Toxicol. 2009;28(1):66–74.

143. Savabieasfahani M, Kannan K, Astapova O, Evans NP, Padmanabhan V. Developmental programming: differential effects of prenatal exposure to bisphenol-A or methoxychlor on reproductive function. Endocrinology. 2006;147(12):5956–66.

144. Amstislavsky SY, Kizilova EA, Golubitsa AN, Vasilkova AA, Eroschenko VP. Preimplantation exposures of murine embryos to estradiol or methoxychlor change postnatal development. Reprod Toxicol. 2004;18(1):103–8.

145. Palanza P, Parmigiani S, vom Saal FS. Effects of prenatal exposure to low doses of diethylstilbestrol, o,p'DDT, and methoxychlor on postnatal growth and neurobehavioral development in male and female mice. Horm Behav. 2001;40(2):252–65.

146. Struve MF, Turner KJ, Dorman DC. Preliminary investigation of changes in the sexually dimorphic nucleus of the rat medial preoptic area following prenatal exposure to fenitrothion. J Appl Toxicol. 2007;27(6):631–6.

147. Okahashi N, Sano M, Miyata K, et al. Lack of evidence for endocrine disrupting effects in rats exposed to fenitrothion in utero and from weaning to maturation. Toxicology. 2005; 206(1):17–31.

148. Turner KJ, Barlow NJ, Struve MF, et al. Effects of in utero exposure to the organophosphate insecticide fenitrothion on androgen-dependent reproductive development in the Crl:CD(SD) BR rat. Toxicol Sci. 2002;68(1):174–83.

149. Quadros PS, Wagner CK. Regulation of progesterone receptor expression by estradiol is dependent on age, sex and region in the rat brain. Endocrinology. 2008;149(6):3054–61.

150. Adamsson A, Simanainen U, Viluksela M, Paranko J, Toppari J. The effects of 2,3,7,8-tetrachlorodibenzo-p-dioxin on foetal male rat steroidogenesis. Int J Androl. 2009;32(5):575–85.

151. Takahashi O, Oishi S. Disposition of orally administered 2,2-Bis(4-hydroxyphenyl)propane (Bisphenol A) in pregnant rats and the placental transfer to fetuses. Environ Health Perspect. 2000;108(10):931–5.
152. Kim CS, Sapienza PP, Ross IA, Johnson W, Luu HM, Hutter JC. Distribution of bisphenol A in the neuroendocrine organs of female rats. Toxicol Ind Health. 2004;20(1–5):41–50.
153. Monje L, Varayoud J, Luque EH, Ramos JG. Neonatal exposure to bisphenol A modifies the abundance of estrogen receptor alpha transcripts with alternative 5′-untranslated regions in the female rat preoptic area. J Endocrinol. 2007;194(1):201–12.
154. Khurana S, Ranmal S, Ben-Jonathan N. Exposure of newborn male and female rats to environmental estrogens: delayed and sustained hyperprolactinemia and alterations in estrogen receptor expression. Endocrinology. 2000;141(12):4512–7.
155. Ramos JG, Varayoud J, Kass L, et al. Bisphenol a induces both transient and permanent histofunctional alterations of the hypothalamic-pituitary-gonadal axis in prenatally exposed male rats. Endocrinology. 2003;144(7):3206–15.
156. Rubin BS, Lenkowski JR, Schaeberle CM, Vandenberg LN, Ronsheim PM, Soto AM. Evidence of altered brain sexual differentiation in mice exposed perinatally to low, environmentally relevant levels of bisphenol A. Endocrinology. 2006;147(8):3681–91.
157. Kwon S, Stedman DB, Elswick BA, Cattley RC, Welsch F. Pubertal development and reproductive functions of Crl:CD BR Sprague-Dawley rats exposed to bisphenol A during prenatal and postnatal development. Toxicol Sci. 2000;55(2):399–406.
158. Gupta C. Reproductive malformation of the male offspring following maternal exposure to estrogenic chemicals. Proc Soc Exp Biol Med. 2000;224(2):61–8.
159. Howdeshell KL, Furr J, Lambright CR, Wilson VS, Ryan BC, Gray Jr LE. Gestational and lactational exposure to ethinyl estradiol, but not bisphenol A, decreases androgen-dependent reproductive organ weights and epididymal sperm abundance in the male long evans hooded rat. Toxicol Sci. 2008;102(2):371–82.
160. Ryan BC, Hotchkiss AK, Crofton KM, Gray Jr LE. In utero and lactational exposure to bisphenol A, in contrast to ethinyl estradiol, does not alter sexually dimorphic behavior, puberty, fertility, and anatomy of female LE rats. Toxicol Sci. 2010;114(1):133–48.
161. Takagi H, Shibutani M, Masutomi N, et al. Lack of maternal dietary exposure effects of bisphenol A and nonylphenol during the critical period for brain sexual differentiation on the reproductive/endocrine systems in later life. Arch Toxicol. 2004;78(2):97–105.
162. Tinwell H, Haseman J, Lefevre PA, Wallis N, Ashby J. Normal sexual development of two strains of rat exposed in utero to low doses of bisphenol A. Toxicol Sci. 2002;68(2):339–48.
163. Kobayashi K, Miyagawa M, Wang RS, Sekiguchi S, Suda M, Honma T. Effects of in utero and lactational exposure to bisphenol A on somatic growth and anogenital distance in F1 rat offspring. Ind Health. 2002;40(4):375–81.
164. Tyl RW, Myers CB, Marr MC, et al. Three-generation reproductive toxicity study of dietary bisphenol A in CD Sprague-Dawley rats. Toxicol Sci. 2002;68(1):121–46.
165. Tanaka M, Nakaya S, Katayama M, et al. Effect of prenatal exposure to bisphenol A on the serum testosterone concentration of rats at birth. Hum Exp Toxicol. 2006;25(7):369–73.
166. Fernandez M, Bianchi M, Lux-Lantos V, Libertun C. Neonatal exposure to bisphenol a alters reproductive parameters and gonadotropin releasing hormone signaling in female rats. Environ Health Perspect. 2009;117(5):757–62.
167. Weltje L, vom Saal FS, Oehlmann J. Reproductive stimulation by low doses of xenoestrogens contrasts with the view of hormesis as an adaptive response. Hum Exp Toxicol. 2005;24(9):431–7.
168. Agency) EUSEP. Bisphenol A, CASRN 80–05–7, IRIS, Integrated Risk Information System. 1993; http://www.epa.gov/iris/subst/0356.htm, 2010.
169. Monje L, Varayoud J, Munoz-de-Toro M, Luque EH, Ramos JG. Neonatal exposure to bisphenol A alters estrogen-dependent mechanisms governing sexual behavior in the adult female rat. Reprod Toxicol. 2009;28(4):435–42.
170. Honma S, Suzuki A, Buchanan DL, Katsu Y, Watanabe H, Iguchi T. Low dose effect of in utero exposure to bisphenol A and diethylstilbestrol on female mouse reproduction. Reprod Toxicol. 2002;16(2):117–22.

3 Reproductive Neuroendocrine Targets of Developmental Exposure...

171. Hyoung UJ, Yang YJ, Kwon SK, et al. Developmental toxicity by exposure to bisphenol A diglycidyl ether during gestation and lactation period in Sprague-Dawley male rats. J Prev Med Public Health. 2007;40(2):155–61.
172. Padmanabhan V, Sarma HN, Savabieasfahani M, Steckler TL, Veiga-Lopez A. Developmental reprogramming of reproductive and metabolic dysfunction in sheep: native steroids vs. environmental steroid receptor modulators. Int J Androl. 2010;33(2):394–404.
173. Singh AR, Lawrence WH, Autian J. Maternal-fetal transfer of 14 C-di-2-ethylhexyl phthalate and 14 C-diethyl phthalate in rats. J Pharm Sci. 1975;64(8):1347–50.
174. Clewell RA, Kremer JJ, Williams CC, et al. Kinetics of selected di-n-butyl phthalate metabolites and fetal testosterone following repeated and single administration in pregnant rats. Toxicology. 2009;255(1–2):80–90.
175. Gray Jr LE, Wolf C, Lambright C, et al. Administration of potentially antiandrogenic pesticides (procymidone, linuron, iprodione, chlozolinate, p, p'-DDE, and ketoconazole) and toxic substances (dibutyl- and diethylhexyl phthalate, PCB 169, and ethane dimethane sulphonate) during sexual differentiation produces diverse profiles of reproductive malformations in the male rat. Toxicol Ind Health. 1999;15(1–2):94–118.
176. Gray Jr LE, Ostby J, Furr J, Price M, Veeramachaneni DN, Parks L. Perinatal exposure to the phthalates DEHP, BBP, and DINP, but not DEP, DMP, or DOTP, alters sexual differentiation of the male rat. Toxicol Sci. 2000;58(2):350–65.
177. Andrade AJ, Grande SW, Talsness CE, Grote K, Chahoud I. A dose-response study following in utero and lactational exposure to di-(2-ethylhexyl)-phthalate (DEHP): non-monotonic dose-response and low dose effects on rat brain aromatase activity. Toxicology. 2006;227(3): 185–92.
178. Nagao T, Ohta R, Marumo H, Shindo T, Yoshimura S, Ono H. Effect of butyl benzyl phthalate in Sprague-Dawley rats after gavage administration: a two-generation reproductive study. Reprod Toxicol. 2000;14(6):513–32.
179. Carruthers CM, Foster PM. Critical window of male reproductive tract development in rats following gestational exposure to di-n-butyl phthalate. Birth Defects Res B Dev Reprod Toxicol. 2005;74(3):277–85.
180. Struve MF, Gaido KW, Hensley JB, et al. Reproductive toxicity and pharmacokinetics of di-n-butyl phthalate (DBP) following dietary exposure of pregnant rats. Birth Defects Res B Dev Reprod Toxicol. 2009;86(4):345–54.
181. Macleod DJ, Sharpe RM, Welsh M, et al. Androgen action in the masculinization programming window and development of male reproductive organs. Int J Androl. 2010;33(2): 279–87.
182. Saillenfait AM, Sabate JP, Gallissot F. Effects of in utero exposure to di-n-hexyl phthalate on the reproductive development of the male rat. Reprod Toxicol. 2009;28(4):468–76.
183. Tyl RW, Myers CB, Marr MC, et al. Reproductive toxicity evaluation of dietary butyl benzyl phthalate (BBP) in rats. Reprod Toxicol. 2004;18(2):241–64.
184. Lee HC, Yamanouchi K, Nishihara M. Effects of perinatal exposure to phthalate/adipate esters on hypothalamic gene expression and sexual behavior in rats. J Reprod Dev. 2006;52(3):343–52.
185. Piepenbrink MS, Hussain I, Marsh JA, Dietert RR. Developmental Immunotoxicology of Di-(2-Ethylhexyl)phthalate (DEHP): Age-Based Assessment in the Female Rat. J Immunotoxicol. 2005;2(1):21–31.
186. Culty M, Thuillier R, Li W, et al. In utero exposure to di-(2-ethylhexyl) phthalate exerts both short-term and long-lasting suppressive effects on testosterone production in the rat. Biol Reprod. 2008;78(6):1018–28.
187. Borch J, Axelstad M, Vinggaard AM, Dalgaard M. Diisobutyl phthalate has comparable antiandrogenic effects to di-n-butyl phthalate in fetal rat testis. Toxicol Lett. 2006;163(3): 183–90.
188. Parks LG, Ostby JS, Lambright CR, et al. The plasticizer diethylhexyl phthalate induces malformations by decreasing fetal testosterone synthesis during sexual differentiation in the male rat. Toxicol Sci. 2000;58(2):339–49.

189. Mylchreest E, Sar M, Wallace DG, Foster PM. Fetal testosterone insufficiency and abnormal proliferation of Leydig cells and gonocytes in rats exposed to di(n-butyl) phthalate. Reprod Toxicol. 2002;16(1):19–28.
190. Fisher JS, Macpherson S, Marchetti N, Sharpe RM. Human 'testicular dysgenesis syndrome': a possible model using in-utero exposure of the rat to dibutyl phthalate. Hum Reprod. 2003;18(7):1383–94.
191. Yamasaki K, Okuda H, Takeuchi T, Minobe Y. Effects of in utero through lactational exposure to dicyclohexyl phthalate and p, p'-DDE in Sprague-Dawley rats. Toxicol Lett. 2009; 189(1):14–20.
192. Zhu YJ, Jiang JT, Ma L, et al. Molecular and toxicologic research in newborn hypospadiac male rats following in utero exposure to di-n-butyl phthalate (DBP). Toxicology. 2009; 260(1–3):120–5.
193. Lee KY, Shibutani M, Takagi H, et al. Diverse developmental toxicity of di-n-butyl phthalate in both sexes of rat offspring after maternal exposure during the period from late gestation through lactation. Toxicology. 2004;203(1–3):221–38.
194. Saillenfait AM, Gallissot F, Sabate JP. Differential developmental toxicities of di-n-hexyl phthalate and dicyclohexyl phthalate administered orally to rats. J Appl Toxicol. 2009;29(6):510–21.
195. Drake AJ, van den Driesche S, Scott HM, Hutchison GR, Seckl JR, Sharpe RM. Glucocorticoids amplify dibutyl phthalate-induced disruption of testosterone production and male reproductive development. Endocrinology. 2009;150(11):5055–64.
196. Chou YY, Huang PC, Lee CC, Wu MH, Lin SJ. Phthalate exposure in girls during early puberty. J Pediatr Endocrinol Metab. 2009;22(1):69–77.
197. Wolff MS, Britton JA, Boguski L, et al. Environmental exposures and puberty in inner-city girls. Environ Res. 2008;107(3):393–400.
198. Buck Louis GM, Gray Jr LE, Marcus M, et al. Environmental factors and puberty timing: expert panel research needs. Pediatrics. 2008;121 Suppl 3:S192–207.
199. Nikaido Y, Yoshizawa K, Danbara N, et al. Effects of maternal xenoestrogen exposure on development of the reproductive tract and mammary gland in female CD-1 mouse offspring. Reprod Toxicol. 2004;18(6):803–11.
200. Kouki T, Okamoto M, Wada S, Kishitake M, Yamanouchi K. Suppressive effect of neonatal treatment with a phytoestrogen, coumestrol, on lordosis and estrous cycle in female rats. Brain Res Bull. 2005;64(5):449–54.
201. Kouki T, Kishitake M, Okamoto M, Oosuka I, Takebe M, Yamanouchi K. Effects of neonatal treatment with phytoestrogens, genistein and daidzein, on sex difference in female rat brain function: estrous cycle and lordosis. Horm Behav. 2003;44(2):140–5.
202. Kubo K, Arai O, Omura M, Watanabe R, Ogata R, Aou S. Low dose effects of bisphenol A on sexual differentiation of the brain and behavior in rats. Neurosci Res. 2003;45(3): 345–56.
203. Wisniewski AB, Cernetich A, Gearhart JP, Klein SL. Perinatal exposure to genistein alters reproductive development and aggressive behavior in male mice. Physiol Behav. 2005;84(2): 327–34.
204. Gunnarsson D, Selstam G, Ridderstrale Y, Holm L, Ekstedt E, Madej A. Effects of dietary phytoestrogens on plasma testosterone and triiodothyronine (T3) levels in male goat kids. Acta Vet Scand. 2009;51:51.
205. Giampietro PG, Bruno G, Furcolo G, et al. Soy protein formulas in children: no hormonal effects in long-term feeding. J Pediatr Endocrinol Metab. 2004;17(2):191–6.
206. Strom BL, Schinnar R, Ziegler EE, et al. Exposure to soy-based formula in infancy and endocrinological and reproductive outcomes in young adulthood. J Am Med Assoc. 2001;286(7): 807–14.
207. Lee BJ, Jung EY, Yun YW, et al. Effects of exposure to genistein during pubertal development on the reproductive system of male mice. J Reprod Dev. 2004;50(4):399–409.
208. Moore TO, Karom M, O'Farrell L. The neurobehavioral effects of phytoestrogens in male Syrian hamsters. Brain Res. 2004;1016(1):102–10.

3 Reproductive Neuroendocrine Targets of Developmental Exposure...

209. Tena-Sempere M. Kisspeptin/GPR54 system as potential target for endocrine disruption of reproductive development and function. Int J Androl. 2010;33(2):360–8.
210. Shirota M, Mukai M, Sakurada Y, et al. Effects of vertically transferred 3,3′,4,4′,5-pentachlorobiphenyl (PCB-126) on the reproductive development of female rats. J Reprod Dev. 2006;52(6):751–61.
211. Faqi AS, Dalsenter PR, Mathar W, Heinrich-Hirsch B, Chahoud I. Reproductive toxicity and tissue concentrations of 3,3′,4,4′-tetrachlorobiphenyl (PCB 77) in male adult rats. Hum Exp Toxicol. 1998;17(3):151–6.
212. Muto T, Imano N, Nakaaki K, et al. Estrous cyclicity and ovarian follicles in female rats after prenatal exposure to 3,3′,4,4′,5-pentachlorobiphenyl. Toxicol Lett. 2003;143(3):271–7.
213. Lyche J, Larsen H, Skaare JU, et al. Effects of perinatal exposure to low doses of PCB 153 and PCB 126 on lymphocyte proliferation and hematology in goat kids. J Toxicol Environ Health A. 2004;67(11):889–904.
214. Oskam IC, Lyche JL, Krogenaes A, et al. Effects of long-term maternal exposure to low doses of PCB126 and PCB153 on the reproductive system and related hormones of young male goats. Reproduction. 2005;130(5):731–42.
215. Sager DB, Girard DM. Long-term effects on reproductive parameters in female rats after translactational exposure to PCBs. Environ Res. 1994;66(1):52–76.
216. Lee CK, Kang HS, Kim JR, et al. Effects of aroclor 1254 on the expression of the KAP3 gene and reproductive function in rats. Reprod Fertil Dev. 2007;19(4):539–47.
217. Gellert RJ. Uterotrophic activity of polychlorinated biphenyls (PCB) and induction of precocious reproductive aging in neonatally treated female rats. Environ Res. 1978;16(1–3):123–30.
218. Stoker TE, Laws SC, Crofton KM, Hedge JM, Ferrell JM, Cooper RL. Assessment of DE-71, a commercial polybrominated diphenyl ether (PBDE) mixture, in the EDSP male and female pubertal protocols. Toxicol Sci. 2004;78(1):144–55.
219. Stoker TE, Cooper RL, Lambright CS, Wilson VS, Furr J, Gray LE. In vivo and in vitro antiandrogenic effects of DE-71, a commercial polybrominated diphenyl ether (PBDE) mixture. Toxicol Appl Pharmacol. 2005;207(1):78–88.
220. Den Hond E, Schoeters G. Endocrine disrupters and human puberty. Int J Androl. 2006;29(1):264–71. discussion 286–290.
221. Blanck HM, Marcus M, Tolbert PE, et al. Age at menarche and tanner stage in girls exposed in utero and postnatally to polybrominated biphenyl. Epidemiology. 2000;11(6):641–7.
222. Gladen BC, Ragan NB, Rogan WJ. Pubertal growth and development and prenatal and lactational exposure to polychlorinated biphenyls and dichlorodiphenyl dichloroethene. J Pediatr. 2000;136(4):490–6.
223. Den Hond E, Roels HA, Hoppenbrouwers K, et al. Sexual maturation in relation to polychlorinated aromatic hydrocarbons: Sharpe and Skakkebaek's hypothesis revisited. Environ Health Perspect. 2002;110(8):771–6.
224. Vasiliu O, Muttineni J, Karmaus W. In utero exposure to organochlorines and age at menarche. Hum Reprod. 2004;19(7):1506–12.
225. Veeramachaneni DN, Palmer JS, Amann RP, Pau KY. Sequelae in male rabbits following developmental exposure to p, p'-DDT or a mixture of p, p'-DDT and vinclozolin: cryptorchidism, germ cell atypia, and sexual dysfunction. Reprod Toxicol. 2007;23(3):353–65.
226. Heinrichs WL, Gellert RJ, Bakke JL, Lawrence NL. DDT administered to neonatal rats induces persistent estrus syndrome. Science. 1971;173(997):642–3.
227. Gellert RJ, Heinrichs WL, Swerdloff RS. DDT homologues: estrogen-like effects on the vagina, uterus and pituitary of the rat. Endocrinology. 1972;91(4):1095–100.
228. Rasier G, Parent AS, Gerard A, et al. Mechanisms of interaction of endocrine-disrupting chemicals with glutamate-evoked secretion of gonadotropin-releasing hormone. Toxicol Sci. 2008;102(1):33–41.
229. Tait S, Ricceri L, Venerosi A, Maranghi F, Mantovani A, Calamandrei G. Long-term effects on hypothalamic neuropeptides after developmental exposure to chlorpyrifos in mice. Environ Health Perspect. 2009;117(1):112–6.

230. Shutoh Y, Takeda M, Ohtsuka R, et al. Low dose effects of dichlorodiphenyltrichloroethane (DDT) on gene transcription and DNA methylation in the hypothalamus of young male rats: implication of hormesis-like effects. J Toxicol Sci. 2009;34(5):469–82.
231. Orikasa C, Kondo Y, Sakuma Y. Transient transcription of the somatostatin gene at the time of estrogen-dependent organization of the sexually dimorphic nucleus of the rat preoptic area. Endocrinology. 2007;148(3):1144–9.
232. Adewale HB, Jefferson WN, Newbold RR, Patisaul HB. Neonatal bisphenol-a exposure alters rat reproductive development and ovarian morphology without impairing activation of gonadotropin-releasing hormone neurons. Biol Reprod. 2009;81(4):690–9.
233. Atanassova N, McKinnell C, Turner KJ, et al. Comparative effects of neonatal exposure of male rats to potent and weak (environmental) estrogens on spermatogenesis at puberty and the relationship to adult testis size and fertility: evidence for stimulatory effects of low estrogen levels. Endocrinology. 2000;141(10):3898–907.
234. Ceccarelli I, Della Seta D, Fiorenzani P, Farabollini F, Aloisi AM. Estrogenic chemicals at puberty change ERalpha in the hypothalamus of male and female rats. Neurotoxicol Teratol. 2007;29(1):108–15.
235. Akingbemi BT, Sottas CM, Koulova AI, Klinefelter GR, Hardy MP. Inhibition of testicular steroidogenesis by the xenoestrogen bisphenol A is associated with reduced pituitary luteinizing hormone secretion and decreased steroidogenic enzyme gene expression in rat Leydig cells. Endocrinology. 2004;145(2):592–603.
236. Della Seta D, Minder I, Belloni V, Aloisi AM, Dessi-Fulgheri F, Farabollini F. Pubertal exposure to estrogenic chemicals affects behavior in juvenile and adult male rats. Horm Behav. 2006;50(2):301–7.
237. Kawai K, Murakami S, Senba E, et al. Changes in estrogen receptors alpha and beta expression in the brain of mice exposed prenatally to bisphenol A. Regul Toxicol Pharmacol. 2007; 47(2):166–70.
238. Kim K, Son TG, Park HR, et al. Potencies of bisphenol A on the neuronal differentiation and hippocampal neurogenesis. J Toxicol Environ Health A. 2009;72(21–22):1343–51.
239. Lomenick JP, Calafat AM, Melguizo Castro MS, et al. Phthalate exposure and precocious puberty in females. J Pediatr. 2010;156(2):221–5.
240. Saillenfait AM, Sabate JP, Gallissot F. Diisobutyl phthalate impairs the androgen-dependent reproductive development of the male rat. Reprod Toxicol. 2008;26(2):107–15.
241. Noriega NC, Howdeshell KL, Furr J, Lambright CR, Wilson VS, Gray Jr LE. Pubertal administration of DEHP delays puberty, suppresses testosterone production, and inhibits reproductive tract development in male Sprague-Dawley and Long-Evans rats. Toxicol Sci. 2009; 111(1):163–78.
242. Ge RS, Chen GR, Dong Q, et al. Biphasic effects of postnatal exposure to diethylhexylphthalate on the timing of puberty in male rats. J Androl. 2007;28(4):513–20.
243. Dalsenter PR, Santana GM, Grande SW, Andrade AJ, Araujo SL. Phthalate affect the reproductive function and sexual behavior of male Wistar rats. Hum Exp Toxicol. 2006; 25(6):297–303.
244. Ma M, Kondo T, Ban S, et al. Exposure of prepubertal female rats to inhaled di(2-ethylhexyl) phthalate affects the onset of puberty and postpubertal reproductive functions. Toxicol Sci. 2006;93(1):164–71.
245. Kurahashi N, Kondo T, Omura M, Umemura T, Ma M, Kishi R. The effects of subacute inhalation of di (2-ethylhexyl) phthalate (DEHP) on the testes of prepubertal Wistar rats. J Occup Health. 2005;47(5):437–44.
246. Faber KA, Hughes Jr CL. Dose-response characteristics of neonatal exposure to genistein on pituitary responsiveness to gonadotropin releasing hormone and volume of the sexually dimorphic nucleus of the preoptic area (SDN-POA) in postpubertal castrated female rats. Reprod Toxicol. 1993;7(1):35–9.
247. Slikker Jr W, Scallet AC, Doerge DR, Ferguson SA. Gender-based differences in rats after chronic dietary exposure to genistein. Int J Toxicol. 2001;20(3):175–9.

248. Jacob DA, Temple JL, Patisaul HB, Young LJ, Rissman EF. Coumestrol antagonizes neuroendocrine actions of estrogen via the estrogen receptor alpha. Exp Biol Med (Maywood). 2001;226(4):301–6.
249. Jefferson WN, Doerge D, Padilla-Banks E, Woodling KA, Kissling GE, Newbold R. Oral exposure to genistin, the glycosylated form of genistein, during neonatal life adversely affects the female reproductive system. Environ Health Perspect. 2009;117(12):1883–9.
250. Salama J, Chakraborty TR, Ng L, Gore AC. Effects of polychlorinated biphenyls on estrogen receptor-beta expression in the anteroventral periventricular nucleus. Environ Health Perspect. 2003;111(10):1278–82.
251. Chung YW, Clemens LG. Effects of perinatal exposure to polychlorinated biphenyls on development of female sexual behavior. Bull Environ Contam Toxicol. 1999;62(6):664–70.
252. Steinberg RM, Juenger TE, Gore AC. The effects of prenatal PCBs on adult female paced mating reproductive behaviors in rats. Horm Behav. 2007;51(3):364–72.
253. Khan IA, Thomas P. Aroclor 1254 inhibits tryptophan hydroxylase activity in rat brain. Arch Toxicol. 2004;78(6):316–20.
254. Eroschenko VP, Cooke PS. Morphological and biochemical alterations in reproductive tracts of neonatal female mice treated with the pesticide methoxychlor. Biol Reprod. 1990;42(3): 573–83.
255. Navarro VM, Gottsch ML, Chavkin C, Okamura H, Clifton DK, Steiner RA. Regulation of gonadotropin-releasing hormone secretion by kisspeptin/dynorphin/neurokinin B neurons in the arcuate nucleus of the mouse. J Neurosci. 2009;29(38):11859–66.
256. Nagao T, Saito Y, Usumi K, Kuwagata M, Imai K. Reproductive function in rats exposed neonatally to bisphenol A and estradiol benzoate. Reprod Toxicol. 1999;13(4):303–11.
257. Salian S, Doshi T, Vanage G. Neonatal exposure of male rats to Bisphenol A impairs fertility and expression of sertoli cell junctional proteins in the testis. Toxicology. 2009;265(1–2): 56–67.
258. Kawai K, Nozaki T, Nishikata H, Aou S, Takii M, Kubo C. Aggressive behavior and serum testosterone concentration during the maturation process of male mice: the effects of fetal exposure to bisphenol A. Environ Health Perspect. 2003;111(2):175–8.
259. Dickerson SM, Cunningham SL, Gore AC. Prenatal PCBs disrupt early neuroendocrine development of the rat hypothalamus. Toxicol Appl Pharmacol 2011; 252:36-46.
260. Dickerson SM, Cunningham SL, Patisaul HB, Woller MJ, Gore AC. Endocrine disruption of brain sexual differentiation by developmental PCB exposure. Endocrinology. 2011;152: 581–94.

Chapter 4
Endocrine Disruptors and Puberty Disorders from Mice to Men (and Women)

Alberto Mantovani

Abstract Puberty disturbances are a serious global issue, which is nested in the worldwide trend of earlier pubertal onset in females. The putative role of endocrine disrupters (ED) has been investigated mainly concerning peripheral precocious puberty in females; however, some data suggests that also male gynecomastia, female central precocious puberty, and delayed puberty should deserve attention.

Epidemiological studies on ED support the involvement of specific widespread pollutants in either precocious [phthalates, dichlorodiphenyltrichloroethane (DDT)] or delayed [polychlorinated biphenyls (PCBs), lead] puberty; more recent evidence points out also dietary agents (mycotoxins, phytoestrogens). Further human studies should focus about multiple exposures and possible additive effects, as well as on other factors that may interact with or potentiate ED action, such as lifestyle, body composition, nutrient intake, and ethnicity.

Toxicological studies are essential to respond to such research needs as etiologic research on environmentally relevant ED levels and different effects at critical developmental windows. Nevertheless, current toxicological studies lack a robust database to define an array of endpoints associated with specific outcomes and modes of action; moreover, a reliable *in vitro* test or battery does not exist to screen chemicals potentially affecting puberty. The two experimental schemes adopted till now include: a "programming" scheme, where exposure is mediated by the dam, either *in utero* or during lactation, and a "direct exposure" scheme, dealing with weaned, immature animals. The two schemes are not mutually exclusive as they represent different susceptibility windows. Some established modes of action may be already associated with outcomes, e.g., ED with estrogenic effects tend to delay male and accelerate female puberty, with long-term histological effects in target tissues. Nevertheless, the spectrum of ED modes of action is going to include new

---

A. Mantovani (✉)
Food and Veterinary Toxicology Unit, Dept. Food Safety and Veterinary Public Health,
Istituto Superiore di Sanità, Rome, Italy
e-mail: alberto.mantovani@iss.it

E. Diamanti-Kandarakis and A.C. Gore (eds.), *Endocrine Disruptors and Puberty*,
Contemporary Endocrinology, DOI 10.1007/978-1-60761-561-3_4,
© Springer Science+Business Media, LLC 2012

mechanisms and targets, such as kisspeptins and N-methyl-D-aspartate receptors. Whereas for some chemicals the available data might be sufficient to trigger precautionary measures, a sound risk analysis needs an interdisciplinary integration between human medicine and experimental toxicology.

Keywords Precocious puberty • Pubertal disturbances

Introduction: Are Puberty Disorders an Increasing Problem?

Puberty is regulated by the endocrine system. Disruption of that system by any factor may, therefore, affect this development profoundly. An ongoing great secular trend is now generally recognized leading to earlier timing of female puberty concerning menarche age as well as the onset of secondary sexual characteristics, such as breast development in girls. The first notice of this phenomenon came from the USA in the 1990s, but then consistent observations came also from Europe and elsewhere. A multidisciplinary expert panel [1] concluded that the girls' data are sufficient to suggest a secular trend toward earlier breast development onset and menarche from 1940 to 1994, but the boys' data are insufficient to suggest a trend during this same period. A change in the timing of puberty markers was considered adverse from a public health perspective. Nutrition, in particular increased body fat, is a major factor involved; however, the weight-of-evidence evaluation of human and animal studies suggested that exposure to endocrine disrupters (ED), particularly the estrogen mimics and antiandrogens, are important factors associated with altered puberty timing [1].

Against the backstage of the general trend toward earlier pubertal timing, clinically evident precocious puberty currently affects 1 in 5,000 children and is ten times more common in girls. Precocious puberty entrains the appearance of secondary sex characteristics before 8 years of age or the onset of menarche before age 9; its incidence is raising in several countries like USA. Besides rare genetic conditions with very early puberty of central origin, identified risk factors for idiopathic precocious puberty include genetic predisposition, ethnicity, with higher frequency in black subjects, and pediatric obesity [2]. Besides precocious puberty with menarche, premature thelarche is a more benign condition, characterized by early breast development and higher estrogen levels, which may be observed also in small girls. Some cases show increased bone age as well and may represent an intermediate state between benign premature thelarche and precocious puberty [3].

Recent studies document the extent of phenomena. A Danish study compared two cohorts of girls aged 5.6–20.0 years that were studied in 1991–1993 (1991 cohort; n = 1100) and 2006–2008 (2006 cohort; n = 995). Onset of puberty, defined as mean age at attainment of glandular breast tissue, occurred 1 year earlier in the 2006 cohort (9.86 years) as compared to the 1991 cohort (10.88 years); the difference remained after adjusting for body mass index (BMI). Estimated mean ages at menarche showed a lesser difference as they were 13.42 years and 13.13 years in the

1991 and 2006 cohorts, respectively. Serum hormones changes were unlikely to account for observed differences: Follicle-stimulating hormone (FSH) and luteinizing hormone (LH) did not differ between the two cohorts at any age interval, whereas significantly lower estradiol levels were found in 8- to 10-year-old girls from the 2006 cohort compared with similarly aged girls from the 1991 cohort [4]. Early breast development may achieve high numbers, although the incidence is clearly different among ethnic groups. A study carried out on more than 1,200 girls at ages 6–8 years in three towns of USA found that at 7 years, 10.4% of white, 14.9% of Hispanic, and 23.4% of black non-Hispanic had glandular breast tissue, whereas at 8 years, the proportions were 18.3%, 30.9%, and 42.9%, respectively. Although the proportion of early breast development was clearly higher in black girls, the authors point out that white girls showed a particularly noticeable increase compared to girls born 10–30 years earlier [5].

The different forms of precocious puberty may lead to psychosocial problems, may affect growth and body composition, and may have long-bearing consequences, yet inadequately known. The early and improper endocrine (e.g., estrogenic) input to the reproductive system and/or to cancer-prone target tissues (e.g., breast) should not be overlooked. A recent study on pediatric ovarian masses demonstrates that precocious puberty is one preoperative indicator that best predicts an ovarian malignancy: Noticeably, patients aged 1–8 years have the greatest incidence of malignancy [6]. Thus, attention is being given to risk factors implicated in earlier pubertal onset.

Genetic background might explain 50–80% of variability in the timing of puberty [7]; genetic polymorphisms have been implicated also in thelarche [3]. Therefore, it may be difficult to find strong associations with individual, nongenetic risk factors; however, the interplay between different factors as well as with genetic predisposition is well worth investigating.

Interestingly, growing evidence has documented the relationship between intrauterine growth retardation (IUGR) and pubertal development [8]. Changes in pubertal development, thus, come to be part of the growing list of IUGR-related health disorders; although underlying mechanisms are still elusive, children born with IUGR may have an altered sensitivity to endocrine and metabolic signaling [8, 9]. Faster progression of puberty was found in girls previously affected by IUGR but not in boys; it may be noteworthy that children with previous IUGR had higher levels of the adrenal-produced steroid dehydroepiandrosterone [10]. The risk of precocious pubarche was associated with IUGR, extensive catch-up growth, and childhood obesity in an Australian study, irrespective of gestational age: The relationship was observed in both sexes, although within cases, girls outnumbered boys – eight to one [9].

Early pubertal development and an increased incidence of sexual precocity have been noticed in children, primarily girls, migrating for foreign adoption in several Western European countries. Factors related to the changing milieu after migration have been considered, but none has been identified as major, manageable contributor. Indeed, in the formerly deprived migrating children, refeeding and catch-up growth may prime maturation; however, precocious puberty has been seen also in some nondeprived migrating children [11].

All the above different factors, from genetic predisposition to increasing prevalence of adiposity, do not rule out the involvement of environmental chemicals; due to biological plausibility as well as widespread presence, ED are suspected to contribute to the trend of earlier pubertal onset as well as to nongenetic precocious puberty and thelarche [1]. The factors regulating the physiological onset of normal puberty are poorly understood, which hampers investigation of the possible role of environmental influences. ED may have more than one mode of action, and the effects may depend on the developmental stage of the exposed individual; moreover, ED typically affect programming, thus, effects of developmental exposures may manifest later in life [12]. According to the hypothesis by Mouritsen et al. [13], the overall features of earlier puberty may indicate an estrogen-like effect without concomitant central activation of the hypothalamic-pituitary axis. Indeed, the mean age at breast development has declined more abruptly than the mean age at menarche [4]; thus, the time span from initiation of breast development to menarche has increased. Moreover, ED modes of action may result in different effects according to sex (e.g., the different effect of an estrogen agonist in males and females); this might be consistent with the different prevalence of precocious puberty between girls and boys. ED might also contribute indirectly to early pubertal onset by altering the endocrine regulation of the feto-placental unit, thus resulting in IUGR which is a recognized risk factor for precocious puberty [8].

The rising global trend has drawn attention mostly toward accelerated puberty; on the other hand, delayed puberty should call for more attention. In particular, male delayed puberty is a commonly presenting problem to pediatricians: This may not always be a benign condition, with adverse impacts on sexual and psychosocial integration, growth, and skeletal maturation. Mutations and polymorphisms in candidate genes are being actively studied, but understanding of other risk factors would be important as well [14]. A relationship of puberty disorders with children exposure to estrogenic or antiandrogenic ED is biologically plausible, thus it should deserve more attention from the epidemiological standpoint. The following sections attempt to review and discuss the recent epidemiological and toxicological evidence on the involvement of ED in puberty disturbances, with special attention to more recent findings.

Human Studies on ED and Puberty Disturbances

Aksglaede et al. [15] point out that fetuses and children may be more sensitive to ED effects on puberty than previously thought: In particular, circulating levels of estradiol in prepubertal children are lower than originally claimed, which may render prepubertal children vulnerable to sex steroid alterations without impact on adults. Besides central puberty, requiring an alteration of the hypothalamic-pituitary axis, ED with estrogen agonist or antagonist activity may plausibly affect peripheral puberty by interacting with estrogen receptor pathways in peripheral tissues. Notwithstanding biological plausibility, the amount of epidemiological studies

4 Endocrine Disruptors and Puberty Disorders from Mice to Men (and Women)

showing convincing associations between specific ED and puberty disorders is comparatively limited. Exposures associated with signs of early female puberty include phthalates, dichlorodiphenyltrichloroethane (DDT), polybrominated biphenyls (PBBs); polychlorinated biphenyls (PCBs) and lead have been associated with delayed puberty in both sexes, and endosulfan, with slower reproductive maturation in boys [7, 16, 17].

In Puerto Rico, a temporal trend toward premature thelarche in girls and gynecomastia in boys has been noted during the early 1980s [18]. Serum samples from cases and controls were analyzed for determining the possible presence of pesticides and/or phthalates. No pesticides or their metabolites could be found in any of the serum samples; however, 68% (28/41) of the girls with premature breast development had measurable levels of phthalates [dimethyl, diethyl, dibutyl and di(2-ethylhexyl)] compared with 17% (6/35) control samples, suggesting an association with the cumulative exposure to these widespread plastic additives [19].

In another study performed in Belgium, plasma of girls with precocious puberty was screened for chlorinated pesticides; data showed an increase in plasma levels of the DDT metabolite p,p'-DDE in foreign children with precocious puberty immigrating from developing countries compared with undetectable levels in Belgian-born girls with idiopathic or organic precocious puberty; since DDT is still used in a number of developing countries, the data suggested a relationship with early exposure to this pesticide [20] together with other risk factors associated with migration [11].

The effect of *in utero* exposure to PBB on sexual maturation was evaluated in Michigan girls whose mothers were accidentally exposed through diet to the flame retardant FireMaster. According to questionnaires sent to mothers of daughters <18 years of age and to the daughters themselves, menarche and pubertal hair growth were highly significantly advanced in girls breast-fed, i.e., with higher perinatal exposure to PBBs; no association was observed with breast development. Interestingly, menarche and breast development are estrogen-dependent events whereas pubertal hair growth is independent from estrogen levels, suggesting that PBBs may interact with different pathways [21].

More recent studies have started tackling other ED that received comparatively little attention till now with respect to puberty disturbances as well as the issue of multiple exposures occurring in real-life situations.

Wolff et al. [22] investigated pubertal status in relation to combined ED exposures among a multiethnic group of 192 healthy 9-year-old girls residing in New York City; namely, the study considered phytoestrogens (enterolactone and daidzein), bisphenol A (BPA), DDE, PCB, and lead. Breast development was present in 53% of girls. The urinary phytoestrogen biomarker concentrations were lower among girls with breast development compared with no development: Main individual effects were attributable to daidzein and genistein, although the effect magnitude was relatively low. Since BMI is a source of peripubertal hormones relevant to breast development, BMI modification of exposure effects was also examined. As expected, stronger associations with delayed breast development were observed among girls with below-median BMI and third tertile of urinary daidzein and/or genistein. On the contrary, high urinary enterolactone was significantly associated with delayed breast development in girls with high BMI. No other associations with breast development were found; no biomarkers were associated with hair development, which was present in 31% of girls. Thus, the data suggest that phytoestrogens can be more significant than other ED in a normal urban setting, and that they may interact with BMI. The same group has recently published a more extended study, investigating the associations of concurrent exposures from three chemical classes (phenols, phthalates, and phytoestrogens) with pubertal stages in a multiethnic longitudinal study of 1,151 girls from New York City, New York, greater Cincinnati, Ohio, and northern California who were 6–8 years of age at enrollment. Breast development was present in 30% of girls, and 22% had pubic hair. This study has involved more centers and younger girls as compared to Wolff et al. [22]; associations of biomarkers of exposure with early puberty signs were less evident than in the former study. Phthalates metabolites showed a weak, but significant, positive trend, namely, both high-molecular-weight and low-molecular-weight metabolites with pubic hair development, whereas only low-molecular-weight ones with breast development. As for phytoestrogens, daidzein confirmed an inverse association with breast stage, but the effect magnitude was small; the effect of enterolactone showed an interaction with BMI partly consistent with that observed in the 2008 paper. In the first enterolactone quintile, high BMI was significantly associated with breast development, whereas there was no association in the fifth, highest quintile, indicating that enterolactone attenuated BMI associations with breast development. The authors suggested that the exposure levels found were too low to lead to but small increases of risk [23].

Another recent Flemish study took a different approach: The sexual maturation of a cohort of adolescents (1,679 subjects aged 14–15 years) was studied in relation to internal exposure to persistent pollutants, including lead and the chlorinated ED hexachlorobenzene (HCB), p,p'-DDE, and more representative PCB congeners (138, 153, and 180). In boys, serum levels of HCB, p,p'-DDE, and PCB were significantly and positively associated with pubertal staging (pubic hair and genital development). Higher levels of serum HCB and blood lead were associated with, respectively, a lower and a higher risk of gynecomastia, i.e., abnormally large breast development in pubertal boys. In girls, significant and negative associations were detected between blood lead and pubic hair development; higher exposure to PCBs was significantly associated with a delay in timing of menarche. Overall, the study provides evidence that environmental exposures to major pollutants at levels actually present in the general population are associated with measurable effects on pubertal development. However, the modes of action seem quite different: Sexual maturation seems to be impaired by in both sexes, accelerated by HCB in males, whereas PCB may accelerate male and delay female development. Of course, the long-term health consequences, if any, of such temporal shifts associated with environmental exposures cannot be inferred presently [24)].

Some recent studies investigated the association of puberty disturbances with agents not previously considered.

Mercury (Hg) is known to be a neurotoxic and nephrotoxic agent, whereas endocrine effects did receive a comparatively limited attention; a potential to alter steroid synthesis and metabolism is worth noting. An association between premature puberty and exposure to Hg from thimerosal-containing vaccines was evaluated in computerized medical records of 1990–1996 birth cohorts; a total of nearly 280,000 subjects were identified. Significantly increased rate ratios were observed for premature puberty for a 100-microgram difference in vaccine-derived Hg exposure in the birth to 7 months (rate ratio = 5.58) and birth to 13 months (rate ratio = 6.45) of age exposure windows [25].

Soy-based milk formulas contain significant amounts of phytoestrogens. Among 694 female infants aged 3–24 months, 92 had consumed soy formulas for more than 3 months and none was referred for abnormal breast development. No differences in the presence of breast tissue buds were observed during the first year of life. However, breast tissue buds were significantly more prevalent in the second year of life in infants fed soy-based formula vs. those that were breast-fed and those fed dairy-based formula (22.0% vs. 10.3%; odds ratio of 2.45, 95% confidence interval 1.11–5.39). Unlike infants on dairy-based formulas and breast-feeding, infants fed a soy-based formula did not demonstrate a decline in the prevalence of breast buds during the second year of life; the authors suggest that phytoestrogens impose a preserving effect on breast tissue that is evolved in early infancy [63]. It is noteworthy that the phytoestrogens daidzein and genistein were associated with a slower breast development by Wolff et al. [22, 23]: The apparent discrepancy may be related to the different exposure window (early infancy vs. childhood) as well as to the different intake pattern (high-level, concentrated intake through milk formula vs. continuous low-level dietary intake).

A hypothesized external agent that induces central precocious puberty would act through the early initiation of the pulsatility of gonadotropin-releasing hormone from the hypothalamus, thereby inducing the cascade of hormonal events that result in pubertal development; however, scant data are available on the possible association between central precocious puberty and environmental agents [15]. Massart et al. [26] performed a multicenter analysis of central precocious puberty in northwest Tuscany, an area of nearly 1,300,000 inhabitants in the period 1998–2003. Similar prevalences were found in the main cities of the areas (mean 30.4 per 100,000 children, standard deviation 18.6), whereas the sea town of Viareggio area (19,219 inhabitants 0–14 years of age) had a prevalence of more than 160 per 100,000. Living in Viareggio area significantly increased the risk; the annual incidence was relatively constant and significantly higher than in other towns of the areas (relative risk 5.04, 95% confidence limits 2.3–11.2). Interestingly, 47% of total cases of the whole northwest Tuscany were distributed in the countryside surrounding Viareggio, with a peak incidence in some villages. The research group hypothesized environmental risk factors and pointed out the dietary exposure to zearalenone, a mycotoxin which is a strong estrogen agonist. High serum levels of zearalenone and its metabolite alpha-zearalenol were found in 6 out of 32 (19%) girls with central precocious puberty, compared with levels found in matched controls. Zearalenone levels correlated with patient's height, weight, and height velocity, but not with BMI, bone age, and gonadal secretion [27]. Zearalenone is a nonsteroidal by-product of *Fusarium* sp. molds that can contaminate grains, nuts, and their based products, thus, dietary exposure may be significant; in addition, alpha-zeranol has been widely used in the USA as a growth promoter to improve fattening rates in cattle. Both zearalenone and alpha-zeranol have higher *in vitro* and *in vivo* estrogen-agonist activity as compared to many man-made ED; accordingly, Massart and Saggese [28] call for more attention toward the mycotoxin as a risk factor for precocious pubertal development.

Experimental Toxicology Studies and Pubertal Onset

A wealth of experimental toxicology studies in rodents indicates potential effects of specific ED on puberty onset. The established markers for onset of puberty are vaginal opening (VO) in female rats and balanopreputial separation (BPS) in male rats; their onset is normally seen starting from PND 35 and PND 42, respectively. Additional parameters may be appearance of estrus and testicular descent in females and males, respectively. Delay of puberty onset might be an unspecific cause of marked toxicity in juvenile rodents; besides, VO and BPS are considered rather characteristic signs of endocrine disruption during the pre- and peripubertal phase [29]. In many cases, changes in puberty-related markers have been detected in studies that specifically target the postweaning, peripubertal phase [postnatal day (PND) 22–55 in the rat] as a sensitive window to ED effects. More subtle, and potentially more sensitive, parameters may also be used concurrently, such as changes in mammary gland histology, hormone balance, or molecular markers, even though a more robust database is needed [30].

The available studies point to signatures of effects typical for distinct modes of action.

Chlorotriazine herbicides are interesting ED since they specifically target neuroendocrine regulation of reproductive development in rats by altering the secretion of luteinizing hormone (LH) and prolactin through a direct effect on the central nervous system [30]. Atrazine is the most known and investigated compound: Several studies indicated that prenatal exposure could delay VO and mammary development in the female offspring of Long-Evans rats [31]. A cross-fostering experiment following exposure on gestation days (GD) 15–19 with 100 mg/kg bw showed that a significant delay in VO was mostly related to exposure from milk; on the other hand, significant delays in epithelial development of mammary glands were related to either intrauterine and/or milk exposure. Delays in pubertal endpoints were not related to body weight or hormone concentrations and appeared to have different susceptibility developmental windows [31]. The late embryonic and fetal window (GD 13–19) is the most sensitive to elicit delayed mammary gland development, without affecting fetal weight; females exposed *in utero* to atrazine may have impaired nursing function, as suggested by the reduced growth of their offspring [32]. Atrazine may delay pubertal development in male rats, too. Stoker et al. [33] showed that BPS was significantly delayed in male Wistar rats orally dosed with dose levels 12.5–200 mg/kg bw along the whole postweaning, peripubertal phase. The weights of accessory reproductive gland was apparently reduced at higher dose levels (50 mg/kg and above), but this effect was partly related to reduced food consumption; testes weights were unaffected by atrazine treatment. Hormone changes were less consistent: A dose-related toward LH reduction was observed, whereas prolactin levels showed a high variability. Prolactin is involved in the maintenance of LH receptors prior to puberty; no difference in LH receptor number was observed in testes. At top dose (200 mg/kg bw) intratesticular testosterone was significantly decreased, and serum estrone and estradiol, significantly increased. Thyroid hormone balance may also be relevant to puberty onset: The only significant change was the increase of triiodothyronine (T3) in the high-dose group. Thus, atrazine produced a clear-cut delay of puberty onset in males that was only partly reflected by hormone levels or organ weight changes at sexual maturation [33]. Further *in vitro* assays showed that atrazine may delay male puberty in rats by directly inhibiting Leydig cell testosterone production [34]. Two studies were recently performed on female Wistar rats orally exposed to another chlorotriazine, simazine: In the first study, rats were treated with 12.5–100 mg/kg bw on PND 22–42 (21 days overall); in the second one, treatment period covered from PND 22 until the first day of estrus after PND 62; and dose levels were 12.5–200 mg/kg bw (at least 41 days overall). In the 21-day exposure, there was a clear delaying effect on puberty onset: VO and the day of first estrus were delayed, and the number of normal cycles was significantly decreased. In the 41-day exposure, VO and the day of first estrus were also delayed, but the number of normal estrous cycles was not different than controls. Both studies showed a significant decrease in serum prolactin, which, together with delayed puberty signs, suggests a neuroendocrine mode of action [35]. While interesting, the findings on chlorotriazine pubertal effects are observed at dose levels much higher than those expected upon environmental exposure. A recent attempt to assess the effects of peri- and postnatal (GD 14 to PND 21) exposure to atrazine at environmentally relevant levels (1 or 100 µg/kg bw) did not achieve convincing results: Subtle behavioral changes of doubtful biological significance and slight changes in sperm and steroid metabolism are reported, the latter being quoted as effects even though they did not reach statistical significance [36]. The study did not explore histological, functional, or molecular markers of target neuroendocrine and reproductive tissues that could have provided more relevant findings: Thus, the actual effects of chlorotriazines at environmentally relevant conditions on pubertal development still await adequate studies.

PCBs have also been reported to exert a gender-specific effect on puberty onset through a neuroendocrine mechanism different than that of chlorotriazines. A reconstituted mixture of four indicator congeners (the dioxin-like PCB 126 and the non-dioxin-like, highly persistent PCB 138, 153, and 180) was injected subcutaneously to dams at the dose of 10 mg/kg daily from GD15 to GD19 and then twice a week till weaning. The developmental PCB exposure produced important changes in the dimorphic hypothalamic expression of both aromatase and the 5 alpha-reductases, which were still evident in adult animals: Female puberty onset occurred earlier but without estrous cycle irregularity, while testicular descent in males was delayed. Neurobehavioral changes (sexual behavior and memory retention) were also noted specifically in males. The study points to more attention toward the endocrine and reproductive impact of altered steroidogenesis at central nervous level; the single, high dose used and the parenteral exposure route, nevertheless, caution toward the actual relevance of these findings for risk assessment [37].

The dicarboximide fungicides, procymidone and vinclozolin, delay male puberty in rodents through direct antagonism with androgen receptor (AR) [38]. Iprodione is a dicarboximide fungicide that does not appear to be an AR antagonist; however, in juvenile rats given 50–200 mg/kg bw iprodione through the whole postweaning-peripubertal phase (PND 23–52), BPS was significantly retarded at ≥ 100 mg/kg bw. Other effects seen at top dose (200 mg/kg) included reduced weight of sex accessory glands and increased adrenal and liver weights. There was a significant impact on steroid production, as evidenced by reduced serum levels of testosterone, 17a-hydroxyprogesterone, and androstenedione as well as by reduced *ex vivo* testis production of testosterone and progesterone; serum LH was unaffected. Overall, these changes suggest an impairment of first steps of testicular steroidogenesis; nevertheless, this different antiandrogenic mode of action leads to outcomes essentially similar to the other dicarboximides [39].

Male puberty may be delayed also by compounds exhibiting estrogenic activity. The ultraviolet filter 4-methyl-benzylidene camphor (4-MBC) is a preferential estrogen receptor (ER)-beta ligand. When 0.7–47 mg/kg bw 4-MBC (/day) was administered in the diet to rats throughout the full developmental cycle (to the parent generation before mating through to offspring's adulthood), male puberty was delayed; in the adult offspring, prostate weight was also significantly decreased. Effects were observable down to the lowest dose level of 0.7 mg/kg; further analysis on prostate tissue revealed a reduced expression of the nuclear receptor coregulator N-CoR [40]. The plasticizer di-n-butyl phthalate (DBP) induces testicular atrophy in prepubertal male rats: Studies upon single and prolonged administration in rats exposed on PND 21–28 indicated this effect being characterized by increased apoptosis of spermatogenic cells, whereas testicular steroidogenesis and serum LH/FSH showed an early but transient reduction. Cotreatment with the antiestrogen ICI 182,780 significantly reduced the apoptosis effect, without reversing the reduced testicular steroidogenesis. Thus, impairment of testicular development in prepubertal rats is likely induced by activating estrogen receptors in the tissue, whereas reduction in testicular steroidogenic function is not associated with spermatogenic cell apoptosis [41].

While delaying male puberty, estrogenic compounds are expected to elicit earlier female puberty. The issue might be somewhat less straightforward: Soy isoflavone supplementation did not alter the histological phenotype of endometrial cells in growing rats, but a hyperplastic response of the endometrium was shown in the adult rats. This age-related difference could have been also related to a higher bioavailability in adults [42]. Dietary soy isoflavones did not potentiate or alter the uterotrophic response to a high-dose level (600 mg/kg bw) of the estrogenic ED bisphenol A in juvenile rats [43]. Diethylstilbestrol (DES) is a recognized model for ER-alpha agonists. Earlier VO and increased uterus weight weaning were observed in the offspring of CD-1 mice given orally 0.01 mg/kg bw during GD 9–16, a critical period for reproductive tract development *in utero*. Histological analysis showed reduced endometrium thickness and ovarian abnormalities (increased polyovular follicles, oocytes with condensed chromatin) at sexual maturity [44]. The persistent chlorinated compound lindane given orally to pregnant CD-1 mice on GD 9–16 (15 mg/kg bw) showed a pattern of effects mainly consistent with that of DES: increased uterus weight in weaned females, earlier VO, and reduced diameters of primary oocytes at sexual maturity, without effects on steroid hormone metabolism enzymes. Further *in vitro* analysis indicated an ER-beta-mediated action [45]. The neonatal (PND 2–20) exposure to the plasticizer benzyl butyl phthalate (BBP) increased the uterine weight/body weight ratio at PND 21 while it decreased the body weight/VO time ratio. BBP did not induce significant changes on the morphology of the mammary gland, but increased proliferative index in terminal end buds at PND 35 days and in lobules till adulthood. Moreover, BBP upregulated many genes related to proliferation and differentiation, communication, and signal transduction, mainly at the end of the exposure (PND 21), becoming less prominent thereafter [46]. Phthalates are not major binders of the ERs and AR; however, taking together the findings by Alam et al. [41] and Moral et al. [46], there are strong indications that some phthalates may elicit estrogen-like effects in target tissues of prepubertal rodents. "Estrogen-like" mechanisms may elicit quite different effects. Masutomi et al. [47] compared three putative estrogenic ED upon perinatal exposure through the maternal diet from GD 15 to PND 10: the chlorinated insecticide methoxychlor (MXC; 24–1200 mg/kg diet), genistein (20–1000 mg/kg diet), and diisononyl phthalate (DINP; 400–20,000 mg/kg diet). Noticeably, soy-free diet was used as a basal diet to eliminate possible estrogenic effects from the standard diet. All chemicals caused signs of maternal toxicity at high doses. The top MXC exposure level (1,200 mg/kg) accelerated VO and delayed BPS in females and males, respectively; genistein did not affect puberty parameters, but reduced body weight in adult offspring at all doses; DINP also did not affect puberty, even though slight but significant histological changes in testes and ovary were seen in the adult offspring at the top exposure of 20,000 mg/kg (corresponding to a very high intake of $\geq 1,000$ mg/kg bw). MXC was the only compound showing apparent "estrogenic" effects; irregular estrous cyclicity and histopathological alterations in the reproductive tract and anterior pituitary were observed in the adult offspring exposed perinatally to highest MXC level. Overall, data on different estrogen-acting chemicals (DES, lindane, BBP, MXC) consistently point out that earlier puberty timing is linked with long-term altered programming of the female reproductive tract [44–47]. Interestingly, estrogenic ED were the topic of the only study attempting at studying additive effects by modeling real-life situations. Two predefined mixtures of phytoestrogens and synthetic chemicals identified as ED were tested in the uterotrophic assay on prepubertal rats: The composition of each mixture (what chemicals and to what amount) was based on human exposure data. The phytoestrogen mixture did elicit an uterotrophic response, whereas the synthetic one has no effect itself nor an additive effect with phytoestrogens, possibly because of exposure levels too low [48].

A single oral prepubertal (PND 29) exposure to 0.01 mg/kg bw of the aryl hydrocarbon receptor (AHR) agonist 2,3,7,8-tetrachlorodibenzo-p-dioxin (TCDD) delayed puberty in female rats as well as induced abnormal cyclicity and premature reproductive senescence, according to vaginal cytology monitoring until 1 year of age. The lifelong exposure to lower TCDD levels (50 or 200 ng/kg bw/week) beginning *in utero* induced a dose- and time-dependent loss of normal cyclicity (mainly prolonged estrous cycles) and significantly hastened the onset of reproductive senescence. The number and size of ovarian follicles and diestrus concentrations of LH were not altered by TCDD, whereas at both dose levels, progesterone levels were increased and estradiol levels were decreased [49]. The relevance of this study resides in the link between AHR agonism, delayed female puberty, and persistent endocrine imbalance with long-bearing consequences on reproductive life.

The biocide tributyltin (TBT) does induce immune and endocrine effects in rodents; one main mode of action may be interference with retinoic acid pathways [50]. Peripubertal (PND 24–45) exposure of male KM mice to 0.05 mg or 0.5 mg TBT chloride/kg bw/3 days decreased serum and intratesticular testosterone levels in mice treated with the top dose level on PND 49; this change was no more present in animals examined at adulthood (PND 84). Adult animals showed a number of other dose-related hormone changes, such as reduced serum and intratesticular estradiol and reduced serum LH. Such changes were not accompanied by evident markers of altered pubertal timing, while unfortunately, no histological analysis was performed; however, it might be noteworthy that mice exposed to the lowest dose level (0.05 mg/kg) exhibited a significant increase in body weight in the late stage of the experiment [51].

A few studies have shown that some thiadiazoles act as ED via disturbance of thyroid hormone homeostasis. Pubertal female rats were orally treated on PND 22–42 with thiadiazoles used as agricultural fungicides in China, namely, thiodiazole copper (4–30 mg/kg bw), bismerthiazole (10–40 mg/kg bw), and saisentong (5–15 mg/kg). All three compounds impaired thyroid function, as evidenced by reduced thyroid hormones and increased TSH concentrations, as well as thyroid gland hyperplasia; however, no histological changes were observed in uterus and ovaries, and VO was unaffected [52–54]. These data suggest that the juvenile female rat is sensitive to thyroid disrupters, but pubertal timing is not a sensitive endpoint.

Finally, attention should be given to studies pointing out an endocrine modulation of puberty by substances and modes of action different from those generally as ED. Diesel exhausts accelerate male puberty in mice exposed *in utero*. Pregnant ICR mice inhaled diesel exhausts at soot concentrations of $0.3–3.0$ mg/m^3 on GD 2–16. When male offspring were examined on PND 28, the weights of the testes and accessory glands and testosterone concentration in serum were significantly increased; moreover, testosterone levels correlated significantly with the expression levels of steroidogenic enzymes and with daily sperm production [55]. A further investigation was conducted on *in utero* exposure to nanoparticle-rich diesel exhaust. Pregnant F344 rats were exposed to either nanoparticle-rich (149 micrograms/m^3) or filtered (3.10 micrograms/m^3) diesel exhaust on GD 1–19. Results in rat male offspring examined on PND 28 were not consistent with those in mice, since a delayed maturation was seen with both nanoparticle-rich and filtered exhausts, namely, weights of accessory glands, serum concentrations of testosterone, progesterone, corticosterone, and FSH and testicular concentrations of steroidogenic acute regulatory

4 Endocrine Disruptors and Puberty Disorders from Mice to Men (and Women) 131

protein, and 17 beta-hydroxysteroid dehydrogenase mRNA were all decreased, whereas serum concentrations of immunoreactive inhibin were increased. The only difference between the two exhausts was the increased transcription of FSH receptor mRNA in the nanoparticle-rich group [56]. The mechanism(s) underlying the effects in both species upon early intrauterine exposure as well as such interspecies difference still await clarification.

Hexavalent chromium (Cr-VI), a recognized industrial pollutant and carcinogenic, has recently raised a renewed attention, owing to potential endocrine-related effects. Cr-VI delayed puberty and impaired female reproductive function thereafter, as evidenced by decreased follicle number and extended estrous cycle. A serious impact on endocrine balance was pointed out by decreased steroidogenesis, GH, and prolactin as well as increased FSH. Cr-VI markedly reduced the expression of steroid metabolism proteins and hormone receptors in rat granulosa cells. Vitamin C pretreatment protected granulosa cells from Cr-VI effects; also, the coadministration of vitamin C in drinking water (500 mg/L) reduced Cr-VI ovarian toxicity *in vivo* [57].

Semicarbazide (SEM) raised concern as contaminant in baby foods, since it is formed from azodicarbonamide present in glass jar–packaged foods. A juvenile toxicity study on rats exposed on PND 22–50 to 40–140 mg/kg bw showed a much higher toxicity in immature animals, than previously reported in adults, with effects on skeletal cartilages, thyroid, and ovary being present also at the lowest dose level, 40 mg/kg bw [58]. SEM-treated female rats showed delayed VO at all tested doses, whereas in males BPS was anticipated at SEM 40 and 75 mg/kg and delayed at 140 mg/kg, the latter effect probably due to the high general toxicity elicited by this dose level. Serum estradiol levels were dose-dependently reduced in females; however, aromatase activity was overall increased, mainly in females. Dehydrotestosterone serum levels were also decreased but without a clear dose response. This multiple and gender-specific array of effect was further explored by an *in vitro* battery of assays: Besides a weak antiestrogenic activity, findings indicated antagonism with N-methyl-aspartate receptors, a rather novel aspect among ED modes of action [59].

The investigation of novel targets and modes of action can, indeed, bring fresh support to risk assessment. Attainment of reproductive capacity at puberty relies on a complex series of maturational events that include sexual differentiation of the brain. Alterations of sex steroid milieu during such critical period may disrupt pubertal maturation and gonadotropic function later in life. Kisspeptins, products of the KiSS-1 gene, have emerged as essential gatekeepers of puberty onset and reproductive function. Neonatal injection of estradiol benzoate or of high dose of the "xenoestrogen" bisphenol A to male and female rats resulted in a dose-dependent decrease in hypothalamic KiSS-1 mRNA levels at the prepubertal stage, linked to lower serum LH concentrations [60]. Female rats injected subcutaneously with the phytoestrogen genistein (10 mg/kg bw) on PND 0–3 showed accelerated VO and lower numbers of KiSS immunoreactive fibers in specific hypothalamic areas (anteroventral, periventricular, and arcuate nuclei). On PND 24, multiple oocyte follicles were no more present in ovaries of untreated animals, but were still detectable in those from treated animals, indicating a disruption in the timing of ovarian development.

It is noteworthy that 10 micrograms/kg bw of estradiol benzoate had the same effects as subcutaneous genistein [61].

Expression and function of hypothalamic KiSS-1 system is sensitive not only to the activational effects but also to the organizing actions of sex steroids during critical stages of development. Studies in rodents have demonstrated that early exposures to androgens, and estrogens are crucial for proper sexual differentiation of the patterns of KiSS-1 mRNA expression, whereas the actions of estrogen along puberty are essential for the rise of hypothalamic kisspeptins during this period [62]. Thus, the investigation of novel ED targets may support the scientific basis of risk assessment, explain age and gender-related differences, and provide new biomarkers to study exposed populations.

Considerations for Risk Assessment

The available evidence points out that puberty disturbances are a serious global issue, which is nested in the worldwide trend of earlier pubertal onset in females, associated with environmental, social, and nutritional factors [1, 3].

Concerning the role of environmental factors, most studies concentrate on peripheral precocious puberty and premature thelarche in females; however, also male gynecomastia [24], female central precocious puberty [28] and delayed puberty [14, 24] should deserve attention.

Epidemiological studies on ED support the involvement of specific widespread pollutants in either precocious (phthalates, DDT) or delayed (PCB, lead) puberty [7, 16, 17]. More recent evidence points out agents that have been somewhat overlooked till now as well as exposure situations that may partly shift the attention from environment to food. The estrogenic mycotoxin zearalenone, a contaminant of grains and nuts, has been associated with central precocious puberty in Italian cluster [27]. Dietary intake of phytoestrogens appears to have a slight but significant puberty-delaying effect in prepubertal girls [22, 23], while soy-based formulas had a preserving effect on breast buds in infants [63]: These data lend further support to the concept that ED effects depend on susceptibility age window as well as on dose and toxicokinetics [12]. Two major issues emerge from human studies on ED and puberty:

- Taking into account the multiple exposures occurring in real-life scenario. While several recent studies assessed concurrently multiple exposures [22–24], very little effort has been made till now to assess a "cocktail effect," i.e., whether low levels of ED with similar modes of action or targets may have an additive action [48].
- Taking into account the interactions with other risk factors involved in puberty disturbances. Phytoestrogens interact with BMI in prepubertal girls; soy phytoestrogens (daidzein and genistein) and lignans (enterolactone) seem to interact with BMI in a different way [22, 23]. Early exposure to persistent pesticides, no more authorized in industrialized countries, may contribute with other factors to

4 Endocrine Disruptors and Puberty Disorders from Mice to Men (and Women) 133

the increased incidence of precocious puberty observed in immigrated girls [11, 20]. Nutrition is an important, yet overlooked, factor modulating vulnerability to ED: A dedicated database on ED-diet interactions, EDID, has been launched in the Italian ED website, http://www.iss.it/inte (also available in english) [64]. Subclinical deficiencies of vitamins or trace elements may increase the vulnerability to specific ED mechanisms, conversely; some ED may directly interact with nutrient-modulated pathways, e.g., retinoic acid [50], hereby increasing the requirements for that nutrient [65]. Finally, ethnicity is a factor itself in puberty onset [2]; ethnic disparities, such as health behaviors associated with poverty, could also result in differences in exposure to particular ED [66].

Toxicological studies are essential to respond to data gaps and research needs identified by Buck Louis et al. [67], such as i) etiologic research focused on ED effects at environmentally relevant levels, ii) different effects at critical developmental windows, and iii) basic research on the association between ED mechanisms and either gonadotropin-dependent central puberty or gonadotropin-independent peripheral puberty. Nevertheless, current toxicological studies suffer from shortcomings, such as the lack of a robust database to define an array of histological, biochemical, and molecular endpoints associated with specific outcomes and modes of action. Moreover, a reliable *in vitro* test or battery does not exist to screen chemicals potentially affecting puberty; however, some papers use ad hoc *in vitro* approaches to clarify and integrate *in vivo* assays [34, 45, 57, 59].

Basically, two schemes have been adopted for toxicological studies on ED effects on puberty:

- The "programming" scheme, where exposure is mediated by the dam, either *in utero* or during lactation, the duration depending on the critical window(s) of interest [31, 32, 37, 44, 45, 55–57]. Such scheme could well fit into the extended one-generation reproductive toxicity study protocol that is currently proposed for adoption by the international testing strategies [68].
- The "direct exposure" scheme, that may be related either to the juvenile toxicity studies already used in pharmacology [44] and/or to the peripubertal assay for ED endorsed by the US Environmental Protection Agency [33; 35]: The former is more similar to subacute toxicity test, albeit with more attention to endpoints relevant to immature animals, and the latter is more of a screening assay targeted to endocrine-relevant endpoints. While risk assessment agencies call for more data on children as vulnerable population group or because of specific exposure situations (foods, toys, etc.), it might be surprising that no international effort is foreseen to develop a robust tool for characterizing chemical hazards in the immature organisms.

Toxicological studies should develop a database that may allow to associate puberty-relevant endpoints with given modes of action. The short review above suggests the following associations:

- Delayed puberty in both sexes with increased prolactin level: inhibition of LH production at central nervous system, as for chlorotriazines [35]

134 A. Mantovani

- Delayed male and accelerated female puberty, without estrous cycle irregularity: altered steroidogenesis at central nervous system [37]
- Impaired male puberty: antiandrogens, such as dicarboximides [40]
- Delayed male and accelerated female puberty, with long-term histological effects in target tissues: estrogenic effects [39, 44, 46]
- Delayed female puberty with long-term disturbances on estral cyclicity: AHR agonists such as TCDD [49]

The above modes of action do not cover at all the spectrum of ED modes of action, which currently is going to include new mechanisms and targets [59, 61].

A wealth of evidence indicates that exposure to ED may be associated with puberty disturbances; whereas for some chemicals the available data might be sufficient to trigger precautionary measures, a sound risk analysis needs an interdisciplinary integration between human medicine and experimental toxicology.

Acknowledgments The paper has been prepared within the frame of the following projects: PREVIENI (htpp://www.iss.it/prvn), funded by the Italian Ministry of Environment; BRIDGE (website in preparation), funded by the EC 7th Framework Programme.
The support of Mrs. Francesca Baldi (Istituto Superiore di Sanità) in the preparation of the manuscript is gratefully acknowledged.

References

1. Euling SY, Selevan SG, Pescovitz OH, Skakkebaek NE. Role of environmental factors in the timing of puberty. Pediatrics. 2008;121(Suppl 3):S167–71.
2. Cesario SK, Hughes LA. Precocious puberty: a comprehensive review of literature. J Obstet Gynecol Neonatal Nurs. 2007;36:263–74.
3. Codner E, Román R. Premature thelarche from phenotype to genotype. Pediatr Endocrinol Rev. 2008;5:760–5.
4. Aksglaede L, Sørensen K, Petersen JH, Skakkebaek NE, Juul A. Recent decline in age at breast development: the Copenhagen puberty study. Pediatrics. 2009;123:932–9.
5. Biro FM, Galvez MP, Greenspan LC, Succop PA, Vangeepuram N, Pinney SM, Teitelbaum S, Windham GC, Kushi LH, Wolff MS. Pubertal assessment method and baseline characteristics in a mixed longitudinal study of girls. Pediatrics. 2010;126:583–90.
6. Oltmann SC, Garcia N, Barber R, Huang R, Hicks B, Fischer A. Can we preoperatively risk stratify ovarian masses for malignancy? J Pediatr Surg. 2010;45:130–4.
7. Toppari J, Juul A. Trends in puberty timing in humans and environmental modifiers. Mol Cell Endocrinol. 2010;324:39–44.
8. Schoeters G, Den Hond E, Dhooge W, van Larebeke N, Leijs M. Endocrine disruptors and abnormalities of pubertal development. Basic Clin Pharmacol Toxicol. 2008;102:168–75.
9. Neville KA, Walker JL. Precocious pubarche is associated with SGA, prematurity, weight gain, and obesity. Arch Dis Child. 2005;90:258–61.
10. van Weissenbruch MM, Engelbregt MJ, Veening MA, Delemarre-van de Waal HA. Fetal nutrition and timing of puberty. Endocr Dev. 2005;8:15–33.
11. Parent AS, Teilmann G, Juul A, Skakkebaek NE, Toppari J, Bourguignon JP. The timing of normal puberty and the age limits of sexual precocity: variations around the world, secular trends, and changes after migration. Endocr Rev. 2003;24:668–93.
12. Mantovani A. Risk assessment of endocrine disrupters. The role of toxicological studies. Ann NY Acad Sci. 2006;1076:239–52.

4 Endocrine Disruptors and Puberty Disorders from Mice to Men (and Women)

13. Mouritsen A, Aksglaede L, Sørensen K, et al. Hypothesis: exposure to endocrine-disrupting chemicals may interfere with timing of puberty. Int J Androl. 2010;33:346–59.
14. Ambler GR. Androgen therapy for delayed male puberty. Curr Opin Endocrinol Diabetes Obes. 2009;16:232–9.
15. Aksglaede L, Juul A, Leffers H, Skakkebaek NE, Andersson AM. The sensitivity of the child to sex steroids: possible impact of exogenous estrogens. Hum Reprod Update. 2006;12: 341–9.
16. Caserta D, Maranghi L, Mantovani A, Marci R, Maranghi F, Moscarini M. Impact of endocrine disruptor chemicals in gynaecology. Hum Reprod Update. 2008;14:59–72.
17. Roy JR, Chakraborty S, Chakraborty TR. Estrogen-like endocrine disrupting chemicals affecting puberty in humans–a review. Med Sci Monit. 2009;15:137–45.
18. Freni-Titulaer LW, Cordero JF, Haddock L, Lebron G, Martinez R, Mills JL. Premature thelarche in Puerto Rico. A search for environmental factors. Am J Dis Child. 1986;140: 1263–7.
19. Colon I, Caro D, Bourdony CJ, Rosario O. Identification of phthalate esters in the serum of young Puerto Rican girls with premature breast development. Environ Health Perspect. 2000;108:895–900.
20. Krstevska-Konstantinove M, Charlier C, Craen M, et al. Sexual precocity after immigration from developing countries to Belgium: Evidence of previous exposure to organochlorine pesticides. Hum Reprod. 2001;16:1020–6.
21. Blanck HM, Marcus M, Tolbert BE, Rubin C, Henderson AK, Hertzberg VS, Zhang RH, Cameron L. Age at menarche and tanner stage in girls exposed in utero and postnatally to polybromonated biphenyl. Epidemiol. 2000;11:641–7.
22. Wolff MS, Britton JA, Boguski L, et al. Environmental exposures and puberty in inner-city girls. Environ Res. 2008;107:393–400.
23. Wolff MS, Teitelbaum SL, Pinney SM, et al. Breast cancer and environment research centers. Investigation of relationships between urinary biomarkers of phytoestrogens, phthalates, and phenols and pubertal stages in girls. Environ Health Perspect. 2010;118:1039–46.
24. Den Hond E, Dhooge W, Bruckers L, et al. Internal exposure to pollutants and sexual maturation in Flemish adolescents. J Expo Sci Environ Epidemiol. 2010 Mar 3. [Epub ahead of print].
25. Geier DA, Young HA, Geier MR. Thimerosal exposure & increasing trends of premature puberty in the vaccine safety datalink. Indian J Med Res. 2010;131:500–7.
26. Massart F, Seppia P, Pardi D, et al. High incidence of central precocious puberty in a bounded geographic area of northwest Tuscany: an estrogen disrupter epidemic? Gynecol Endocrinol. 2005;20:92–8.
27. Massart F, Meucci V, Saggese G, Soldani G. High growth rate of girls with precocious puberty exposed to estrogenic mycotoxins. J Pediatr. 2008;152:690–5.
28. Massart F, Saggese G. Oestrogenic mycotoxin exposures and precocious pubertal development. Int J Androl. 2010;33:369–76.
29. Laffin B, Chavez M, Pine M. The pyrethroid metabolites 3-phenoxybenzoic acid and 3-phenoxybenzyl alcohol do not exhibit estrogenic activity in the MCF-7 human breast carcinoma cell line or Sprague-Dawley rats. Toxicology. 2010;267:39–44.
30. Calamandrei G, Maranghi F, Venerosi A, Alleva E, Mantovani A. Efficient testing strategies for evaluation of xenobiotics with neuroendocrine activity. Reprod Toxicol. 2006;22:164–74.
31. Rayner JL, Wood C, Fenton SE. Exposure parameters necessary for delayed puberty and mammary gland development in Long-Evans rats exposed in utero to atrazine. Toxicol Appl Pharmacol. 2004;195:23–34.
32. Rayner JL, Enoch RR, Fenton SE. Adverse effects of prenatal exposure to atrazine during a critical period of mammary gland growth. Toxicol Sci. 2005;87:255–66.
33. Stoker TE, Laws SC, Guidici DL, Cooper RL. The effect of atrazine on puberty in male wistar rats: an evaluation in the protocol for the assessment of pubertal development and thyroid function. Toxicol Sci. 2000;58:50–9.
34. Friedmann AS. Atrazine inhibition of testosterone production in rat males following peripubertal exposure. Reprod Toxicol. 2002;16:275–9.

35. Zorrilla LM, Gibson EK, Stoker TE. The effects of simazine, a chlorotriazine herbicide, on pubertal development in the female Wistar rat. Reprod Toxicol. 2010;29:393–400.
36. Belloni V, Dessì-Fulgheri F, Zaccaroni M, Di Consiglio E, De Angelis G, Testai E, Santochirico M, Alleva E, Santucci D. Early exposure to low doses of atrazine affects behavior in juvenile and adult CD1 mice. Toxicology. 2011;279(1–3):19–26.
37. Colciago A, Casati L, Mornati O, Vergoni AV, Santagostino A, Celotti F, Negri-Cesi P. Chronic treatment with polychlorinated biphenyls (PCB) during pregnancy and lactation in the rat Part 2: Effects on reproductive parameters, on sex behavior, on memory retention and on hypothalamic expression of aromatase and 5alpha-reductases in the offspring. Toxicol Appl Pharmacol. 2009;239:46–54.
38. Blystone CR, Lambright CS, Cardon MC, et al. Cumulative and antagonistic effects of a mixture of the antiandrogens vinclozolin and iprodione in the pubertal male rat. Toxicol Sci. 2009;111:179–88.
39. Blystone CR, Lambright CS, Furr J, Wilson VS, Gray Jr LE. Iprodione delays male rat pubertal development, reduces serum testosterone levels, and decreases ex vivo testicular testosterone production. Toxicol Lett. 2007;174(1–3):74–81.
40. Durrer S, Ehnes C, Fuetsch M, Maerkel K, Schlumpf M, Lichtensteiger W. Estrogen sensitivity of target genes and expression of nuclear receptor co-regulators in rat prostate after pre- and postnatal exposure to the ultraviolet filter 4-methylbenzylidene camphor. Environ Health Perspect. 2007;115(Suppl 1):42–50.
41. Alam MS, Ohsako S, Matsuwaki T, et al. Induction of spermatogenic cell apoptosis in prepubertal rat testes irrespective of testicular steroidogenesis: a possible estrogenic effect of di(n-butyl) phthalate. Reproduction. 2010;139:427–37.
42. Chang MJ, Nam HK, Myong N, Kim SH. Age-related uterotrophic response of soy isoflavone intake in rats. J Med Food. 2007;10:300–7.
43. Wade MG, Lee A, McMahon A, Cooke G, Curran I. The influence of dietary isoflavone on the uterotrophic response in juvenile rats. Food Chem Toxicol. 2003;41:1517–25.
44. Maranghi F, Tassinari R, Moracci G, Macrì C, Mantovani A. Effects of a low oral dose of diethylstilbestrol (DES) on reproductive tract development in F1 female CD-1 mice. Reprod Toxicol. 2008;26:146–50.
45. Maranghi F, Rescia M, Macrì C, et al. Lindane may modulate the female reproductive development through the interaction with ER-beta: an in vivo-in vitro approach. Chem Biol Interact. 2007;169:1–14.
46. Moral R, Wang R, Russo IH, Mailo DA, Lamartiniere CA, Russo J. The plasticizer butyl benzyl phthalate induces genomic changes in rat mammary gland after neonatal/prepubertal exposure. BMC Genomics. 2007;8:453.
47. Masutomi N, Shibutani M, Takagi H, Uneyama C, Takahashi N, Hirose M. Impact of dietary exposure to methoxychlor, genistein, or diisononyl phthalate during the perinatal period on the development of the rat endocrine/reproductive systems in later life. Toxicology. 2003;192:149–70.
48. van Meeuwen JA, van den Berg M, Sanderson JT, Verhoef A, Piersma AH. Estrogenic effects of mixtures of phyto- and synthetic chemicals on uterine growth of prepubertal rats. Toxicol Lett. 2007;170:165–76.
49. Franczak A, Nynca A, Valdez KE, Mizinga KM, Petroff BK. Effects of acute and chronic exposure to the aryl hydrocarbon receptor agonist 2,3,7,8-tetrachlorodibenzo-p-dioxin on the transition to reproductive senescence in female Sprague-Dawley rats. Biol Reprod. 2006;74:125–30.
50. Yonezawa T, Hasegawa S, Ahn JY, et al. Tributyltin and triphenyltin inhibit osteoclast differentiation through a retinoic acid receptor-dependent signaling pathway. Biochem Biophys Res Commun. 2007;355:10–5.
51. Si J, Wu X, Wan C, Zeng T, Zhang M, Xie K, Li J. Peripubertal exposure to low doses of tributyltin chloride affects the homeostasis of serum T, E2, LH, and body weight of male mice. Environ Toxicol. 2010 Jan 5. [Epub ahead of print]

4 Endocrine Disruptors and Puberty Disorders from Mice to Men (and Women) 137

52. Zhang L, Wang J, Zhu GN. Pubertal exposure to bismerthlazol inhibits thyroid function in juvenile female rats. Exp Toxicol Pathol. 2009;61:453–9.
53. Zhang L, Wang J, Zhu GN. Pubertal exposure to saisentong: effects on thyroid and hepatic enzyme activity in juvenile female rats. Exp Toxicol Pathol. 2010;62:127–32.
54. Zhang L, Wang J, Zhu GN, Su L. Pubertal exposure to thiodiazole copper inhibits thyroid function in juvenile female rats. Exp Toxicol Pathol. 2010;62:163–9.
55. Yoshida S, Ono N, Tsukue N, Oshio S, Umeda T, Takano H, Takeda K. In utero exposure to diesel exhaust increased accessory reproductive gland weight and serum testosterone concentration in male mice. Environ Sci. 2006;13:139–47.
56. Li C, Taneda S, Taya K, et al. Effects of in utero exposure to nanoparticle-rich diesel exhaust on testicular function in immature male rats. Toxicol Lett. 2009;185:1–8.
57. Banu SK, Samuel JB, Arosh JA, Burghardt RC, Aruldhas MM. Lactational exposure to hexavalent chromium delays puberty by impairing ovarian development, steroidogenesis and pituitary hormone synthesis in developing Wistar rats. Toxicol Appl Pharmacol. 2008;232:180–9.
58. Maranghi F, Tassinari R, Lagatta V, et al. Effects of the food contaminant semicarbazide following oral administration in juvenile Sprague-Dawley rats. Food Chem Toxicol. 2009; 47:472–9.
59. Maranghi F, Tassinari R, Marcoccia D, et al. The food contaminant semicarbazide acts as an endocrine disrupter: evidence from an integrated in vivo/in vitro approach. Chem Biol Interact. 2010;183:40–8.
60. Navarro VM, Sánchez-Garrido MA, Castellano JM, et al. Persistent impairment of hypothalamic KiSS-1 system after exposures to estrogenic compounds at critical periods of brain sex differentiation. Endocrinology. 2009;150:2359–67.
61. Losa SM, Todd KL, Sullivan AW, Cao J, Mickens JA, Patisaul HB. Neonatal exposure to genistein adversely impacts the ontogeny of hypothalamic kisspeptin signaling pathways and ovarian development in the peripubertal female rat. Reprod Toxicol. 2010 Oct 15. [Epub ahead of print].
62. Tena-Sempere M. Kisspeptin/GPR54 system as potential target for endocrine disruption of reproductive development and function. Int J Androl. 2010;33:360–8.
63. Zung A, Glaser T, Kerem Z, Zadik Z. Breast development in the first 2 years of life: an association with soy-based infant formulas. J Pediatr Gastroenterol Nutr. 2008;46:191–5.
64. Baldi F, Mantovani A. A new database for food safety: EDID (endocrine disrupting chemicals – diet interaction database). Ann Ist Super Sanita. 2008;44:57–63.
65. Latini G, Knipp G, Mantovani A, Marcovecchio ML, Chiarelli F, Söder O. Endocrine disruptors and human health. Mini Rev Med Chem. 2010;10:846–55.
66. Ranjit N, Siefert K, Padmanabhan V. Bisphenol-A and disparities in birth outcomes: a review and directions for future research. J Perinatol. 2010;30(1):2–9.
67. Buck Louis GM, Gray Jr LE, Marcus M, et al. Environmental factors and puberty timing: expert panel research needs. Pediatrics. 2008;121(Suppl 3):S192–07.
68. Piersma AH, Rorije E, Beekhuijzen ME et al. Combined retrospective analysis of 498 rat multi-generation reproductive toxicity studies: on the impact of parameters related to F1 mating and F2 offspring. Reprod Toxicol. 2010 Dec 3. [Epub ahead of print].

Chapter 5
Thyroid Hormone Regulation of Mammalian Reproductive Development and the Potential Impact of Endocrine-Disrupting Chemicals

**Kara Renee Thoemke*, Thomas William Bastian*,
and Grant Wesley Anderson**

Abstract Thyroid hormones are essential for growth and development of all vertebrate organisms. A growing literature reveals that thyroid hormones are centrally important in the development of reproductive tissues. In males, the predominant target is the developing testis with aberrant Sertoli and Leydig cell development and accompanying testicular dysfunction. Hypothyroidism has also been associated with precocious puberty in a distinct subpopulation of adolescent girls. The hypothalamic–pituitary–thyroid axis is the target of numerous endocrine-disrupting chemicals. Intriguingly, many of these chemicals also perturb the development and function of mammalian reproductive tissues. We now review the key literature detailing the role of thyroid hormones in mammalian reproductive development and provide an analysis of the evidence linking thyroidal disruption by polychlorinated biphenyl, polybrominated diphenyl ether, phthalate, triclosan, and bisphenol A exposure, with maldevelopment of mammalian reproductive systems.

Keywords Childhood • Development • Endocrine disruptors • Hyperthyroid • Hypothyroid • Leydig cell • Puberty • Sertoli cell • Testes • Thyroid hormone

Introduction

Thyroid hormones are essential for growth and development of all vertebrate organisms. The role of thyroid hormones in controlling the development and adult function of many organs including the brain, heart, and liver, is well described.

---

* Contributed equally

K.R. Thoemke
Department of Biology, The College of St. Scholastica, Duluth, USA

T.W. Bastian • G.W. Anderson (✉)
Department of Pharmacy Practice and Pharmaceutical Sciences,
College of Pharmacy Duluth, University of Minnesota, Duluth, USA
e-mail: ander163@d.umn.edu

E. Diamanti-Kandarakis and A.C. Gore (eds.), *Endocrine Disruptors and Puberty*,
Contemporary Endocrinology, DOI 10.1007/978-1-60761-561-3_5,
© Springer Science+Business Media, LLC 2012

Significantly, less attention has been historically paid to the role of thyroid hormones on mammalian reproductive development; however, a growing literature reveals that thyroid hormones are also centrally important in the development of reproductive tissues. The hypothalamic–pituitary–thyroid axis is the target of numerous endocrine-disrupting chemicals. Intriguingly, many of these chemicals also perturb the development and function of mammalian reproductive tissues. Might disruption of the thyroid axis contribute to the observed effects of these chemicals on mammalian reproduction and reproductive potential? Research testing this hypothesis is summarized in the following literature review.

Thyroid Hormone Synthesis, Metabolism, and Action

Environmental chemicals can adversely affect thyroid hormone synthesis, metabolism, transport, excretion, and other aspects of thyroid biology. Therefore, it is prudent to give a brief overview of thyroid hormone physiology before discussing the impact of thyroid hormones on the development of reproductive tissues. Thyroxine (T4) and 3,5,3'-triiodothyronine (T3) production by the thyroid gland require iodine, which is typically obtained from dietary sources. Most of the world's iodine is found in seawater and is concentrated by marine plants and animals. In mountainous regions and areas with frequent flooding, soils and food products are often iodine deficient. In these regions, dietary iodine supplementation, in the form of iodized salts or oils, must be relied upon for adequate human iodine consumption. Absorption of iodine, in the form of iodide, in the stomach and duodenum is mediated by the sodium/iodine symporter (NIS), which is expressed on the apical surface of enterocytes [1]. NIS is also expressed at the basolateral membrane of the thyroid gland follicular cells and utilizes the energy released by transporting sodium down its concentration gradient to concentrate iodide in the thyroid (Fig. 5.1b, #1) [2]. Upon entering the thyroid follicular cell, iodide is transported across the apical membrane into the colloid by pendrin, a chloride/iodide transporter (Fig. 5.1b, #2) [3]. In the colloid space, hydrogen peroxide and the heme-containing enzyme thyroid peroxidase (TPO) oxidize iodide and subsequently iodinate tyrosyl residues on thyroglobulin (Tg) in a process termed iodide organification (Fig. 5.1b, #3) [4]. TPO also catalyzes the coupling of two diiodotyrosine (DIT) residues to form T4 and the coupling of monoiodotyrosine (MIT) and DIT residues to form T3 (Fig. 5.1b, #4) [4]. The newly synthesized T4 and T3 remain attached to Tg (henceforth denoted T4-Tg or T3-Tg) and are stored in the colloid until needed [4]. Secretion of T4 and T3 begins with endocytosis of T4-Tg and T3-Tg from the colloid into the follicular cells (Fig. 5.1b, #5). Within the thyroid follicular cell, T4-Tg and T3-Tg are incorporated into phagolysosomes, and T4 and T3 are released from Tg into the circulation by proteolytic hydrolysis (Fig. 5.1b, #6, 7). The prohormone T4 is the major secreted iodothyronine product secreted from the thyroid gland. The secreted T4:T3 ratio is approximately 11:1 in humans and 5:1 in rodents [5].

Thyroid hormone synthesis is regulated by the hypothalamic–pituitary–thyroid (HPT) negative feedback system (Fig. 5.1a). Hypothalamic neurons in the paraventricular nucleus (PVN) express thyrotropin-releasing hormone (TRH) and have axonal processes extending into the median eminence. The median eminence connects to the anterior pituitary through a series of blood vessels collectively known as the hypophyseal portal system, allowing direct delivery of TRH to its site of action in the anterior pituitary. TRH is first synthesized as a large inactive peptide (pro-TRH) and is converted to the active TRH tripeptide (PyroGlu-His-Pro-NH$_2$) through several posttranslational processing steps [6]. In the final processing step, TRH-gly, the immediate precursor to TRH, is α(alpha)-amidated on the carboxy-terminus by the copper-containing peptidylglycine α(alpha)-amidating monooxygenase (PAM) enzyme [6]. TRH binding to TRH receptors in the anterior pituitary thyrotrope cells stimulates production of the thyroid stimulating hormone (TSH)

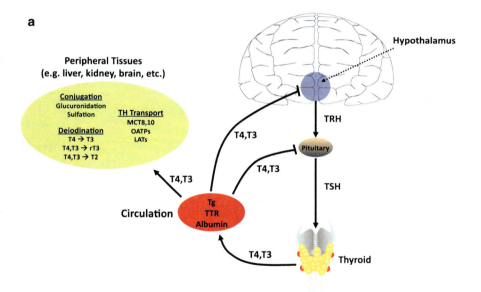

Fig. 5.1 Environmental disruption of thyroid hormone synthesis and metabolism. (**a**) The hypothalamic–pituitary–thyroid axis including control points of thyroid hormone metabolism and availability. (**b**) Thyroid hormone synthesis in the thyroid gland. (1) The Na$^+$/I$^-$ symporter (NIS) controls entry of iodide into the thyroid follicular cell. (2) Pendrin facilitates exit of iodide into the follicular lumen (colloid space). (3) Thyroid peroxidase (TPO) utilizes hydrogen peroxide to oxidize iodide and, subsequently, iodinate tyrosyl residues on thyroglobulin (Tg). (4) TPO also catalyzes the coupling of iodinated tyrosines on Tg to produce T4 and T3. (5) Thyroglobulin containing T4 and T3 is brought into the thyroid follicular cell through endocytosis. (6) Endosomes fuse with lysosomes, and T4 and T3 are released from Tg via proteolytic hydrolysis. (7) T4 and T3 are transported out of the thyroid follicular cell into circulation. (8) Iodotyrosine dehalogenase enzymes recycle iodide from uncoupled iodotyrosines. (9) TSH binding to the TSH receptor (TSHR) stimulates many aspects of TH synthesis and release. (**c**) Molecular basis of thyroid hormone action in a thyroid hormone-responsive cell. T3 and/or T4 are transported into a thyroid hormone-responsive cell via specific transport proteins. T3 and T4 are deiodinated by deiodinase enzymes to produce additional T3, rT3, and T2. T3 enters the nucleus and binds to the thyroid hormone receptor to regulate transcription of TH-responsive genes

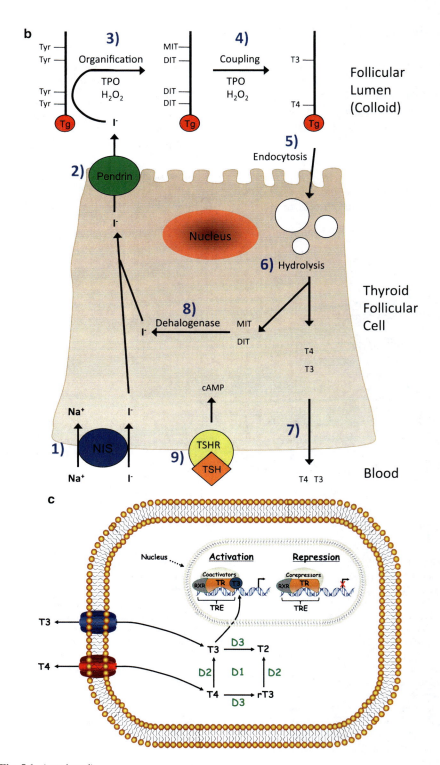

Fig. 5.1 (continued)

β(beta)-subunit [7], posttranslational glycosylation of TSH [8], and release of TSH into circulation [7]. TSH binding to TSH receptors on the basolateral membrane of thyroid follicle cells stimulates adenylate cyclase activity and increases intracellular cyclic adenosyl monophosphate (AMP) (Fig. 5.1b, #9). Cyclic AMP activates downstream signaling pathways resulting in increased iodine uptake, Tg synthesis, TPO activity, TH synthesis, and TH release [4]. T4 and T3 negatively regulate the expression and activity of hypothalamic PAM and TRH and pituitary TSH [6,9,10]. Type II deiodinase (D2), which converts T4 to T3, is expressed in both the hypothalamus and the anterior pituitary and is thought to play an important role in the negative feedback of T4 levels on TRH and TSH [5].

Circulating T4 and T3 are bound to carrier proteins such as thyroxine binding globulin (TBG), transthyretin (TTR), and albumin (Fig. 5.1a). Approximately 0.03% of T4 and 0.3% of T3 in the circulation remains unbound, and it is this "free" hormone that enters cells and exerts a molecular response [11]. The circulating concentration of T4 in humans is approximately 100 nM and in rats is approximately 44 nM [5]. The circulating concentration of T3 in humans is approximately 1.8 nM and in rats is approximately 750 nM [5]. Therefore, the molar ratio of circulating total T4 to total T3 is approximately 50–60:1 in both humans and rodents. The circulating T4:T3 ratio is higher than the ratio of T4:T3 that is secreted from the thyroid, which can be accounted for by the higher clearance rate of T3 (65% per day for T3 compared to 10% per day for T4) [5].

Transport of thyroid hormones across cell membranes is facilitated by carrier-mediated transport systems (Figs. 5.1a and 5.1c). Thyroid hormone transporters identified to date include the monocarboxylic acid transporters (MCT) 8 and 10, the system L amino acid transporters, and the organic anion-transporting polypeptides (Oatps) [12]. All these transporters transport thyroid hormones bidirectionally along concentration gradients. Thyroid hormone transporters are expressed in a wide variety of tissues including the brain, intestine, kidney, liver, and testes. Unique among the thyroid hormone transporters, Oatps mediate transmembrane transport of a wide variety of amphipathic organic anions in addition to thyroid hormones including steroid conjugates, prostanoids, bile salts, oligopeptides, drugs, toxins, and other xenobiotics [13,14]. The transport mechanism of Oatps is sodium independent and likely anion exchange. One of the Oatps, Oatp1c1, is expressed in the luminal and abluminal membranes of BBB endothelial cells, the basolateral membrane of choroid plexus epithelial cells, human ciliary body epithelial cells, and Leydig cells of the testes [15–17]. Compared to other Oatps, Oatp1c1 has a relatively narrow range of substrates. However, Oatp1c1 is a high-affinity thyroxine transporter and has the lowest identified K_m for T4 of all known transporters [16]. Other Oatp1c1 substrates include 3,3',5'-triiodothyronine (reverse T3), cerivastatin, and several glucuronidated sterols [16].

THs are predominately metabolized by three different deiodinase enzymes (Figs. 5.1a and 5.1c). Type II deiodinase (D2) catalyzes the removal of an outer ring iodine on T4 and T3. Type III deiodinase (D3) catalyzes the removal of an inner ring iodine on T4 and T3. Type I deiodinase (D1) performs both inner ring and outer ring deiodination reactions. Thus, D1 and D2 activate T4 by conversion to T3, and D1 and D3 inactivate T4 by conversion to reverse T3 (rT3). In humans, approximately 20% of circulating T3 comes from thyroidal production and the remaining 80% comes from peripheral conversion of T4 to T3 by D1 and D2 [5]. In rodents, the contribution of peripheral D1 and D2 conversion of T4 to T3 to the circulating T3 pool is approximately 60% and the remaining 40% comes from thyroidal production of T3 [5]. Further metabolism of T3 and rT3 by the deiodinases produces other inactive metabolites such as T2 and T1. In addition to deiodination, THs also undergo sulfation by sulfotransferases (SULTs) and glucuronidation by uridine diphosphate glucuronosyl transferases (UDPGTs) (Fig. 5.1a). These reactions primarily take place in the liver and kidney but can also occur in other tissues such as the choroid plexus. TH sulfation and glucuronidation serve to make THs more water-soluble and allow their concentration and excretion in the bile and urine [18,19]. D3 contributes to the clearance of THs from circulation by conversion to inactive metabolites, which are subsequently conjugated and excreted.

Intracellular T3 levels are controlled in a tissue-specific manner by both TH transporters, as described above, and deiodinase activity. In humans, D2 is expressed in the brain, thyroid, heart, skeletal muscle, spinal cord, and placenta [5]. D2 is localized to the endoplasmic reticulum membrane and thus converts T4 to T3 intracellularly. The proximity of D2 to the nucleus provides easy access of D2-generated T3 to its site of action. In addition, in the brain, it is thought that D2-expressing glial cells provide T3 to the surrounding neurons [20]. T3 negatively regulates D2 mRNA expression, while T4 and rT3 decrease D2 activity by stimulating ubiquitination and proteasomal degradation of D2 [20]. In humans, D1 is predominately expressed in the liver, kidney, and thyroid and is localized to the cell membrane [5]. Therefore, T3 generated by D1 rapidly equilibrates with circulating T3 and is not a major source of intracellular T3 in these tissues. D3 is expressed in the brain, pituitary, fetal liver, lung, placenta, and skin and is also localized to the plasma membrane [21]. In contrast to D2, both D1 and D3 expression are positively regulated by T3 [20]. Together, the relative expression and activity of D1, D2, and D3 within a given cell helps to control the intracellular T3 concentration.

The molecular basis of TH action is predominately at the nuclear level where T3 regulates gene transcription through binding to nuclear thyroid hormone receptors (TRs) (Fig. 5.1c). Two separate genes, alpha- and beta-c-erbA, encode four TH-binding TR isoforms including TRα(alpha)-1, TRβ(beta)-1, TRβ(beta)-2, and TRβ(beta)-3. These four TRs regulate transcription by binding, as a heterodimer with retinoid X receptor (RXR), to specific thyroid hormone response elements (TREs) in the upstream promoter region of a TH-responsive gene. In the absence of T3, TRs bound to the TRE typically recruit corepressors such as N-CoR or SMRT, resulting in deacetylation of chromatin and repression of gene transcription. TR binding of T3 alters TR protein conformation and signals for the recruitment of co-activators such as SRC-1, CBP, or PCAF, resulting in acetylation of chromatin and activation of TH-responsive gene transcription. The relative expression of TRα(alpha) and TRβ(beta) isoforms differs considerably across tissue and cell types and throughout development.

5 Thyroid Hormone Regulation of Mammalian Reproductive Development... 145

THs can also mediate molecular responses outside of their role in gene activation. These non-genomic effects may occur through interactions of T4, T3, or other TH metabolites with cytosolic or cell surface proteins. As discussed above, T4 binding to D2 at the ER membrane results in ubiquitination and proteasomal degradation of D2, contributing to tight control of intracellular T3 levels [20]. Recently, integrin α(alpha)Vβ(beta)3 was identified as a cell surface receptor for iodothyronines [22]. T4, the predominate ligand, binding to integrin α(alpha)Vβ(beta)3 results in the phosphorylation and activation of mitogen-activated protein kinase (MAPK). PhosphoMAPK (pMAPK) mediates several downstream events including phosphorylation of TRβ(beta)-1 and estrogen receptor-α(alpha), activating these receptors in the absence of T3 or estrogen [23,24]. In addition, T3 activates phosphatidylinositol 3-kinase (PI3-K) via Src kinase, resulting in increased cell membrane insertion and activity of the Na-K-ATPase [25]. Both T4 and rT3, but not T3, contribute to actin polymerization in astrocytes, without altering actin gene expression [26,27]. T3 also causes a rapid increase in cellular glucose uptake in the absence of new protein synthesis, but it also affects the expression of glucose transporters [28,29]. Therefore, TH exerts both genomic and non-genomic effects on cellular glucose uptake. Several additional non-genomic effects of TH have been described and are extensively reviewed elsewhere [30–32].

Disruption of the Thyroid Axis

Disruption of the physiologic effects of thyroid hormone can occur through two main mechanisms: alteration of circulating and tissue thyroid hormone levels, and disruption of thyroid hormone control over gene expression in the nucleus. We shall first consider the mechanisms identified as targets for disruption of thyroid hormone synthesis and metabolism. As demonstrated in Fig. 5.1a, b, there are many theoretical targets for disruption resulting in altered thyroid hormone levels including the thyroid hormone producing thyrocyte, the thyroid hormone-binding proteins responsible for carrying thyroid hormone in the blood and cerebrospinal fluid, and the enzymes and transporters involved in thyroid hormone metabolism and excretion. Indeed, numerous drugs and toxicants have been identified that interact with the thyroid axis at the level of each of these targets. In the thyrocyte, the uptake of iodide from the blood into the thyrocyte via sodium-iodide symporter NIS is inhibited by perchlorates, nitrates, and thiocyanates [33,34]. These toxicants compete with iodide for binding to the symporter resulting in decreased uptake of iodide and concomitant reduction in thyroid hormone production. Environmental exposure to combinations of NIS inhibitors such as perchlorate and thiocyanates, in the face of low iodine intake, has been shown to correlate with reductions in serum T4 in the human population [35]. Phthalates may also affect NIS activity through regulating NIS mRNA levels although the directionality of regulation varies between individual phthalate esters [36]. The drugs methimazole (MMI) and propylthiouracil (PTU), and disruptors such as soy isoflavones and the sunscreen agent benzophenone, target the enzyme thyroperoxidase responsible for the organification and conjugation reactions of thyroid hormone synthesis [37–39].

After release from the thyrocyte, the thyroid hormones circulate predominantly bound to carrier proteins such as thyroid binding globulin, transthyretin, or albumin. The free hormone levels are significantly less than the total hormone levels. A large number of compounds compete with the thyroid hormones for carrier protein binding resulting in marked changes in circulating hormone concentrations [39]. However, through classical endocrine feedback mechanisms, significant changes in total hormone concentration are often not associated with changes in free hormone levels. For example, the perfluorinated toxicant perfluorooctane sulfonate (PFOS) has been shown to significantly reduce circulating thyroid hormone levels in mammals [40]. However, no differences in free thyroid hormone levels have been observed in PFOS-dosed animals or in the hypothalamic–pituitary–thyroid axis [41]. In keeping with the observed unchanged free hormone levels in treated animals, expression of the TH-regulated target gene malic enzyme in the liver is not different between the treatment groups [41]. Not surprisingly, dosed animals also show no changes in the development or function of other thyroid hormone target tissues. This example demonstrates that environmental toxicants may exert significant effects on the total levels of circulating hormones without affecting the free hormone concentrations. A number of disruptors in addition to the perfluorinated compounds, including hydroxyl-PCBs and pentachlorophenol, affect thyroid hormone binding to carrier proteins and, presumably, also alter circulating TH levels [39].

Thyroid hormone disruptors also target enzymes involved in the metabolism of thyroid hormones such as the deiodinases and the conjugating sulfotransferases and UDP-glucuronosyltransferases (UGTs). The goitrogen propylthiouracil is a well-described inhibitor of type 1 deiodinase in the peripheral tissues [33]. PCBs also inhibit the action of type 2 deiodinase (D2) with potential implications for neural development as D2 serves as a critical responder to T3 levels in the developing brain [42]. Reductions in brain T3 result in robust activation of D2 expression and activity [43]. Thus, inhibition of deiodinase activity interferes with the regulatory roles played by these enzymes in the production of T3 in the periphery and in the degradation of the active forms of the thyroid hormones. Inhibition of sulfotransferases by hydroxy-PCBs, triclosan, and pentachlorophenol results in the reduced sulfation of thyroid hormones while PCBs, dioxins, the polybrominated diphenol ether (PBDE) DE-71, and bisphenol A upregulate UGT expression and activity resulting in increased glucuronidated thyroid hormones [44]. As glucuronidation facilitates the elimination of thyroid hormones in the bile, increased glucuronidation likely results in reduced thyroid hormone levels in animals exposed to these toxicants.

New to our understanding of mechanisms of disruption of thyroid hormone action is the concept of toxicant disruption of thyroid hormone transport. Thyroid hormone transporters serve to facilitate entry of thyroid hormones into target tissues and excretion of thyroid hormones and their metabolites from the body [45]. Relatively little attention has been focused on mechanisms controlling the expression of thyroid hormone transporters. However, expression of certain thyroid hormone transporters such as the Oatps is regulated by the xenobiotic sensing nuclear receptors [46], suggesting a novel mechanism of disruption for certain classes of thyroid hormone disruptors. For example, Oatp expression in the liver is transcriptionally regulated by the thyroid hormone disrupters [47] and aryl hydrocarbon receptor (AhR) ligands TCDD and PCB 126 [48]. In addition, expression of the brain barrier thyroid hormone transporter Oatp1c1 is upregulated in response to reduced circulating TH levels [16], suggesting a compensatory mechanism employed by the brain to ensure adequate THs are delivered to this tissue, especially during brain development. Direct competition of toxicants and thyroid hormones for transporter binding sites provides another potential mechanism of disruption. Such a mechanism is unlikely however, to disrupt thyroid hormone transport by the MCT or LAT transporters as they exhibit an extremely narrow range of substrate specificities [45]. Organic anion-transporting polypeptides, on the other hand, are exceptionally promiscuous and transport a wide variety of endo- and xenobiotics [49] including drugs such as the fenamate class of nonsteroidal anti-inflammatory drugs, iopanoic acid and phenytoin [50]. Each of these agents have been shown capable of inhibiting thyroid hormone transport in vitro at levels similar to what is achieved by therapeutic dosing indicating the potential of these xenobiotics in inhibiting thyroid hormone transport in vivo. The perfluorinated carboxylate compounds such as perfluorooctanoate (PFOA) [51,52] are also transported by Oatps although the in vivo effects of these compounds on thyroid hormone transport and metabolism have not been reported.

Similar to many nuclear hormone receptors, the thyroid hormone receptors exhibit a high degree of ligand specificity. Indeed, triiodothyronine (T3) exhibits an over 100-fold increased affinity for TR compared to the T3 isomer reverse T3 (rT3) [53]. Thus, it is not surprising that few endocrine disruptors have been shown to directly interact with the thyroid hormone receptors. Some substrates with demonstrated affinity for TR include hydroxy-PCBs, hydroxylated polybrominated diphenyl ethers (PBDEs), phthalates, and bisphenol A [33,39,54–57]. The most extensively studied of the TR-binding endocrine disruptors are the hydroxylated PCBs [55–57]. In vitro studies from the Zoeller laboratory demonstrated that a mixture of hydroxylated PCB congeners acts as rat TR agonists [58], inducing promoter activity in a TR-dependent fashion. Interestingly, none of the individual congeners induced transcription, suggesting that the mixture induced enzymatic modification of specific congeners that in turn act as TR agonists. In contrast, the Koibuchi laboratory has demonstrated that T3-dependent activation of human TRβ1 is inhibited by the hydroxylated form of mono-ortho PCB (4(OH)-2',3,3',4'5'-PCB) as well as a non-coplanar hydroxylated PCB and the mixed congener Aroclor 1254 [59]. The hydroxylated PCBs inhibited co-activator-dependent T3-dependent activation of TR. Interestingly, hydroxylated PCB congeners may exert promoter-specific effects as congeners may inhibit T3-dependent transactivation of some, but not all, tested promoters [60]. Several studies from this group suggest that the observed inhibition is the result of PCB interaction with the TR DNA-binding domain but not the ligand-binding domain [60–62]. Interaction with the DNA-binding domain ultimately results in congener-induced dissociation of TR from the TRE and subsequent inhibition of T3-dependent transactivation. These findings were unexpected as the chemical structures of PCBs are strikingly similar to the thyroid hormones, and thus, it would seem likely that the structurally similar chemicals would recognize the same ligand-binding pocket in the TR ligand-binding domain. Further work is necessary to determine whether the PCBs are binding directly to discrete sites within the DNA-binding domain or to sites outside the domain that when occupied, modify TR structure sufficiently enough that the DNA binding domain is unable to remain bound to the TRE.

In addition to the hydroxylated PCBs, several other chemicals affect TH-dependent signaling through the thyroid hormone receptor. Phthalates reduce circulating thyroid hormone levels through unknown mechanisms; however, results from a recent in vitro study performed with reporter gene assays suggest that phthalates may possess TR antagonist activity [54]. However, the in vivo relevance of this finding is unclear as the IC_{50} concentrations observed were between 10^{-6} and 10^{-4} M, suggesting that phthalate concentrations in vivo would have to be very high before thyroid hormone binding is inhibited. Polybrominated diphenyl ethers may also disrupt thyroid hormone signaling by thyroid hormone receptors. In vitro studies suggest that hydroxylated PBDEs directly bind to the thyroid hormone receptors [63] and inhibit TR-mediated TH-dependent transactivation [63,64]. Interestingly, as was observed for hydroxylated PCBs, studies performed with chimeric receptors suggest that these compounds may exert indirect antagonist action via allosteric binding to the DNA-binding domain of the thyroid hormone receptor, possibly disrupting TR interactions with the TRE [64]. Finally, the thyroid hormone disrupter bisphenol A (BPA) may also directly bind to the thyroid hormone receptor and alter TH-dependent gene activation [39,57]. Moriyama and colleagues reported that BPA displaces radiolabeled T3 from endogenous TR and inhibits T3-mediated transcription by recruiting the nuclear corepressor N-CoR to human TR [65]. Another group however, was unable to detect inhibition of BPA binding to human TR, although brominated BPA derivatives were determined to inhibit T3 binding to the TR [66]. These compounds also inhibited T3-dependent transactivation of T3-responsive promoters. The direct antagonistic action of BPA on TR-mediate transcription has also been observed in Xenopus [67]. Additional work in *Xenopus* suggests that bisphenol A may also repress expression of both TRα and TRβ [68] further contributing to the antagonistic effects of BPA on thyroid hormone action observed in this species [67].

Dietary Effects on the Thyroid Axis

In addition to anthropogenic environmental chemicals, naturally occurring chemicals are also capable of altering thyroid homeostasis. The presence of dietary and environmental thyroid disruptors, even in low concentrations, may be particularly detrimental in regions where other insults to the thyroid axis coexist. Thiocyanate precursors such as thioglucosides and cyanogenic glucosides are present in many common staple foods. Thioglucosides are found in vegetables belonging to the *Brassica* genus of plants such as cabbage, broccoli, cauliflower, kale, turnips,

Brussel sprouts, rutabaga, rapeseed, and mustard [69]. Upon digestion, thioglucosides are metabolized and thiocyanate is released. Cassava, lima beans, maize, bamboo shoots, sweet potatoes, and linseed are staple foods in African and Asian nations and contain cyanogenic glucosides [69]. After ingestion, cyanogenic glucosides are hydrolyzed in the gut to release cyanide, which is metabolized to thiocyanate. Thiocyanate alters thyroid physiology by interacting with NIS transporters in the thyroid follicular cells, thereby inhibiting iodine uptake [70]. In addition, thiocyanate inhibits thyroid TPO activity resulting in reduced iodine organification [71,72]. Thiocyanate also displaces T4 from serum thyroid hormone binding proteins leading to increased serum free T4 levels [73]. Studies in rodents have demonstrated that a diet high in goitrogenous foods, such as bamboo shoot or radish, increases thyroid weight and decreases thyroid TPO activity and serum total T4 and T3 levels even after supplementation with adequate iodine [74,75]. Another recent study demonstrated that giving thiocyanate to pregnant mice from late gestation through weaning alters several thyroid function tests in the offspring [76]. Thiocyanate treatment reduced thyroid iodine content and plasma-free T4 and T3 levels and increased plasma TSH levels and thyroid weights in the postnatal day 25 offspring. Cassava is the main staple food for millions of people in developing countries. Soaking cassava roots in water for several days removes cyanogenic glucosides and detoxifies cassava prior to ingestion. However, these practices are not always followed. Several epidemiological studies demonstrate that thiocyanate from poorly detoxified cassava contributes to a hypothyroid state and goiter development [77–79]. Another study demonstrated that consumption of *Brassica* vegetables leads to thiocyanate overload contributing to goiter development in an area of moderate to severe iodine deficiency [80]. Together these data suggest that chronic consumption of thiocyanate-containing foods contributes to thyroid dysfunction in regions of iodine insufficiency.

Thiocyanate is also generated from cigarette smoking, through metabolism of cyanide. Smokers have increased thyroid volume and goiter prevalence compared to nonsmokers, especially in the presence of iodine deficiency [81,82]. In addition, a recent study in healthy American women reported a decrease in serum total T4 and T3 levels in active and passive smokers compared to nonsmokers [83]. The effects of smoking are particularly important during pregnancy and infancy (see Chap. 2) as maternal smoking reduces iodine levels in the breast milk [84] and neonates from smoking mothers have an increased risk of hypothyroidism [85]. These findings are consistent with the antithyroid effects of thiocyanate on NIS transport proteins and TPO [86].

Flavonoids are another group of naturally occurring dietary goitrogens and are found in soy and millet. The isoflavonoids genistein and daidzein are structurally similar to THs and are major components of soy. In vitro studies indicated that genistein and daidzein inhibit TPO activity [87]. An in vivo study in rats demonstrated that dietary genistein reduces TPO activity in a dose-dependent manner [88]. The same study showed that rats consuming a diet containing soy had reduced thyroidal TPO activity compared to those consuming a soy-free diet. However, other thyroidal parameters including serum T4, T3, and TSH levels and thyroid weight and histology did not change in the treated rats. Additional aspects of thyroid physiology are altered by genistein, daidzein, and other flavonoids. Flavonoids decrease thyroid iodide uptake [89] inhibit type I and type II deiodinase activities [90,91] inhibit thyroidal SULT activity [92] and bind to plasma TTR [93]. Human infants consuming soy formula were shown to develop goiters, which can be reversed by switching to milk or through iodine supplementation [94,95]. Infant consumption of soy formula is also associated with increased risk of developing autoimmune thyroid disease during adolescence [96]. In addition, women of childbearing age receiving 2 mg soy per day for 3 months have reduced free T3 levels [97].

Dietary micronutrient deficiencies are a significant health problem throughout the world, especially in developing countries. Several micronutrients including iron (Fe), copper (Cu), and selenium (Se) are required in addition to iodine for normal thyroid function. The World Health Organization (WHO) estimates that two billion people worldwide, including 31.5% of school-age children, have insufficient iodine intake [98]. Iodine deficiency is most prevalent in developing countries where soils are iodine depleted and access to iodized salt or oil is limited. Iodine deficiency impairs thyroid hormone (TH) synthesis and results in a spectrum of disorders including goiter, cretinism, cognitive impairment, and growth retardation. It is also becoming clear that even mild iodine deficiency alters thyroid homeostasis and impairs neurological development [99,100].

The WHO estimates that over 1.6 billion people suffer from anemia or Fe deficiency, including over 500 million women of childbearing age [101]. Studies in both humans and rodents have demonstrated that Fe deficiency adversely affects thyroid function. Post-weanling rats fed an Fe-deficient diet have reduced circulating total T4 and T3 levels [102–105]. Hess et al. recently showed that Fe deficiency in post-weanling rats reduces activity of the Fe-containing TPO enzyme [105]. Other studies indicate that rat hepatic type I deiodinase activity is reduced in response to Fe deficiency [103,104]. Fe deficiency may also impair central control of the HPT axis as Fe-deficient rats have reduced plasma TSH levels and a blunted TSH response to exogenous TRH injection [103]. Several recent epidemiological studies have demonstrated an interaction between Fe and thyroid function in humans. An estimated 20–44% of iodine-deficient goitrous children in North and West Africa also suffer from iron-deficient anemia (IDA) [106–108]. In these children, Fe repletion improves the efficacy of iodine supplementation resulting in decreased thyroid volume and goiter prevalence [106,108,109]. Another recent study in pregnant Swiss women showed that low maternal body Fe stores are correlated with increased circulating TSH levels and decreased circulating T4 levels, suggesting that Fe deficiency may compound the deleterious effects of iodine deficiency during development [110].

Cu deficiency is less prevalent than Fe or iodine deficiencies but likely contributes to the millions of people suffering from non-iron-deficient anemia [101]. Cu deficiency reduces circulating total T4 and T3 in post-weanling rats [111,112]. This may be due to decreased TPO activity as Cu deficiency reduces serum Fe levels [113]. Additionally, Cu deficiency reduces hepatic type I deiodinase and brown adipose tissue type II deiodinase activities in rats [111,112]. Cu is also necessary for

5 Thyroid Hormone Regulation of Mammalian Reproductive Development... 151

full α(alpha)-amidating activity of PAM [114]. Therefore, the TSH-stimulating activity of TRH may be impaired during Cu deficiency. Interestingly, maternal Cu or Fe deficiency during gestation and lactation reduces circulating and brain TH concentrations in neonatal rats [43,115].

Selenium deficiency is most prevalent in Southeast Asia, Russia, and African countries and often occurs in combination with iodine deficiency [116]. Selenium is essential to the activity of all three iodothyronine deiodinases involved in thyroid hormone metabolism. In addition, selenium is essential for glutathione peroxidase (GPX) and thioredoxin reductase protection against oxidative damage in the thyroid gland and other tissues. Rat studies demonstrate that selenium deficiency compounds the effect of iodine deficiency leading to lower plasma T4 and T3, higher plasma TSH, larger thyroid glands, reduced thyroid iodine levels, and increased cerebral D2 activity compared to iodine deficiency alone [117,118]. Several human studies have implicated selenium deficiency as a compounding factor in the incidence of myxedematous cretinism [119,120]. In iodine- and selenium-deficient school children, selenium supplementation reduces serum total T4, free T4, and rT3 levels [121]. Subsequent iodine supplementation was unable to return T4 levels to normal, suggesting that iodine deficiency should be corrected prior to selenium supplementation. A study in West Africa demonstrated that severe selenium deficiency impaired the thyroidal response of children to iodine supplementation, suggesting that both iodine and severe selenium deficiency should be corrected [122]. Individually, Se, Fe, Cu, and other micronutrient deficiencies may have only minor effects on the thyroid axis. However, in combination with iodine deficiency, dietary goitrogens, and/or environmental thyroid disruptors, these effects may be exacerbated resulting in a more severe insult to the thyroid axis and more severe developmental impairments.

TH and Mammalian Sexual Maturation

In mammals, thyroid hormone has its most profound effect on fetal and neonatal development. Much of what is known about the effects of TH on mammalian development comes from inducing TH deficiency (hypothyroidism) or excess (hyperthyroidism) in rodent model systems.

Hypo- and Hyperthyroidism in Males

In rodents, transient gestational-onset hypothyroidism impairs fertility and decreases sperm forward motility [123]. In humans, congenital hypothyroidism diagnosed and treated neonatally leads to normal sexual development, suggesting that defects in testicular development due to gestational hypothyroidism in humans can be reversed by early treatment [124]. However, the role of transient gestational-onset hypothyroidism in human fertility has not been intensively investigated.

Neonatal hypothyroidism prior to puberty in both humans and rodents leads to testicular atrophy, delayed sexual maturation, altered gonadotropin secretion, and impaired gonadal function [125,126]. Juvenile/transient neonatal hypothyroidism in humans and rodents induces precocious spermatogenesis and macroorchidism with increased Sertoli, Leydig, and germ cell numbers [127–134]. In humans, prolonged prepubertal hypothyroidism extending into adulthood is associated with decreased testicular weight, impaired fertility, and hypogonadotropic hypogonadism [135]. Testicular biopsies from adult patients affected by untreated, juvenile-onset hypothyroidism indicate testicular atrophy and involution [136]. Short-term postpubertal hypothyroidism leads to minor decreases in seminal volume and sperm motility with a significant effect on sperm morphology [137].

In rodents, transient neonatal-prepubertal hyperthyroidism leads to a reduction in Sertoli cell and germ cell numbers, an increase in the adult Leydig cell populations, and an overall decrease in testicular weight [138,139]. Mild neonatal-induced hyperthyroidism extending into adulthood has no impact on Sertoli and germ cell proliferation and development but leads to a slight, transient delay in Leydig cell development and spermatogenesis [140].

In humans, hyperthyroidism leads to increased sex hormone binding globulin (SHBG) with a concomitant increase in circulating total testosterone, reduced testosterone bioavailability, and reduced metabolic clearance rate while free testosterone levels remain normal [141–143]. Total and free estradiol concentrations may be elevated and may explain the decreased libido and gynecomastia experienced in some patients [144].

Both hypo- and hyperthyroidism in males have been associated with erectile dysfunction, decreased libido, delayed ejaculation, and premature ejaculation. Restoration to the euthyroid state leads to significant improvement of symptoms for most patients [145–147].

Hypo- and Hyperthyroidism in Females

In rodents, hypothyroidism induced in neonatal females leads to delayed vaginal opening and sexual maturation – markers of puberty – along with reduction in size of ovaries and ovarian follicles and underdeveloped uterine and vaginal tissues [148,149]. Postpubertal-induced hypothyroidism results in irregular estrus cycles and ovarian atrophy [150,151]. Congenitally hypothyroid female *hyt* mice, carrying a spontaneous mutation of the TSH receptor, are infertile and do not reach estrus unless treated with exogenous gonadotrophins [152]. Finally, studies from the 1950s show that treatment of mice with thyroid protein, presumably rendering the animals hyperthyroid, is associated with advanced vaginal opening [153].

Much less is known about the effects of thyroidal dysfunction on the developing human female reproductive system. Female congenitally hypothyroid children treated within the first month after birth exhibit normal onset and progression of puberty [124]. However, in 1960, Van Wyk and Grumbach described a syndrome

5 Thyroid Hormone Regulation of Mammalian Reproductive Development... 153

characterized by juvenile hypothyroidism accompanied by precocious puberty and ovarian enlargement [154]. Several small studies confirming these findings have been reported over the past five decades and demonstrate gonadal stimulation in the face of gonadotropin levels that are generally not elevated [155–158]. Importantly, the clinical signs revert back to normal after treating with thyroxine replacement. Interestingly, such gonadal stimulation is only observed in a distinct minority of hypothyroid adolescent girls, suggesting that another contributing factor is necessary for presentation of premature pubertal onset. It has been hypothesized that the TSH, which is circulating at high levels, may directly interact with the follicle-stimulating hormone (FSH) receptor and cause the observed gonadal stimulation [127]. Sequence analysis of the FSH receptor in four children with Van Wyk and Grumbach syndrome failed to identify mutations of the gene that could be linked to increased sensitivity of the liganded receptor [159] however, suggesting that abnormal FSH receptor signaling is likely not the molecular basis for the observed clinical findings. Hypothyroidism in women is also associated with decreased rates of metabolic clearance of androstenedione and estrone, and menstrual disturbances, most commonly oligomenorrhea [160]. Hyperthyroidism is associated with delayed sexual maturation, menstrual disturbances (most commonly hypomenorrhea and polymenorrhea), and delayed onset of menses [161,162]. Hyperthyroid patients have increased SHBG, elevated estrogen, testosterone, and androstenedione levels, and elevated androgen metabolism [163].

Thyroid Hormone and Male Sexual Development and Maturation – Cellular and Molecular Studies

The mammalian testes consist of seminiferous tubules containing germ cells surrounded by Sertoli cells and the supporting interstitial connective tissue between seminiferous cords consisting of steroidogenic Leydig cells, endothelial cells, fibroblasts, macrophages, and lymphocytes. The testes function to produce spermatozoa and synthesize androgens.

The mammalian gonad develops from a bipotential precursor tissue in the embryo, the genital ridges. Initiation of testis development is triggered by expression of the Y chromosomal testis-determining gene, *SRY*, which encodes a high mobility group (HMG) DNA-binding domain [164]. SRY acts as both a transcriptional activator [165] and repressor [166], but the molecular mechanisms of its action remain unknown. Some evidence indicates that SRY acts to prevent female development by repressing a negative regulator of male testis development [166] and, in addition, repressing transcription of genes required for ovarian development [167] SRY also acts to initiate testis development by activating the transcription of *SOX9*, an HMG box transcription factor [165].

A subset of somatic cells within the indifferent, bipotential gonad express *SRY* at approximately 41 days post-ovulation (dpo) in humans [168] and at approximately gestational day 10.5 in mice [169]. *SRY* expression triggers the differentiation into Sertoli cells and subsequent upregulation of steroidogenic factor-1 (*SF-1*) and anti-Mullerian hormone (*AMH*) gene expression [170,171].

Sertoli cell number must reach a critical threshold to ensure testis development [172]. The number of Sertoli cells present at puberty determines adult testicular size and daily sperm production and therefore also determines reproductive function and fertility in rat [173]. Sertoli cell proliferation normally occurs during the fetal or neonatal period, and in the peripubertal period, in all species examined [174]. These two proliferative periods may overlap as occurs in the rat [175], or may be temporally separated as in the human [176]. The number of Sertoli cells produced is coordinately controlled by genetic and hormonal factors. In particular, FSH functions to increase the rate of Sertoli cell proliferation. In neonatal rats, suppression of FSH results in an approximate 40% decrease in the final Sertoli cell number [175,177]. Conversely, increasing FSH concentrations either by injection of recombinant FSH or neonatal hemicastration increases the final Sertoli cell number by up to 49% [178].

Thyroid hormone is a critical regulator of the transition from the proliferative phase to the maturation phase of Sertoli cell development [179,180]. TRα1 expression in the rat is maximal in Sertoli cell nuclei through the proliferative phase, approximately postnatal day 15. Expression declines as proliferation is inhibited and maturation is induced [181].

In transiently hypothyroid neonatal rodents, the Sertoli cell proliferation phase is extended, leading to an increase in Sertoli cell number, testicular size, and sperm production compared to control animals [129,130]. When transient juvenile hyperthyroidism was induced in rats, Sertoli cell proliferation arrested prematurely and Sertoli cell maturation commenced prematurely, resulting in a 50% reduction in adult testis size and reduction in sperm production [138]. The mechanisms by which thyroid hormone inhibits Sertoli cell proliferation while stimulating functional maturation remain unclear. One mechanism may involve synergistic downregulation of *AMH* transcription and concomitant upregulation of androgen receptor transcription by T3 and FSH leading to a progressive Sertoli cell maturation [182,183].

Potential targets of TH action in the Sertoli cell include the cyclin-dependent kinase inhibitors (CDKIs) p27Kip1 (p27) and p21Cip1 (p21) and the gap junction protein, Connexin43 (Cx43). TH may inhibit Sertoli cell proliferation by upregulating the expression of p27 and p21 [184,185]. Testicular phenotypes of p27 or p21 single knockout mice and the p27/p21 double knockout mice support this hypothesis; the knockout mice have enlarged testes, increased Sertoli cell numbers, and increased daily sperm production compared with wild type controls [186]. In addition, the transition to functional maturation may be achieved, in part, through TH upregulation of Cx43, which is required for spermatogenesis and is a highly abundant gap junction protein in testicular cells [187]. Consistent with a role for Cx43 in Sertoli cell transition to functional maturation, the proliferative phase is extended and maturation is delayed in Sertoli cell-specific Cx43 knockout mice. Knockout mice also fail to initiate spermatogenesis and are infertile [188].

During Sertoli cell differentiation, the gonad increases in size due to proliferation and migration of cells from adjacent mesonephros. The mesonephric cells give

5 Thyroid Hormone Regulation of Mammalian Reproductive Development...

rise to peritubular myoid cells and to steroidogenic Leydig cells [189,190]. During gonadogenesis, Sertoli cells function to organize the testis cords and initiate differentiation of other cell types, including steroidogenic Leydig cells. During puberty, the function of the Sertoli cell switches from an organizing center to a physical and metabolic support cell for germ cells.

In rodents, Sertoli cells secrete Desert hedgehog (Dhh), as a paracrine signal to induce the differentiation of both the fetal Leydig cells and of the peritubular myoid cells [191]. Mutant mouse gonads lacking functional Dhh fail to develop Leydig cells, and the lack of androgen production leads to feminization of the external genitalia [192].

The peritubular myoid cells surround the seminiferous tubules to provide structural support and participate in the regulation of spermatogenesis and testicular function [193]. The steroidogenic Leydig cells occupy the gonad interstitium, often in clusters close to blood vessels. In mammals, two types of Leydig cells are necessary for sexual development. The fetal Leydig cell population is comprised of cells from the coelomic epithelium, and the migrating mesonephric cell lineages [194–196]. Adult Leydig cells differentiate from a pool of spindle-shaped, interstitial stem Leydig cells. The stem Leydig cells commit to the Leydig cell lineage and give rise to the progenitor Leydig cells, spindle-shaped cells expressing luteinizing hormone receptors (LHRs) and possess 3αHSD and 3βHSD enzymatic activities required for steroid synthesis. Progenitor Leydig cells enlarge, decrease proliferative rate, and transition to the round morphology of the immature Leydig cells (day 28–56 postpartum). Immature Leydig cells undergo a final round of division and mature into adult Leydig cells. The adult Leydig cells are the most common interstitial cell type, but small populations of stem Leydig cells, progenitor Leydig cells, and immature Leydig cells persist throughout adulthood. In rodents, fetal Leydig cells arise by day 12.5 of gestation, and adult Leydig cells by day 56 postpartum. Adult Leydig cells differentiate from their precursor cells during the second week of postnatal age in the rat [197].

Human Leydig cell development consists of three stages based on the triphasic increase in plasma testosterone levels. The initial testosterone peak occurs at 14–18 weeks gestation due to the functional maturation of the fetal Leydig cells. The second testosterone peak occurs after the second or third postnatal month, and the third peak coincides with the establishment of puberty [198].

In humans, defects in fetal Leydig cell specification or differentiation can result in incomplete masculinization of male fetuses [199]. Fetal Leydig cells differentiate from precursor lineages at 7–14 weeks gestation with maturation occuring at 14–18 weeks gestation. The neonatal Leydig cell population is comprised of mature adult Leydig cells and dedifferentiating or apoptotic fetal Leydig cells. At puberty, luteinizing hormone stimulates the growth, proliferation, differentiation, and maturation of Leydig cells. The pubertal surge of testosterone is required for initiation of spermatogenesis, development of accessory sex glands, and development of secondary sexual characteristics.

In addition to a role in Sertoli cell maturation, TH has been shown to regulate the development of Leydig cells and, based on TR expression, is likely involved in other aspects of testicular development and function. TR isoforms have been identified in Sertoli cells, Leydig cells, peritubular cells, germ cells, and sperm [181–183,200–202]. Importantly, TR expression is maximal in the neonatal–prepubertal ages in human and rat testis indicating a critical window for thyroid hormone action [203,204].

Transient neonatal hypothyroidism arrests Leydig cell differentiation, allowing continuous proliferation of mesenchymal Leydig cell precursors. Following recovery to euthyroidism, adult Leydig cell populations and testis size are increased compared to controls [205,206]. Neonatal–prepubertal hyperthyroidism leads to increased numbers of differentiated adult Leydig cells by increasing the number of mesenchymal cells produced and recruited into the differentiation pool [139]. Prenatal-induced chronic dietary hypothyroidism leads to a 40–60% reduction in adult Leydig cell progenitor formation and proliferation at postpartum days 16 and 28 with increased proliferation from day 42 to day 63 postpartum. Fetal Leydig cell populations are not adversely affected by hypothyroidism [140].

T3 also regulates testicular steroidogenesis. Steroid hormone synthesis is initiated and regulated by the transport of cholesterol from the outer to the inner mitochondrial membrane. The phosphoprotein, steroidogenic acute regulatory protein (StAR), is involved in cholesterol transport to the inner membrane. In mouse Leydig tumor cells, acute T3 treatment (37.5 mmol/l) increases StAR protein levels and steroid production [207].

In rats, transient neonatal hypothyroidism leads to a 69% increase in numbers of Leydig cells, a 20% decline in Leydig cell volume, and a permanent decrease in steroidogenic function. Peripheral testosterone levels remained unchanged. In adult Leydig cells isolated from transient neonatal hypothyroid rats, a 55% reduction in LH-stimulated testosterone production was observed. The addition of steroid precursor 22(R)-hydroxycholesterol failed to increase testosterone secretion and results in a 73% decrease in testosterone production [132].

Disruption of the Thyroid Axis and Sexual Development

The effects of endocrine disruption on pubertal development are an important, growing focus of contemporary medical research as recent industrialization has increased the exposure of children to a wide variety of chemicals with poorly understood toxicities [208–210]. Additionally, epidemiologic evidence reveals coincident increases in gonadal maldevelopment and changes in pubertal timing. Determining whether these coincident changes are related to cause and effect through hypothesis-driven inquiry requires detailed knowledge of the physiologic and molecular effects of endocrine-disrupting chemicals on reproductive development and health in both model organisms and humans. A small but growing literature reveals that several well-described disruptors of gonadal development and puberty also disrupt thyroid hormone synthesis, metabolism, and action including polychlorinated biphenyls, bisphenol A, and phthalates [33,56,57,211]. In theory, any compound

5 Thyroid Hormone Regulation of Mammalian Reproductive Development... 157

capable of disrupting thyroid hormone synthesis, metabolism, or action should be considered capable of disrupting mammalian gonadal development. However, as detailed previously, a minimum level of thyroidal dysfunction will need to be achieved before these aberrant biologic effects are manifested. In addition, the thyroidal insult must occur during a specific window of time during development. With these caveats in mind, we will next assess the literature implicating thyroidal disruption as a potential contributor to gonadal maldevelopment and altered onset of puberty.

Triclosan

The chlorinated phenolic compound triclosan has recently been shown to impair female pubertal development and thyroid hormone homeostasis [212]. In this single study, three doses of triclosan (1, 10, and 50 mg/kg/day) were given to rat dams throughout gestation and lactation. Circulating total T4 levels were reduced throughout gestation and lactation at the two highest doses given to the dams. Circulating T3 levels were reduced by all three doses but only from midgestation to postnatal day 10. Free hormone or TSH levels were not measured. The authors observed that vaginal opening was significantly delayed in female rats exposed to all doses of triclosan. No effects on first estrus or uterine and ovary weights were observed. Repletion of neonates with T3 was not performed. In contrast to the developing female, another study revealed that triclosan did not alter the timing of puberty onset in male rats [213]. In this study, male rats were exposed to triclosan from weaning until necropsy with no chemical exposure to the developing fetus or nursing neonate. The authors noted significant declines in circulating total T4 and T3 with no effects on TSH levels at doses above 30 mg/kg. Triclosan exposure at any dose did not affect onset of preputial separation or gonadal tissue weights. Sex-dependent differences in triclosan exposure may explain the disparate results of these two studies, as may the differences in study design where fetuses and nursing neonates were exposed to triclosan in the female rat study.

Bisphenol A

Bisphenol A (BPA) has been shown to disrupt thyroid hormone signaling through the thyroid hormone receptor as indicated earlier. BPA also alters pubertal timing in rodents [214–216]. Howdeshell and colleagues observed reduced time to first estrus in prenatal exposure of mice to 2.4 micrograms/kg BPA [216]. No effects were noted on age of vaginal opening. In another mouse study, mice treated with low doses of BPA from gestational days 11–17 exhibited both earlier vaginal opening and time of first estrus [215]. No changes in anogenital distance were noted in either males or females in this study. Similar negative results were observed in a separate study in male rats dosed from weaning through postnatal day 53, although thyroid weights were elevated in BPA-treated animals [217]. Thyroid hormone levels have only been assessed in one study to date in BPA-dosed animals [218]. In this study, postnatal day 15 pups dosed during gestation and for two weeks postnatally exhibited elevated T4 levels with normal TSH values. However, as BPA likely disrupts thyroid hormone signaling at the level of the thyroid hormone receptor, it is perhaps not surprising that BPA has minimal to no effect on circulating thyroid hormone levels. No studies have yet assessed the potential linkage between BPA exposure, pubertal timing, and thyroidal disruption.

Phthalates

Phthalates and phthalate metabolites have been shown to alter thyroid hormone levels in both humans and rodents [219]. A group of 18–55-year-old male partners of subfertile couples were assessed for urinary phthalate metabolites and serum thyroid hormones and TSH [220]. An inverse association was noted between mono(2-ethylhexyl) phthalate (MEHP) and free T4 and T3 serum levels and a suggested positive correlation with TSH. Similar findings were noted in a separate study of phthalate monoesters and thyroid hormones in pregnant Taiwanese women [221]. Negative correlations were noted between total and free T4 and urinary monobutyl phthalate (MBP). Importantly, phthalate metabolites were also found to be negatively associated with serum-free and total T3 in both male and female Danish children of 4–9 years of age [222]. Together, these data indicate that phthalate exposure in both children and adults is correlated with reduced serum thyroid hormone levels. Similar findings have been noted in rodent studies. Adult male rats fed high doses of the phthalate esters di-2-ethylhexyl phthalate (DEHP), di-n-hexyl phthalate (DnHP), or di-n-oxtyl phthalate (DnOP) for 21 days demonstrated significant reductions in serum total T3 and T4 [223]. Interestingly, T3 levels were rapidly reduced within 7 days of initiating phthalate feeding but then increased above control values by 21 days. T4 levels were reduced throughout. Thyroid histology was noted as abnormal, suggesting thyroidal hyperactivity and a possible direct impact of phthalates on the thyroid gland [223,224]. These data suggest that phthalate exposure results in rapid reduction in circulating thyroid hormones by unknown mechanisms. However, the feedback control mechanisms of the hypothalamic–pituitary–thyroid axis remain functional; they can sense reduced circulating thyroid hormones and respond by increasing activity of the thyroid gland. Other rodent studies have not revealed changes in thyroid hormone levels after several weeks of phthalate exposure [225]; however, it is likely that changes in thyroidal status would be observed at earlier time points of treatment if time course experiments would be conducted.

Decreased anogenital distance has recently been correlated with phthalate exposure in male human infants aged 2–36 months [226]. These findings are consistent with studies performed in pre- and postnatally exposed rodents where altered sexual differentiation has been noted in both developing male and female rodents. DEHP dosing as low as 10 mg/kg reduces testes weight and anogenital distance when given from approximately midgestation to two weeks after birth [227]. Similar results were

5 Thyroid Hormone Regulation of Mammalian Reproductive Development... 159

seen in male rat fetuses when dams were administered with di-n-propyl phthalate (DnPP) during gestational days 6–20 [228]. Multiple studies have also noted effects of phthalate dosing on preputial separation [229–231]. Di-n-butyl phthalate (DBP) administered during gestational days 10–19 resulted in reduced weights of testes and accessory organs at postnatal day 31, and delayed preputial separation [229]. Similar results have been observed for male rodents dosed perinatally with DEHP [230] and benzyl butyl phthalate (BBP) [231]. In female rodents exposed to DEHP during gestation through weaning, anogenital distance changes and delayed first estrus are not observed [232]. However, significant delays in vaginal opening have been observed at all doses over 15 mg DEHP/kg/day [232]. Together, these data indicate that multiple phthalates have the potential to negatively impact both male and female pubertal onset and gonadal development if the animals are exposed during critical periods of gonadal development. The observed correlations of phthalate exposure with circulating thyroid hormones in both humans and rodents suggest the possibility that at least some of the observed effects of phthalates on the developing reproductive system may be attributed to altered thyroidal status. Assessment of the effects of phthalates on thyroid hormone levels and action in the developing rodent are warranted.

Polychlorinated Biphenyls (PCBs)

PCBs are a group of persistent industrial compounds that have been linked to thyroid hormone disruption and also impact normal mammalian gonadal development. In vitro studies generally suggest that PCBs inhibit normal thyroid hormone signaling through the nuclear receptor; however, there are also human and animal data suggesting that PCBs may affect the levels of circulating thyroid hormones [56]. The effects of PCBs on mammalian reproductive development are varied depending on the congener assessed and the developmental time period of exposure [210,233]. Developmental exposure to certain PCB congeners has been associated with impaired neurodevelopment [57]. Human studies have generally not shown an association with pubertal timing in girls with the exception of a study associating four estrogenic PCB congeners with the onset of menarche in Akwesasne Mohawk girls [209]. Although the findings are mixed, at least two studies reveal a possible relationship between PCB exposure and delayed puberty in boys [210]. In rodents, findings are also mixed as dosing with mixed congeners (PCB-126, 138, 153, and 180) during gestation and early postnatal life has been associated with accelerated vaginal opening and delayed testicular descent [234] while dosing with PCB-126 alone for approximately the same period of time but at a lower dose, has been associated with delayed vaginal opening [235]. In contrast, exposure of late gestation rat embryos to the PCB mixture Aroclor 1221 was not associated with alterations in female pubertal timing [236]. Similarly, midgestational rat exposure to 1–4 mg/kg/day PCB-153 was not associated with changes in testes weight, ovary weight, or anogenital distance or changes in plasma concentrations of T4, T3, or TSH [237]. Exposure of midgestation embryos to higher doses of PCB-153 did, however, result in decreases in T4 and T3 accompanied by nonsignificant reductions in testes weight at 1 and 3 weeks of age [238]. Treatment of early postnatal male rat pups with PCB-153 from postnatal days 3–7 surprisingly resulted in reductions in circulating total and free T4 several weeks later at postnatal day 77 [239]. Although no changes in testes weight were observed, PCB dosing was associated with significant reductions in expression of TH-regulated markers of Sertoli cell proliferation and differentiation, connexin43 and P27kip1, suggesting possible effects of altered thyroid hormone status. Finally, a recent study from the Akwesasne Mohawk community shows that a grouping of eight persistent PCBs was positively associated with TSH and inversely associated with free T4 [240]. Associations with these PCBs and onset of menarche have not been reported.

Polybrominated Diphenyl Ethers

Studies of polybrominated diphenyl ethers (PBDEs) provide some of the most intriguing data, suggesting a possible linkage between thyroidal disruption and disruption of mammalian puberty [44,241,242]. In the first of these studies, pups born to rat dams treated with the commercial PBDE mixture DE-71 from gestational days 6 through weaning developed significant hypothyroxinemia from birth to at least postnatal day 20 in both males and females [242]. DE-71 dosing was also correlated with delayed preputial separation at a dose of 30.6 mg/kg/day. Anogenital distance on postnatal day 7 and testis weights were not changed although a nonsignificant increase in testis weight was noted at the highest DE-71 dose. In female rats, a significant reduction in mammary gland epithelial outgrowth, lateral branching, and terminal end bud development at postnatal day 21 was associated with DE-71 dosing. This comprehensive study suggests that PBDEs can cause substantial changes in TH homeostasis and reproductive development of both male and female rodents. Similar results for DE-71 have been reported when rats were dosed from weaning through postnatal days 41–53 [44]. Treatment resulted in decreases in both T4 and T3 and increases in TSH for both male and female rats. DE-71 dosing was also correlated with delayed preputial separation in males and delayed vaginal opening in females. Finally, in a study assessing the effects of DE-71 in the lactating rat, dosing dams with DE-71 from gestation day 6 through lactation day 18 led to reductions in T4, increases in TSH, and increased testis weights at postnatal day 31 [241]. Together, these studies suggest a possible linkage between PBDE-mediated thyroidal disruption and pubertal timing.

Summary

Thyroid hormone deficiency and excess during development are associated with abnormalities in reproductive development in both male and female vertebrates. In both humans and rodents, male testicular development is particularly susceptible

5 Thyroid Hormone Regulation of Mammalian Reproductive Development... 161

to perturbations in thyroid hormone levels and action. Exposure to several endocrine disruptors targeting the thyroid results in alterations in pubertal timing. A key unanswered question is related to cause and effect. To date, few if any rodent or human studies directly test the thyroid hormone disrupting hypothesis by conducting thyroid hormone repletion studies. Such studies are warranted but would only be informative for disruptors that alter circulating and tissue thyroid hormone levels. Such studies would determine whether the tested chemical disruptors exert their action through thyroidal interference or through unrelated mechanisms. This question is especially pertinent for disruptors with known impact on other endocrine systems such as the androgens and gonadotropins. Other key unanswered questions relate to the timing of thyroidal perturbation and the extent of thyroid hormone disruption necessary to impact reproductive development. These issues are paramount as agencies work to develop guidelines for acceptable endocrine-disrupting chemical exposure. In addition, many of the studies cited in this chapter did not assess the effects of endocrine disruptors on TSH or free T4, considered the key indicators of biologically significant thyroid hormone status. While the accumulated evidence points to severe thyroidal insult as necessary for detectable perturbation of reproductive development, there are a clear lack of studies assessing the impact of multiple mild to moderate insults on the thyroid axis on the reproductive tract. These potential secondary insults include insufficient dietary micronutrient intake (e.g., iodine and iron), exposure to goitrogens in the diet and tobacco smoke, and the impact of endobiotics and xenobiotics on thyroid hormone metabolism and transport into target tissues. Finally, relatively little research has been conducted on the impact of thyroid hormone disruption and female reproductive development. Further studies assessing both the role of thyroid hormones on mammalian reproductive development and the potential impact of endocrine-disrupting chemicals on these processes are clearly justified.

References

1. Nicola JP, Basquin C, Portulano C, Reyna-Neyra A, Paroder M, Carrasco N. The Na+/I– symporter mediates active iodide uptake in the intestine. Am J Physiol Cell Physiol. 2009;296(4):C654–662.
2. Eskandari S, Loo DD, Dai G, Levy O, Wright EM, Carrasco N. Thyroid Na+/I– symporter. Mechanism, stoichiometry, and specificity. J Biol Chem. 1997;272(43):27230–8.
3. Royaux IE, Suzuki K, Mori A, et al. Pendrin, the protein encoded by the Pendred syndrome gene (PDS) is an apical porter of iodide in the thyroid and is regulated by thyroglobulin in FRTL-5 cells. Endocrinology. 2000;141(2):839–45.
4. Dunn JT, Dunn AD. Update on intrathyroidal iodine metabolism. Thyroid. 2001;11(5): 407–14.
5. Bianco AC, Salvatore D, Gereben B, Berry MJ, Larsen PR. Biochemistry, cellular and molecular biology, and physiological roles of the iodothyronine selenodeiodinases. Endocr Rev. 2002;23(1):38–89.
6. Nillni EA. Regulation of the hypothalamic Thyrotropin Releasing Hormone (TRH) neuron by neuronal and peripheral inputs. Front Neuroendocrinol. 2010;31:134.

7. Haisenleder DJ, Ortolano GA, Dalkin AC, Yasin M, Marshall JC. Differential actions of thyrotropin (TSH)-releasing hormone pulses in the expression of prolactin and TSH subunit messenger ribonucleic acid in rat pituitary cells in vitro. Endocrinology. 1992;130(5):2917–23.
8. Weintraub BD, Gesundheit N, Taylor T, Gyves PW. Effect of TRH on TSH glycosylation and biological action. Ann N Y Acad Sci. 1989;553:205–13.
9. Koller KJ, Wolff RS, Warden MK, Zoeller RT. Thyroid hormones regulate levels of thyrotropin-releasing-hormone mRNA in the paraventricular nucleus. Proc Natl Acad Sci USA. 1987;84(20):7329–33.
10. Shupnik MA, Ridgway EC. Thyroid hormone control of thyrotropin gene expression in rat anterior pituitary cells. Endocrinology. 1987;121(2):619–24.
11. Yen PM. Physiological and molecular basis of thyroid hormone action. Physiol Rev. 2001;81(3):1097–142.
12. Visser WE, Friesema EC, Visser TJ. Minireview: thyroid hormone transporters: the knowns and the unknowns. Mol Endocrinol. 2011;25(1):1–14.
13. Hagenbuch B, Meier PJ. Organic anion transporting polypeptides of the OATP/ SLC21 family: phylogenetic classification as OATP/ SLCO superfamily, new nomenclature and molecular/functional properties. Pflugers Arch. 2004;447(5):653–65.
14. Mikkaichi T, Suzuki T, Tanemoto M, Ito S, Abe T. The organic anion transporter (OATP) family. Drug Metab Pharmacokinet. 2004;19(3):171–9.
15. Gao B, Huber RD, Wenzel A, et al. Localization of organic anion transporting polypeptides in the rat and human ciliary body epithelium. Exp Eye Res. 2005;80(1):61–72.
16. Sugiyama D, Kusuhara H, Taniguchi H, et al. Functional characterization of rat brain-specific organic anion transporter (Oatp14) at the blood-brain barrier: high affinity transporter for thyroxine. J Biol Chem. 2003;278(44):43489–95.
17. Tohyama K, Kusuhara H, Sugiyama Y. Involvement of multispecific organic anion transporter, Oatp14 (Slc21a14), in the transport of thyroxine across the blood-brain barrier. Endocrinology. 2004;145(9):4384–91.
18. Hood A, Klaassen CD. Differential effects of microsomal enzyme inducers on in vitro thyroxine (T(4)) and triiodothyronine (T(3)) glucuronidation. Toxicol Sci. 2000;55(1):78–84.
19. Kester MH, Kaptein E, Roest TJ, et al. Characterization of human iodothyronine sulfotransferases. J Clin Endocrinol Metab. 1999;84(4):1357–64.
20. Gereben B, Zavacki AM, Ribich S, et al. Cellular and molecular basis of deiodinase-regulated thyroid hormone signaling. Endocr Rev. 2008;29(7):898–938.
21. Huang SA. Physiology and pathophysiology of type 3 deiodinase in humans. Thyroid. 2005;15(8):875–81.
22. Bergh JJ, Lin HY, Lansing L, et al. Integrin alphaVbeta3 contains a cell surface receptor site for thyroid hormone that is linked to activation of mitogen-activated protein kinase and induction of angiogenesis. Endocrinology. 2005;146(7):2864–71.
23. Davis PJ, Shih A, Lin HY, Martino LJ, Davis FB. Thyroxine promotes association of mitogen-activated protein kinase and nuclear thyroid hormone receptor (TR) and causes serine phosphorylation of TR. J Biol Chem. 2000;275(48):38032–9.
24. Tang HY, Lin HY, Zhang S, Davis FB, Davis PJ. Thyroid hormone causes mitogen-activated protein kinase-dependent phosphorylation of the nuclear estrogen receptor. Endocrinology. 2004;145(7):3265–72.
25. Bhargava M, Lei J, Mariash CN, Ingbar DH. Thyroid hormone rapidly stimulates alveolar Na, K-ATPase by activation of phosphatidylinositol 3-kinase. Curr Opin Endocrinol Diabetes Obes. 2007;14(5):416–20.
26. Siegrist-Kaiser CA, Juge-Aubry C, Tranter MP, Ekenbarger DM, Leonard JL. Thyroxine-dependent modulation of actin polymerization in cultured astrocytes. A novel, extranuclear action of thyroid hormone. J Biol Chem. 1990;265(9):5296–302.
27. Farwell AP, Dubord-Tomasetti SA, Pietrzykowski AZ, Leonard JL. Dynamic nongenomic actions of thyroid hormone in the developing rat brain. Endocrinology. 2006;147(5):2567–74.

5 Thyroid Hormone Regulation of Mammalian Reproductive Development... 163

28. Segal J. A rapid, extranuclear effect of 3,5,3'-triiodothyronine on sugar uptake by several tissues in the rat in vivo. Evidence for a physiological role for the thyroid hormone action at the level of the plasma membrane. Endocrinology. Jun 1989;124(6):2755–64.

29. Pickard MR, Sinha AK, Ogilvie LM, Leonard AJ, Edwards PR, Ekins RP. Maternal hypothyroxinemia influences glucose transporter expression in fetal brain and placenta. J Endocrinol. 1999;163(3):385–94.

30. Cheng SY, Leonard JL, Davis PJ. Molecular aspects of thyroid hormone actions. Endocr Rev. 2010;31:139.

31. Davis PJ, Leonard JL, Davis FB. Mechanisms of nongenomic actions of thyroid hormone. Front Neuroendocrinol. 2008;29(2):211–8.

32. Bassett JH, Harvey CB, Williams GR. Mechanisms of thyroid hormone receptor-specific nuclear and extra nuclear actions. Mol Cell Endocrinol. 2003;213(1):1–11.

33. Patrick L. Thyroid disruption: mechanism and clinical implications in human health. Altern Med Rev. 2009;14(4):326–46.

34. Brucker-Davis F. Effects of environmental synthetic chemicals on thyroid function. Thyroid. 1998;8(9):827–56.

35. Steinmaus C, Miller MD, Howd R. Impact of smoking and thiocyanate on perchlorate and thyroid hormone associations in the 2001–2002 national health and nutrition examination survey. Environ Health Perspect. 2007;115(9):1333–8.

36. Breous E, Wenzel A, Loos U. The promoter of the human sodium/iodide symporter responds to certain phthalate plasticisers. Mol Cell Endocrinol. 2005;244(1–2):75–8.

37. Doerge DR, Sheehan DM. Goitrogenic and estrogenic activity of soy isoflavones. Environ Health Perspect. 2002;110 Suppl 3:349–53.

38. Schmutzler C, Bacinski A, Gotthardt I, et al. The ultraviolet filter benzophenone 2 interferes with the thyroid hormone axis in rats and is a potent in vitro inhibitor of human recombinant thyroid peroxidase. Endocrinology. 2007;148(6):2835–44.

39. Miller MD, Crofton KM, Rice DC, Zoeller RT. Thyroid-disrupting chemicals: interpreting upstream biomarkers of adverse outcomes. Environ Health Perspect. 2009;117(7):1033–41.

40. Yu WG, Liu W, Jin YH, et al. Prenatal and postnatal impact of perfluorooctane sulfonate (PFOS) on rat development: a cross-foster study on chemical burden and thyroid hormone system. Environ Sci Technol. 2009;43(21):8416–22.

41. Chang SC, Thibodeaux JR, Eastvold ML, et al. Thyroid hormone status and pituitary function in adult rats given oral doses of perfluorooctanesulfonate (PFOS). Toxicology. 2008;243(3):330–9.

42. Courtin F, Zrouri H, Lamirand A, et al. Thyroid hormone deiodinases in the central and peripheral nervous system. Thyroid. 2005;15(8):931–42.

43. Bastian TW, Prohaska JR, Georgieff MK, Anderson GW. Perinatal iron and copper deficiencies alter neonatal rat circulating and brain thyroid hormone concentrations. Endocrinology. 2010;151(8):4055–65.

44. Stoker TE, Laws SC, Crofton KM, Hedge JM, Ferrell JM, Cooper RL. Assessment of DE-71, a commercial polybrominated diphenyl ether (PBDE) mixture, in the EDSP male and female pubertal protocols. Toxicol Sci. 2004;78(1):144–55.

45. Heuer H, Visser TJ. Minireview: pathophysiological importance of thyroid hormone transporters. Endocrinology. 2009;150(3):1078–83.

46. Meyer zu Schwabedissen HE, Kim RB. Hepatic OATP1B transporters and nuclear receptors PXR and CAR: interplay, regulation of drug disposition genes, and single nucleotide polymorphisms. Mol Pharm. 2009;6(6):1644–61.

47. Martin L, Klaassen CD. Differential effects of polychlorinated biphenyl congeners on serum thyroid hormone levels in rats. Toxicol Sci. 2010;117(1):36–44.

48. Cheng X, Maher J, Dieter MZ, Klaassen CD. Regulation of mouse organic anion-transporting polypeptides (Oatps) in liver by prototypical microsomal enzyme inducers that activate distinct transcription factor pathways. Drug Metab Dispos. 2005;33(9):1276–82.

49. Westholm DE, Rumbley JN, Salo DR, Rich TP, Anderson GW. Organic anion-transporting polypeptides at the blood-brain and blood-cerebrospinal fluid barriers. Curr Top Dev Biol. 2008;80:135–70.

50. Westholm DE, Stenehjem DD, Rumbley JN, Drewes LR, Anderson GW. Competitive inhibition of organic anion-transporting polypeptide 1c1-mediated thyroxine transport by the fenamate class of nonsteroidal antiinflammatory drugs. Endocrinology. 2009;150(2):1025–32.

51. Weaver YM, Ehresman DJ, Butenhoff JL, Hagenbuch B. Roles of rat renal organic anion transporters in transporting perfluorinated carboxylates with different chain lengths. Toxicol Sci. 2010;113(2):305–14.

52. Yang CH, Glover KP, Han X. Characterization of cellular uptake of perfluorooctanoate via organic anion-transporting polypeptide 1A2, organic anion transporter 4, and urate transporter 1 for their potential roles in mediating human renal reabsorption of perfluorocarboxylates. Toxicol Sci. 2010;117(2):294–302.

53. Koenig RJ, Warne RL, Brent GA, Harney JW, Larsen PR, Moore DD. Isolation of a cDNA clone encoding a biologically active thyroid hormone receptor. Proc Natl Acad Sci USA. 1988;85(14):5031–5.

54. Shen O, Du G, Sun H, et al. Comparison of in vitro hormone activities of selected phthalates using reporter gene assays. Toxicol Lett. 2009;191(1):9–14.

55. Koibuchi N, Iwasaki T. Regulation of brain development by thyroid hormone and its modulation by environmental chemicals. Endocr J. 2006;53(3):295–303.

56. Jugan ML, Levi Y, Blondeau JP. Endocrine disruptors and thyroid hormone physiology. Biochem Pharmacol. 2010;79(7):939–47.

57. Zoeller TR. Environmental chemicals targeting thyroid. Hormones (Athens). 2010;9(1): 28–40.

58. Gauger KJ, Giera S, Sharlin DS, Bansal R, Iannacone E, Zoeller RT. Polychlorinated biphenyls 105 and 118 form thyroid hormone receptor agonists after cytochrome P4501A1 activation in rat pituitary GH3 cells. Environ Health Perspect. 2007;115(11):1623–30.

59. Iwasaki T, Miyazaki W, Takeshita A, Kuroda Y, Koibuchi N. Polychlorinated biphenyls suppress thyroid hormone-induced transactivation. Biochem Biophys Res Commun. 2002; 299(3):384–8.

60. Amano I, Miyazaki W, Iwasaki T, Shimokawa N, Koibuchi N. The effect of hydroxylated polychlorinated biphenyl (OH-PCB) on thyroid hormone receptor (TR)-mediated transcription through native-thyroid hormone response element (TRE). Ind Health. 2010;48(1): 115–8.

61. Miyazaki W, Iwasaki T, Takeshita A, Kuroda Y, Koibuchi N. Polychlorinated biphenyls suppress thyroid hormone receptor-mediated transcription through a novel mechanism. J Biol Chem. 2004;279(18):18195–202.

62. Miyazaki W, Iwasaki T, Takeshita A, Tohyama C, Koibuchi N. Identification of the functional domain of thyroid hormone receptor responsible for polychlorinated biphenyl-mediated suppression of its action in vitro. Environ Health Perspect. 2008;116(9):1231–6.

63. Li F, Xie Q, Li X, et al. Hormone activity of hydroxylated polybrominated diphenyl ethers on human thyroid receptor-beta: in vitro and in silico investigations. Environ Health Perspect. 2010;118(5):602–6.

64. Ibhazehiebo K, Iwasaki T, Kimura-Kuroda J, Miyazaki W, Shimokawa N, Koibuchi N. Disruption of thyroid hormone receptor-mediated transcription and thyroid hormone-induced purkinje cell dendrite arborization by polybrominated diphenylethers. Environ Health Perspect. 2011;119:168.

65. Moriyama K, Tagami T, Akamizu T, et al. Thyroid hormone action is disrupted by bisphenol A as an antagonist. J Clin Endocrinol Metab. 2002;87(11):5185–90.

66. Kitamura S, Kato T, Iida M, et al. Anti-thyroid hormonal activity of tetrabromobisphenol A, a flame retardant, and related compounds: affinity to the mammalian thyroid hormone receptor, and effect on tadpole metamorphosis. Life Sci. 2005;76(14):1589–601.

67. Heimeier RA, Das B, Buchholz DR, Shi YB. The xenoestrogen bisphenol A inhibits postembryonic vertebrate development by antagonizing gene regulation by thyroid hormone. Endocrinology. 2009;150(6):2964–73.

68. Iwamuro S, Yamada M, Kato M, Kikuyama S. Effects of bisphenol A on thyroid hormone-dependent up-regulation of thyroid hormone receptor alpha and beta and down-regulation of retinoid X receptor gamma in Xenopus tail culture. Life Sci. 2006;79(23):2165–71.

5 Thyroid Hormone Regulation of Mammalian Reproductive Development...

69. Vanderpas J. Nutritional epidemiology and thyroid hormone metabolism. Annu Rev Nutr. 2006;26:293–322.
70. Saito K, Yamamoto K, Takai T, Yoshida S. Inhibition of iodide accumulation by perchlorate and thiocyanate in a model of the thyroid iodide transport system. Acta Endocrinol (Copenh). 1983;104(4):456–61.
71. Virion A, Deme D, Pommier J, Nunez J. Opposite effects of thiocyanate on tyrosine iodination and thyroid hormone synthesis. Eur J Biochem. 1980;112(1):1–7.
72. Chandra AK, Mukhopadhyay S, Lahari D, Tripathy S. Goitrogenic content of Indian cyanogenic plant foods & their in vitro anti-thyroidal activity. Indian J Med Res. 2004;119(5): 180–5.
73. Michajlovskij N, Langer P. Increase of serum free thyroxine following the administration of thiocyanate and other anions in vivo and in vitro. Acta Endocrinol (Copenh). 1974;75(4): 707–16.
74. Chandra AK, Ghosh D, Mukhopadhyay S, Tripathy S. Effect of bamboo shoot, Bambusa arundinacea (Retz.) Willd. on thyroid status under conditions of varying iodine intake in rats. Indian J Exp Biol. 2004;42(8):781–6.
75. Chandra AK, Mukhopadhyay S, Ghosh D, Tripathy S. Effect of radish (Raphanus sativus Linn.) on thyroid status under conditions of varying iodine intake in rats. Indian J Exp Biol. 2006;44(8):653–61.
76. Ghorbel H, Fetoui H, Mahjoubi A, Guermazi F, Zeghal N. Thiocyanate effects on thyroid function of weaned mice. C R Biol. 2008;331(4):262–71.
77. Abuye C, Kelbessa U, Wolde-Gebriel S. Health effects of cassava consumption in south Ethiopia. East Afr Med J. 1998;75(3):166–70.
78. Akindahunsi AA, Grissom FE, Adewusi SR, Afolabi OA, Torimiro SE, Oke OL. Parameters of thyroid function in the endemic goitre of Akungba and Oke-Agbe villages of Akoko area of southwestern Nigeria. Afr J Med Med Sci. 1998;27(3–4):239–42.
79. Vanderpas J, Bourdoux P, Lagasse R, et al. Endemic infantile hypothyroidism in a severe endemic goitre area of central Africa. Clin Endocrinol (Oxf). 1984;20(3):327–40.
80. Erdogan MF, Erdogan G, Sav H, Gullu S, Kamel N. Endemic goiter, thiocyanate overload, and selenium status in school-age children. Biol Trace Elem Res. 2001;79(2):121–30.
81. Hegedus L, Karstrup S, Veiergang D, Jacobsen B, Skovsted L, Feldt-Rasmussen U. High frequency of goitre in cigarette smokers. Clin Endocrinol (Oxf). 1985;22(3):287–92.
82. Brix TH, Hansen PS, Kyvik KO, Hegedus L. Cigarette smoking and risk of clinically overt thyroid disease: a population-based twin case-control study. Arch Intern Med. 2000;160(5): 661–6.
83. Soldin OP, Goughenour BE, Gilbert SZ, Landy HJ, Soldin SJ. Thyroid hormone levels associated with active and passive cigarette smoking. Thyroid. 2009;19(8):817–23.
84. Laurberg P, Nohr SB, Pedersen KM, Fuglsang E. Iodine nutrition in breast-fed infants is impaired by maternal smoking. J Clin Endocrinol Metab. 2004;89(1):181–7.
85. Chanoine JP, Toppet V, Bourdoux P, Spehl M, Delange F. Smoking during pregnancy: a significant cause of neonatal thyroid enlargement. Br J Obstet Gynaecol. 1991;98(1):65–8.
86. Fukayama H, Nasu M, Murakami S, Sugawara M. Examination of antithyroid effects of smoking products in cultured thyroid follicles: only thiocyanate is a potent antithyroid agent. Acta Endocrinol (Copenh). 1992;127(6):520–5.
87. Divi RL, Doerge DR. Inhibition of thyroid peroxidase by dietary flavonoids. Chem Res Toxicol. 1996;9(1):16–23.
88. Chang HC, Doerge DR. Dietary genistein inactivates rat thyroid peroxidase in vivo without an apparent hypothyroid effect. Toxicol Appl Pharmacol. 2000;168(3):244–52.
89. Schroder Van der Elst JP, Smit JW, Romijn HA, van der Heide D. Dietary flavonoids and iodine metabolism. Biofactors. 2003;19(3–4):171–6.
90. Ferreira AC, Lisboa PC, Oliveira KJ, Lima LP, Barros IA, Carvalho DP. Inhibition of thyroid type 1 deiodinase activity by flavonoids. Food Chem Toxicol. 2002;40(7):913–7.
91. Mori K, Stone S, Braverman LE, Devito WJ. Involvement of tyrosine phosphorylation in the regulation of 5′-deiodinases in FRTL-5 rat thyroid cells and rat astrocytes. Endocrinology. 1996;137(4):1313–8.

92. Ebmeier CC, Anderson RJ. Human thyroid phenol sulfotransferase enzymes 1A1 and 1A3: activities in normal and diseased thyroid glands, and inhibition by thyroid hormones and phytoestrogens. J Clin Endocrinol Metab. 2004;89(11):5597–605.

93. Green NS, Foss TR, Kelly JW. Genistein, a natural product from soy, is a potent inhibitor of transthyretin amyloidosis. Proc Natl Acad Sci USA. 2005;102(41):14545–50.

94. Chorazy PA, Himelhoch S, Hopwood NJ, Greger NG, Postellon DC. Persistent hypothyroidism in an infant receiving a soy formula: case report and review of the literature. Pediatrics. 1995;96(1 Pt 1):148–50.

95. Hydovitz JD. Occurrence of goiter in an infant on a soy diet. N Engl J Med. 1960;262: 351–3.

96. Fort P, Moses N, Fasano M, Goldberg T, Lifshitz F. Breast and soy-formula feedings in early infancy and the prevalence of autoimmune thyroid disease in children. J Am Coll Nutr. 1990; 9(2):164–7.

97. Duncan AM, Merz BE, Xu X, Nagel TC, Phipps WR, Kurzer MS. Soy isoflavones exert modest hormonal effects in premenopausal women. J Clin Endocrinol Metab. 1999;84(1):192–7.

98. de Benoist B, McLean E, Andersson M, Rogers L. Iodine deficiency in 2007: global progress since 2003. Food Nutr Bull. 2008;29(3):195–202.

99. Zimmermann MB. The adverse effects of mild-to-moderate iodine deficiency during pregnancy and childhood: a review. Thyroid. 2007;17(9):829–35.

100. Berbel P, Mestre JL, Santamaria A, et al. Delayed neurobehavioral development in children born to pregnant women with mild hypothyroxinemia during the first month of gestation: the importance of early iodine supplementation. Thyroid. 2009;19(5):511–9.

101. McLean E, Cogswell M, Egli I, Wojdyla D, de Benoist B. Worldwide prevalence of anaemia, WHO Vitamin and Mineral Nutrition Information System, 1993–2005. Public Health Nutr. 2009;12(4):444–54.

102. Beard J, Green W, Miller L, Finch C. Effect of iron-deficiency anemia on hormone levels and thermoregulation during cold exposure. Am J Physiol. 1984;247(1 Pt 2):R114–119.

103. Beard J, Tobin B, Green W. Evidence for thyroid hormone deficiency in iron-deficient anemic rats. J Nutr. 1989;119(5):772–8.

104. Beard JL, Tobin BW, Smith SM. Effects of iron repletion and correction of anemia on norepinephrine turnover and thyroid metabolism in iron deficiency. Proc Soc Exp Biol Med. 1990;193(4):306–12.

105. Hess SY, Zimmermann MB, Arnold M, Langhans W, Hurrell RF. Iron deficiency anemia reduces thyroid peroxidase activity in rats. J Nutr. 2002;132(7):1951–5.

106. Hess SY, Zimmermann MB, Adou P, Torresani T, Hurrell RF. Treatment of iron deficiency in goitrous children improves the efficacy of iodized salt in Cote d'Ivoire. Am J Clin Nutr. 2002;75(4):743–8.

107. Zimmermann M, Adou P, Torresani T, Zeder C, Hurrell R. Persistence of goiter despite oral iodine supplementation in goitrous children with iron deficiency anemia in Cote d'Ivoire. Am J Clin Nutr. 2000;71(1):88–93.

108. Zimmermann MB, Zeder C, Chaouki N, Torresani T, Saad A, Hurrell RF. Addition of microencapsulated iron to iodized salt improves the efficacy of iodine in goitrous, iron-deficient children: a randomized, double-blind, controlled trial. Eur J Endocrinol. 2002;147(6): 747–53.

109. Zimmermann M, Adou P, Torresani T, Zeder C, Hurrell R. Iron supplementation in goitrous, iron-deficient children improves their response to oral iodized oil. Eur J Endocrinol. 2000;142(3):217–23.

110. Zimmermann MB, Burgi H, Hurrell RF. Iron deficiency predicts poor maternal thyroid status during pregnancy. J Clin Endocrinol Metab. 2007;92(9):3436–40.

111. Lukaski HC, Hall CB, Marchello MJ. Body temperature and thyroid hormone metabolism of copper-deficient rats. J Nutr Biochem. 1995;6(8):445–51.

112. Olin KL, Walter RM, Keen CL. Copper deficiency affects selenoglutathione peroxidase and selenodeiodinase activities and antioxidant defense in weanling rats. Am J Clin Nutr. 1994; 59(3):654–8.

5 Thyroid Hormone Regulation of Mammalian Reproductive Development... 167

113. Pyatskowit JW, Prohaska JR. Multiple mechanisms account for lower plasma iron in young copper deficient rats. Biometals. 2008;21(3):343–52.
114. Kulathila R, Consalvo AP, Fitzpatrick PF, et al. Bifunctional peptidylglycine alpha-amidating enzyme requires two copper atoms for maximum activity. Arch Biochem Biophys. 1994; 311(1):191–5.
115. Bastian TW, Lassi KC, Anderson GW, Prohaska JR. Maternal iron supplementation attenuates the impact of perinatal copper deficiency but does not eliminate hypotriiodothyroninemia nor impair sensorimotor development. J Nutr Biochem. (In Press).
116. Combs Jr GF. Selenium in global food systems. Br J Nutr. 2001;85(5):517–47.
117. Arthur JR, Nicol F, Beckett GJ. The role of selenium in thyroid hormone metabolism and effects of selenium deficiency on thyroid hormone and iodine metabolism. Biol Trace Elem Res. 1992;33:37–42.
118. Beckett GJ, Nicol F, Rae PW, Beech S, Guo Y, Arthur JR. Effects of combined iodine and selenium deficiency on thyroid hormone metabolism in rats. Am J Clin Nutr. 1993;57(2 Suppl):240S–3S.
119. Vanderpas JB, Contempre B, Duale NL, et al. Iodine and selenium deficiency associated with cretinism in northern Zaire. Am J Clin Nutr. 1990;52(6):1087–93.
120. Goyens P, Golstein J, Nsombola B, Vis H, Dumont JE. Selenium deficiency as a possible factor in the pathogenesis of myxoedematous endemic cretinism. Acta Endocrinol (Copenh). 1987;114(4):497–502.
121. Contempre B, Duale NL, Dumont JE, Ngo B, Diplock AT, Vanderpas J. Effect of selenium supplementation on thyroid hormone metabolism in an iodine and selenium deficient population. Clin Endocrinol (Oxf). 1992;36(6):579–83.
122. Zimmermann MB, Adou P, Torresani T, Zeder C, Hurrell RF. Effect of oral iodized oil on thyroid size and thyroid hormone metabolism in children with concurrent selenium and iodine deficiency. Eur J Clin Nutr. 2000;54(3):209–13.
123. Anbalagan J, Sashi AM, Vengatesh G, Stanley JA, Neelamohan R, Aruldhas MM. Mechanism underlying transient gestational-onset hypothyroidism-induced impairment of posttesticular sperm maturation in adult rats. Fertil Steril. 2010;93(8):2491–7.
124. Salerno M, Micillo M, Di Maio S, et al. Longitudinal growth, sexual maturation and final height in patients with congenital hypothyroidism detected by neonatal screening. Eur J Endocrinol. 2001;145(4):377–83.
125. Ahuja MM, Chopra IJ, Sridhar CB. Sporadic cretinism and juvenile hypothyroidism. Metabolism. 1969;18(6):488–96.
126. Jannini EA, Ulisse S, D'Armiento M. Thyroid hormone and male gonadal function. Endocr Rev. 1995;16(4):443–59.
127. Anasti JN, Flack MR, Froehlich J, Nelson LM, Nisula BC. A potential novel mechanism for precocious puberty in juvenile hypothyroidism. J Clin Endocrinol Metab. 1995;80(1): 276–9.
128. Bruder JM, Samuels MH, Bremner WJ, Ridgway EC, Wierman ME. Hypothyroidism-induced macroorchidism: use of a gonadotropin-releasing hormone agonist to understand its mechanism and augment adult stature. J Clin Endocrinol Metab. 1995;80(1):11–6.
129. Cooke PS, Hess RA, Porcelli J, Meisami E. Increased sperm production in adult rats after transient neonatal hypothyroidism. Endocrinology. 1991;129(1):244–8.
130. Cooke PS, Meisami E. Early hypothyroidism in rats causes increased adult testis and reproductive organ size but does not change testosterone levels. Endocrinology. 1991;129(1): 237–43.
131. Cooke PS, Porcelli J, Hess RA. Induction of increased testis growth and sperm production in adult rats by neonatal administration of the goitrogen propylthiouracil (PTU): the critical period. Biol Reprod. 1992;46(1):146–54.
132. Hardy MP, Kirby JD, Hess RA, Cooke PS. Leydig cells increase their numbers but decline in steroidogenic function in the adult rat after neonatal hypothyroidism. Endocrinology. 1993; 132(6):2417–20.
133. Hardy MP, Sharma RS, Arambepola NK, et al. Increased proliferation of Leydig cells induced by neonatal hypothyroidism in the rat. J Androl. 1996;17(3):231–8.

134. Hess RA, Cooke PS, Bunick D, Kirby JD. Adult testicular enlargement induced by neonatal hypothyroidism is accompanied by increased Sertoli and germ cell numbers. Endocrinology. 1993;132(6):2607–13.
135. De La Balze FA, Arrillaga F, Mancini RE, Janches M, Davidson OW, Gurtman AI. Male hypogonadism in hypothyroidism: a study of six cases. J Clin Endocrinol Metab. 1962; 22:212–22.
136. Hoffman WH, Kovacs KT, Gala RR, et al. Macroorchidism and testicular fibrosis associated with autoimmune thyroiditis. J Endocrinol Invest. 1991;14(7):609–16.
137. Krassas GE, Papadopoulou F, Tziomalos K, Zeginiadou T, Pontikides N. Hypothyroidism has an adverse effect on human spermatogenesis: a prospective, controlled study. Thyroid. 2008;18(12):1255–9.
138. van Haaster LH, de Jong FH, Docter R, de Rooij DG. High neonatal triiodothyronine levels reduce the period of Sertoli cell proliferation and accelerate tubular lumen formation in the rat testis, and increase serum inhibin levels. Endocrinology. 1993;133(2):755–60.
139. Ariyaratne HB, Chamindrani Mendis-Handagama S. Changes in the testis interstitium of Sprague Dawley rats from birth to sexual maturity. Biol Reprod. 2000;62(3):680–90.
140. Rijntjes E, Wientjes AT, Swarts HJ, de Rooij DG, Teerds KJ. Dietary-induced hyperthyroidism marginally affects neonatal testicular development. J Androl. 2008;29(6):643–53.
141. Ruder H, Corvol P, Mahoudeau JA, Ross GT, Lipsett MB. Effects of induced hyperthyroidism on steroid metabolism in man. J Clin Endocrinol Metab. 1971;33(3):382–7.
142. Vermeulen A, Vandeweghe M, Verdonck L. Determination of cortisol and testosterone, using the protein binding method. Z Klin Chem Klin Biochem. 1969;7(1):111.
143. Abalovich M, Levalle O, Hermes R, et al. Hypothalamic-pituitary-testicular axis and seminal parameters in hyperthyroid males. Thyroid. 1999;9(9):857–63.
144. Krassas GE, Pontikides N, Deligianni V, Miras K. A prospective controlled study of the impact of hyperthyroidism on reproductive function in males. J Clin Endocrinol Metab. 2002;87(8):3667–71.
145. Krassas GE, Tziomalos K, Papadopoulou F, Pontikides N, Perros P. Erectile dysfunction in patients with hyper- and hypothyroidism: how common and should we treat? J Clin Endocrinol Metab. 2008;93(5):1815–9.
146. Meikle AW. The interrelationships between thyroid dysfunction and hypogonadism in men and boys. Thyroid. 2004;14 Suppl 1:S17–25.
147. Carani C, Isidori AM, Granata A, et al. Multicenter study on the prevalence of sexual symptoms in male hypo- and hyperthyroid patients. J Clin Endocrinol Metab. 2005;90(12): 6472–9.
148. Dijkstra G, de Rooij DG, de Jong FH, van den Hurk R. Effect of hypothyroidism on ovarian follicular development, granulosa cell proliferation and peripheral hormone levels in the prepubertal rat. Eur J Endocrinol. 1996;134(5):649–54.
149. Gellert RJ, Bakke JL, Lawrence NL. Delayed vaginal opening in the rat following pharmacologic doses of T4 administered during the neonatal period. J Lab Clin Med. 1971;77(3):410–6.
150. Mattheij JA, Swarts JJ, Lokerse P, van Kampen JT, Van der Heide D. Effect of hypothyroidism on the pituitary-gonadal axis in the adult female rat. J Endocrinol. 1995;146(1):87–94.
151. Tohei A, Imai A, Watanabe G, Taya K. Influence of thiouracil-induced hypothyroidism on adrenal and gonadal functions in adult female rats. J Vet Med Sci. 1998;60(4):439–46.
152. Jiang JY, Imai Y, Umezu M, Sato E. Characteristics of infertility in female hypothyroid (hyt) mice. Reproduction. 2001;122(5):695–700.
153. Soliman FA, Reineke EP. Influence of variations in environmental temperature and thyroid status on sexual function in young female mice. Am J Physiol. 1952;168(2):400–5.
154. Van Wyk JJ, Grumbach MM. Syndrome of precocious menstruation and galactorrhea in juvenile hypothyroidism: An example of hormonal overlap in pituitary feedback. J Pediatr. 1960;57:416–35.
155. Browne LP, Boswell HB, Crotty EJ, O'Hara SM, Birkemeier KL, Guillerman RP. Van Wyk and Grumbach syndrome revisited: imaging and clinical findings in pre- and postpubertal girls. Pediatr Radiol. 2008;38(5):538–42.

156. Lee PA, Blizzard RM. Serum gonadotropins in hypothyroid girls with and without sexual precocity. Johns Hopkins Med J. 1974;135(1):55–60.
157. Ozgen T, Guven A, Aydin M. Precocious puberty in a girl with down syndrome due to primary hypothyroidism. Turk J Pediatr. 2009;51(4):381–3.
158. Sanjeevaiah AR, Sanjay S, Deepak T, Sharada A, Srikanta SS. Precocious puberty and large multicystic ovaries in young girls with primary hypothyroidism. Endocr Pract. 2007;13(6):652–5.
159. Suarez EA, D'Alva CB, Campbell A, et al. Absence of mutation in the follicle-stimulating hormone receptor gene in severe primary hypothyroidism associated with gonadal hyperstimulation. J Pediatr Endocrinol Metab. 2007;20(8):923–31.
160. Longcope C, Abend S, Braverman LE, Emerson CH. Androstenedione and estrone dynamics in hypothyroid women. J Clin Endocrinol Metab. 1990;70(4):903–7.
161. Saxena KM, Crawford JD, Talbot NB. Childhood thyrotoxicosis: a long-term perspective. Br Med J. 1964;2(5418):1153–8.
162. Krassas GE, Pontikides N, Kaltsas T, Papadopoulou P, Batrinos M. Menstrual disturbances in thyrotoxicosis. Clin Endocrinol (Oxf). 1994;40(5):641–4.
163. Ridgway EC, Longcope C, Maloof F. Metabolic clearance and blood production rates of estradiol in hyperthyroidism. J Clin Endocrinol Metab. 1975;41(3):491–7.
164. Giese K, Pagel J, Grosschedl R. Distinct DNA-binding properties of the high mobility group domain of murine and human SRY sex-determining factors. Proc Natl Acad Sci USA. 1994;91(8):3368–72.
165. Sekido R, Lovell-Badge R. Sex determination involves synergistic action of SRY and SF1 on a specific Sox9 enhancer. Nature. 2008;453(7197):930–4.
166. McElreavey K, Vilain E, Abbas N, Herskowitz I, Fellous M. A regulatory cascade hypothesis for mammalian sex determination: SRY represses a negative regulator of male development. Proc Natl Acad Sci USA. 1993;90(8):3368–72.
167. Peng H, Ivanov AV, Oh HJ, Lau YF, Rauscher 3rd FJ. Epigenetic gene silencing by the SRY protein is mediated by a KRAB-O protein that recruits the KAP1 co-repressor machinery. J Biol Chem. 2009;284(51):35670–80.
168. Hanley NA, Hagan DM, Clement-Jones M, et al. SRY, SOX9, and DAX1 expression patterns during human sex determination and gonadal development. Mech Dev. 2000;91(1–2):403–7.
169. Hacker A, Capel B, Goodfellow P, Lovell-Badge R. Expression of Sry, the mouse sex determining gene. Development. 1995;121(6):1603–14.
170. Parker KL, Rice DA, Lala DS, et al. Steroidogenic factor 1: an essential mediator of endocrine development. Recent Prog Horm Res. 2002;57:19–36.
171. De Santa Barbara P, Bonneaud N, Boizet B, et al. Direct interaction of SRY-related protein SOX9 and steroidogenic factor 1 regulates transcription of the human anti-Mullerian hormone gene. Mol Cell Biol. 1998;18(11):6653–65.
172. Palmer SJ, Burgoyne PS. In situ analysis of fetal, prepubertal and adult XX—XY chimaeric mouse testes: Sertoli cells are predominantly, but not exclusively, XY. Development. 1991;112(1):265–8.
173. Orth JM, Gunsalus GL, Lamperti AA. Evidence from Sertoli cell-depleted rats indicates that spermatid number in adults depends on numbers of Sertoli cells produced during perinatal development. Endocrinology. 1988;122(3):787–94.
174. Plant TM, Marshall GR. The functional significance of FSH in spermatogenesis and the control of its secretion in male primates. Endocr Rev. 2001;22(6):764–86.
175. Sharpe RM, Turner KJ, McKinnell C, et al. Inhibin B levels in plasma of the male rat from birth to adulthood: effect of experimental manipulation of Sertoli cell number. J Androl. Jan–Feb 1999;20(1):94–101.
176. Cortes D, Muller J, Skakkebaek NE. Proliferation of Sertoli cells during development of the human testis assessed by stereological methods. Int J Androl. 1987;10(4):589–96.
177. Atanassova N, McKinnell C, Walker M, et al. Permanent effects of neonatal estrogen exposure in rats on reproductive hormone levels, Sertoli cell number, and the efficiency of spermatogenesis in adulthood. Endocrinology. 1999;140(11):5364–73.

178. Simorangkir DR, de Kretser DM, Wreford NG. Increased numbers of Sertoli and germ cells in adult rat testes induced by synergistic action of transient neonatal hypothyroidism and neonatal hemicastration. J Reprod Fertil. 1995;104(2):207–13.
179. Cooke PS, Zhao YD, Bunick D. Triiodothyronine inhibits proliferation and stimulates differentiation of cultured neonatal Sertoli cells: possible mechanism for increased adult testis weight and sperm production induced by neonatal goitrogen treatment. Biol Reprod. 1994; 51(5):1000–5.
180. Palmero S, Prati M, Bolla F, Fugassa E. Tri-iodothyronine directly affects rat Sertoli cell proliferation and differentiation. J Endocrinol. 1995;145(2):355–62.
181. Buzzard JJ, Morrison JR, O'Bryan MK, Song Q, Wreford NG. Developmental expression of thyroid hormone receptors in the rat testis. Biol Reprod. 2000;62(3):664–9.
182. Arambepola NK, Bunick D, Cooke PS. Thyroid hormone and follicle-stimulating hormone regulate Mullerian-inhibiting substance messenger ribonucleic acid expression in cultured neonatal rat Sertoli cells. Endocrinology. 1998;139(11):4489–95.
183. Arambepola NK, Bunick D, Cooke PS. Thyroid hormone effects on androgen receptor messenger RNA expression in rat Sertoli and peritubular cells. J Endocrinol. 1998;156(1): 43–50.
184. Buzzard JJ, Wreford NG, Morrison JR. Thyroid hormone, retinoic acid, and testosterone suppress proliferation and induce markers of differentiation in cultured rat sertoli cells. Endocrinology. 2003;144(9):3722–31.
185. Holsberger DR, Jirawatnotai S, Kiyokawa H, Cooke PS. Thyroid hormone regulates the cell cycle inhibitor p27Kip1 in postnatal murine Sertoli cells. Endocrinology. 2003;144(9): 3732–8.
186. Holsberger DR, Kiesewetter SE, Cooke PS. Regulation of neonatal Sertoli cell development by thyroid hormone receptor alpha1. Biol Reprod. 2005;73(3):396–403.
187. Gilleron J, Nebout M, Scarabelli L, et al. A potential novel mechanism involving connexin 43 gap junction for control of sertoli cell proliferation by thyroid hormones. J Cell Physiol. 2006;209(1):153–61.
188. Brehm R, Zeiler M, Ruttinger C, et al. A sertoli cell-specific knockout of connexin43 prevents initiation of spermatogenesis. Am J Pathol. 2007;171(1):19–31.
189. Schmahl J, Eicher EM, Washburn LL, Capel B. Sry induces cell proliferation in the mouse gonad. Development. 2000;127(1):65–73.
190. Capel B, Albrecht KH, Washburn LL, Eicher EM. Migration of mesonephric cells into the mammalian gonad depends on Sry. Mech Dev. 1999;84(1–2):127–31.
191. Yao HH, Whoriskey W, Capel B. Desert Hedgehog/Patched 1 signaling specifies fetal Leydig cell fate in testis organogenesis. Genes Dev. 2002;16(11):1433–40.
192. Park SY, Tong M, Jameson JL. Distinct roles for steroidogenic factor 1 and desert hedgehog pathways in fetal and adult Leydig cell development. Endocrinology. 2007;148(8):3704–10.
193. Maekawa M, Kamimura K, Nagano T. Peritubular myoid cells in the testis: their structure and function. Arch Histol Cytol. 1996;59(1):1–13.
194. Karl J, Capel B. Sertoli cells of the mouse testis originate from the coelomic epithelium. Dev Biol. 1998;203(2):323–33.
195. Byskov AG. Differentiation of mammalian embryonic gonad. Physiol Rev. 1986;66(1): 71–117.
196. Brennan J, Tilmann C, Capel B. Pdgfr-alpha mediates testis cord organization and fetal Leydig cell development in the XY gonad. Genes Dev. 2003;17(6):800–10.
197. Mendis-Handagama SM, Risbridger GP, de Kretser DM. Morphometric analysis of the components of the neonatal and the adult rat testis interstitium. Int J Androl. 1987;10(3): 525–34.
198. Forest MG, Cathiard AM, Bertrand JA. Evidence of testicular activity in early infancy. J Clin Endocrinol Metab. 1973;37(1):148–51.
199. Geissler WM, Davis DL, Wu L, et al. Male pseudohermaphroditism caused by mutations of testicular 17 beta-hydroxysteroid dehydrogenase 3. Nat Genet. 1994;7(1):34–9.

5 Thyroid Hormone Regulation of Mammalian Reproductive Development... 171

200. Canale D, Agostini M, Giorgilli G, et al. Thyroid hormone receptors in neonatal, prepubertal, and adult rat testis. J Androl. Mar–Apr 2001;22(2):284–8.
201. Falcone M, Miyamoto T, Fierro-Renoy F, Macchia E, DeGroot LJ. Antipeptide polyclonal antibodies specifically recognize each human thyroid hormone receptor isoform. Endocrinology. 1992;131(5):2419–29.
202. Macchia E, Nakai A, Janiga A, et al. Characterization of site-specific polyclonal antibodies to c-erbA peptides recognizing human thyroid hormone receptors alpha 1, alpha 2, and beta and native 3,5,3'-triiodothyronine receptor, and study of tissue distribution of the antigen. Endocrinology. 1990;126(6):3232–9.
203. Jannini EA, Carosa E, Rucci N, Screponi E, D'Armiento M. Ontogeny and regulation of variant thyroid hormone receptor isoforms in developing rat testis. J Endocrinol Invest. 1999;22(11):843–8.
204. Jannini EA, Dolci S, Ulisse S, Nikodem VM. Developmental regulation of the thyroid hormone receptor alpha 1 mRNA expression in the rat testis. Mol Endocrinol. 1994;8(1): 89–96.
205. Maran RR, Arunakaran J, Jeyaraj DA, Ravichandran K, Ravisankar B, Aruldhas MM. Transient neonatal hypothyroidism alters plasma and testicular sex steroid concentration in puberal rats. Endocr Res. 2000;26(3):411–29.
206. Mendis-Handagama C, Ariyaratne S. Prolonged and transient neonatal hypothyroidism on Leydig cell differentiation in the postnatal rat testis. Arch Androl. 2004;50(5):347–57.
207. Manna PR, Kero J, Tena-Sempere M, Pakarinen P, Stocco DM, Huhtaniemi IT. Assessment of mechanisms of thyroid hormone action in mouse Leydig cells: regulation of the steroidogenic acute regulatory protein, steroidogenesis, and luteinizing hormone receptor function. Endocrinology. 2001;142(1):319–31.
208. Buck Louis GM, Gray Jr LE, Marcus M, et al. Environmental factors and puberty timing: expert panel research needs. Pediatrics. 2008;121 Suppl 3:S192–207.
209. Jacobson-Dickman E, Lee MM. The influence of endocrine disruptors on pubertal timing. Curr Opin Endocrinol Diabetes Obes. 2009;16(1):25–30.
210. Toppari J, Juul A. Trends in puberty timing in humans and environmental modifiers. Mol Cell Endocrinol. 2010;324(1–2):39–44.
211. Zoeller RT. Environmental neuroendocrine and thyroid disruption: relevance for reproductive medicine? Fertil Steril. 2008;89(2 Suppl):e99–e100.
212. Rodriguez PE, Sanchez MS. Maternal exposure to triclosan impairs thyroid homeostasis and female pubertal development in Wistar rat offspring. J Toxicol Environ Health A. 2010; 73(24):1678–88.
213. Zorrilla LM, Gibson EK, Jeffay SC, et al. The effects of triclosan on puberty and thyroid hormones in male Wistar rats. Toxicol Sci. 2009;107(1):56–64.
214. Markey CM, Wadia PR, Rubin BS, Sonnenschein C, Soto AM. Long-term effects of fetal exposure to low doses of the xenoestrogen bisphenol-A in the female mouse genital tract. Biol Reprod. 2005;72(6):1344–51.
215. Honma S, Suzuki A, Buchanan DL, Katsu Y, Watanabe H, Iguchi T. Low dose effect of in utero exposure to bisphenol A and diethylstilbestrol on female mouse reproduction. Reprod Toxicol. 2002;16(2):117–22.
216. Howdeshell KL, Hotchkiss AK, Thayer KA, Vandenbergh JG, Vom Saal FS. Exposure to bisphenol A advances puberty. Nature. 1999;401(6755):763–4.
217. Tan BL, Kassim NM, Mohd MA. Assessment of pubertal development in juvenile male rats after sub-acute exposure to bisphenol A and nonylphenol. Toxicol Lett. 2003;143(3): 261–70.
218. Zoeller RT, Bansal R, Parris C. Bisphenol-A, an environmental contaminant that acts as a thyroid hormone receptor antagonist in vitro, increases serum thyroxine, and alters RC3/neurogranin expression in the developing rat brain. Endocrinology. 2005;146(2):607–12.
219. Sathyanarayana S. Phthalates and children's health. Curr Probl Pediatr Adolesc Health Care. 2008;38(2):34–49.

220. Meeker JD, Calafat AM, Hauser R. Di(2-ethylhexyl) phthalate metabolites may alter thyroid hormone levels in men. Environ Health Perspect. 2007;115(7):1029–34.
221. Huang PC, Kuo PL, Guo YL, Liao PC, Lee CC. Associations between urinary phthalate monoesters and thyroid hormones in pregnant women. Hum Reprod. 2007;22(10):2715–22.
222. Boas M, Frederiksen H, Feldt-Rasmussen U, et al. Childhood exposure to phthalates: associations with thyroid function, insulin-like growth factor I, and growth. Environ Health Perspect. 2010;118(10):1458–64.
223. Hinton RH, Mitchell FE, Mann A, et al. Effects of phthalic acid esters on the liver and thyroid. Environ Health Perspect. 1986;70:195–210.
224. Price SC, Chescoe D, Grasso P, Wright M, Hinton RH. Alterations in the thyroids of rats treated for long periods with di-(2-ethylhexyl) phthalate or with hypolipidaemic agents. Toxicol Lett. 1988;40(1):37–46.
225. Botelho GG, Golin M, Bufalo AC, Morais RN, Dalsenter PR, Martino-Andrade AJ. Reproductive effects of di(2-ethylhexyl)phthalate in immature male rats and its relation to cholesterol, testosterone, and thyroxin levels. Arch Environ Contam Toxicol. 2009;57(4): 777–84.
226. Swan SH, Main KM, Liu F, et al. Decrease in anogenital distance among male infants with prenatal phthalate exposure. Environ Health Perspect. 2005;113(8):1056–61.
227. Christiansen S, Boberg J, Axelstad M, et al. Low-dose perinatal exposure to di(2-ethylhexyl) phthalate induces anti-androgenic effects in male rats. Reprod Toxicol. 2010;30(2):313–21.
228. Saillenfait AM, Roudot AC, Gallissot F, Sabate JP, Chagnon MC. Developmental toxic potential of di-n-propyl phthalate administered orally to rats. J Appl Toxicol. 2011;31(1): 36–44.
229. Kim TS, Jung KK, Kim SS, et al. Effects of in utero exposure to DI(n-Butyl) phthalate on development of male reproductive tracts in Sprague-Dawley rats. J Toxicol Environ Health A. 2010;73(21–22):1544–59.
230. Andrade AJ, Grande SW, Talsness CE, et al. A dose-response study following in utero and lactational exposure to di-(2-ethylhexyl) phthalate (DEHP): effects on androgenic status, developmental landmarks and testicular histology in male offspring rats. Toxicology. 2006; 225(1):64–74.
231. Gray Jr LE, Ostby J, Furr J, Price M, Veeramachaneni DN, Parks L. Perinatal exposure to the phthalates DEHP, BBP, and DINP, but not DEP, DMP, or DOTP, alters sexual differentiation of the male rat. Toxicol Sci. 2000;58(2):350–65.
232. Grande SW, Andrade AJ, Talsness CE, Grote K, Chahoud I. A dose-response study following in utero and lactational exposure to di(2-ethylhexyl)phthalate: effects on female rat reproductive development. Toxicol Sci. 2006;91(1):247–54.
233. Abaci A, Demir K, Bober E, Buyukgebiz A. Endocrine disrupters – with special emphasis on sexual development. Pediatr Endocrinol Rev. 2009;6(4):464–75.
234. Colciago A, Casati L, Mornati O, et al. Chronic treatment with polychlorinated biphenyls (PCB) during pregnancy and lactation in the rat Part 2: Effects on reproductive parameters, on sex behavior, on memory retention and on hypothalamic expression of aromatase and 5alpha-reductases in the offspring. Toxicol Appl Pharmacol. 2009;239(1):46–54.
235. Shirota M, Mukai M, Sakurada Y, et al. Effects of vertically transferred 3,3',4,4',5-pentachlorobiphenyl (PCB-126) on the reproductive development of female rats. J Reprod Dev. 2006;52(6):751–61.
236. Steinberg RM, Walker DM, Juenger TE, Woller MJ, Gore AC. Effects of perinatal polychlorinated biphenyls on adult female rat reproduction: development, reproductive physiology, and second generational effects. Biol Reprod. 2008;78(6):1091–101.
237. Kobayashi K, Miyagawa M, Wang RS, Suda M, Sekiguchi S, Honma T. Effects of in utero exposure to 2,2',4,4',5,5'-hexachlorobiphenyl on postnatal development and thyroid function in rat offspring. Ind Health. 2009;47(2):189–97.
238. Kobayashi K, Miyagawa M, Wang RS, Suda M, Sekiguchi S, Honma T. Effects of in utero exposure to 2,2',4,4',5,5'-hexachlorobiphenyl (PCB 153) on somatic growth and endocrine status in rat offspring. Congenit Anom (Kyoto). 2008;48(4):151–7.

239. Xiao W, Li K, Wu Q, Nishimura N, Chang X, Zhou Z. Influence of persistent thyroxine reduction on spermatogenesis in rats neonatally exposed to 2,2',4,4',5,5'-hexa-chlorobiphenyl. Birth Defects Res B Dev Reprod Toxicol. 2010;89(1):18–25.
240. Schell LM, Gallo MV. Relationships of putative endocrine disruptors to human sexual maturation and thyroid activity in youth. Physiol Behav. 2010;99(2):246–53.
241. Ellis-Hutchings RG, Cherr GN, Hanna LA, Keen CL. Polybrominated diphenyl ether (PBDE)-induced alterations in vitamin A and thyroid hormone concentrations in the rat during lactation and early postnatal development. Toxicol Appl Pharmacol. 2006;215(2): 135–45.
242. Kodavanti PR, Coburn CG, Moser VC, et al. Developmental exposure to a commercial PBDE mixture, DE-71: neurobehavioral, hormonal, and reproductive effects. Toxicol Sci. 2010; 116(1):297–312.

Part II
Developmental Exposure to Endocrine Disruptors and Adverse Reproductive Outcomes

Chapter 6
Developmental Exposure to Endocrine Disruptors and Ovarian Function

Evanthia Diamanti-Kandarakis, Eleni Palioura, Eleni A. Kandaraki

Abstract The developing organism is particularly sensitive to environmental insults that can permanently reprogram normal physiology leading to long-term functional aberrations. With regard to the female gonad, any disruption of the complex regulatory mechanisms that control ovarian development/function could provoke disorders of normal steroidogenesis, folliculogenesis, ovulation, and, finally, female fertility. Endocrine-disrupting chemicals are among the factors that are incriminated to have an adverse impact on several aspects of ovarian biology. Through their capacity to mimic or antagonize natural hormones, this diverse class of chemicals is believed to interfere in the early stages of gonadal development. This chapter summarizes experimental evidence linking pre/peripubertal exposure (prenatal and/or early life until puberty exposure) to environmental chemical contaminants and perturbations in ovarian development and function.

Keywords 1,1-Dichloro-2,2-bis(p-chlorophenyl)ethylene (DDE) • 2,2-Bis (p-hydroxyphenyl)- 1,1,1-trichloroethane (HPTE) • Antiandrogenic endocrine disruptors • Bisphenol A (BPA) • Diethylstilbestrol (DES) • Endocrine disruptors (EDs) • Estrogenic endocrine disruptors • Genistein • methoxychlor (MXC) • Ovary • Puberty • Vinclozolin

---

E. Diamanti-Kandarakis (✉) • E. Palioura
Third Department of Medicine, Sotiria Hospital, Medical School
University of Athens, Athens, Greece
e-mail: e.diamanti.kandarakis@gmail.com

E.A. Kandaraki
Department of Medicine, Huddersfield Royal Infirmary Hospital, Huddersfield, UK

E. Diamanti-Kandarakis and A.C. Gore (eds.), *Endocrine Disruptors and Puberty*,
Contemporary Endocrinology, DOI 10.1007/978-1-60761-561-3_6,
© Springer Science+Business Media, LLC 2012

Introduction

During the last decades, systematic research on the field of endocrine disruption has revealed the vulnerability of all hormone-sensitive physiological systems to exogenous agents that interfere with various aspects of natural hormone physiology. The possible reproductive hazards of these environmental chemicals have generated concern as it remains controversial whether they can alter ovarian development and function, purportedly through estrogenic, antiestrogenic, and/or antiandrogenic effects. Mammalian oocyte maturation and follicle physiology are shown to be impaired by persistent environmental pollutants [1]; therefore, the female gonad could be characterized as a potential target of disruption. Indeed, ovarian development and function are regulated by a milieu of endogenous factors as well as exogenous agents that may interfere even from the early stages of gonadal development.

The range of biological effects varies depending not only on the nature of the chemical and dose but on the susceptibility of the individual to the compound. The timing of exposure to environmental insults is a determining factor since the developmental periods from periconception during pregnancy, infancy, childhood, and puberty are considered critical and sensitive windows of susceptibility [2]. This is derived from the developmental programming hypothesis, which proposes that exposure of the developing tissues/organs to an adverse stimulus or insult during critical or sensitive times of development can permanently reprogram normal physiological responses leading to hormonal disorders later in life [3, 4]. Therefore, the concept of developmental reprogramming includes permanent changes related to a particular function as a result of some events that occur during the perinatal period.

A representative example of this developmental programming in human is provided by women exposed to diethylstilbestrol (DES) in utero. Diethylstilbestrol is a potent estrogen given in the mid-twentieth century under the erroneous assumption that it would prevent miscarriages and other pregnancy complications. Exposure to this estrogenic substance during the critical period of prenatal development has been associated with reproductive tract abnormalities as well as increased incidence of cervical–vaginal adenocarcinoma in the female offspring [5]. Interestingly, preliminary data from a unique cohort of young women, born to mothers prenatally exposed to diethylstilbestrol, provide evidence of possible transmissible effects in third-generation female descendants with irregular menstrual cycles and possible infertility in DES granddaughters. However, further research is required to draw any safe conclusion [6]. Overall, there is substantial evidence that the female reproductive tract is a target of disruption after exposure in critical developmental windows resulting in permanent reproductive aberrations later in life. The possibility that these aberrations may be transgenerational is of particular concern.

A rational query is whether the mammalian ovary could also represent a target for developmental programming as a result of environmental hormone exposure. Indeed, when considering the exquisitely balanced hormonal mechanisms that control ovarian development and function, it is reasonable to assume that any disturbance from the early stages of mammalian gonad and follicular genesis could permanently reprogram the normal ovarian function leading to long-term disorders.

Susceptibility of the ovaries to different classes of endocrine-disrupting chemicals depends on the stage of development at which exposure occurs. At the oocyte level, the stage of development at which the follicle is impaired determines the impact that the exposure to the chemical will have on reproduction. Disruption of primordial follicles would have a long-term effect on reproduction as these cells represent the pool of oocytes available for a lifetime of fertility. In contrast, any damage of large growing or antral follicles would lead to a temporary impairment given that these cells will be replaced by recruitment from the pool of primordial follicles.

Theoretically, potential "ovarian disruptors" could act through two general routes: directly on the ovary by affecting oocyte maturation and steroid hormone production or indirectly on the hypothalamus and pituitary by altering gonadotropins' secretion, or by causing abnormalities of female reproductive tract. Disruption of any of these sites can ultimately manifest as a disruption of ovarian function. At the level of the ovary multiple processes could be disrupted, including folliculogenesis, ovarian steroidogenesis, and intraovarian regulatory factors as well as fecundity/reproductive capacity. The present text will discuss available evidence that environmental chemicals disrupt pre- and postpubertal normal ovarian development and function, with particular attention to chemicals with estrogenic, antiandrogenic, and combined effects.

The Developing Ovary: Physiology

The mammalian ovary performs two major functions: folliculogenesis, defined as the development and maturation of ovarian follicles, and steroidogenesis, defined as the production of ovarian hormones. Both processes ensure normal reproductive function and species preservation. Ovarian biology is one of the most complex processes as its cellular and molecular aspects are not yet absolutely understood, and moreover, little is known about its control mechanisms.

Follicles are the functional units of the ovary, and each follicle consists of an oocyte surrounded by one or more layers of somatic cells. Mammalian ovary undergoes a tightly regulated biological process from germ line development and gonad formation to oogenesis, ovulatory process, and steroidogenesis. During embryogenesis, germ cell development involves primordial germ cell formation and migration into the urogenital ridge and mitotic division of germ cell nuclei to form a multinucleated syncytia or germ cell nest [7, 8]. Germ cells subsequently enter meiosis, becoming oocytes, and remain connected to each other through cytoplasmic bridges. Meiotically active oocytes progress through the stages of prophase I and arrest in diplotene until ovulation, at which time meiosis is resumed (Fig. 6.1). The majority of the oocytes within these nests will undergo apoptosis as the germ cell clusters break down to give rise to primordial follicles (reviewed in [7, 8]). Primordial follicle

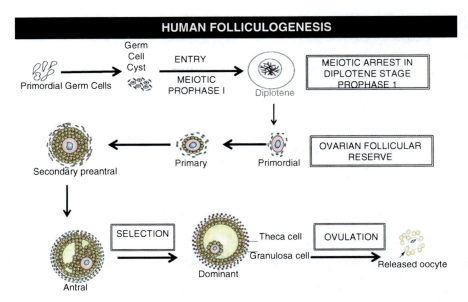

Fig. 6.1 This figure is a schematic representation of the stages of folliculogenesis from primordial germ cell migration into the genital ridge and oocytes' arrest, to diplotene stage of meiotic prophase I, to selection of the dominant follicle and ovulation. Primordial cells are formed prenatally and represent the ovarian follicular reserve with a pool of oocytes available for a lifetime of fertility

formation occurs when oocytes that survive the process of germ cell cluster breakdown are individually surrounded with squamous pregranulosa cells (Fig. 6.1). This represents the first stage of folliculogenesis, and in humans, it is initiated in utero around 21 weeks of gestation [9]. The population of these primordial follicles serves as a resting and finite pool of oocytes available for a lifetime of fertility. These oocytes are subject to environmental exposures, thus being vulnerable to the effects of endocrine-disrupting chemicals during reproductive life.

Following primordial follicle formation, the major stages of ovarian folliculogenesis are primordial-to-primary follicle transition with a morphological change in granulosa cells from squamous to cuboidal and the subsequent formation of secondary and tertiary follicle. Lastly, ovulation and formation of a corpus luteum occur. Preantral follicles begin to develop prenatally in humans and do not require stimulation by the pituitary gonadotropins [7]. The precise mechanisms involved in early ovarian follicle formation are not completely understood, but are essential in organizing the fetal ovary.

The ovum inside the developing follicle is surrounded by layers of granulosa cells followed by theca cells, where steroidogenesis takes place. Steroid hormone synthesis is controlled by the activity of several highly substrate-selective cytochrome P450 enzymes. Several environmental chemicals (e.g., Genistein, Bisphenol A, Methoxychlor) have been shown to exert direct inhibitory and/or stimulatory effects on key ovarian steroidogenic enzymes (e.g., 17β-hydroxysteroid dehydrogenase

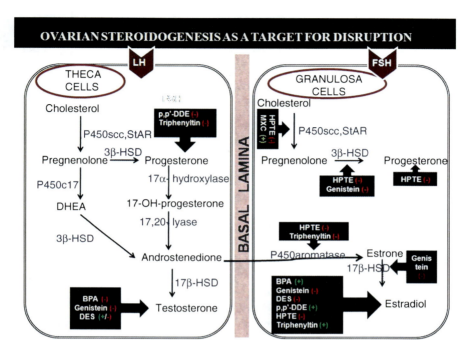

Fig. 6.2 This figure illustrates that ovarian synthesis of steroid hormones is subject to direct inhibitory (−) and/or stimulatory (−/+) modulation by several environmental chemicals. Note that key ovarian steroidogenic enzymes also represent a target for disruption. Abbreviations: *BPA* bisphenol A, *DES* diethylstilbestrol, HPTE, DDE, 1,1-dichloro-2,2-bis(*p*-chlorophenyl)ethylene, *MXC* methoxychlor, *P450scc* cholesterol side-chain cleavage cytochrome P450, *StAR* steroidogenic acute regulatory protein, *3β-HSD* 3β-hydroxysteroid dehydrogenase, *17β-HSD* 17β-hydroxysteroid dehydrogenase, *P450c17* 17 α-hydroxylase/C17–20 lyase cytochrome P450, *DHEA* dehydroepiandrosterone

(17β-HSD), 3β-hydroxysteroid dehydrogenase (3β-HSD), cytochrome P450 side-chain cleavage (CYP11A)) [10, 11] and therefore could disrupt follicular development and maturation (Fig. 6.2).

Since the developmental stages of mammalian ovary commence in intrauterine life and continue after birth until the initiation of ovulation and reproductive capacity postpubertal, any disruption during this wide life period could adversely affect normal ovarian physiology. Therefore, developmental exposure to chemicals that have profound negative effects on ovarian components could reprogram normal gonad biology.

It is widely recognized that some synthetic chemicals have the potential to disrupt or interfere with the endocrine system by mimicking or inhibiting the action of sex steroid hormones, estradiol (E2) and testosterone. This is particularly important given that hormones play a prominent role in regulating ovarian follicle development and function. In a broader context, steroid hormones can guide ovarian development. The importance of estrogens has been elucidated in estrogen receptor

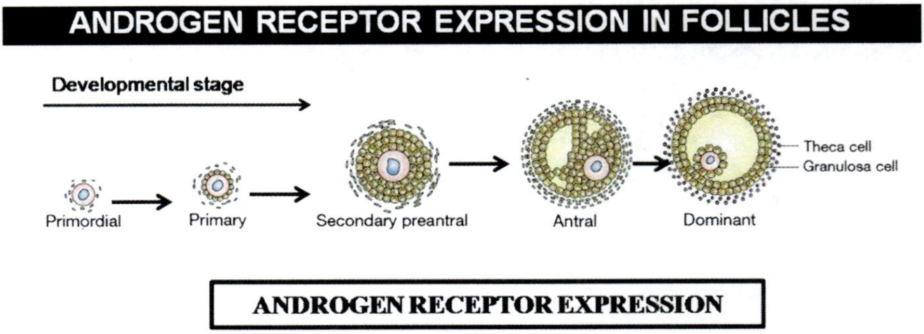

ANDROGEN RECEPTOR EXPRESSION IN FOLLICLES

Fig. 6.3 AR expression is present throughout most stages of follicular development

(ER)-α, ER-β, or ER-α and ER-β knockout mouse models, all of which converge upon various defects in ovulation and/or follicle development [12]. Furthermore, in the aromatase knockout (ArKO) mice which are characterized by E2 deficiency, a permanent reduction of primordial and primary follicles is observed implying the regulatory effect of estrogen on fetal ovarian folliculogenesis [13]. Similarly, a reduction in the number of primordial follicles is observed in estrogen-depleted fetal baboon ovaries [14]. Additionally, rodents' early neonatal exposure to E2 is shown to inhibit oocyte nest breakdown and primordial follicle assembly [15] as well as primordial-to-primary follicle transition [16].

In general, as articulated by Britt and Findlay, "estrogen is obligatory for normal folliculogenesis beyond the antral stage and for the maintenance of the female phenotype of the somatic cells within the ovaries" [17]. From the above, it seems obvious that estrogen plays a major role in fetal and early postnatal ovarian development; thus, environmental agents that mimic or antagonize the actions of ovarian steroids could disrupt follicle development and/or ovarian function.

Apart from the well-studied role of estrogens on female reproductive physiology, relatively fewer effects are known concerning the role of other steroid hormones in mammalian ovary development. Androgens, except from being the precursor of estradiol biosynthesis, have also been shown to play a role in regulating follicle development and ovulation. Indeed, the androgen receptor (AR) expression is present throughout most stages of follicular development in the oocyte, granulosa, and theca cells [18] (Fig. 6.3). AR-mediated actions are believed to play an important role in maintaining female fertility, as the AR-null mice shows premature ovarian failure (POF)-like symptoms [19]. Experiments have suggested that progesterone also mediates early follicle development. In ovaries from newborn rat, progesterone was shown to inhibit primordial follicle assembly and the primordial-to-primary follicle transition, [16] and in mice, progesterone was shown to inhibit oocyte nest breakdown [15]. Similarly, in cultured fetal bovine ovaries in which follicle assembly occurs prior to birth as in humans, treatment with high progesterone doses significantly decreased follicle assembly [20]. Given that many chemicals possess androgenic/antiandrogenic properties, an interruption of the normal androgenic control over ovarian function could represent another possible mechanism of endocrine disruptors–related ovarian disorders.

In addition to extrinsic regulation of ovarian development by sex steroid hormones, emerging evidence supports the significant role of a plethora of locally produced growth factors as intraovarian regulators of folliculogenesis. The proper formation of ovarian follicles in the fetus depends on a balance between those factors and systemic sex steroid levels. Many of these factors expressed by ovarian somatic cells and oocytes belong to the transforming growth factor-β (TGF-β) superfamily. For example, activin and its functional antagonist inhibin are believed to exert local autocrine–paracrine actions to modulate follicle growth, gonadotropin responsiveness, steroidogenesis, oocyte maturation, ovulation, and corpus luteum function [21]. In particular, activin has been shown to regulate ovarian granulosa cell proliferation and differentiation, to modulate ovarian follicle atresia, to increase FSH receptor expression in secondary follicles, and stimulate oocyte maturation [21]. Inhibin has been shown to affect follicle development by exerting a negative feedback effect on FSH synthesis and secretion [21]. Another TGF-β superfamily member, anti-Mullerian hormone (AMH), has been shown to have negative effect in the initiation of primordial follicle growth and on preantral follicle development [22]. Environmental factors that may either inhibit or stimulate the production of these local factors during the critical time frame of follicle formation could alter follicle dynamics. This may represent another possible mechanism of ovarian disorders related to endocrine disruptor exposure.

Developmental Interference by EDs on the Programming of Prepubertal–Pubertal Ovary

As previously discussed, ovarian development in different stages is regulated by numerous intraovarian growth factors and steroid hormones working in concert with the pituitary gonadotropins' control. Human and wildlife are exposed at critical periods of development to endocrine-disrupting chemicals that may be responsible for reproductive disorders. Prenatal or early postnatal exposure to some synthetic chemicals may affect the reproductive system of the offspring in the long term. Therefore, the complex regulatory system of gonad development can be disrupted by exogenous agents that mimic or antagonize endogenous hormone-mediated mechanisms. Most of the endocrine-disrupting chemicals are traditionally believed to exert estrogenic, antiestrogenic, and/or antiandrogenic affect, and their developmental interference with ovarian maturation progress may lead to long-term adverse effects on reproduction. The following sections provide a review of the literature on the interference of known endocrine disruptors with the mammalian ovary during critical and sensitive windows of susceptibility mainly focused on the period from prenatal to pre- and pubertal period (Fig. 6.4).

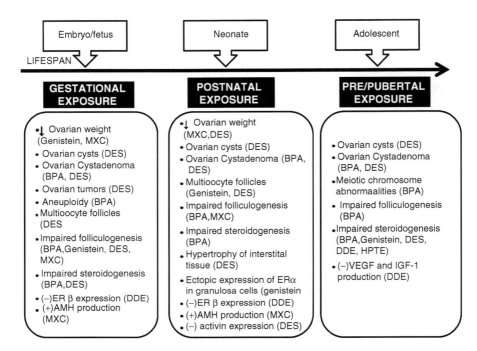

Fig. 6.4 Ovarian disorders following developmental exposure to select EDC during critical and sensitive windows of susceptibility (experimental data).Note that many of these effects are not apparent at the exposed period but later in life. The stimulatory (+) and inhibitory (−) effects of the environmental endocrine disruptors reviewed are shown. Abbreviations: *BPA* bisphenol A, *DES* diethylstilbestrol, *p,p´-DDE* p,p′ -1,1-dichloro-2,2-bis(p-chlorophenyl)ethylene, *MXC* methoxychlor, *HPTE* 2,2-bis(p-hydroxyphenyl)- 1,1,1-trichloroethane, *ER β* estrogen receptor β, *AMH* anti-Mullerian hormone, *VEGF* vascular endothelial growth factor, *IGF-1* insulin growth factor 1. *Symbols*: ↓: decreased

Estrogenic Endocrine Disruptors

Ovarian development and folliculogenesis depend on a number of endocrine, paracrine, and autocrine signals regulated by a milieu of endogenous factors including natural sex steroid hormones. Since several endocrine disruptors may possess estrogenic properties, their impact on the developing ovary may involve the sex steroid-dependent mechanisms controlling the gonads and the reproductive neuroendocrine axis. This is pathophysiologically plausible, given that several studies have demonstrated the significant role of estrogens even from the earliest stages of follicle development. Indeed, many estrogenic EDs have a high affinity to ER-β that appears to be the predominant receptor type mediating the estrogenic control of folliculogenesis [11] as it is expressed in the human ovary, mainly in granulosa cells [23, 24].

Bisphenol A

Bisphenol A (BPA) is a high-volume production chemical used primarily in the manufacture of polycarbonate plastic and epoxy resins [25, 26]. With a broad spectrum of applications including use in certain food and drink packaging, in many plastic consumer products as well as dental materials, [27]. BPA is now considered one of the most ubiquitous estrogen-mimicking industrial compounds. BPA has been shown to leach from food and beverage containers, and some dental sealants and composites, under normal conditions of use [28]. Concentrations of BPA have been measured directly in maternal and fetal plasma, in breast milk, and in placental and adipose tissue in humans [28–30]. Biochemical assays have determined that BPA mimics the action of estradiol by binding both ER-α and ER-β, with approximately tenfold higher affinity to ER-β [31, 32]. Due to its estrogenic properties and its ubiquitous detection in the environment, scientific concern has arisen regarding its potency to interfere with mammalian development and adversely affect reproduction or development.

Developmental exposure to BPA is associated with numerous reproductive abnormalities in experimental animals, suggesting that the ovary is a particularly sensitive target for the effects of this chemical. BPA can easily pass through the placenta, as shown in rodents, [33] and does not bind to α-fetoprotein, the liver-produced glycoprotein that normally binds circulating estrogen in the developing fetus. Thus, it is reasonable to assume that exposure to even low doses of BPA during fetal and neonatal period, when endogenous estrogens are low, could adversely affect ovarian development [34]. Mice exposed to BPA during fetal life displayed changes in ovarian follicular development, manifested as an excess of antral follicles at the expense of corpora lutea formation and a persistent state of estrus and/or metestrus with impaired estrous cyclicity [35]. Moreover, 18-month-old female mice exposed to BPA during a critical period of gonad differentiation (prenatally) exhibited several benign abnormalities including increased incidence of ovarian cysts formations and ovarian cyst adenomas compared to the control group [36].

In a reptile model, *Caiman latirostris*, ovum exposure to BPA provides evidence of the adverse effects of this estrogenic chemical on gonadal differentiation, folliculogenesis, and steroidogenesis [37]. Juvenile caimans prenatally exposed to BPA exhibit abnormal ovarian morphology with a high incidence of multioocyte follicles (MOFs), while in neonatal life, the exposed animals have higher estrogen levels and decreased testosterone levels compared to the controls (Fig. 6.2) [37]. With relevance to humans, it should be noted that MOFs have also been observed in humans; however, the pathogenetic connection is yet unidentified. However, as indicated in a mice study by Iguchi et al. [38] the oocytes derived from MOFs have reduced capacity for in vitro fertilization (IVF) compared to single oocyte follicles. Therefore, in reproductive life, this may translate into compromised fertility capacity.

Apart from the disturbance in steroidogenesis, folliculogenesis was also impaired in caimans exposed ovum to BPA [37]. In particular, developmental exposure to BPA has been associated with altered follicular dynamics with an almost total absence of type III follicles in neonates but a higher proportion of type III follicles in juvenile animals compared to controls [37].

BPA has also been incriminated as a potent disruptor of meiosis in the fetal ovary based on a series of animal studies assessing the possible effect of this estrogenic chemical on oocyte maturation. In mice, low-dose BPA exposure during pregnancy was shown to disturb oocyte development in unborn female fetuses, while in adulthood, these perturbations were translated into an increase in chromosomally abnormal (aneuploid) eggs and embryos [39]. In humans, aneuploidy is considered as one of the leading causes of miscarriages and mental retardation [40]. Currently, there are no data on in utero exposure to BPA and aneuploidy in humans, but the possibility of an interaction is compelling. Recently, a link was suggested between BPA exposure and recurrent miscarriages, as BPA levels were higher in the blood of women who miscarried than in women who carried their pregnancies to term [41].

Neonatal exposure to BPA has been associated with the induction of several abnormalities in the ovary, including an increased incidence of ovarian cysts [42] and disorders of follicle development manifested as large antral-like follicles and fewer corpora lutea in ovaries of exposed rodents [43].

In prepubertal female mice, short-term, oral exposure to low doses of BPA resulted in increased incidence of meiotic chromosome abnormalities indicated by an increase in the congression failure (defect of alignment of chromosomes during the first meiotic division) [44] These findings amplify the potency of BPA to act as meiotic aneugen [44].

Apart from these effects on oocyte meiosis, experimental data indicate that prepubertal exposure to BPA disrupts normal ovulation with absence of corpora lutea at 4 weeks of age in almost all of the exposed female rats. However, this dysfunction was only transient as at 8 weeks of age, rats were ovulatory [45]. As previously mentioned, the stage of development at which the impact on the follicle occurs determines the dysregulatory effects that the exposure to the chemical will have on reproduction. Hence, in the aforementioned experimental study by Nikaido et al. [45] the consequences on ovulation should be investigated in long-term studies, as it was shown in this study that these effects were transient as the pool of primordial follicles seemed not to be affected.

Genistein

Over the last few years, public and scientific interest in phytoestrogens such as genistein has increased because of these compounds' proposed beneficial effects. Human exposure to genistein is predominantly from soy products in the diet or soy-derived products including soy-based infant formulas. Given that human infants are exposed to high levels of genistein in soy-based foods [46], concern arises regarding the potential effects of such exposure on the developing organism. Genistein is a major soy isoflavone acting as a phytoestrogen mainly by binding to the ER, with a higher binding affinity (20- to 30-fold more) to ER-β than to ER-α, [32] displaying both weak estrogenic and antiestrogenic activity, depending on its concentrations and on the organ involved.

Several animal experiments have shown adverse effects on the developing ovary following exposure to genistein. In utero exposure to high doses of genistein has been associated with impaired ovarian follicular development manifested as an absence of corpora lutea at 4-week-old treated mice, but in the adult ovary (in ≥8 weeks of age), corpora lutea were present, suggesting a transient delay of ovulation caused by genistein administration [47]. Furthermore, in prepubertal rats exposed to genistein from gestation through lactation, ovarian weight and sex steroid levels (estrogen and progesterone concentrations) were significantly reduced compared to control values with unaltered gonadotropin values (Fig. 6.2). These findings imply a possible inhibitory effect of genistein on granulosa cell growth and maturation [48]. When these animals reached adulthood, the genistein-treated ovaries exhibited more frequent follicular atresia and secondary interstitial cells (androgen-productive cells derived from the theca cells of atretic follicles) [48].

Neonatal exposure to genistein has also been associated with adverse consequences on ovarian development and function. In a series of studies conducted by Jefferson et al., it was shown that mice neonatally exposed to genistein exhibit altered ovarian differentiation manifested as a dose-related increase in the incidence of multioocyte follicles (MOFs) in the prepubertal [49] and in the mature ovary, [50] via an ER-β-mediated mechanism [49]. Interestingly, when these genistein-treated female mice mated with control males, their female offspring also exhibited increased MOFs, suggesting that these developmental effects could be transmitted to subsequent generations [51] Regarding the process linking neonatal genistein exposure and MOFs in mice, Jefferson et al. demonstrated that this estrogenic substance inhibits oocyte nest breakdown and reduces the number of apoptotic oocytes, [52] thereby distorting normal primordial follicle assembly [53]. This is in accordance with other studies that have also demonstrated that genistein inhibits nest breakdown and primordial follicle assembly in the neonatal mouse ovary both in vitro and in vivo [15].

Functional ovarian defects have also been reported in laboratory animals exposed to genistein during neonatal life. Anovulation has been observed in 4-month-old mice that were neonatally exposed to high doses of genistein (50 mg/kg/day) on days 1–5 after birth [50]. On the contrary, lower doses of genistein are shown to enhance ovulation rates, indicating an increased ovulatory capacity of the neonatally exposed rodent ovary [49]. Disrupted estrous cyclicity and impaired fertility have also been reported in rodents exposed to genistein in postnatal life [51].

Furthermore, neonatal exposure to genistein was shown to stimulate the ectopic expression of ER-α in granulosa cells of genistein-treated mice. This biochemical effect on the maturing ovary was partly attributed to tyrosine-kinase inhibitory actions of the phytoestrogen [49].

Prepubertal exposure to genistein has been associated with a higher incidence of transient anovulation in exposed rats compared to the untreated group [45]. Moreover, steroidogenesis appears to be partly affected by genistein exposure. In isolated immature rat ovarian follicles, phytoestrogen administration resulted to a prominent decrease in testosterone levels (Fig. 6.2) without significant changes in secreted estradiol [54].

To sum up, developmental exposure of laboratory animals to genistein has been associated with morphological, biochemical, and functional abnormalities of the ovaries. The list of the ovarian defects reported includes multioocyte follicles, impaired folliculogenesis and steroidogenesis, anovulation/superovulation, as well as ectopic expression of ERα in granulosa cells.

Diethylstilbestrol (DES)

Diethylstilbestrol (DES) represents the most well-studied estrogenic chemical with known deleterious effects on female reproductive system. In utero DES exposure has been incriminated in the development of clear cell vaginal and cervical adenocarcinoma as well as structural reproductive tract abnormalities, impaired fertility, poor pregnancy outcomes, and menstrual irregularities in prenatally exposed women [5, 55].

Biochemical assays have determined that DES is a potent estrogenic substance as it can bind to both ER-α and ER-β with higher affinity than 17β-estradiol itself, the natural circulating estrogen [31].

Rodent models have been utilized to replicate and predict many of the adverse effects seen in DES-exposed humans. Gestational exposure to DES has been shown to exert a promoting effect on follicular maturation. In utero exposure to DES was associated with an increased percentage of primary and secondary follicles and a reduced percentage of primordial follicles in 3-week-old female mice [56]. At an earlier embryonic stage of development, DES has been shown to promote somatic cells' proliferation given that the number of proliferating somatic cells in the DES-treated mice ovary was significantly higher than that in the control ovary [57]. Morphological ovarian abnormalities are also reported in animals exposed to DES during gestation including polyovular follicles, oocytes with irregular shape, oocytes with condensed chromatin along the nuclear membrane, [58] ovarian cysts, and ovarian tumors [59]. Similar morphological and functional changes in the ovaries (absence of corpora lutea, hypertrophy of interstitial tissue, [60] polyovular follicles, [38] and decreased body weight [61]) have also been reported in animals exposed to DES in the neonatal period. Finally, female mice in utero exposed to DES exhibit elevated testosterone levels produced by the ovary, a hormonal disorder not observed in humans [59]. Although the aforementioned disorders have not been reported in women, women prenatally exposed to DES commonly exhibit paraovarian cysts which contribute significantly to the infertility that characterizes this population [62].

An inhibitory effect on follicle formation and development (both primary follicle progression from primordial follicles and secondary follicle progression) has been reported in the neonatal DES-treated (C57BL/6 J) mice ovary, via an ER-α-mediated mechanism [63]. Furthermore, in the mouse ovary, DES is shown to suppress activin expression and signaling pathways. Given that activin is believed to regulate follicle development, its inhibition may represent another pathway that enables estrogenic chemicals to disrupt normal folliculogenesis [61]. Indeed, exposure to DES was associated with a decreased number of small antral follicles in experimental animals [61].

Interestingly, prepubertal exposure to this pharmaceutical xenoestrogen was associated with anovulation at the age of 4 weeks in exposed female rats; however, this effect was not permanent [45]. Moreover, according to an in vitro study, administration of DES to isolated immature rat ovarian follicles resulted to a decline in estradiol and testosterone levels, implying an inhibitory effect of DES on ovarian steroidogenesis (Fig. 6.2) [54].

In summary, developmental exposure to DES in laboratory animals has been associated with oocyte morphological abnormalities (polyovular follicles, oocytes with irregular shape, oocytes with condensed chromatin along nuclear membrane, ovarian cysts), ovarian tumors, anovulation, inhibitory/stimulatory effect on follicle development as well as a promoting effect on somatic cells, and hypertrophy of interstitial tissue. Additionally, DES neonatal exposure has been shown to inhibit activin expression and signaling.

Antiandrogenic Endocrine Disruptors

Given that several endocrine disruptors (EDs) have been shown to possess antiandrogenic traits, these chemicals could impair the normal androgen-orchestrated events that control folliculogenesis and ovulation. As previously described, androgens support ovarian development and function; thus, chemicals that target the ovarian AR could adversely affect reproduction. Therefore, the role of antiandrogenic endocrine disruptors such as DDE and vinclozolin on the female gonad will be discussed.

1,1-Dichloro-2,2-bis(p-chlorophenyl)ethylene (DDE)

DDE (1,1-dichloro-2,2-bis(p-chlorophenyl)ethylene), the primary and most stabile metabolite of DDT (2,2-bis(4-Chlorophenyl)-1,1,1-trichloroethane), is a known persistent organic pollutant with antiandrogenic properties, as it inhibits androgen action by binding to the androgen receptor [64] However, a number of studies have found that DDE also binds to estrogen receptor, though this effect is relatively weak [65]. Although the agricultural and industrial use of DDT has been banned in developed countries, it remains abundant in the environment and detectable in many human tissues due to its lipophilic and bioaccumulative characteristics [66] and also because of its capacity to easily cross the placental barrier [67]. Indeed, DDE continues to be detected in human serum and in follicular fluid of women [68], indicating that this toxicant has direct access to the ovary.

With regard to female reproductive function, DDE has been shown to negatively affect ovarian function in healthy reproductive-aged women, as it was associated with shorter luteal phase length and decreased luteal phase progesterone levels [69]. Shorter cycle length has also been reported in women with high levels of serum DDE [70]. The mechanism by which DDE may affect ovarian function is not known; however, several studies have shown that DDE exposure alters ovarian steroidogenesis.

In prepubertal porcine granulosa and theca cells, the p,p′-DDE isomer was shown to affect sex steroid synthesis by stimulating estradiol secretion and inhibiting progesterone production, confirming its action as a disruptor of steroidogenesis [71]. Conversion of testosterone to estradiol was significantly increased presumably through a direct stimulatory effect of DDE on aromatase activity, the cytochrome P450 enzyme regulating the conversion of androgens to estrogens [71]. Finally, cultures of prepubertal porcine granulosa and theca cells with o,p′-DDE resulted in a decrease of ER-β expression showing DDE's potency to modulate ER-β expression on ovarian follicular cells [71].

A similar inhibitory affect on basal and FSH-stimulated progesterone synthesis was observed in primary cultures of prepubertal porcine granulosa cells, though the mechanism appeared to be cAMP-mediated [72]. Apart from the impact of DDE on ovarian steroidogenesis, studies report an effect on other aspects of ovarian physiology. In immature female rats, DDE was shown to increase the expression of the ovarian growth factors, namely, vascular endothelial growth factor (VEGF) and insulin growth factor (IGF)-1. As these factors are necessary for folliculogenesis and ovarian function, their dysregulation could lead to adverse reproductive disorders [73].

To sum up, with regard to DDE's effects on female ovary, experimental studies indicate that this chemical alters ovarian steroidogenesis (Fig. 6.2) as well as folliculogenesis directly or indirectly through modulation of intraovarian regulators. Moreover, DDE is shown to modulate expression of ER-β in ovarian follicles.

Vinclozolin

Vinclozolin, a dicarboximide fungicide extensively used in fruit and vegetables, has known antiandrogenic properties acting as an androgen receptor (AR) antagonist through its two main metabolites, M1 and M2 [74]. With regard to its effects on female reproduction, vinclozolin has been shown to provoke changes in the ovary of adult mice exposed to the antiandrogen from embryonic life to lactation. In a two-generation reproductive toxicity study, dietary vinclozolin was shown to cause histopathological alterations in the ovaries of exposed rodents, in particular hyperplasia of ovarian interstitial cells and vacuolation of lutein cells [75]. No alterations were reported in sex steroid levels, estrous cycle, and reproductive capacity among treated animals [75].

Furthermore, gestational exposure to this fungicide is associated with morphogenic abnormalities of the embryos manifested as virilization of female rodents and feminization of males [76]. At the molecular level, vinclozolin has an effect on the expression of estrogen and progesterone receptors in genital tubercles from exposed fetuses; however, the ovary was not examined in this experiment [76].

In 7-week-old female rats, exposure to vinclozolin for 28 days affected sex hormone status, inducing changes in the estrogen to testosterone ratio and luteinizing hormone (LH) levels [77].

Therefore, although information on effects of developmental exposure to vinclozolin in laboratory animals is limited, the results suggest that such exposure is associated with functional and histopathological alterations in the developing ovary.

Combined Endocrine Disruptors Effects on the Ovary

One of the most complicated issues related to the effects of EDs on female reproduction is the complexity and diversity of mechanisms of actions that characterize some chemicals. Indeed, several EDs may exert both estrogenic and antiestrogenic activities, depending on the dosage, timing of exposure, receptor subtype, or the tissue target, while others may be naturally metabolized to subproducts with different properties. The balance between estrogenic and androgenic (among other) properties of EDs can be biologically significant, given that reproduction of both sexes involves an interplay of androgens and estrogens.

Another key issue regarding the mechanisms of action and consequences of exposure to EDs is the importance of mixtures of chemicals in which organisms are exposed. The combined effect of different substances that may act synergistically or antagonistically will determine the biological outcome. Humans and animals are more likely to be exposed to a mixture of EDs rather than a single ED, and this interference should be considered when investigating the impact on the developing ovary. Therefore, the role of EDs and their metabolites with diverse mechanisms of action as well as the combined effect of different classes of chemicals on the developing female gonad will be subsequently examined, using methoxychlor as an example.

Methoxychlor [1,1,1-trichloro-2,2-bis(4-methoxyphenyl)ethane; MXC] is an organochlorine pesticide used as a replacement for DDT. MXC acts as a weak estrogen agonist at the level of the uterus and oviduct but as an antiestrogen in the ovary [78]. Additionally, its major metabolite 2,2-bis(p-hydroxyphenyl)- 1,1,1-trichloroethane (HPTE) has multiple modes of action by displaying estrogenic, antiestrogenic, or antiandrogenic properties, depending on the receptor subtype [79, 80]. Since methoxychlor is rapidly converted by the liver to its active metabolite, HPTE, its biological effects on the developing organism may be partly determined by the action of this metabolite.

Exposure to MXC from gestation to early postnatal life has been associated with molecular and morphological alterations in the ovary, impaired follicular development, and disrupted ovarian function possibly leading to multiple reproductive abnormalities in adult female rats [81]. In particular, high-dose MXC exposure was linked to smaller-sized ovaries containing more antral follicles and very few corpora lutea compared to controls. Moreover, at 4 months of age, animals exposed to high-dose MXC had irregular estrous cyclicity, and prepubertal females exhibited reduced superovulatory response to exogenous gonadotropins. At a molecular level, high-dose MXC resulted in a reduction of ER-β, while anti-Mullerian hormone (AMH) was upregulated by both low- and high-dose MXC in preantral and early antral follicles [81].

Early postnatal exposure to MXC has been shown to provoke alterations in ovarian morphology and size, follicle number, and AMH production in the prepubertal female rat gonad [82]. More specifically, the exposed animals exhibited a statistically significant reduction in ovarian weight and size compared to control group.

Moreover, MXC was reported to have a direct inhibitory effect on folliculogenesis during the later preantral stages with a dose-dependent reduction in the number of antral follicles and a concomitant increase in preantral follicles. In addition, MXC as well as its major metabolite, HPTE, had a stimulatory effect on AMH production in rat granulosa cells in vitro [82]. Given that AMH inhibits initial primordial follicle growth in mice as well as in humans [83] and may also modify preantral and small antral follicle growth [84], these findings indicate a potential direct inhibitory effect of MXC on follicle development in the prepubertal ovary via local paracrine actions of AMH [82].

Apart from the stimulatory impact of HPTE in AMH production of rat ovarian granulosa cells, the predominant MXC metabolite is reported to possess a direct inhibitory effect on ovarian steroidogenesis (Fig. 6.2). When granulosa cells from prepubertal rats were treated with HPTE, FSH-dependent steroidogenesis was inhibited, as indicated by the suppression of FSH-stimulated progesterone and 17β-estradiol production compared to control cultures [85]. Moreover, in the presence of HPTE, FSH-stimulated mRNA levels of steroid biosynthetic enzymes (P450 side-chain cleavage enzyme, 3-beta-hydroxysteroid dehydrogenase, and P450 aromatase) were impaired, while mRNA levels of sterol acute regulatory protein (StAR) were not significantly altered by HPTE. Interestingly, when investigators examined the impact of HPTE on cAMP-induced steroid production, a decrease in progesterone levels was observed, while estrogen production was not affected [85]. In evaluating the level of steroidogenic enzyme, a diversity of effects were noted, suggesting a complex mechanism of action. Zachow and Uzumcu demonstrated that HPTE clearly inhibits FSH-stimulated sex hormone production in a culture of primary granulosa cells [85].

In adult female rodents, MXC is shown to inhibit growth and induce atresia of antral follicles through an oxidative stress pathway [86]. To the best of our knowledge, in an earlier period of life, such effects have not been reported.

In conclusion, developmental exposure to methoxychlor and its major metabolite HPTE has been associated with alterations in ovarian morphology and size, folliculogenesis, as well as steroidogenesis. In addition, they were shown to affect AMH production, ovulation, as well as ER-β expression.

Dysregulation of Ovarian Function at the Peri- and Postpubertal Period in Humans and Experimental Animals

Human puberty is a complex neuroendocrine process leading to the maturation of secondary sex characteristics, accelerated linear growth, and attainment of reproductive capacity. At the onset of puberty, the stimulation of pituitary gonadotropins by the increased levels of hypothalamic gonadotropin-releasing hormone (GnRH) activates the female gonad. Mammalians undergo their first estrous cycle (menarche), ovulation is initiated, and the ovaries produce and release steroids hormones. This progressive transition from amenorrhea to ovulatory cycles represents a crucial stage in the maturation process of the reproductive system and is regulated by complex hormonal mechanisms.

As previously reported, developmental exposure to diverse classes of endocrine-disrupting chemicals is associated with adverse effects on ovarian development and function. Given that puberty is a sensitive developmental stage, exposure to environmental toxicants may affect ovary during this phase. According to experimental data, immediately after the onset of puberty, mice prenatally exposed to the estrogenic chemicals DES, BPA, and genistein exhibit transient anovulation [47]. A similar lack of corpora lutea was observed in female mice exposed to same substances in prepubertal life [45]. It should be noted that anovulatory cycles are usual in first postpubertal years; however, in both studies, the percentage of animals with lack of corpora lutea was higher compared to the control group, implying the adverse effect of endocrine disruptors' administration. However, the impact on ovulation was not permanent given that at the end of the experimental period (16 and 24 weeks of age) corpura lutea was present at the exposed animals. Nevertheless, the exposed population demonstrated abnormal estrous cycle, with an increase in the percentage of time spent in estrus phase [45, 47].

In a culture of ovarian granulosa cells from prepubertal rats, investigators demonstrated that FSH-stimulated steroidogenesis was inhibited in the presence of HPTE, the major metabolite of methoxychlor [85]. A similar inhibitory effect on FSH-stimulated progesterone synthesis was observed in primary cultures of prepubertal porcine granulosa cells with DDE [72]. Both experimental findings converge to an aggravating impact of endocrine disruptors on FSH-stimulated steroid hormones' synthesis. The activation of ovarian steroidogenic activity by pituitary gonadotropins is a central endocrine component of puberty; therefore, its dysregulation after exposure to these exogenous agents could lead to impaired ovarian function at puberty.

In prepubertal porcine granulosa and theca cells, the p,p'-DDE isomer was shown to disrupt ovarian steroid production by stimulating estradiol secretion and inhibiting progesterone production [71]. Additionally, peripubertal rodents' exposure to the organotin compound triphenyltin coincides with early postpubertal alterations in steroid hormone concentration, decreased ovarian aromatase activity, and increased ovarian size, [87] revealing the susceptibility of the pubertal ovary to this chemical. Common denominator in both studies is the observation that pre/pubertal exposure to environmental contaminants impairs normal steroidogenesis in the pre/pubertal follicles in vivo and in vitro.

From the above experimental studies, it seems plausible that exposure to several classes of environmental chemicals in the peripubertal period or in an earlier developmental stage may exert a negative effect on normal ovarian function at puberty or at early postpubertal years.

Concerning human puberty, data is limited. Since the nineteenth century, there have been significant modifications in puberty timing with earlier age of thelarche and menarche in girls [88]. This puberty timing alteration has been associated, apart from apparent improvements in general health and nutrition, with a potential impact of endocrine-disrupting chemicals, particularly the estrogen mimics and antiandrogens [89]. There is no direct evidence linking impaired ovarian function at puberty with exposure to endocrine disruptors. However, low birth weight appears to be associated with premature pubarche, ovulatory disorders (low ovulatory rates), and ovarian hyperandrogenism in girls, [90, 91] implying a link between disorders of ovarian function manifested at puberty with alterations in intrauterine conditions. It is known that environmental factors have a significant impact on the developing organism; therefore, it is possible that endocrine-disrupting chemicals may also represent another factor that impairs ovarian functionality during puberty manifested as delayed ovulatory cycles and/or enhanced androgen production.

Conclusions

Studies in experimental animals and in vitro models demonstrate several defects of ovarian development and function in response to ED exposure. Developmental exposure of laboratory animals to several classes of chemicals has been associated with alterations at multiple levels of ovarian development and function (Fig. 6.5).

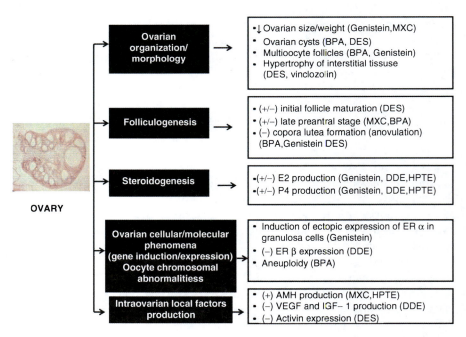

Fig. 6.5 Summary of the effects of developmental exposure to select endocrine disruptors in ovarian physiology. The stimulatory (+) and inhibitory (−) effects of the environmental endocrine disruptors reviewed are shown. Abbreviations: *BPA* bisphenol A, *DES* diethylstilbestrol, *DDE* 1,1-dichloro-2,2-bis(p-chlorophenyl)ethylene, *DCB*, *MXC* methoxychlor, *HPTE* 2,2-bis(p-hydroxyphenyl)- 1,1,1-trichloroethane, *ER β* estrogen receptor β, *ER α* estrogen receptor α, *AMH* anti-Mullerian hormone, *VEGF* vascular endothelial growth factor, *IGF-1* insulin growth factor 1, *E2* estradiol, *P4* progesterone. *Symbols*: ↓: decreased

As analyzed earlier in the text, several substances have been demonstrated to adversely affect ovarian differentiation/development, folliculogenesis, and steroidogenesis. Moreover, the local production of intraovarian factors that regulate ovarian follicle development is shown to be modulated by select environmental chemicals, while at a molecular level, some studies indicate that ovarian estrogen receptor type α and β expression is also modulated.

The timing of exposure appears to be a key factor as developmental exposure during critical windows of susceptibility is shown to disrupt ovarian development and function. This interaction does not only affect the developing embryo but also may alter adult ovarian function by targeting steroidogenesis. Moreover, in many of the aforementioned studies, the observed disorders are caused after exposure to low doses of a chemical or a combination of different classes of chemicals. This is particularly important given that human disorders are more likely the additive result of chronic exposure to low amounts of mixtures of EDC. Furthermore, the consequences of developmental exposure to an ED often remain insidious for several years and become manifest at a subsequent developmental stage. Therefore, early life exposures may lead to disorders manifested in adulthood.

However, caution should be exerted whether the data derived from experimental animal studies and in vitro models could be extrapolated to human population. Since the cellular biology of estrogens is conserved across vertebrate species, it is biologically plausible that endocrine-disrupting chemicals that affect laboratory animals could also have a similar impact on the human ovary. These effects may only be apparent in adult life with no clinically evident disorder at earlier stages of life. For instance, it is possible that some adult female reproductive disorders such as infertility/subfertility, premature ovarian failure, polycystic ovary syndrome, or decreased ovarian reserve may be precipitated by exposure to chemicals that are shown to alter ovarian physiology in laboratory animals. This environmental interaction could occur during the stages of major organogenesis and folliculogenesis, but the biological outcome could only be clinically manifested at adulthood. This concept awaits confirmation by studies in humans.

References

1. Pocar P, Brevini TAL, Fischer B, et al. The impact of endocrine disruptors on oocyte competence. Reproduction. 2003;125:313–25.
2. Woodruff TJ, Carlson A, Schwartz JM, et al. Proceedings of the summit on environmental challenges to reproductive health and fertility: executive summary. Fertil Steril. 2008;89: 281–300.
3. Barker DJ, Eriksson JG, Forsén T, et al. Fetal origins of adult disease: strength of effects and biological basis. Int J Epidemiol. 2002;31:1235–9.
4. Gluckman PD, Hanson MA, Pinal C. The developmental origins of adult disease. Matern Child Nutr. 2005;1(3):130–41.
5. Schrager S, Potter BE. Diethylstilbestrol exposure. Am Fam Physician. 2004;69:2395–400.
6. Titus-Ernstoff L, Troisi R, Hatch EE, et al. Menstrual and reproductive characteristics of women whose mothers were exposed in utero to diethylstilbestrol (DES). Int J Epidemiol. 2006;35(4):868–70.

7. Edson MA, Nagaraja AK, Matzuk MM. The mammalian ovary from genesis to revelation. Endocr Rev. 2009;30:624–712.
8. Hartshorne GM, Lyrakou S, Hamoda H, et al. Oogenesis and cell death in human prenatal ovaries: what are the criteria for oocyte selection? Mol Hum Reprod. 2009;15(12):805–19.
9. Kurilo LF. Oogenesis in antenatal development in man. Hum Genet. 1981;57:86–92.
10. Sanderson JT. The steroid hormone biosynthesis pathway as a target for endocrine-disrupting chemicals. Toxicol Sci. 2006;94(1):3–21.
11. Uzumcu M, Zachow R. Developmental exposure to environmental endocrine disruptors: consequences within the ovary and on female reproductive function. Reprod Toxicol. 2007;23: 337–52.
12. Dupont S, Krust A, Gansmuller A, et al. Effect of single and compound knockouts of estrogen receptors α (ERα) and β (ERβ) on mouse reproductive phenotypes. Development. 2000;127:4277–91.
13. Britt KL, Saunders PK, McPherson SJ, et al. Estrogen actions on follicle formation and early follicle development. Biol Reprod. 2004;71:1712–23.
14. Zachos NC, Billiar RB, Albrecht ED, et al. Developmental regulation of baboon fetal ovarian maturation by estrogen. Biol Reprod. 2002;67:1148–56.
15. Chen Y, Jefferson WN, Newbold RR, et al. Estradiol, progesterone, and genistein inhibit oocyte nest breakdown and primordial follicle assembly in the neonatal mouse ovary in vitro and in vivo. Endocrinology. 2007;148:3580–90.
16. Kezele P, Skinner MK. Regulation of ovarian primordial follicle assembly and development by estrogen and progesterone: endocrine model of follicle assembly. Endocrinology. 2003;144: 3329–37.
17. Britt KL, Findlay JK. Estrogen actions in the ovary revisited. J Endocrinol. 2002;175: 269–76.
18. Walters KA, Allan CM, Handelsman DJ. Androgen actions and the ovary. Biol Reprod. 2008;78:380–9.
19. Shiina H, Matsumoto T, Sato T, et al. Premature ovarian failure in androgen receptor-deficient mice. Proc Natl Acad Sci USA. 2006;103(1):224–9.
20. Nilsson EE, Skinner MK. Progesterone regulation of primordial follicle assembly in bovine fetal ovaries. Mol Cell Endocrinol. 2009;313:9–16.
21. Knight PG, Glister C. Potential local regulatory functions of inhibins, activins and follistatin in the ovary. Reproduction. 2001;121:503–12.
22. Knight PG, Glister C. Focus on TGF-b Signalling.TGF-β superfamily members and ovarian follicle development. Reproduction. 2006;132:191–206.
23. Bradenberger A, Tee M, Lee J, et al. Tissue distribution of estrogen receptors alpha (ER-alpha) and beta (ER-beta) mRNA in the midgestational human fetus. J Clin Endocrinol Metab. 1997;82:3509–12.
24. Enmark E, Pelto-Huikko M, Grandien K, et al. Human estrogen receptor beta-gene structure, chromosomal localization, and expression pattern. J Clin Endocrinol Metab. 1997;82: 4258–65.
25. Biles JE, McNeal TP, Begley TH, et al. Determination of bisphenol A in reusable polycarbonate food-contact plastics and migration to food-simulating liquids. J Agric Food Chem. 1997;45(9):3541–4.
26. Feldman D. Editorial: estrogens from plastic – are we being exposed? Endocrinology. 1997;138(5):1777–9.
27. Noda M, Komatsu H, Sano H. HPLC analysis of dental resin composites component. J Biomed Mater Res. 1999;47:374–8.
28. Vandenberg LN, Hauser R, Marcus M, et al. Human exposure to bisphenol A (BPA). Reprod Toxicol. 2007;24(2):139–77.
29. Ikezuki Y, Tsutsumi O, Takai Y, et al. Determination of bisphenol A concentrations in human biological fluids reveals significant early prenatal exposure. Hum Reprod. 2002;17(11): 2839–41.

6 Developmental Exposure to Endocrine Disruptors and Ovarian Function

30. Fernandez MF, Arrebola JP, Taoufiki J, et al. Bisphenol-A and chlorinated derivatives in adipose tissue of women. Reprod Toxicol. 2007;24:259–64.

31. Kuiper GGJM, Carlsson B, Grandien K, et al. Comparison of ligand binding specificity and transcript tissue distribution of estrogen receptors α and β. Endocrinology. 1997;138:863–70.

32. Kuiper GG, Lemmen JG, Carlsson B, et al. Interaction of estrogenic chemicals and phytoestrogens with estrogen receptor beta. Endocrinology. 1998;139(10):4252–63.

33. Takahashi O, Oishi S. Disposition of orally administered 2,2-bis(4-hydroxyphenyl) propane (bisphenol A) in pregnant rats and the placental transfer to fetuses. Environ Health Perspect. 2000;108:931–5.

34. Vandenberg LN, Maffini MV, Sonnenschein C, et al. Bisphenol-A and the great divide: a review of controversies in the field of endocrine disruption. Endocr Rev. 2009;30(1):75–95.

35. Markey CM, Coombs MA, Sonnenschein C, et al. Mammalian development in a changing environment: exposure to endocrine disruptors reveals the developmental plasticity of steroid-hormone target organs. Evol Dev. 2003;5(1):67–75.

36. Newbold RR, Jefferson WN, Padilla-Banks E. Prenatal exposure to bisphenol A at environmentally relevant doses adversely affects the murine female reproductive tract later in life. Environ Health Perspect. 2009;117:879–85.

37. Stoker C, Beldomenico PM, Bosquiazzo VL, et al. Developmental exposure to endocrine disruptor chemicals alters follicular dynamics and steroid levels in *Caiman latirostris*. Gen Comp Endocrinol. 2008;156:603–12.

38. Iguchi T, Fukazawa Y, Uesugi Y, et al. Polyovular follicles in mouse ovaries exposed neonatally to diethylstilbestrol in vivo and in vitro. Biol Reprod. 1990;43:478–84.

39. Susiarjo M, Hassold TJ, Freeman E, et al. Bisphenol A exposure in utero disrupts early oogenesis in the mouse. PLoS Genet. 2007;3(1):63–70.

40. Hassold T, Hunt P. To err (meiotically) is human: to assess the effect of damaged polycarbonate caging materials, the genesis of human aneuploidy. Nat Rev Genet. 2001;2:280–91.

41. Sugiura-Ogasawara M, Ozaki Y, Sonta S, et al. Exposure to bisphenol A is associated with recurrent miscarriage. Hum Reprod. 2005;20:2325–9.

42. Newbold RR, Jefferson WN, Padilla-Banks E. Long-term adverse effects of neonatal exposure to bisphenol A on the murine female reproductive tract. Reprod Toxicol. 2007;24(2):253–8.

43. Adewale HB, Jefferson WN, Newbold RR, et al. Neonatal bisphenol-a exposure alters rat reproductive development and ovarian morphology without impairing activation of gonadotropin-releasing hormone neurons. Biol Reprod. 2009;81(4):690–9.

44. Hunt PA, Koehler KE, Susiarjo M, et al. Bisphenol A exposure causes meiotic aneuploidy in the female mouse. Curr Biol. 2003;13:546–53.

45. Nikaido Y, Danbara N, Tsujita-Kyutoku M. Effects of prepubertal exposure to xenoestrogen on development of estrogen target organs in female CD-1 mice. In Vivo. 2005;19(3):487–94.

46. Setchell KD, Zimmer-Nechemias L, Cai J, et al. Isoflavone content of infant formulas and the metabolic fate of these phytoestrogens in early life. Am J Clin Nutr. 1998;68:1453–61.

47. Nikaido Y, Yoshizawa K, Danbara N, et al. Effects of maternal xenoestrogen exposure on development of the reproductive tract and mammary gland in female CD-1 mouse offspring. Reprod Toxicol. 2004;18(6):803–11.

48. Awoniyi CA, Roberts D, Veeramachaneni DN, et al. Reproductive sequelae in female rats after in utero and neonatal exposure to the phytoestrogen genistein. Fertil Steril. 1998;70:440–7.

49. Jefferson WN, Couse JF, Padilla-Banks E, et al. Neonatal exposure to genistein induces estrogen receptor (ER)α expression and multioocyte follicles in the maturing mouse ovary: evidence for ERβ-mediated and nonestrogenic actions. Biol Reprod. 2002;67(4):1285–96.

50. Jefferson WN, Padilla-Banks E, Newbold RR. Disruption of the developing female reproductive system by phytoestrogens: genistein as an example. Mol Nutr Food Res. 2007;51:832–44.

51. Jefferson WN, Padilla-Banks E, Newbold RR. Disruption of the female reproductive system by the phytoestrogen genistein. Reprod Toxicol. 2007;23:308–16.

52. Jefferson W, Newbold R, Padilla-Banks E, et al. Neonatal genistein treatment alters ovarian differentiation in the mouse: inhibition of oocyte nest breakdown and increased oocyte survival. Biol Reprod. 2006;74(1):161–8.

53. Pepling ME, Spradling AC. Mouse ovarian germ cell cysts undergo programmed breakdown to form primordial follicles. Dev Biol. 2001;234:339–51.
54. Myllymaki S, Haavistoa T, Vainioa M, et al. In vitro effects of diethylstilbestrol, genistein, 4-tert-butylphenol, and 4-tert-octylphenol on steroidogenic activity of isolated immature rat ovarian follicles. Toxicol Appl Pharmacol. 2005;204:69–80.
55. Herbst AL, Ulfelder H, Poskanzer DC. Adenocarcinoma of the vagina. N Engl J Med. 1971;284:878–81.
56. Yamamoto M, Shirai M, Sugita K, et al. Effects of maternal exposure to diethylstilbestrol on the development of the reproductive system and thyroid function in male and female rat offspring. J Toxicol Sci. 2003;28(5):385–94.
57. Ikeda Y, Tanaka H, Esaki M. Effects of gestational diethylstilbestrol treatment on male and female gonads during early embryonic development. Endocrinology. 2008;149(8):3970–9.
58. Maranghi F, Tassinari R, Moracci G, et al. Effects of a low oral dose of diethylstilbestrol (DES) on reproductive tract development in F1 female CD-1 mice. Reprod Toxicol. 2008;26(2): 146–50.
59. Newbold R. Cellular and molecular effects of developmental exposure to diethylstilbestrol: implications for other environmental estrogens. Environ Health Perspect. 1995;103(7):83–7.
60. Tenenbaum A, Forsberg JG. Structural and functional changes in ovaries from adult mice treated with diethylstilboestrol in the neonatal period. J Reprod Fertil. 1985;73:465–77.
61. Kipp JL, Kilen SM, Bristol-Gould S, et al. Neonatal exposure to estrogens suppresses activin expression and signaling in the mouse ovary. Endocrinology. 2007;148:1968–76.
62. Haney AF, Newbold RR, Fetter BF, et al. Paraovarian cysts associated with prenatal diethylstilbestrol exposure. Comparison of the human with a mouse model. Am J Pathol. 1986;124:405–11.
63. Kim H, Hayashia S, Chambonb P, et al. Effects of diethylstilbestrol on ovarian follicle development in neonatal mice. Reprod Toxicol. 2009;27:55–62.
64. Kelce WR, Stone CR, Laws SC, et al. Persistent DDT metabolite p, p'-DDE is a potent androgen receptor antagonist. Nature. 1995;375(6532):581–5.
65. Matthews J, Celius T, Halgren R, et al. Differential estrogen receptor binding of estrogenic substances: a species comparison. J Steroid Biochem Mol Biol. 2000;74:223–34.
66. Jaga K, Dharmani C. Global surveillance of DDT and DDE levels in human tissues. Int J Occup Med Environ Health. 2003;16(1):7–20.
67. Dorea JG, Cruz-Granja AC, Lacayo-Romero ML, et al. Perinatal metabolism of dichlorodiphenyldichloroethylene in Nicaraguan mothers. Environ Res. 2001;86:229–37.
68. Jarrell JF, Villeneuve D, Franklin C, et al. Contamination of human ovarian follicular fluid and serum by chlorinated organic compounds in three Canadian cities. Can Med Assoc J. 1993; 148(8):1321–7.
69. Windham GC, Lee D, Mitchell P, et al. Exposure to organochlorine compounds and effects on ovarian function. Epidemiology. 2005;16:182–90.
70. Ouyang F, Perry MJ, Xu X, et al. Serum DDT, age at menarche, and abnormal menstrual cycle length. Occup Environ Med. 2005;62:878–84.
71. Wójtowicz AK, Kajta M, Gregoraszczuk EŁ. DDT- and DDE-induced disruption of ovarian steroidogenesis in prepubertal porcine ovarian follicles: a possible interaction with the main steroidogenic enzymes and estrogen receptor beta. J Physiol Pharmacol. 2007;58(4):873–85.
72. Chedrese PJ, Feyles F. The diverse mechanism of action of dichlorodiphenyldichloroethylene (DDE) and methoxychlor in ovarian cells in vitro. Reprod Toxicol. 2001;15:693–8.
73. Holloway AC, Petrik JJ, Younglai EV. Influence of dichlorodiphenylchloroethylene on vascular endothelial growth factor and insulin-like growth factor in human and rat ovarian cells. Reprod Toxicol. 2007;24(3–4):359–64.
74. Kelce WR, Monosson E, Gamcsik MP, et al. Environmental hormone disruptors: evidence that vinclozolin developmental toxicity is mediated by antiandrogenic metabolites. Toxicol Appl Pharmacol. 1994;126:276–85.
75. Matsuura I, Saitoh T, Ashina M, et al. Evaluation of a two-generation reproduction toxicity study adding endpoints to detect endocrine disrupting activity using vinclozolin. J Toxicol Sci. 2005;30(Spec No):163–88.

6 Developmental Exposure to Endocrine Disruptors and Ovarian Function 199

76. Buckley J, Willingham E, Agras K, et al. Embryonic exposure to the fungicide vinclozolin causes virilization of females and alteration of progesterone receptor expression in vivo: an experimental study in mice. Environ Health. 2006;5:4.
77. Shin JH, Moon HJ, Kim TS, et al. Repeated 28-day oral toxicity study of vinclozolin in rats based on the draft protocol for the "Enhanced OECD Test Guideline No. 407" to detect endocrine effects. Arch Toxicol. 2006;80(9):547–54.
78. Hall DL, Payne LA, Putnam JM, et al. Effect of methoxychlor on implantation and embryo development in the mouse. Reprod Toxicol. 1997;11:703–8.
79. Maness SC, McDonnell DP, Gaido KW. Inhibition of androgen receptor-dependent transcriptional activity by DDT isomers and methoxychlor in HepG2 human hepatoma cells. Toxicol Appl Pharmacol. 1998;151:135–42.
80. Gaido KW, Maness SC, McDonnell DP, et al. Interaction of methoxychlor and related compounds with estrogen receptor alpha and beta, and androgen receptor: structure-activity studies. Mol Pharmacol. 2000;58:852–8.
81. Armenti AE, Zama AM, Passantino L, et al. Developmental methoxychlor exposure affects multiple reproductive parameters and ovarian: folliculogenesis and gene expression in adult rats. Toxicol Appl Pharmacol. 2008;233(2):286–96.
82. Uzumcu M, Kuhn PE, Marano JE, et al. Early postnatal methoxychlor exposure inhibits folliculogenesis and stimulates anti-Mullerian hormone production in the rat ovary. J Endocrinol. 2006;191:549–58.
83. Carlsson IB, Scott JE, Visser JA, et al. Anti-Müllerian hormone inhibits initiation of growth of human primordial ovarian follicles in vitro. Hum Reprod. 2006;21(9):2223–7.
84. Durlinger AL, Visser JA, Themmen AP. Regulation of ovarian function: the role of anti-Müllerian hormone. Reproduction. 2002;124:601–9.
85. Zachow R, Uzumcu M. The methoxychlor metabolite, 2,2-bis-(p-hydroxyphenyl)-1,1,1-trichloroethane, inhibits steroidogenesis in rat ovarian granulosa cells in vitro. Reprod Toxicol. 2006;22(2006):659–65.
86. Gupta RK, Miller KP, Babus JK, et al. Methoxychlor inhibits growth and induces atresia of antral follicles through an oxidative stress pathway. Toxicol Sci. 2006;93(2):382–9.
87. Grote K, Andrade AJ, Grande SW, et al. Effects of peripubertal exposure to triphenyltin on female sexual development of the rat. Toxicology. 2006;222(1–2):17–24.
88. Euling SY, Herman-Giddens ME, Lee PA, et al. Examination of US puberty-timing data from 1940 to 1994 for secular trends: panel findings. Pediatrics. 2008;121(3):172–91.
89. Euling SY, Selevan SG, Pescovitz OH, et al. Role of environmental factors in the timing of puberty. Pediatrics. 2008;121(Suppl 3):167–71.
90. Ibanez L, de Zegher F. Puberty and prenatal growth. Mol Cell Endocrinol. 2006;254–255: 22–5.
91. Ibanez L, Potau N, Francois I, de Zegher F. Precocious pubarche, hyperinsulinism, and ovarian hyperandrogenism in girls: relation to reduced fetal growth. J Clin Endocrinol Metab. 1998;83:3558–62.

Chapter 7
Developmental Exposure to Environmental Endocrine Disruptors and Adverse Effects on Mammary Gland Development

Suzanne E. Fenton, Lydia M. Beck, Aditi R. Borde, and Jennifer L. Rayner

Abstract The breast or mammary gland is an organ highly dependent upon hormones and other endogenous growth catalysts for normal development. Environmental chemical exposures have been associated with altered breast developmental timing in populations of girls, and several chemicals and dietary agents are known to induce delayed or accelerated mammary gland development in rodent models. These alterations in development are more likely to occur if exposure to endocrine disruptors coincides with periods of rapid cellular proliferation. These periods of growth include prenatal, peripubertal, and pregnant/lactational mammary development. This chapter will outline the studies that have shown significant effects

---

Disclaimer: The information in this document has been subjected to review by the National Institute for Environmental Health Sciences and approved for publication. The interpretations and conclusions in this review are those of the authors. Approval does not signify that the contents reflect the views of the Institute, nor does mention of trade names or commercial products constitute endorsement or recommendation for use.

Funding Information: Support for L. Beck and A. Borde was provided through the NIH Summers of Discovery Program, with funding by the National Toxicology Program and the American Recovery and Reinvestment Act.

S.E. Fenton (✉)
National Toxicology Program, National Institute of Environmental
Health Sciences, Research Triangle Park, NC, USA
e-mail: fentonse@niehs.nih.gov

L.M. Beck
Summers of Discovery Internship Program, National Institute of Environmental
Health Sciences, Research Triangle Park, NC, USA

School of Medicine, University of Missouri, Columbia, MO, USA

A.R. Borde
Summers of Discovery Internship Program, National Institute of Environmental
Health Sciences, Research Triangle Park, NC, USA

J.L. Rayner
SRC Inc., Arlington, VA, USA

E. Diamanti-Kandarakis and A.C. Gore (eds.), *Endocrine Disruptors and Puberty*,
Contemporary Endocrinology, DOI 10.1007/978-1-60761-561-3_7,
© Springer Science+Business Media, LLC 2012

of environmental chemicals on mammary gland development in rodent models and discuss the relationship of these data to later life adverse health repercussions.

Keywords Breast • Endocrine disruptor • Lactation • Mammary gland • Terminal end buds • Tumor development

Introduction

Mammary gland or breast development is a critical step in pubertal progression in all mammals and this novelty helps define an entire class of vertebrates [1]. The timing of breast or mammary gland development in human adolescents is characterized in stages with Tanner stage II [2] being an easily detected end point used to define the beginnings of puberty in girls. The median age for US and European girls to initiate breast development has significantly decreased from that reported several decades ago [3], and experts in the puberty field agree that obesity and environmental exposures have played a prominent role in this shift [4]. Environmental chemical exposures have been associated with altered breast developmental timing in populations of girls, and several chemicals and dietary agents are known to induce delayed or accelerated mammary gland development in rodent models for human health [5]. Because of the concomitant increase in the incidence of breast cancer cases in recent decades (now one in eight women are at risk for diagnosis [6]), the mammary gland has been an important target for research on the adverse effects of environmental compounds. In order to study the effects of environmental chemicals and carcinogens on breast tissue in later life, it is important to understand the developmental process of the mammary gland and how morphological changes caused by the fetal or neonatal environment can increase the vulnerability to external threats such as carcinogens. This chapter outlines the studies that have shown significant effects of environmental chemicals on mammary gland development in rodent models and discusses the relationship of those data to later life adverse health repercussions.

Mammary Gland Development

The developmental process of mammary gland formation is fairly well defined in rodents, differing slightly between mice and rats, but less well described in humans [5]. The epithelial bud and ductal outgrowth of the mouse and rat mammary gland begin to form during late gestation, approximately 6–7 days before birth [7]. The little information available on fetal breast epithelial development suggests that the human tissue begins ductal development early in gestation, about embryonic week 12–14 [7, 8]. Both the rodent and human mammary epithelia grow at an isometric rate (at the same rate as the body) until just before puberty, although there is a short burst of growth just before birth [7, 8]. Peripubertal growth of the mammary tissue, controlled by the rapidly changing hormonal milieu, is exponential, and the fat pad rapidly fills

7 Developmental Exposure to Environmental Endocrine Disruptors... 203

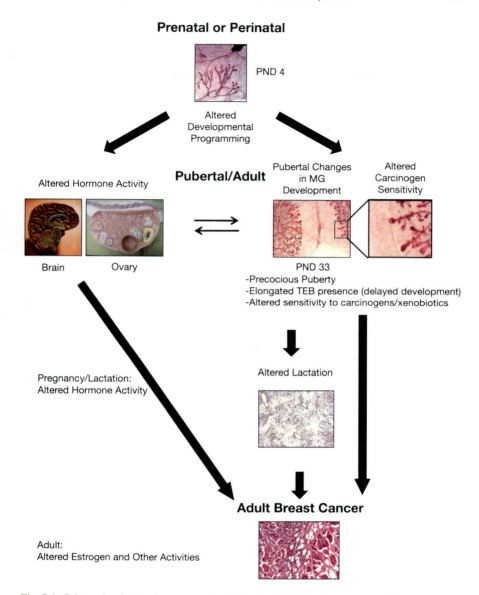

Fig. 7.1 Schematic of altered mammary gland development following exposure to EDCs

with epithelium to manifest the adult form of the gland. The gland will stay in this form, with minor changes depending on the stage of the estrous (rodent) or menstrual (human) cycle, or will undergo dramatic differentiation during pregnancy [7].

There are three phases of mammary gland growth that are suggested as critical because paramount developmental events occur during these times (Fig. 7.1). The prenatal development of the mammary epithelial sprout, when the primary ducts form at the bud site coincident with the nipple, serves as a critical event. Prior to birth, the mammary epithelial bud receives signals from the surrounding fat pad to form primary ducts and begin extension into the fat [7]. Interference in this occurrence could lead to altered timing of mammary development or abnormal formation of glandular structures (i.e., altered number of primary ducts, blind ducts, or unusual presence of nipple/areola), leaving lasting effects on the gland. Another critical interval of mammary gland development is the peripubertal period, when mammary growth is exponential in nature (Fig. 7.1). This span of time, several weeks in rodents or years in girls, features unique highly proliferative structures, terminal end buds (TEBs), present throughout the gland [7–9]. The ends of these teardrop-shaped structures are multiple cell layers thick and are the sites of further ductal branching [7]. These structures eventually disappear from the mature gland as they differentiate into lobulo-alveolar units. Several studies have determined that TEBs are sensitive to chemical carcinogens in rodent models, and in fact, the presence of TEBs at the time of carcinogen exposure is positively associated with tumor multiplicity [9, 10]. Potentially, any compound that prolongs the period of TEB presence in the developing gland or slows differentiation could affect the sensitivity of the gland to chemical carcinogen action. Alternatively, environmental exposures can cause precocious TEB differentiation of the breast which may decrease breast cancer risk.

Finally, the mammary gland undergoes a third critical period of development during pregnancy (Fig. 7.1). During this time, the gland prepares itself for functional lactation. Interruption of this process can lead to mortality or malnutrition of the offspring. This is particularly important in wildlife and domestic species that rely solely on maternal nourishment for survival. Pregnancy has been shown to be protective of breast tissue to later-life disease (cancer) [11].

Endocrine-Disrupting Compounds (EDCs) Alter Mammary Gland Development

Most mammary gland development occurs during the peripubertal period, so this is an essential stage of development to monitor. In female rodents, puberty can be determined by vaginal opening, first estrus, and mammary gland development (differentiation of TEBs). In humans, the pubertal end points used are menarche (begin menstruation) and thelarche (breast bud development). The correlation between mammary gland development in rodents and humans has made the rodent mammary gland an optimal research model of the human breast [5]. Rats and humans produce similar types of tumors in their mammary tissue [10, 12], and the rodent and human undergo mammary gland development at a similar biological pace (albeit the absolute time is quite different) with a few exceptions [13].

Environmental endocrine-disrupting compounds (EDCs) including xenoestrogens and xenobiotics, naturally occurring hormones, organochlorines, industrial chemicals, metals, and lipids have been shown to cause altered puberty timing, delayed mammary development, permanent morphological changes to the mammary gland, and heightened or decreased susceptibility to environmental carcinogens (Table 7.1)

7 Developmental Exposure to Environmental Endocrine Disruptors… 205

Table 7.1 Environmental compounds shown to alter mammary gland development

Compound	Animal model	Precocious MG development	Impaired MG development	Reported putative consequences
Atrazine	Rat		X	Lactation Transgenerational effects Persistent MG growth effects Increased susceptibility to MG tumors
Bisphenol A (BPA)	Rat and mouse	X	X	Development effects limited to early life exposure Neoplasia Increased susceptibility to MG tumors
Cadmium	Rat		X	Estrogen-like effects
Diethylstilbestrol (DES)	Rat and mouse	X		Lactation Spontaneous MG tumors
Dioxin or TCDD	Rat and mouse		X	Prenatal exposure needed for effects Lactation Increased susceptibility to MG tumors
Estradiol	Monkey and mouse	X	X	Delayed MG growth after treatment followed by accelerated growth
Genistein	Rat and mouse	X	X	MG development dependent on time and route of exposure Hyperplasia (male and female) Increased and decreased susceptibility to MG tumors
High-fat diet (n-3 PUFA)	Rat and mouse		X	Effects only when treated during active growth period Decreased susceptibility to MG tumors
High-fat diet (n-6 PUFA)	Rat and mouse	X		Persistent MG growth changes Increased susceptibility to MG tumors

(continued)

Table 7.1 (continued)

Compound	Animal model	Precocious MG development	Impaired MG development	Reported putative consequences
Leptin	Mouse	X		Increased susceptibility to MG tumors
Nonylphenol	Rat	X		Development effects limited to early life exposure
Phthalates (DBP)	Rat		X	Hypoplasia in female MG
				Male nipple retention
Perfluorooctanoic acid (PFOA)	Mouse	X	X	Persistent MG growth inhibition
				Stimulated MG growth in one strain
Polybrominated diphenyl ether (PBDE)	Rat		X	Possible persistent MG growth inhibition
Testosterone	Rat and mouse	X	X	Female virilization
				Enhanced MG development

MG mammary gland, *PUFA* polyunsaturated fatty acids, *DBP* dibutyl phthalate

following both in utero and postpartum exposures. Exposures during pregnancy have caused an interruption in the differentiation of mammary gland development including stunted or delayed growth, decreased branching, and decreased formation of alveoli, all important structural components needed for lactation. Although difficulties with nursing in humans can be remedied with infant formulas, wildlife and domestic species do not have this luxury. Studies have shown that improper lactation as a result of interrupted mammary gland development can cause high morbidity among wildlife offspring [14]. Precocious or delayed mammary development is sometimes associated with timing of vaginal opening and first estrus. The effects of EDCs on these events are addressed in more detail in other chapters of this book.

EDCs Causing Mixed Effects on Mammary Gland Development

Xenoestrogens

This group of EDCs is particularly interesting, as the effects for many of the compounds depend on the timing, dose, and route of exposure. Consistent results have been shown across strains, within a species, and across species, in many cases. In some instances, effects at high doses are not seen at low doses, and low doses may manifest significant effects through unique modes of action. Furthermore, at some time points evaluated, precocious development may be detected, and at others, the effects may be interpreted as delayed development. The idiosyncrasies of these exposure/effect scenarios are detailed here.

Genistein

Genistein, a phytoestrogen present in soy products, is currently a widely studied EDC because some humans may rely solely on soy infant formula for sustenance during their first year of life and because of its varying effects on development in rodents, which are dependent on the time and method of exposure as well as the dose administered. A notable developmental outcome of genistein exposure is the alteration of mammary gland development, in addition to ovarian dysfunction.

In a study using CD-1 mice, changes in TEB differentiation and ductal branching were observed, with results depending on the amount of genistein given between postnatal days (PND) 1 and 5. By postnatal week 6, the highest dose, 50 mg/kg (s.c.), caused a decrease in the number of TEBs and decreased branching, suggesting delayed mammary gland development. However, mice exposed to the lowest dose, 0.5 mg/kg genistein, displayed increased branching and ductal elongation when compared to the controls during the same time period. Additionally, mammary gland morphological differences including reduced lobulo-alveolar development and dilated ducts in mice treated neonatally with 5 and 50 mg/kg appeared to be permanent as they were observed at 9 months of age [15]. Prepubertal exposure to 0.5 mg/kg genistein in Sprague–Dawley rats caused a significant decrease in the number of TEBs at 50 days compared to controls. Additionally, the number of terminal ducts in the genistein-treated rats was considerably greater than controls. These results suggest an enhanced rate of differentiation of TEBs into terminal ducts after genistein treatment, thus signifying precocious development [16].

Numerous studies conducted to assess a possible relationship between genistein exposure and cancer susceptibility resulted in a variety of conclusions ranging from increased susceptibility to protective effects dependent on timing of exposure during the developmental period. In two studies which included at least 5 days of postnatal exposure, TEBs appeared much earlier in treated animals than in control animals (precocious mammary development). However, these TEBs also differentiated into mature structures earlier in development compared to the controls. In the mammary gland, TEBs have been shown to be sensitive to carcinogens, thus a reduction in TEBs present may indicate decreased susceptibility to cancer [17]. The results of both studies showed a correlation with a decreased risk of developing cancer. One study observed a decrease in multiplicity of tumors in rats treated with genistein [18], and the other found a significant increase in the density of lobulo-alveolar structures [19]. The development of these differentiated structures is correlated with a decreased susceptibility to environmental carcinogens [19].

Conversely, some genistein and cancer susceptibility studies have found results consistent with an increased risk of mammary tumorigenesis in rats and mice. In fact, hyperplasia was shown in multiple generations of male rats fed genistein. This finding enforces other reports that mammary gland hyperplasia in the male rat is a sensitive marker of estrogenic EDCs [20]. The studies in females, which utilized gestational exposures, showed precocious development of TEBs with decreased differentiation with age, indicating a longer period of TEB exposure to environmental toxins. Administration of DMBA, a known carcinogen, on PND 50 increased mammary tumor incidence in these studies [21, 22]. These results were associated with an increased risk of mammary gland tumor development.

Bisphenol A

The widespread commercial use of the xenoestrogen bisphenol A (BPA) in sealants and plastics has put it in the forefront of current toxicology studies. Research has shown that BPA causes adverse mammary gland development by altering mammary gland cell proliferation, ductal elongation, and ductal differentiation. The effects are accelerated or delayed depending on the treatment dose. A low-dose BPA treatment (25 µg/kg) to CD-1 mice administered from gestation day (GD) 9 to the end of pregnancy caused the treated offspring to develop greater ductal elongation than the control, while the high-dose treatment (250 µg/kg) caused less ductal elongation. Additionally, this study found significant increases in the number of alveolar and ductal structures of both groups of BPA-treated animals at 6 months of age [23]. Environmentally relevant doses (25–250 ng/kg/day) of BPA administered to CD-1 mouse dams gestationally and lactationally resulted in female offspring having a significantly higher number of TEBs relative to the ductal area and decreased apoptotic TEB cells at PND 30 compared to the control offspring. In low-dose exposed offspring, the number of lateral branches was significantly increased [24]. The histoarchitecture and tissue organization of the fetal mammary gland was shown to be altered after in utero exposure (GD 8–18) to BPA. Researchers observed increased ductal growth with smaller epithelial cells, accelerated fat pad maturation resulting in a less dense fat pad, increased ductal collagen density, and decreased collagen density in the stroma following exposure to 250 ng/kg BPA [25]. Importantly, changes in fat pad density and stromal composition may be involved in determining the risk for tumor development over a lifetime and further work on the role of EDCs in these end points is ongoing.

Additional research has made a case for heightened susceptibility to spontaneous hyperplasia and neoplastic lesions with increasing age. In CD-1 mice exposed to 0–25 µg/kg from GD 8 through PND 16, an increased incidence of beaded ducts, a spontaneous lesion that is thought to be precursory to ductal carcinoma in situ, was observed at significantly increased incidence in the mammary glands of 0.25, 2.5, and 25 µg/kg BPA-exposed offspring at 9 months of age. The beaded appearance of the ducts was due to bridges being formed between the ductal walls by actively proliferating epithelial cells in the ductal lumen (intraductal hyperplasia) or at locations where the ducts branched, merged together, and branched again. The cells in the beaded ducts had proliferation indices much higher than normal ducts [26]. In a study conducted by Durando et al. [27], pregnant Wistar rats were exposed to 25 µg/kg/day from GD 8 to GD 23. After puberty, effects including decreased apoptosis, structures resembling hyperplastic ducts, and increased stromal nuclear density were observed. By PND 110, the number of mast cells surrounding the hyperplastic ducts was increased in BPA-treated rats, and fibroblastic stroma had replaced adipose tissue. A significant increase in hyperplastic lesions was observed at PND 180 in the BPA-treated animals following a subcarcinogenic dose of N-nitroso-N-methylurea (NMU), and an enhanced response was observed by the presence of neoplastic lesions (ductal carcinomas in situ) [27]. Similar results were observed in a study in which pregnant Wistar rats were treated with 2.5–1,000 µg/kg BPA from GD 9 to PND 1. By PND 50, a significant threefold to fourfold increase in ductal hyperplasia was observed in all animals exposed to BPA, and by PND 95, a significant increase in ductal hyperplasia was observed only in animals exposed to 2.5 µg/kg BPA. Ductal hyperplasias are believed to be precursors to carcinomas, and these results suggest that BPA induces not only an elevated susceptibility to carcinogens [28] but also an inherent ability to induce neoplasias.

Xenobiotics

Nonylphenol

Nonylphenol is a xenobiotic used in industry as a surfactant, emulsifier, and detergent. It is often found in, or lining, food containers and wraps, in cleaning compounds,

210 S.E. Fenton et al.

and spermicides. Colerangle and Roy [29] administered 0.01 mg nonylphenol/day via minipump for 11 days to Noble rats which resulted in a 200% increase in mammary proliferative effects as measured by PCNA staining. At a higher dose, 7.1 mg/day, a 400% increase in mammary cell proliferation was observed. They also found that nonylphenol altered mammary epithelial cell cycle kinetics by increasing movement from G0 and G1 phases to the S phase [29]. Because nonylphenol alters cell-cycle kinetics, it can produce DNA mutation and chromosomal abnormality [30]. These changes within the basic structure of the body cause genetic instability, leading to an increased risk for the development of neoplastic lesions and mammary gland tumorigenesis. It is noted that the mammary gland results observed by Colerangle and Roy [29] were not reproducible in a later study [31]; however, differences in laboratory quantification of mammary gland differentiation and cell proliferation may have played a role in the differences between the two studies.

Gestational exposure (GD 15–19, oral gavage) of Long–Evans rat dams to 100 mg nonylphenol/kg resulted in proliferative mammary epithelial branching and budding observable by PND 4, using a developmental scoring system. By PND 33, the nonylphenol-treated rats displayed more alveolar buds and increased TEB differentiation into mature lobules compared to the controls. These differences between nonylphenol treated and control offspring were evident for the duration of the experiment but were not significant at 10 mg/kg [32]. Two of three studies of early life nonylphenol exposure demonstrated the ability to cause precocious mammary epithelial development.

Phthalates

Di (*n*-butyl) phthalate (DBP) is a commercially important chemical used as a plasticizing agent; DBP and related phthalates are used to make plastics soft and to contain and disperse scent (in perfumes and health and beauty aids). Their presence in medical tubing used in neonatal care wards and children's toys has been especially disconcerting. Phthalate environmental contamination is widespread, making studies on its toxicity imperative to current toxicology and epidemiology studies. In the case of DBP, studies have shown effects on the female mammary gland as well as on male mammary tissues. A study testing the effects of 0–10,000 ppm DBP administered to dams during both gestation and lactation (GD 15 to PND 21) reported both male and female mammary gland development changes. At PND 14, nipple retention was observed in males at the highest dose. Adverse mammary gland effects were evident into adulthood; in males, this included the dilation of alveolar buds and ducts, and poor alveolar branching and hypoplasia in females [33].

Nipple retention in males has been observed in other studies of DBP. Nipple and areola retention on PND 14 was observed in 91% of the male rat offspring dosed with 500 mg/kg/day from GD 12 to 21. The nipple buds on these males were similar to those on females of the same age. Nipple retention was also observed in males exposed to a lower dose, 100 mg/kg/day [34]. At 250 mg/kg/day, 56% of male offspring and at 500 mg/kg/day, 87% of pups retained their thoracic nipples on PND

7 Developmental Exposure to Environmental Endocrine Disruptors... 211

14 [34]. Combining and analyzing the results of these studies on the effects of phthalate on male rodents highlights the extremely sensitive end point of nipple retention and disrupted mammary gland development after phthalate exposure.

Sex Hormones

The sex hormones testosterone and estradiol, along with the synthetic estrogen diethylstilbestrol (DES), have been shown to cause altered mammary development in animals. While not environmental EDCs per se, aberrations in the timing or the amount of exposure to endogenous sex hormones, or administration of exogenous hormones, disrupt endocrine systems, including the mammary gland.

Testosterone

Exposure to testosterone has been shown to defeminize the female rodent mammary gland. Female rats exposed in utero to 1 and 10 mg testosterone/kg on GD 13–20 showed no nipple development – similar to that in male control rats of the same age. The results of this study suggested that gestational exposure to critical levels of testosterone could severely inhibit the development of nipples in female offspring, rendering them unable to properly nurse their offspring later in life [35]. Other studies have confirmed the virilization of female rats exposed to testosterone during critical periods of development. Prepubertal treatment with testosterone caused the defeminization of female rats offspring, delayed age of vaginal opening by nearly 6 days, and diminished circulating levels of follicle-stimulating hormone (involved in follicular growth) and luteinizing hormone (ovulation trigger) [36]. The permanent changes in endocrine function of the female rat caused by prenatal testosterone could have adverse effects on reproduction as well as stunted mammary development and decreased lactational function.

Early postnatal treatment with testosterone stimulated ductal branching in the mammary gland of BALB/c mice. Mice treated with 20 µg/day on PND 1–5 displayed normal ductal branching on PND 6; however, a precocious increase in ductal branching was observed by PND 26 and PND 33, resulting in an increased amount of ductal branching compared to the control mice [37].

Estradiol

Postnatal treatment with estradiol causes accelerated mammary gland development in mice. Early postnatal (PND 1–5) estradiol exposure to BALB/c mice resulted in no differences in mammary gland development between control and dosed animals at 3 weeks of age. However, a dose-dependent response was observed in the total duct junctions and total mammary growth area by the time the mice had reached

5 weeks of age. As the dose increased from 25 to 70 µg/d, ductal branching increased, but the ratio between branching and mammary area remained constant. This study was among the first to discover the precocious developmental effects of estradiol treatment on mammary epithelial [38]. Tamooka and Bern [37] found two distinct effects of estradiol administered on PND 1–5 to BALB/c mice: inhibition of growth immediately after treatment and stimulation of growth by 4 weeks of age. Treatment with estradiol resulted in prominent inhibition of ductal branching in the first 6 days of life. Ductal branching remained relatively stagnant for the first 3 weeks of life, but the number of ducts branching nearly tripled by PND 33, surpassing growth in control animals [37]. Treatment with 2–4 µg estradiol on PND 1–3 caused the enhanced development of the mammary parenchyma on PND 25–35. This accelerated development after estradiol exposure also correlated with a greater number of undifferentiated TEBs in the mammary gland throughout the study observed as late as PND 114. While TEBs presented earlier in the estradiol-treated animals than the control animals, density of the epithelial ducts was comparable on PND 35. At PND 50, the TEBs in control animals differentiated into lobulo-alveolar units, while the TEBs in the treated mice remained unchanged, making the animal more susceptible to neoplastic lesions [39].

In rhesus monkeys, estradiol increased the epithelium proliferation of the mammary gland sixfold compared to normal proliferation in control animals. As previously mentioned, increased cell proliferation may play a role in cancer susceptibility [40].

Diethylstilbestrol (DES)

Altered mammary development caused by DES in Alderley Park rats was observed in a study following exposure to 0.055 mg/kg DES via subcutaneous minipumps for 11 days. The number of lobules present was markedly increased over control rats at the same age of development. This increase in lobules was coupled with a decrease in TEBs and terminal ducts [41]. Nipple morphology has also been shown to be altered after gestational treatment of DES. In one study, pregnant Sprague–Dawley rats were injected with varying doses of DES during GD 10 and 13; GD 15, 17, and 19; or GD 15 and 18. In dams dosed on GD 10 and 13, excessive nipple development was observed in female ($\geq$12 µg/kg/day) and male (1,200 µg/kg/day) offspring. In dams dosed on GD 15 and afterward and at doses greater than 120 µg/kg/day, nipples were present in female and male offspring on PND 1, an event usually occurring around PND 5 in females. Visible nipple development was also observed in female offspring exposed to 1.2 and 120 µg/kg/day. However, regardless of the precocious appearance of nipples, the DES-exposed female pups were unable to support their own offspring after birth. This was most likely due to the absence of the nipple sheath. The lack of this structure indicated the failure to form a pathway between the mammary gland and the nipple, rendering the female unable to nurse [42].

Hovey and colleagues [43] investigated mammary gland development from PND 21 to postpuberty in BALB/c mice dosed within 36 h of birth with 0.0125–50 µg DES (s.c.). Mice treated with 12.5 µg DES had significantly increased ductal outgrowth at PND 33 compared to control animals, but mammary gland morphological differences were not observed until 12 weeks of age. At this time, dilated mammary ducts filled with proteinaceous secretion and lined with cells filled with lipid droplets were observed in mice treated with 25 µg. Alveolar development was significantly increased after treatment with 0.0125 µg and significantly reduced in mice at ≥12.5 µg. These morphological changes in DES-treated mice (12.5 µg) were accompanied by precocious lactation [43].

DES has also been shown to increase tumorigenesis among rats exposed gestationally (0, 0.8, or 8 µg on GD 15 and 18), postnatally (0 or 2.5 mg at 12 weeks of age), or both. In addition to morphological changes seen in the exposed rats (elongated nipples and extensive lobulo-alveolar proliferation), decreased tumor latency and greater multiplicity of tumors were observed at low and high doses of DES following gestational or lactational exposure. Furthermore, treating rats both gestationally and postnatally yielded increased tumor multiplicity and decreased tumor latency compared to all other exposures and the control groups [44].

Alteration of mammary development timing, changes in mammary morphology, and increased incidence of tumorigenesis are well-noted adverse effects of DES on rodents. However, DES has an impact on human breast cancer risk as well. Women exposed to DES, including those exposed in utero during its widespread use in the 1950s and 1960s, have an increased risk for breast cancer. Women over the age of 40 exposed to DES in utero had an estimated 1.9 times the risk of developing breast cancer compared to unexposed women of the same age. Additionally, the highest risks could be correlated with the highest cumulative doses to the synthetic estrogen in utero [45].

Metals

Naturally occurring metals have the potential to be classified as EDCs especially if they mimic or perturb the normal hormonal milieu. Cadmium is a good example of a metal that has resulted in altered MG development in mice and rats following exposure.

Cadmium

One study demonstrated that cadmium mimicked the effects of estrogen in rats and disrupted normal mammary development [46]. At a low in utero exposure (0.5 µg/kg), the mammary gland of female offspring had more TEBs than controls at PND 35 and less alveolar buds in the treated offspring than in control offspring at 5 µg/kg [46]. Increased TEBs and decreased differentiation into alveolar buds signals cause for concern because of the heightened susceptibility of TEBs to carcinogenic activity. A study investigating cadmium in MCF-7 cells, human breast cancer cells, showed that cadmium decreased estrogen receptor protein and mRNA, stimulated the estrogen response element, and induced cell growth. This activity suggests that cadmium plays a role in modulating and promoting growth of breast cancer cells [47].

Altered mammary gland development was also observed in a study in which lactating NMRI mice were injected (s.c.) with 0–2,000 µg cadmium/kg body weight from PND 8 through 10. Glands from mice given ≥ 5 µg/kg had a greater number of fat globules and degenerated alveolar epithelial cells. When further analyzed, a negative dose–response relationship was observed between cadmium and β-casein expression. At the highest dose, litter weights were significantly reduced compared to control offspring body weight. These results suggested that alterations of the lactating mammary gland caused by cadmium may perturb its function and impair offspring development [48].

Lipids and Adipose-Derived Hormones

Lipids and adipose-derived hormones, when fed or administered at high levels or during sensitive developmental periods, have been studied for their effects on pubertal development. Maternal diet and health can have a large effect on fetal development, and assessing the effects of dietary changes on developing mammary glands is an important end point.

n-6 Polyunsaturated Fatty Acid (PUFA)

One study monitored the effects of a maternal rodent diet high in n-6 polyunsaturated fatty acid (PUFA). Gestational exposure (GD 0–birth) to high n-6 PUFA levels correlated with a larger mammary fat pad area in the offspring as well as an increased epithelial cell density at 4 weeks of age. An increased density of TEBs was also apparent throughout the experiment, and these effects were observed as late as 11 weeks of age concurrently with reduced differentiation into alveolar buds [39], indicating delayed development and an increased risk of tumorigenesis [49].

n-3 Polyunsaturated Fatty Acid (PUFA)

Maternal diets high in n-6 PUFA have been shown to cause precocious mammary gland development among offspring, yet delayed differentiation. A different type of fat, n-3 polyunsaturated fatty acid (n-3 PUFA) has also been studied to determine if its effects are similar or different to those of a high n-6 PUFA maternal diet.

There are many different types of n-3 PUFA oils, but a study addressing the effect of menhaden (fish) oil on mammary gland development in BALB/c mice found significant differences in development between the treated and control glands following exposure to a high n-3 PUFA diet. Mice fed the fish-oil diets

(15.5–19% Menhaden oil) experienced considerable suppression of mammary gland development. Most importantly, expansive ductal growth was reduced in all immature mice that had been exposed to diets of varying high amounts of n-3 PUFA. These suppressive effects were only found in mice undergoing states of intense proliferation, namely those in early developmental and pubertal stages. Menhaden oil did not have an effect on mammary gland development in mature mice suggesting mouse pubertal mammary gland development is a sensitive window to a high n-3 PUFA diet [50].

Most studies regarding n-3 PUFA have assessed its role in mammary gland tumorigenesis. Upon exposure to a known carcinogen, a diet high in n-3 PUFA was found to decrease the incidence of mammary tumors and increase the tumor latency period compared to rodents with a high n-6 PUFA diet which displayed enhanced tumor development and decreased tumor latency [51]. This may be related to the increased number of TEB in the n-6 PUFA diet offspring and significantly reduced TEB numbers in the n-3 PUFA diet females.

Leptin

The adipose-derived hormone leptin is an essential hormone to the development and survival of animals, including humans and rodents. Although primarily studied for its role in energy balance through its actions on leptin receptors in the hypothalamus, leptin has lesser-known roles in mammary gland development and function. Leptin-deficient mice are not able to support pups after birth because of undeveloped mammary glands [52]. In addition to promoting normal mammary development, leptin has been shown to promote mammary tumor development. It was reported that HBL100 and T-47D breast epithelial cells treated with leptin exhibited greater proliferation than controls [53]. Transgenic mice overexpressing leptin and fed a high-fat diet (and thus developed a high serum concentration of leptin) developed mammary tumors at a younger age than lean mice of the same genetic line [53]. These findings show that leptin plays a key role in mammary gland development and, subsequently, susceptibility to tumorigenesis. Taken together, there is no single role for lipids or adipose-derived hormones, such as leptin in mammary gland development, but they clearly influence the tissue and affect the susceptibility to tumor formation depending on the type of PUFA.

EDCs Causing Delayed Development

The adverse effects of precocious or accelerated development on the functional and morphological structure of the mammary gland have been widely studied. Another result of exposure to chemicals present in the environment is delayed development which can have detrimental effects on the function and morphology of mammary glands and may also predict further problems in later life.

Organochlorines

Atrazine

The pesticide atrazine is a common herbicide used on crops in the USA and has been the target for scientific research on its toxicity because of its ability to permeate waterways, soil, and drinking-water supplies. Different paradigms of atrazine exposure were tested to assess effects on mammary gland development [54]. By using a cross-fostering design, it was possible to observe the effects on Long–Evans rat pups exposed to atrazine in utero (GD 15–19), through suckling (residual atrazine in the milk), or both. All rats exposed to atrazine, in utero, through milk, or both, exhibited delayed mammary gland development (decreased branching, delayed migration into the fat pad, and fewer TEBs) as early as PND 4 and as late as PND 40. This altered development was especially prominent in the offspring that were exposed to atrazine both in utero and through nursing. By PND 40, most TEBs on control animals had differentiated, while the atrazine-exposed animals still retained large numbers of TEBs. These results suggested that atrazine-delayed mammary gland development is mediated transplacentally as well [54].

In another study [55], the effects of restricted gestational exposures as well as nursing capabilities of atrazine-exposed offspring were examined. The most persistent delays in mammary development occurred in rats exposed to atrazine between GD 17 and 19 and were equivalent to effects seen after longer exposures (GD 13–19). Delayed mammary gland development was observed in these animals well after they had reached sexual maturity. Studies to determine lactational ability of those first-generation females exhibiting delayed mammary development revealed that second-generation pups born to mothers who had been exposed in utero to atrazine had significantly lower body weights than those born to control dams. This observation suggested decreased lactational capacity of mammary glands in rats that display significantly delayed mammary gland development [55].

The toxicity of atrazine and its metabolites in an environmentally based mixture was studied in pregnant Long–Evans rats dosed on GD 15–19 with 0.09–8.73 mg/kg. Delayed mammary gland development in metabolite-exposed offspring was observed as early as PND 4 and persisted through PND 60. At PND 4, these adverse effects included delayed migration of epithelium into the mammary fat pad and decreased branching. Lack of differentiation of the TEBs was still apparent at PND 60 in all rats exposed to the mixture, while control animals had fully differentiated TEBs. These results suggested that metabolites of atrazine are biologically active and may play a role in the effects observed [56]. This study also suggests that the metabolite mixture, which is how this class of herbicide is found in nature, is effective at lower doses than atrazine alone.

The effect of atrazine exposure on mammary gland tumorigenesis is an important field of study because of its widespread use in the environment. A lifetime administration of atrazine to Sprague–Dawley rats in the diet caused an earlier onset of mammary tumor formation than the controls, while the incidence of tumors was

7 Developmental Exposure to Environmental Endocrine Disruptors... 217

not affected. These results supported the hypothesis that exposure to atrazine causes acceleration of endocrine-changing effects within the body, resulting in an earlier onset of tumorigenesis [57].

Dioxins

2, 3, 7, 8-Tetrachlorodibenzo-p-dioxin (TCDD)

The endocrine disruptor TCDD disrupts both development and reproduction. From the molecular signaling level to organ function, dioxin exposure can result in permanent adverse changes in bodily function. Mammary glands are not excluded from dioxin's widespread perturbation of hormone systems, metabolism, tissue and organ function, and cellular communication. The carcinogenic potential has also been well demonstrated [59].

One adverse effect of dioxin on the mammary gland is the interruption of proper differentiation of the mammary gland during pregnancy. Interrupting this process can severely decrease the dam's ability to nurse her young. Exposing pregnant C57BL/6 mouse dams to 5 µg/kg TCDD during pregnancy was shown to cause severe defects in development and differentiation of the mammary gland apparent as soon as 9 days after exposure. These defects included stunted growth, decreased amounts of lobulo-alveolar structures, and a smaller network of ductal branching. Poor differentiation in these structures caused impaired lactational ability in the dams, and all pups born to these TCDD-exposed dams died within 24 h of birth [14]. Although the inability to lactate does not raise mortality concerns among humans, wildlife and domestic animal survival is contingent on the ability to nurse their young until weaning.

The effect of TCDD exposure on the mammary development of offspring has also been studied. The female offspring of Long–Evans rats exposed to 1 mg/kg TCDD on GD 15 exhibited severe delays in differentiation of the mammary gland. Stunted development was observed from PND 4 to PND 68. Adverse effects included decreased ductal branching, delayed epithelial migration into the fat pad, and fewer differentiated terminal structures. Exposure on GD 20 or during lactation did not result in altered mammary gland development in the offspring [60]. Later studies by Lewis et al. [61] confirmed and further described the mammary developmental defects induced by a single prenatal TCDD exposure in an additional rat strain.

A study observing mammary gland differentiation and its effects on tumorigenesis later in life showed that Sprague–Dawley rats exposed gestationally to 1 µg/kg TCDD on GD 15 exhibited permanent mammary gland alteration. On PND 21, the number of TEBs and terminal ducts were similar between treated and control offspring. By PND 50, there were significantly more TEBs present in the treated animals and significantly fewer lobulo-alveolar units. To test the susceptibility of these lobulo-alveolar units to chemically induced tumorigenesis, rats from both exposed and control groups were treated with DMBA on PND 50 and observed for tumor formation. Animals exposed to TCDD developed twice as many tumors, had a higher tumor incidence, and developed tumors at a younger age when compared to control animals. These results suggest that permanent changes to mammary glands during gestational TCDD exposure result in a heightened risk for the development of tumors later in life [62].

An epidemiological study conducted in Seveso, Italy, made the correlation between TCDD exposure and breast cancer among women. A manufacturing plant released up to 30 kg of TCDD during an industrial explosion that exposed the surrounding population to abnormally high levels of TCDD. Increased risk of breast cancer was correlated with higher serum TCDD levels in study participants that were young when exposed. The hazard ratio for breast cancer associated with a tenfold increase in serum TCDD levels was significantly increased to 2.1 (95% confidence interval, 1.0–4.6). This observation was strongly supported by the results of animal studies that examined TCDD exposure and mammary tumorigenesis. Because the women in the study who had the greatest TCDD exposure are not yet between the ages of 40 and 55, the age of highest breast cancer risk [63], it will be important to follow these women as they age and continue to observe breast cancer incidence among this population [64]. Studies evaluating early life exposures to dioxins/PCBs have found delayed breast development in adolescents with the highest circulating (Seveso, Italy) [65] or prenatal/lactational dioxin levels (The Netherlands) [66].

Brominated Compounds

Polybrominated Diphenyl Ether (PBDE)

Polybrominated diphenyl ethers (PBDEs) are a chemical widely used for commercial and industrial products, including flame retardants in textiles, construction materials, and polymers used in electronics. Their widespread use and suspected toxicity has resulted in studies testing for health effects following developmental exposure. In one study, Long–Evans rat dams were treated from GD 6 through PND 21 with 0–30.6 mg/kg of a PBDE mixture (DE-71), giving offspring both gestational and lactational exposure to the compound. At PND 21, the mammary glands in the two highest dose groups were significantly delayed in their development. Effects included decreased epithelial growth, limited TEB development, and a decrease in the number of lateral branches [67]. These findings are consistent with other altered reproductive end points in rodents upon exposure to PBDEs [68].

Perfluorinated Compounds

Perfluorooctanoic Acid (PFOA)

Perfluorooctanoic acid (PFOA) is a chemical used in fire-fighting foams, electronics, and making commercial products grease and waterproof, and has been found to be the final degradation product of other > 8-carbon perfluorinated materials. Persistent traces of the compound have been found in humans and wildlife alike, making it a prime target for toxicity studies, especially developmental effects [69]. Various studies on the mammary gland rodent model have indeed found altered development after gestational PFOA exposure.

Research has shown significant delays in gland development after the gestational exposure of mice to PFOA, both in dams and offspring. A dose of 5 mg PFOA/kg/day administered to CD-1 mouse dams during gestation altered the lactational ability of dams. The dams dosed during GD 8–17 and 1–17 displayed severely delayed mammary gland development on PND 10, the peak of lactation. This dose administered from GD 12 to 17, GD 8 to 17, or GD 1 to 17 caused significant retardation of mammary gland development, proliferation, and differentiation in the offspring. Pups also had decreased body weight compared to their vehicle-treated counterparts, but it was not found to be a significant variable in their mammary growth disparity [70].

Other studies utilizing cross-fostering or restricted gestational exposure designs reported that gestational and/or residual lactational (in milk) exposure to PFOA caused detrimental effects to the developing mammary gland. In control mice offspring cross-fostered to dams exposed on GD 1–17, developmental deficits were apparent from the initial sacrifice day of PND 21 and up to 9 weeks of age. These lactationally exposed pups had low serum PFOA levels comparable in magnitude to levels found in humans exposed to PFOA through contaminated drinking water or occupational hazards, making these findings much more human relevant. Similar mammary gland developmental delays were observed in the exposed offspring cross-fostered to control dams as well as those receiving gestational and lactational exposure. The study also found that developmental effects of gestational-only and/or lactational-only exposed mice could be observed as early as 12 h after parturition _ENREF_65 [71, 72]. These immediate effects are proposed to be a result of an interruption of the rapid growth and development of the mammary gland parenchyma that takes place in the first day of postnatal development [7]. Recent studies [73, 74] define even lower levels of PFOA (5 ppb in water; 0.01 mg/kg/d from GD 10 to 17, respectively) as effective in delaying mammary gland development, and the internal doses overlap with those reported in humans living in PFOA-contaminated communities in OH and WV, USA [69].

Young adult PFOA exposure also altered mammary gland development. In BALB/c mice, a dose of 5 or 10 mg PFOA/kg administered for 4 weeks, starting after weaning, caused reduced ductal length, decreased number of terminal end buds, and decreased stimulated terminal ducts in the mammary glands. The number of proliferating cells in the ducts and terminal end buds were significantly lower than those in controls. In C57BL/6 mice, the same doses caused a different response. At 5 mg/kg, a stimulatory effect was observed with a significant increase in the number of terminal end buds and stimulated terminal ducts. At 10 mg/kg, mammary growth was delayed [75]. Further examination of the mechanism behind the stimulation observed in C57BL/6 mice revealed that PFOA may alter hormones, growth factors, and cellular receptors that promote mammary gland cell proliferation [76].

Conclusions and Data Needs

Breast cancer is one of the leading killers of women around the world, and the need to understand the environmental basis of the disease is pivotal. There are known links between pubertal timing, breast density, endocrine disruption, and breast cancer risk. Childhood obesity and precocious puberty, both of which may indeed influence breast cancer risk, are at epidemic levels in many countries. Understanding the link between environmental exposures and the breast cancer risk factors may be the missing link that will lead to eventual prevention of the disease. Young women today begin breast development earlier than did their mothers, take longer to attain full breast size, thus lengthening the time that they are susceptible to EDCs in their environment. Studies of rodent models exposed to specific EDCs or classes of EDCs may help define those chemicals or xenoestrogens most suspect to influence breast cancer risk in women.

Although research in the last 20 years has made great strides toward understanding the effects of a limited subset of environmental EDCs and how they affect mammary gland development, there is still room for advancing knowledge. This chapter covers most of the known environmental EDCs that have been shown to alter mammary gland development. However, few of these studies report numerous end points in addition to mammary effects, or a dose range for mammary and other effects, so it is often difficult or impossible to assess the sensitivity of mammary tissue compared to other tissues. As the knowledge of the toxicity of environmental endocrine-disrupting compounds becomes more widespread, applying this knowledge to human-relevant toxicity and end points is crucial. This can be achieved by creating a cooperative effort between toxicological and epidemiological studies and including human-relevant data and end points into animal studies, such as internal dosimetry. Some studies identified in this chapter have used doses relevant to humans, whether they achieved blood serum levels similar to those of humans or dosed animals at levels comparable to proven human exposure rates.

Many toxicology studies on rodents today address pubertal end points such as first estrous, estrous cyclicity, and vaginal opening. Although these are important markers in determining rodent puberty, they do not directly apply to human pubertal mechanisms and end points. Including mammary gland development as an end point in these studies creates a vital bridge between animal studies and human relevancy.

As sections of this chapter have pointed out, EDCs have been shown to affect male mammary gland development as well. These findings shed light on a major gap in mammary gland development research. Including male development in studies and determining which environmental compounds affect this sensitive end point should be a key goal as mammary gland development research continues to discover and identify compounds negatively affecting this tissue.

References

1. Oftedal O. The mammary gland and its origin during Synapsid evolution. J Mammary Gland Biol Neoplasia. 2002;7:225–252.
2. Tanner J. Growth and adolescence. 2nd ed. Oxford: Blackwell Scientific Publications; 1962.
3. Euling SY, Herman-Giddens ME, Lee PA, et al. Examination of US puberty-timing data from 1940 to 1994 for secular trends: panel findings. Pediatrics. 2008;121(Suppl 3):S172–191.
4. Euling SY, Selevan SG, Pescovitz OH, Skakkebaek NE. Role of environmental factors in the timing of puberty. Pediatrics. 2008;121(Suppl 3):S167–171.
5. Fenton SE. Endocrine-disrupting compounds and mammary gland development: early exposure and later life consequences. Endocrinology. 2006;147(6 Suppl):S18–24.
6. ACS. Cancer facts and figures. Atlanta: American Cancer Society; 2010.
7. Neville MC, Daniel CW. The mammary gland: development, regulation, and function. New York: Plenum Press; 1987.
8. Elston CW, Ellis IO. The breast. 3rd ed. Edinburgh/New York: Churchill Livingstone; 1998.
9. Russo IH, Russo J. Developmental stage of the rat mammary gland as determinant of its susceptibility to 7,12-dimethylbenz[a]anthracene. J Natl Cancer Inst. 1978;61(6):1439–1449.
10. Russo J, Russo IH. Experimentally induced mammary tumors in rats. Breast Cancer Res Treat. 1996;39(1):7–20.
11. Russo J, Russo IH. Differentiation and breast cancer. Medicina (B Aires). 1997;57(Suppl 2): 81–91.
12. Cardiff RD, Anver MR, Gusterson BA, et al. The mammary pathology of genetically engineered mice: the consensus report and recommendations from the Annapolis meeting. Oncogene. 2000;19(8):968–988.
13. Fenton SE, Condon M, Ettinger AS, et al. Collection and use of exposure data from human milk biomonitoring in the United States. J Toxicol Environ Health A. 2005;68(20):1691–1712.
14. Vorderstrasse BA, Fenton SE, Bohn AA, Cundiff JA, Lawrence BP. A novel effect of dioxin: exposure during pregnancy severely impairs mammary gland differentiation. Toxicol Sci. 2004;78(2):248–257.
15. Padilla-Banks E, Jefferson WN, Newbold RR. Neonatal exposure to the phytoestrogen genistein alters mammary gland growth and developmental programming of hormone receptor levels. Endocrinology. 2006;147(10):4871–4882.
16. Murrill WB, Brown NM, Zhang JX, Manzolillo PA, Barnes S, Lamartiniere CA. Prepubertal genistein exposure suppresses mammary cancer and enhances gland differentiation in rats. Carcinogenesis. 1996;17(7):1451–1457.
17. Russo IH, Koszalka M, Russo J. Effect of human chorionic gonadotropin on mammary gland differentiation and carcinogenesis. Carcinogenesis. 1990;11(10):1849–1855.
18. Hilakivi-Clarke L, Onojafe I, Raygada M, et al. Prepubertal exposure to zearalenone or genistein reduces mammary tumorigenesis. Br J Cancer. 1999;80(11):1682–1688.
19. Cabanes A, Wang M, Olivo S, et al. Prepubertal estradiol and genistein exposures up-regulate BRCA1 mRNA and reduce mammary tumorigenesis. Carcinogenesis. 2004;25(5):741–748.
20. John R, Latendresse TJ, Bucci, GO, Paul M, Constance C. Weis B, Thorn RR, Newbold K, Barry D. Genistein and ethinyl estradiol dietary exposure in multigenerational and chronic studies induce similar proliferative lesions in mammary gland of male Sprague–Dawley rats, Reprod Toxicol. 2009;28(3):342–353.
21. Hilakivi-Clarke L, Cho E, Clarke R. Maternal genistein exposure mimics the effects of estrogen on mammary gland development in female mouse offspring. Oncol Rep. 1998;5(3):609–616.
22. Hilakivi-Clarke L, Cho E, Onojafe I, Raygada M, Clarke R. Maternal exposure to genistein during pregnancy increases carcinogen-induced mammary tumorigenesis in female rat offspring. Oncol Rep. 1999;6(5):1089–1095.
23. Markey CM, Luque EH, Munoz De Toro M, Sonnenschein C, Soto AM. In utero exposure to bisphenol A alters the development and tissue organization of the mouse mammary gland. Biol Reprod. 2001;65(4):1215–1223.

24. Munoz-de-Toro M, Markey CM, Wadia PR, et al. Perinatal exposure to bisphenol-A alters peripubertal mammary gland development in mice. Endocrinology. 2005;146(9):4138–4147.
25. Vandenberg LN, Maffini MV, Wadia PR, Sonnenschein C, Rubin BS, Soto AM. Exposure to environmentally relevant doses of the xenoestrogen bisphenol-A alters development of the fetal mouse mammary gland. Endocrinology. 2007;148(1):116–127.
26. Vandenberg LN, Maffini MV, Schaeberle CM, et al. Perinatal exposure to the xenoestrogen bisphenol-A induces mammary intraductal hyperplasias in adult CD-1 mice. Reprod Toxicol. 2008;26(3–4):210–219.
27. Durando M, Kass L, Piva J, et al. Prenatal bisphenol A exposure induces preneoplastic lesions in the mammary gland in Wistar rats. Environ Health Perspect. 2007;115(1):80–86.
28. Murray TJ, Maffini MV, Ucci AA, Sonnenschein C, Soto AM. Induction of mammary gland ductal hyperplasias and carcinoma in situ following fetal bisphenol A exposure. Reprod Toxicol. 2007;23(3):383–390.
29. Colerangle JB, Roy D. Exposure of environmental estrogenic compound nonylphenol to noble rats alters cell-cycle kinetics in the mammary gland. Endocrine. 1996;4(2):115–122.
30. Roy D, Colerangle JB, Singh KP. Is exposure to environmental or industrial endocrine disrupting estrogen-like chemicals able to cause genomic instability? Front Biosci. 1998;3:d913–921.
31. Odum J, Pyrah ITG, Foster JR, Van Miller JP, Joiner RL, Ashby J. Comparative activities of p-nonylphenol and diethylstilbestrol in noble rat mammary gland and uterotrophic assays. Regul Toxicol Pharmacol. 1999;29(2):184–195.
32. Moon HJ, Han SY, Shin JH, et al. Gestational exposure to nonylphenol causes precocious mammary gland development in female rat offspring. J Reprod Dev. 2007;53(2):333–344.
33. Lee KY, Shibutani M, Takagi H, et al. Diverse developmental toxicity of di-n-butyl phthalate in both sexes of rat offspring after maternal exposure during the period from late gestation through lactation. Toxicology. 2004;203(1–3):221–238.
34. Mylchreest E, Wallace DG, Cattley RC, Foster PM. Dose-dependent alterations in androgen-regulated male reproductive development in rats exposed to Di(n-butyl) phthalate during late gestation. Toxicol Sci. 2000;55(1):143–151.
35. Goldman AS, Shapiro B, Neumann F. Role of testosterone and its metabolites in the differentiation of the mammary gland in rats. Endocrinology. 1976;99(6):1490–1495.
36. Bloch GJ, Mills R. Prepubertal testosterone treatment of neonatally gonadectomized male rats: defeminization and masculinization of behavioral and endocrine function in adulthood. Neurosci Biobehav Rev. 1995;19(2):187–200.
37. Tomooka Y, Bern HA. Growth of mouse mammary glands after neonatal sex hormone treatment. J Natl Cancer Inst. 1982;69(6):1347–1352.
38. Warner MR. Effect of various doses of estrogen to BALB/cCrgl neonatal female mice on mammary growth and branching at 5 weeks of age. Cell Tissue Kinet. 1976;9(5):429–438.
39. Hilakivi-Clarke L, Cho E, Raygada M, Kenney N. Alterations in mammary gland development following neonatal exposure to estradiol, transforming growth factor alpha, and estrogen receptor antagonist ICI 182,780. J Cell Physiol. 1997;170(3):279–289.
40. Zhou J, Ng S, Adesanya-Famuiya O, Anderson K, Bondy CA. Testosterone inhibits estrogen-induced mammary epithelial proliferation and suppresses estrogen receptor expression. FASEB J. 2000;14(12):1725–1730.
41. Odum J, Pyrah IT, Foster JR, Van Miller JP, Joiner RL, Ashby J. Comparative activities of p-nonylphenol and diethylstilbestrol in noble rat mammary gland and uterotrophic assays. Regul Toxicol Pharmacol. 1999;29(2 Pt 1):184–195.
42. Boylan ES. Morphological and functional consequences of prenatal exposure to diethylstilbestrol in the rat. Biol Reprod. 1978;19(4):854–863.
43. Hovey RC, Asai-Sato M, Warri A, et al. Effects of neonatal exposure to diethylstilbestrol, tamoxifen, and toremifene on the BALB/c mouse mammary gland. Biol Reprod. 2005;72(2):423–435.
44. Rothschild TC, Boylan ES, Calhoon RE, Vonderhaar BK. Transplacental effects of diethylstilbestrol on mammary development and tumorigenesis in female ACI rats. Cancer Res. 1987;47(16):4508–4516.

7 Developmental Exposure to Environmental Endocrine Disruptors... 223

45. Palmer JR, Wise LA, Hatch EE, et al. Prenatal diethylstilbestrol exposure and risk of breast cancer. Cancer Epidemiol Biomarkers Prev. 2006;15(8):1509–1514.
46. Johnson MD, Kenney N, Stoica A, et al. Cadmium mimics the in vivo effects of estrogen in the uterus and mammary gland. Nat Med. 2003;9(8):1081–1084.
47. Garcia-Morales P, Saceda M, Kenney N, et al. Effect of cadmium on estrogen receptor levels and estrogen-induced responses in human breast cancer cells. J Biol Chem. 1994;269(24): 16896–16901.
48. Ohrvik H, Yoshioka M, Oskarsson A, Tallkvist J. Cadmium-induced disturbances in lactating mammary glands of mice. Toxicol Lett. 2006;164(3):207–213.
49. Hilakivi-Clarke L, Cho E, Cabanes A, et al. Dietary modulation of pregnancy estrogen levels and breast cancer risk among female rat offspring. Clin Cancer Res. 2002;8(11):3601–3610.
50. Welsch CW, O'Connor DH. Influence of the type of dietary fat on developmental growth of the mammary gland in immature and mature female BALB/c mice. Cancer Res. 1989;49(21): 5999–6007.
51. Jurkowski JJ, Cave Jr WT. Dietary effects of menhaden oil on the growth and membrane lipid composition of rat mammary tumors. J Natl Cancer Inst. 1985;74(5):1145–1150.
52. Malik NM, Carter ND, Murray JF, Scaramuzzi RJ, Wilson CA, Stock MJ. Leptin requirement for conception, implantation, and gestation in the mouse. Endocrinology. 2001;142(12): 5198–5202.
53. Hu X, Juneja SC, Maihle NJ, Cleary MP. Leptin–a growth factor in normal and malignant breast cells and for normal mammary gland development. J Natl Cancer Inst. 2002;94(22): 1704–1711.
54. Rayner JL, Wood C, Fenton SE. Exposure parameters necessary for delayed puberty and mammary gland development in Long-Evans rats exposed in utero to atrazine. Toxicol Appl Pharmacol. 2004;195(1):23–34.
55. Rayner JL, Enoch RR, Fenton SE. Adverse effects of prenatal exposure to atrazine during a critical period of mammary gland growth. Toxicol Sci. 2005;87(1):255–266.
56. Enoch RR, Stanko JP, Greiner SN, Youngblood GL, Rayner JL, Fenton SE. Mammary gland development as a sensitive end point after acute prenatal exposure to an atrazine metabolite mixture in female Long-Evans rats. Environ Health Perspect. 2007;115(4):541–547.
57. Wetzel LT, Luempert 3rd LG, Breckenridge CB, et al. Chronic effects of atrazine on estrus and mammary tumor formation in female Sprague-Dawley and Fischer 344 rats. J Toxicol Environ Health. 1994;43(2):169–182.
58. Birnbaum LS, Fenton SE. Cancer and developmental exposure to endocrine disruptors. Environ Health Perspect. 2003;111(4):389–394.
59. Fenton SE, Hamm JT, Birnbaum LS, Youngblood GL. Persistent abnormalities in the rat mammary gland following gestational and lactational exposure to 2,3,7,8-tetrachlorodibenzo-p-dioxin (TCDD). Toxicol Sci. 2002;67(1):63–74.
60. Lewis BC, Hudgins S, Lewis A, et al. In utero and lactational treatment with 2,3,7,8-tetrachlorodibenzo-p-dioxin impairs mammary gland differentiation but does not block the response to exogenous estrogen in the postpubertal female rat. Toxicol Sci. 2001;62(1):46–53.
61. Brown NM, Manzolillo PA, Zhang JX, Wang J, Lamartiniere CA. Prenatal TCDD and predisposition to mammary cancer in the rat. Carcinogenesis. 1998;19(9):1623–1629.
62. Lipworth L. Epidemiology of breast cancer. Eur J Cancer Prev. 1995;4(1):7–30.
63. Warner M, Eskenazi B, Mocarelli P, et al. Serum dioxin concentrations and breast cancer risk in the Seveso Women's Health Study. Environ Health Perspect. 2002;110(7):625–628.
64. Den Hond E, Roels HA, Hoppenbrouwers K, et al. Sexual maturation in relation to polychlorinated aromatic hydrocarbons: Sharpe and Skakkebaek's hypothesis revisited. Environ Health Perspect. 2002;110(8):771–776.
65. Leijs MM, Koppe JG, Olie K, et al. Delayed initiation of breast development in girls with higher prenatal dioxin exposure; a longitudinal cohort study. Chemosphere. 2008;73(6):999–1004.
66. Kodavanti PR, Coburn CG, Moser VC, et al. Developmental exposure to a commercial PBDE mixture, DE-71: neurobehavioral, hormonal, and reproductive effects. Toxicol Sci. 2010; 116(1):297–312.

67. Talsness CE, Kuriyama SN, Sterner-Kock A, et al. In utero and lactational exposures to low doses of polybrominated diphenyl ether-47 alter the reproductive system and thyroid gland of female rat offspring. Environ Health Perspect. 2008;116(3):308–314.
68. White SS, Fenton SE, Hines EP. Endocrine disrupting properties of perfluorooctanoic acid. J Steroid Biochem Mol Biol. 2011; In press. Accepted March 4, 2011; doi: 10.1016/j.jsbmb.2011.03.011.
69. White SS, Calafat AM, Kuklenyik Z, et al. Gestational PFOA exposure of mice is associated with altered mammary gland development in dams and female offspring. Toxicol Sci. 2007;96(1):133–144.
70. Wolf CJ, Fenton SE, Schmid JE, et al. Developmental toxicity of perfluorooctanoic acid in the CD-1 mouse after cross-foster and restricted gestational exposures. Toxicol Sci. 2007; 95(2):462–473.
71. White SS, Kato K, Jia LT, et al. Effects of perfluorooctanoic acid on mouse mammary gland development and differentiation resulting from cross-foster and restricted gestational exposures. Reprod Toxicol. 2009;27(3–4):289–298.
72. White SS, Stanko JP, Kato K, et al. Investigating the multigenerational effects of prenatal PFOA exposure on mouse mammary gland development, function, and tumor susceptibility. Environ Health Perspect. 2011;119(8):1070–1076.
73. Macon MB, Villanueva L, Tatum-Gibbs K, et al. Prenatal perfluorooctanoic acid exposure in CD-1 mice: low dose developmental effects and internal dosimetry. Toxicol Sci. 2011; In press; doi: 10.1093/toxsci/kfr076 First published online: April 11, 2011 .
74. Yang C, Tan YS, Harkema JR, Haslam SZ. Differential effects of peripubertal exposure to perfluorooctanoic acid on mammary gland development in C57Bl/6 and Balb/c mouse strains. Reprod Toxicol. 2009;27(3–4):299–306.
75. Zhao Y, Tan YS, Haslam SZ, Yang C. Perfluorooctanoic acid effects on steroid hormone and growth factor levels mediate stimulation of peripubertal mammary gland development in C57BL/6 mice. Toxicol Sci. 2010;115(1):214–224.

Chapter 8
Developmental Exposure to Endocrine Disruptors and Male Urogenital Tract Malformations

Mariana F. Fernandez and Nicolas Olea

Abstract Discrepancies in adverse trends in male reproductive health around the world are due in part to the difficulty of comparing studies from different time periods with distinct study populations and varied clinical definitions and diagnostic criteria for these processes. In this context, the apparent increase in the incidence of cryptorchidism and hypospadias in the Western world over recent decades, with a leveling off in hypospadias incidence in most European countries during the 1980s, can be questioned due to differences in case definitions, age at diagnosis, examination techniques, and study populations, e.g., registry versus cohort studies. Experimental investigations have supported the hypothesis of a link between environmental factors and urogenital tract malformations, suggesting that cryptorchidism and hypospadias are associated with exposure to environmental chemicals, especially those identified as endocrine-disrupting chemicals (EDCs). Exposure to EDCs during pregnancy may play a role in the development of male sexual disorders because sexual differentiation and reproductive functioning are critically dependent on the ratio of androgens to estrogens, and an imbalance in this ratio may be responsible for male congenital anomalies. Epidemiology studies have also indicated a link between EDCs and malformations. The accumulated evidence appears sufficient to endorse a precautionary approach, with the implementation of measures to reduce community exposure to EDCs, especially in women of childbearing age, both before and during pregnancy.

M.F. Fernandez (✉)
Biomedical Research Centre, University of Granada. San Cecilio
University Hospital, Granada, CIBERESP, Spain
e-mail: marieta@ugr.es

N. Olea
School of Medicine, University of Granada. San Cecilio University Hospital,
Granada, CIBERESP, Spain
e-mail: nolea@ugr.es

E. Diamanti-Kandarakis and A.C. Gore (eds.), *Endocrine Disruptors and Puberty*,
Contemporary Endocrinology, DOI 10.1007/978-1-60761-561-3_8,
© Springer Science+Business Media, LLC 2012

Keywords Congenital malformations • Cryptorchidism • Endocrine-disrupting chemicals (EDCs) • Hypospadias

Introduction

In Chap. 9, it is reported that most disorders related to male reproductive health, including reduced semen quality, testicular cancer, and congenital reproductive malformations, share common risk factors and can be caused by disturbance of testes development during fetal life. It has been suggested that exposure of the developing male fetus to environmental pollutants is responsible for sexual maturation anomalies and reproductive malfunction in adult life [1]. In particular, recent human data have focused attention on human exposure to endocrine-disrupting chemicals (EDCs) during pregnancy as having a possible role in the origin of male sexual disorders [2,3]. Male sexual differentiation and reproductive functioning are critically dependent on the ratio of androgens to estrogens, and an imbalance in this ratio may be responsible for male congenital anomalies, supporting the EDC hypothesis.

Urogenital Tract Malformations

Male congenital malformations include hypospadias, a urethral opening on the underside of the penis or on the perineum, and cryptorchidism, a failure of one or both testicles to descent into the scrotum. Following the proposal of their separate genetic but common environmental components, they have been considered etiologically comparable entities [4,5]. Highly variable prevalences of cryptorchidism and hypospadias have been reported in Western countries on the basis of registry studies [6–9]. However, intercountry comparisons in the incidence rate of urogenital malformations may be limited by differences in study populations (registry versus cohort studies), case definitions, age at diagnosis, and examination techniques.

The incidence of these malformations appears to have been increasing in the Western world over recent decades but with an apparent leveling off in the incidence of hypospadias in most European countries during the 1980s [10,11]. However, trends in urogenital malformations sometimes follow a more complex pattern. Thus, the Spanish Collaborative Study of Congenital Malformations (ECEMC) reported that the frequency of hypospadias in Spain had remained at 1 in 284 male births (0.35%) during the past few decades until after 1996, when a decreasing frequency of severe forms was recorded [10]. Surprisingly, although the authors suggested that this effect was probably caused by a radical change in exposure that affected the whole country, this change was not observed in the prevalence of any other congenital malformations. In contrast, the prevalence of hypospadias in Denmark, for instance, has consistently risen over recent decades from 0.24% in 1977 to 0.52% in 2005, i.e., an annual increase of 2.40% [7].

Recent prospective cohort studies suggested that the incidence of cryptorchidism at birth may be 9% in Denmark and 6% in UK, higher than previously thought [6,12] and similar to findings (5.7%) in male Lithuanian newborns [13]. However, a much lower incidence (2.4%) was found in Finland, where the incidence of hypospadias was lower than in Denmark (0.27% vs. 1.05%) [6].

The prevalence of cryptorchidism and/or hypospadias was 1.8% in a case-control study nested in a mother–child cohort ($n = 702$) established at the San Cecilio University Hospital of Granada, Southern Spain. This prevalence and that of each malformation (prevalence of cryptorchidism at 1 month of age of 1.1% and of hypospadias at delivery of 0.74%) fell within the range reported by studies in other geographic areas [3,6,8,13,14].

These changes in temporal trends and major geographical variations in prevalence suggest a role for environmental factors in the etiology of male urogenital tract malformations, as discussed in the next sections.

Endocrine-Disrupting Chemicals (EDCs) Hypothesis

Humans are exposed to more than 100,000 synthetic chemicals, some of which have the ability to disrupt normal endocrine homeostasis and alter endocrine system function, thereby producing adverse health effects in an intact organism or its progeny or in subpopulations [15]. Many of these EDCs are persistent compounds that accumulate in human tissues; e.g., certain pesticides [16]. They can affect hormonal activity by a variety of mechanisms; some compounds with estrogenic effects bind to estrogen receptors [e.g., DDT and endosulfan [17]], some with antiandrogenic properties bind to but do not activate the androgen receptor (AR) [e.g., p,p´-DDE and vinclozolin [18,19]] while others affect endocrine homeostasis via multiple mechanisms and complex interactions [20]. Because of their ubiquity and persistence in the environment, EDC pesticides can be found in soil, water, wildlife, and in the adipose tissue of mothers, reaching children during pregnancy and breastfeeding [21]. The growing list of chemicals identified as having endocrine-disrupting properties in animal studies includes also numerous substances found in household and consumer products with shorter half-life; e.g., phthalates and phenols.

Human exposure to EDCs at certain times of life influences normal maturation and may result in impaired male reproductive development and function. Pregnancy is a period of rapid growth, cell differentiation, immaturity of metabolic pathways, and development of vital organ systems. Each normal developmental process occurs during a specific period of a few days or weeks, and it is in this window of particular susceptibility that exposure may have adverse health effects, such as urogenital anomalies. These sensitive periods largely occur during the first trimester and sometimes before the mother knows she is pregnant. An embryo may also be affected by chemicals with a long biological half-life to which the mother was exposed before its conception [22,23].

The EDC hypothesis poses several challenges to epidemiological research [24]. First, the exposure scenario: humans are exposed to a complex cocktail of very heterogeneous classes of environmental chemicals [25], hampering a correct exposure assessment and identification of the individual chemicals responsible for effects. In this regard, several experimental studies exposed animals to mixtures of 3–7 chemicals with antiandrogenic properties at doses at which each chemical alone had no significant effect, finding a major impairment of masculinization and the development of hypospadias [26,27]. Importantly, these mixture effects are usually observed at lower levels of the individual chemicals than those to which humans are generally exposed.

Second, the dose–effect relationship based on a threshold model may underestimate the activity of environmental chemicals. EDCs can induce an inverted-U dose–response curve, and risk assessment in this area cannot rely solely on linear measurements of effects [28,29]. In addition, identification of pathways between the source of the chemicals and human exposure is very complex. The evaluation process is further complicated by the fact that many EDCs are persistent and accumulate in human tissues and that there can be long latency periods between exposure and the manifestation of a response. For instance, in utero exposure may lead to developmental effects that only become evident when the offspring reaches the age of sexual maturity. Furthermore, although some persistent EDCs (e.g., many organochlorine pesticides) are no longer used in many countries, areas that were heavily polluted in the past may serve as redistribution sources for these compounds in the present.

With this complexity in mind, it is perhaps to be expected that epidemiological studies may struggle to provide a definitive association or lack of association between exposure to individual environmental chemicals and any male reproductive disorder in humans [30].

Epidemiological Studies on Congenital Malformations of the Male Reproductive Tract and EDCs

Some studies have provided support for the hypothesis of a link between environmental factors and urogenital tract malformations by suggesting that human cryptorchidism and hypospadias are associated with exposure to environmental chemicals. Cryptorchidism and hypospadias are birth defects of exclusively prenatal origin with an easy and early diagnosis, making them a good choice as biomarkers of chemical exposure and its effects. Unfortunately, studies of exposure are not abundant, especially in relation to EDCs and urogenital malformations.

Epidemiological Studies Related to Occupational Exposure

The occupational activities of parents have sometimes been associated with increased risk, as is the case of farmers and gardeners [5,30–37], drivers [38], foresters, carpenters, and service station attendants [39].

8 Developmental Exposure to Endocrine Disruptors... 229

Parental involvement in agricultural work and/or parental exposure to pesticides has been associated with higher risk of a wide range of congenital malformations. A register-based case-control study in Denmark investigated parental occupation in farming or gardening during the year of conception [5]. All live-born males between 1983 and 1992 with a diagnosis of cryptorchidism (6,177) or hypospadias (1,345) were compared with a randomized control group of 23,273 boys born in the same period. The risk of cryptorchidism, adjusted for year of birth and birth weight, was greater in the sons of women working in farming or gardening (OR 1.38; 95% CI, 1.10–1.73). The effect was even stronger when the analysis was restricted to mothers working in gardening (OR 1.67; 95% CI, 1.14–2.47). However, maternal occupation in farming and gardening had no significant effect on the risk of hypospadias (OR 1.27, 95% CI, 0.81–1.99). Interestingly, the occupation of the father had no effect on the risk of either cryptorchidism or hypospadias.

Kristensen et al. [31] adopted a different approach, studying 192,417 children born to farming families in Norway from 1967 to 1991 and relating different birth defects to the purchase of pesticides and the amount of pesticide spraying equipment on the farm, with no consideration of parental occupation. They found a significant positive relationship with cryptorchidism and a nonsignificant association with hypospadias. The prevalence of specific defects at birth was retrospectively determined, and the purchases of pesticides in 1968 (from the 1969 census) and tractor-drawn pesticide spraying equipment present on the farm in 1979 (from the 1979 census) were used as proxies of exposure. Odds ratios were calculated from contingency tables, and multiple logistic regression analysis was used to control for potential confounders such as maternal age or birth order. The adjusted OR (95% CI) for cryptorchidism in association with pesticide purchase on the farm was 1.70 (1.16–2.50). The association was stronger when the analysis was limited to vegetable farms (adjusted OR 2.32; 95% CI, 1.34–4.01). The OR for hypospadias was only moderately increased in association with the presence of spraying equipment (adjusted OR 1.38; 95% CI, 0.95–1.99). A small increase in statistical significance was noted when the analysis was limited to the presence of tractor spraying equipment on grain farms (adjusted OR 1.51; 95% CI, 1.00–2.26).

In Spain, maternal involvement in agricultural activity during the month before conception and the first trimester of pregnancy was associated with a threefold increase in the risk of bearing a child with a malformation [32]. Moreover, an ecological investigation into variations in the rates of orchidopexy (surgery to move an undescended testicle down into the base of the scrotum), in the Spanish province of Granada, found an association between exposure to pesticides and the risk of cryptorchidism [40]. A later retrospective case-control study in the same geographical area suggested that cryptorchidism was related to the father's employment in agriculture [33].

Similar findings had been published by other studies. The prevalence of both cryptorchidism and hypospadias was studied in an agricultural area of Sicily that includes 12 municipalities with intense production of fruits and vegetables in open-air cultivation and greenhouses [34]. Between 1998 and 2002, 60 cases of cryptorchidism and 55 of hypospadias were identified in the Sicilian Registry of

Congenital Malformations and confirmed by a pediatric consultant. The municipalities were classified by the degree of "pesticide impact," which was rated as high, medium, or low according to the intensity of their agriculture activity. The study showed a higher prevalence of the two malformations in communities with greater "pesticide impact," finding a closer and significant association for hypospadias.

Parental occupational exposure to EDCs was also assessed by questionnaire in the same agricultural area. The interview focused on possible occupational exposures to EDCs by both parents during the critical exposure period, i.e., from 3 months before to 3 months after conception. A clear tendency was observed for an increased risk of cryptorchidism among the sons of women working in agriculture and exposed to pesticides [35] but not among the sons of fathers involved in agricultural work. A major limitation in all of these studies is that none of the data concerning exposure were based on actual measurements of exposure.

The hypothesis was also confirmed among female greenhouse workers exposed to pesticides during pregnancy. A prospective cohort study was designed to investigate possible associations between occupational exposure to pesticides in greenhouses during pregnancy and adverse effects on the reproductive development of male infants [36]. Pregnant women referred to the Department of the Occupational and Environmental Medicine at Odense University Hospital were consecutively recruited between 1996 and 2000, finally enrolling 113 mother–son pairs. It was not possible to estimate individual exposure to endocrine-disrupting chemicals; therefore, the classification of mothers as exposed or unexposed to pesticides was performed independently by two toxicologists with expertise in working conditions in greenhouse horticultures, based on a detailed questionnaire-assisted interview. The prevalence of congenital cryptorchidism among the sons of greenhouse workers was significantly and more than threefold higher (6.2%; 95% CI 3.0–12.4) than among boys born in the same area (1.9%; 95% CI 1.2–3.0) [36]. In addition, the group of boys whose mothers were exposed to pesticides had decreased penile length and testicular volume.

In contrast to the above studies, other authors found no association between cryptorchidism and maternal exposure to pesticides [5,37–41]. Restrepo and coworkers selected cases and controls from among the offspring of 8,867 floriculture workers and their spouses included in a prevalence survey in 1982–1983 for adverse reproductive outcomes in Colombia. They compared 222 cases of congenital malformations with 443 randomly selected controls matched by maternal age and birth order, determining exposure to pesticides during the pregnancy by means of a questionnaire. The exposure variable was treated as categorical (working or not working in floriculture during pregnancy), and the Mantel–Haenszel test was used for statistical analysis of the different types of congenital malformations, followed by multivariate analysis to control for confounders. Although the relative risk of cryptorchidism after maternal exposure to pesticides during pregnancy was 4.6, the confidence interval (not specified) and p value (>0.05) indicate imprecision and uncertainty in the statistical estimate. The authors concluded that there was no significant association between the 16 cases of cryptorchidism and maternal exposure to pesticides during pregnancy [41].

8 Developmental Exposure to Endocrine Disruptors...

Zhu and coworkers conducted a prospective cohort study to examine whether exposure to pesticides influenced pregnancy outcomes among female gardeners and farmers in Denmark [37]. Using data from the National Birth Cohort in Denmark, they identified 226 pregnancies of gardeners and 214 pregnancies of farmers between 1997 and 2003. In two interviews (before the pregnancy and at around 16 weeks of gestation), data were gathered on their work activities and exposure to pesticides. No significant association was found between exposure to pesticides and congenital malformations among these gardeners or farmers in comparison to other workers in the cohort.

Occupational exposure of the father was strongly associated with cryptorchidism in a nested case-control study conducted by Pierik et al. [42] in the city of Rotterdam. Based on a cohort of 8,695 boys recruited at child health-care centers during the first days of life, the authors compared 78 cryptorchidism cases and 56 hypospadias cases with 313 age-matched controls. Their main aim was to identify risk factors for cryptorchidism and hypospadias, focusing on exposure to EDCs. A parental questionnaire, completed within a few weeks of delivery, collected data on personal characteristics, health, and occupation. Occupational exposure was assessed by generic questions to elicit a description of jobs held in the year before pregnancy and by a checklist for self-reported exposure to ionizing radiation, to physical exposures, and to chemicals with endocrine activity or previously described as male reproductive toxicants, such as pesticides. They assessed occupational exposure by applying a job-exposure matrix for potential EDCs [43]. The study concluded that paternal exposure should be included in studies on risk factors for urogenital malformations because the pesticide exposure of fathers was associated with cryptorchidism in their progeny (expressed as odds ratio [OR]) (OR 3.8, 95% CI 1.1–13.4), and their smoking habit was associated with hypospadias (OR 3.8, 95% CI 1.8–8.2). These authors found no association between the occupational exposure of mothers and any abnormality.

The Vrijheid study [44] analyzed the relation between risk of hypospadias and maternal occupation. It included hypospadias ($n = 3,471$) diagnosed at birth and recorded in the National Congenital Anomaly System in England between 1980 and 1996. The potential exposure to EDCs was measured as a surrogate for exposure, using the same job-exposure matrix as the above-mentioned Pierik study, categorizing 348 job titles according to their unlikely, possible, or probable exposure to seven groups of EDCs: pesticides, polychlorinated organic compounds, phthalates, alkylphenolic compounds, biphenolic compounds, heavy metals, and other hormone-disrupting chemicals [43]. Hairdressers, cleaners, and painters were the largest occupational groups with probable exposure to EDCs. There was some indication of an increased risk of hypospadias in the offspring of hairdressers and of some workers with reported exposure to phthalates. Although the job-exposure matrix developed by the authors was unable to distinguish among exposure levels, compound potencies, action mechanisms, possible interactions, or changes in exposure over time, this is a very interesting approach that deserves further exploration. Vrijheid et al. [44] concluded that their results should be interpreted with caution and insisted on the need for validation of the job-exposure matrix by the assessment of real exposure in human tissues.

A case-control study in 1997–1998 examined the risk of hypospadias associated with maternal occupational exposure to EDCs during the first 3 months of pregnancy in an area that included the health regions of North Thames, South Thames, and Anglia (part of the Anglia and Oxford Health Authority), comprising 120 districts [45]. Out of 731 patients identified, 471 agreed to participate and were compared with 490 randomly selected controls. Authors found a two- to threefold increased risk of hypospadias among children of mothers exposed to hair spray (OR = 2.39; 95% CI 1.40–4.17) and phthalates (OR = 3.12; 95% CI 1.04–11.46) in the workplace during pregnancy, after controlling for potential confounders. To assess the occupational exposure to EDCs during the first 3 months of pregnancy, mothers were asked about their job title, department, company, their five main tasks, possible exposure to a list of 26 occupational exposures, and the hours per week that they were in contact with these exposures while at work. Finally, the authors classified all of the occupational exposure data by applying a similar job exposure matrix to that used by Pierik et al. 2004 [43].

Nonoccupational exposure to environmental estrogens can occur through the domestic use of pesticides in homes and gardens, the consumption of treated foodstuffs or contaminated drinking water, or residence in agricultural areas associated with contaminated air, water, or soil, among other routes. Studies of residential exposure to pesticides have used proxy measures of exposure, including pesticide usage at home, residence in or near pesticide application areas, or residence near agricultural crops [46]. Thus, Dolk et al. [46] investigated 1,089 cases of congenital anomaly and 2366 control births without malformation in a collaborative European (Eurohazcon) study on the risk of congenital anomaly among people living near hazardous-waste landfill sites. The study was set up to analyze the risk of congenital anomaly among individuals exposed to chemicals present in hazardous-waste landfill sites. Hypospadias was one of the outcomes examined in this study. Twenty-one such sites were included, and the distance of the mother's place of residence from the nearest site served as a surrogate measurement of exposure to the chemical contaminants it contained. An area of 7-km radius around each landfill was considered, and a zone of 3-km radius around each site was defined as the "proximate zone" within which most exposure to chemical contaminants would occur. Socioeconomic status and maternal age were recorded as confounding variables. Residence within 3 km of a landfill site was associated (borderline significance) with a higher risk of hypospadias (OR 1.96, 95% CI 0.98–3.92).

Epidemiological Studies Reporting Actual Exposure Data

It has been suggested that epidemiological studies would benefit from a more accurate method for measuring exposure to EDCs, e.g., by determining their presence in tissues and fluids of the mother–child pair. In this regard, Hosie et al. [47] quantified a large number of chemical residues in children, including DDT and metabolites, polychlorinated biphenyls, toxaphene, hexachlorocyclohexane, chlorinated cyclodienes, and chlorinated benzenes, examining the fatty tissue of 48 boys undergoing

8 Developmental Exposure to Endocrine Disruptors... 233

surgical procedures, 18 with cryptorchidism and 30 controls. All of the substances were detected in all boys studied. The statistical analysis only revealed highly significant differences in the presence of heptachlor epoxide (HE) and hexachlorobenzene (HCB) between controls and patients with cryptorchidism, with both HE and HCB showing higher mean concentrations in patients with cryptorchidism than in controls.

Two published studies have compared maternal blood levels of DDE and DDT with cryptorchidism and hypospadias in the offspring based on two large birth cohorts conducted in the USA in the 1960s [48, 49]. Although both studies were based on biological samples collected in a period during which DDT was still in use (from 1959 to 1966), neither study found firm associations. In the study by Longnecker et al., maternal exposure was assessed by determining serum DDE levels in the third trimester of pregnancy in a large sample of mother–child pairs. The authors used a nested case-control design with subjects selected from the Collaborative Perinatal Project, a birth cohort study in the USA. Among the male offspring, 219 boys with cryptorchidism and 199 with hypospadias were eligible for the study and were matched with 599 male controls randomly selected from the same cohort [48]. Exposure data were categorized into five DDE concentration strata, and the odds of having a birth defect in relation to DDE level were estimated using logistic regression. After examining stratum-specific effects for several possible confounders and comparing the highest category of serum DDE with the lowest, the adjusted ORs were 1.3 (95% CI, 0.7–2.4) for cryptorchidism, and 1.2 (95% CI, 0.6–2.4) for hypospadias. Bhatia et al. (2005) found no association [49].

Humans may also be simultaneous exposed to other environmental chemicals that contribute to the effect on testicular descent, and pesticide measurements may possibly represent a proxy (sentinel) marker for other exposures, such as brominated flame retardants. A longitudinal birth cohort study in Finland (Turku University Hospital) and Denmark (National University Hospital at Copenhagen) from 1997 to 2001 [50] explored any association between polybrominated diphenyl ethers (PBDEs) and testicular maldescent. Exposure measurements were prospectively planned, including the analysis of 14 PBDEs in samples of placenta (95 cryptorchid/185 healthy boys) and breast milk (62 cryptorchid /68 healthy boys) as a reflection of the accumulated body burden of the mother [51]. The authors found no significant differences between boys with and without cryptorchidism for individual congeners, but the concentration of PBDEs in breast milk was again significantly higher in boys with testicular maldescent, finding an association with the sum of seven PBDEs and with several PBDE congeners. Assessment of PBDE exposure in placenta samples yielded lower concentrations and did not support the above association [51].

Exposure to PBDEs cannot explain the reported difference in cryptorchidism between Denmark and Finland, which differed in the pattern of PBDE congener distribution but not in the total amount found in milk or placenta samples; hence, different sources and/or timings of exposure may be responsible. However, it should be borne in mind that the breast milk of Danish mothers contains higher concentrations of other persistent chemicals in comparison to Finnish mothers [52]. Besides quantitative differences in exposure levels, Danish and Finnish children show qualitatively distinct exposure patterns and typical chemical signatures, possibly indicating a higher environmental exposure for Danish fetuses and infants in comparison to their Finnish counterparts (European Science Foundation: htpp://www.esf.org).

The longitudinal birth cohort study of Main and coworkers [51] was also used to describe regional prevalence rates and risk factors (lifestyle and exposure) for cryptorchidism and to investigate whether individual in utero exposure to organochlorine pesticides, estimated by measurements of maternal breast milk concentrations, was associated with cryptorchidism [50]. They nested a case-control study in a prospective birth cohort, including 62 milk samples from mothers of cryptorchid boys and 68 from mothers of healthy boys. Concentrations of compounds in breast milk are considered a suitable proxy for fetal exposure during pregnancy. Milk was collected as individual pools between 1 and 3 months postpartum and analyzed for 27 organochlorine pesticides. Most of the persistent organochlorine pesticides were found in higher concentrations in boys with cryptorchidism than in controls, although no individual compound was significantly correlated with cryptorchidism. Specifically, 17 of 21 organochlorine pesticides [p,p'-DDT, p,p'-DDE, p,p'-DDD, o,p'-DDT, HCH (α, β, γ), HCB, PCA, α-endosulfan, cis-HE, chlordane (cis-, $trans$-) oxychlordane, methoxychlor, OCS, and dieldrin] showed higher median concentrations in case versus control milk samples while eight organochlorine pesticides were measurable in all samples. However, combined statistical analysis of the eight most abundant persistent pesticides showed that pesticide levels in breast milk were significantly higher in boys with cryptorchidism ($p=0.032$), suggesting that exposure to more than one chemical at low concentrations represents a risk factor for congenital cryptorchidism.

A case-control study nested in a mother–child cohort ($n=702$) established at Granada University Hospital revealed an increased risk of male urogenital malformations related to the combined effect of environmental estrogens in placenta samples. The total effective xenoestrogen burden (TEXB) and the levels of 16 organochlorine pesticides were measured in placenta tissues. Compared with controls, cases had an OR for detectable versus nondetectable TEXB of 2.82 (95% CI, 1.10–7.24), and more pesticides were detected in cases than in controls (9.34±3.19 vs. 6.97±3.93). After adjusting for potential confounders in the conditional regression analysis, ORs for cases with detectable levels of pesticides were 2.25 (95% CI, 1.03–4.89) for o,p'-DDT, 2.63 (95% CI, 1.21–5.72) for p,p'-DDT, 3.38 (95% CI, 1.36–8.38) for lindane, 2.85 (95% CI, 1.22–6.66) for mirex, and 2.19 (95% CI, 0.99–4.82) for endosulfan alpha. Maternal engagement in agriculture (OR=3.47; 95% CI, 1.33–9.03), paternal occupational exposure to xenoestrogens (OR=2.98; 95% CI, 1.11–8.01), and history of previous stillbirths (OR=4.20; 95% CI, 1.11–16.66) were also associated with risk of malformations [30].

Similar findings were recently published [53]. A prospective case-control study was performed to assess the incidence of cryptorchidism and fetal exposure to selected chemicals in the Nice area, a region better known for tourism than for agricultural or industrial activities. It included 151 cord blood samples from 67 cryptorchid cases and 84 closely matched controls and 125 colostrum samples from 56 cryptorchid cases and 69 controls recruited between April 2002 and April 2005.

8 Developmental Exposure to Endocrine Disruptors... 235

Several xenobiotics selected for their known antiandrogenic and/or estrogenic effects were measured in placenta and maternal milk, including seven nonplanar polychlorinated biphenyl (PCB) congeners (PCB 28, PCB 52, PCB 101, PCB 118, PCB138, PCB 153, and PCB 180), one plasticizer (dibutyl phthalate, DBP) and its metabolite monobutyl phthalate (MBP), and the pesticide DDE [52]. Brucker-Davis et al. confirmed that in utero and lactational exposure of fetuses and infants to the EDCs was ubiquitous. Although the difference between cryptorchid cases and controls was modest, boys with cryptorchidism were more likely to be classified in colostrum groups most contaminated with DDE, $\sum$PCBs, and $\sum$PCBs + DDE combined. The OR for cryptorchidism was significantly higher at the highest $\sum$PCB score, finding only a trend for DDE and $\sum$PCBs + DDE versus the lowest score of these components [53].

The Biological Plausibility

Male sexual differentiation and reproductive functioning are critically dependent on a balanced androgen/estrogen ratio. Estrogens are known to have marked effects on the male reproductive system. Thus, a collaborative follow-up study of three cohorts of US men exposed in utero to diethylstilbestrol (DES), a synthetic estrogen widely prescribed to pregnant women during the 1940s to the 1970s, was designed to examine any association between this exposure and the prevalence of abnormalities. The study included 3,067 men (1,638 exposed, 1,429 unexposed). The DES exposure status was determined by reviewing the mothers' medical records and gathering data on the cumulative DES dose and timing of the first exposure to this compound. A cutoff point of 11 weeks gestational age was selected because male genitalia are most susceptible to teratogens during this early period (the first 9 weeks of gestation) [54]. Exposed men had an increased prevalence of cryptorchidism (relative risk [RR]: 1.9; 95% CI 1.1–3.4). The association was strongest for exposure before the 11th week of gestation and for a cumulative DES dose of ≥ 5 g. Authors could not determine the most important factor involved [54]. The same research group found no substantially increased risk of hypospadias among the sons of women exposed to DES in utero [55].

The importance of these findings lies in their applicability to the question of whether environmental factors acting as endocrine disruptors have a detrimental effect on the male reproductive system. The confirmation of an increased risk of these conditions in men prenatally exposed to high doses of DES lends credence to the endocrine disruption hypothesis.

Conclusion

Higher concentrations of selected xenobiotics in boys with cryptorchidism and/or hypospadias may be a marker of a higher global exposure and/or decreased capacity to metabolize and eliminate xenobiotics in general. Studies in animals exposed to mixtures of three to seven EDCs, administering doses at which each chemical alone was without significant effect, caused major impairment of masculinization and development of hypospadias [26,56]. Given that a single, isolated compound is unlikely to be responsible for cryptorchidism and hypospadias, research should focus on assessing the effects of mixtures of chemicals.

Despite the above evidence, epidemiological studies to date are not considered to provide adequate grounds to confirm the hypothesis that environmental endocrine disruptors contribute to urogenital anomalies (e.g., cryptorchidism and hypospadias). Discrepancies in the scientific community about adverse trends in male reproductive health could be related in part to the difficulties in comparing studies from different time periods with distinct study populations and varied clinical definitions and diagnostic criteria. Moreover, epidemiologists traditionally analyze the incidence and risk factors separately for each disorder. Skakkebaek et al. [1] proposed that trends in sperm counts, demand for assisted reproduction, testicular cancer, hypospadias, and undescended testes are components of a single underlying entity, the so-called testicular dysgenesis syndrome (TDS), and that environment and lifestyle factors are the most likely causes, with a lesser participation of the genomic background [1].

Nevertheless, we believe that the evidence of a link between EDCs and congenital malformation is sufficient for a precautionary approach to be taken, implementing measures to reduce community exposure to these chemicals. Importantly, the planning of care for women of childbearing age should take account of these environment concerns, not only during pregnancy but also even before conception.

Acknowledgments We are indebted to Richard Davies for editorial assistance. This study was supported by grants from the Junta de Andalucía (grant numbers P09-CTS-5488 and SAS 07/0133), Spanish Ministry of Science and Innovation (Ramón y Cajal Programme to MF. Fernandez), Spanish Ministry of Health (CIBERESP), and the European Union Commission (Environ. Reprod. Health, QLK4-1999-01422; EDEN, QLRT-2001-00603; CASCADE, Food-CT-2003-506319; and CONTAMED, 2009-212502, projects).

References

1. Skakkebæk NE, Rajpert-De Meyts E, Main KM. Testicular dysgenesis syndrome: an increasingly common developmental disorder with environmental aspects. Hum Reprod. 2001; 16(5):972–8.
2. Itoh N, Kayama F, Tatsuki J, Tsukamoto T. Have sperm counts deteriorated over the past 20 years in healthy, young Japanese men? Results from the Sapporo area. J Androl. 2001; 22:40–4.
3. Toppari J, Larsen JC, Christiansen P, Giwercman A, Grandjean P, Guillette Jr LJ, et al. Male reproductive health and environmental xenoestrogens. Environ Health Perspect. 1996;104: 741–803.
4. Akre O, Lipworth L, Cnattingius S, Sparen P, Ekbom A. Risk factor patterns for cryptorchidism and hypospadias. Epidemiology. 1999;10(4):364–9.
5. Weidner IS, Moller H, Jensen TK, Skakkebaek N. Cryptorchidism and hypospadias in sons of gardeners and farmers. Environ Health Perspect. 1998;106:793–6.

8 Developmental Exposure to Endocrine Disruptors...

6. Boisen KA, Kaleva M, Main KM, Virtanen HE, Haavisto A-M, Schmidt IM, et al. Difference in prevalence of congenital cryptorchidism in infants between two Nordic countries. Lancet. 2004;363:1264–9.
7. Lund L, Engebjerg MC, Pedersen L, Ehrenstein V, Nørgaard M, Sørensen HT. Prevalence of hypospadias in Danish boys: a longitudinal study, 1977–2005. Eur Urol. 2009;55(5):1022–6.
8. Paulozzi LJ. International trends in rates of hypospadias and cryptorchidism. Environ Health Perspect. 1999;107:297–302.
9. Toppari J, Kaleva M, Virtanen HE. Trends in the incidence of cryptorchidism and hypospadias, and methodological limitations of registry-based data. Hum Reprod Update. 2001;7(3): 282–6.
10. Martinez-Frias ML, Prieto D, Prieto L, Bermejo E, Rodriguez-Pinilla E, Cuevas L. Secular decreasing trend of the frequency of hypospadias among newborn male infants in Spain. Birth Defects Res A Clin Mol Teratol. 2004;70:75–81.
11. Virtanen HE, Toppari J. Epidemiology and pathogenesis of cryptorchidism. Hum Reprod Update. 2008;14(1):49–58.
12. Acerini CL, Miles HL, Dunger DB, Ong KK, Hughes IA. The descriptive epidemiology of congenital and acquired cryptorchidism in a UK infant cohort. Arch Dis Child. 2009;94: 868–72.
13. Preiksa RT, Zilaitien B, Matulevicius V, Skakkebaek NE, Petersen JH, Jorgensen N, et al. Higher than expected prevalence of congenital cryptorchidism in Lithuania: a study of 1204 boys at birth and 1 year follow-up. Hum Reprod. 2005;20:1928–32.
14. Boisen KA, Chellakooty M, Schmidt IM, Kai CM, Damgaard IN, Suomi AM, et al. Hypospadias in a cohort of 1072 Danish newborn boys: prevalence and relationship to placental weight, anthropometrical measurements at birth, and reproductive hormone levels at 3 months of age. J Clin Endocrinol Metab. 2005;90:4041–6.
15. Colborn T, Clement C. Wingspread consensus statement. Chemically induced alterations in sexual, functional development: the wildlife/human connection. Princeton: Princeton Scientific; 1992. p. 1–8.
16. Porta M, Puigdomènech E, Ballester F, Selva J, Ribas-Fitó N, Domínguez-Boada L, et al. Studies conducted in Spain on concentrations in humans of persistent toxic compounds. Gac Sanit. 2008;22(3):248–66. Review.
17. Soto AM, Sonnenschein C, Chung KL, Fernandez MF, Olea N, Olea-Serrano MF. The E-Screen as a tool to identify estrogens: an update on estrogenic environmental pollutants. Environ Health Perspect. 1995;103:113–22.
18. Gray LE, Ostby J, Monosson E, Klece WR. Environmental antiandrogens: low doses of the fungicide vinclozolin alter sexual differentiation of the male rat. Toxicol Ind Health. 1999;15:48–64.
19. Sohoni P, Sumpter JP. Several environmental oestrogens are also antiandrogens. J Endocrinol. 1998;158:327–39.
20. Sultan C, Balaguer P, Terouanne B, Georget V, Paris F, Jeandel C, et al. Environmental xeno-estrogens, antiandrogens and disorders of male sexual differentiation. Mol Cell Endocrinol. 2001;178:99–105.
21. Olea N, Olea-Serrano F, Lardelli-Claret P, Rivas A, Barba-Navarro A. Inadvertent exposure to xenoestrogens in children. Toxicol Ind Health. 1999;15:151–8.
22. Dolk H, Vrijheid M. The impact of environmental pollution on congenital anomalies. Br Med Bull. 2003;68:25–45.
23. Swan SH, Main KM, Liu F, Stewart SL, Kruse RL, Calafat AM, et al. Decrease in anogenital distance among male infants with prenatal phthalate exposure. Environ Health Perspect. 2005;113:1056–61.
24. Krimsky S. Hormonal chaos: the scientific and social origins of the environmental endocrine hypothesis. Baltimore: Johns Hopkins University; 2000.
25. Lopez-Espinosa MJ, Granada A, Carreno J, Salvatierra M, Olea-Serrano F, et al. Organochlorine pesticides in placentas from Southern Spain and some related factors. Placenta. 2007;28(7): 631–8.

26. Christiansen S, Scholze M, Axelstad M, Boberg J, Kortenkamp A, Hass U. Combined exposure to anti-androgens causes markedly increased frequencies of hypospadias in the rat. Int J Androl. 2008;31(2):241–8.
27. Kortenkamp A. Low dose mixture effects of endocrine disrupters: implications for risk assessment and epidemiology. Int J Androl. 2008;31(2):233–40.
28. Almstrup K, Fernandez MF, Petersen JH, Olea N, Skakkebaek NE, Leffers H. Dual effects of phytoestrogens result in u-shaped dose-response curves. Environ Health Perspect. 2002;110: 743–8.
29. Weltje L, vom Saal FS, Oehlmann J. Reproductive stimulation by low doses of xenoestrogens contrasts with the view of hormesis as an adaptive response. Hum Exp Toxicol. 2005;24: 431–7.
30. Fernandez MF, Olmos B, Granada A, López-Espinosa MJ, Molina-Molina JM, Fernandez JM, et al. Human exposure to endocrine-disrupting chemicals and prenatal risk factors for cryptorchidism and hypospadias: a nested case-control study. Environ Health Perspect. 2007;115 Suppl 1:8–14.
31. Kristensen P, Irgens LM, Andersen A, Bye AS, Sundheim L. Birth defects among offspring of Norwegian farmers. Epidemiology. 1997;8:537–44.
32. Garcia AM, Flecher T, Benavides FG, Orts E. Parental agricultural work and selected congenital malformations. Am J Epidemiol. 1999;149:64–74.
33. Rueda-Domingo MT, Lopez-Navarrete E, Nogueras-Ocaña M, Lardelli-Claret P. Factores de riesgo de criptorquidia. Gac Sanit. 2001;15:398–405.
34. Carbone P, Giordano F, Nori F, Mantovani A, Taruscio D, Lauria L, et al. Cryptorchidism and hypospadias in the Sicilian district of Ragusa and the use of pesticides. Reprod Toxicol. 2006;22(1):8–12.
35. Carbone P, Giordano F, Nori F, Mantovani A, Taruscio D, Lauria L, et al. The possible role of endocrine disrupting chemicals in the aetiology of cryptorchidism and hypospadias: a population-based case-control study in rural Sicily. Int J Androl. 2007;30(1):3–13.
36. Andersen HR, Schmidt IM, Grandjean P, Jensen TK, Budtz-Jørgensen E, Kjaerstad MB, et al. Impaired reproductive development in sons of women occupationally exposed to pesticides during pregnancy. Environ Health Perspect. 2008;116(4):566–72.
37. Zhu JL, Hjollund NH, Andersen AM, Olsen J. Occupational exposure to pesticides and pregnancy outcomes in gardeners and farmers: a study within the Danish National Birth Cohort. J Occup Environ Med. 2006;48(4):347–52.
38. Irgens A, Kruger K, Skorve AH, Irgens LM. Birth defects and paternal occupational exposure. Hypotheses tested in a record linkage based data set. Acta Obstet Gynecol Scand. 2000;79:465–70.
39. Olshan AF, Teschke K, Baird PA. Paternal occupation and congenital anomalies in offspring. Am J Ind Med. 1991;20:447–75.
40. Garcia-Rodriguez J, Garcia-Martin M, Nogueras-Ocaña M, de Dios Luna-del-Castillo J, Espigares Garcia M, Olea N, et al. Exposure to pesticides and cryptorchidism: geographical evidence of a possible association. Environ Health Perspect. 1996;104:1090–5.
41. Restrepo M, Muñoz N, Day N, Parra JE, Hernandez C, Blettner M, et al. Birth defects among children born to a population occupationally exposed to pesticides in Colombia. Scand J Work Environ Health. 1990;16:239–46.
42. Pierik FH, Burdorf A, Deddens JA, Juttmann RE, Weber RFA. Maternal and paternal risk factors for cryptorchidism and hypospadias: a case-control study in newborn boys. Environ Health Perspect. 2004;112:1570–6.
43. Van Tongeren M, Nieuwenhuijsen MJ, Gardiner K, Armstrong B, Vrijheid M, Dolk H, et al. A job-exposure matrix for potential endocrine-disrupting chemicals developed for a study into the association between maternal occupation and hypospadias. Ann Occup Hyg. 2002;46: 465–77.
44. Vrijheid M, Armstrong B, Dolk H, van Tongeren M, Botting B. Risk of hypospadias in relation to maternal occupational exposure to potential endocrine disrupting chemicals. Occup Environ Med. 2003;60:543–50.

8 Developmental Exposure to Endocrine Disruptors... 239

45. Ormond G, Nieuwenhuijsen MJ, Nelson P, Toledano MB, Iszatt N, Geneletti S, et al. Endocrine disruptors in the workplace, hair spray, folate supplementation, and risk of hypospadias: case-control study. Environ Health Perspect. 2009;117(2):303–7.
46. Dolk H, Vrijheid M, Armstrong B, Abramsky L, Bianchi F, Garne E, et al. Risk of congenital anomalies near hazardous-waste landfill sites in Europe: the EUROHAZCON study. Lancet. 1998;352:423–7.
47. Hosie S, Loff S, Witt K, Niessen K, Waag KL. Is there a correlation between organochlorine compounds and undescended testes? Eur J Pediatr Surg. 2000;10:304–9.
48. Longnecker MP, Klebanoff MA, Brock JW, Zhou H, Gray KA, Needham LL, et al. Maternal serum level of 1,1-dichloro-2,2-bis(p-chlorophenyl)ethylene and risk of cryptorchidism, hypospadias, and polythelia among male offspring. Am J Epidemiol. 2002;155:313–22.
49. Bhatia R, Shiau R, Petreas M, Weintraub JM, Farhang L, Eskenazi B. Organochlorine pesticides and male genital anomalies in the child health and development studies. Environ Health Perspect. 2005;113:220–4.
50. Damgaard IN, Skakkebaek NE, Toppari J, Virtanen HE, Shen H, Schramm KW, et al. Persistent pesticides in human breast milk and cryptorchidism. Environ Health Perspect. 2006;114(7):1133–8.
51. Main KM, Kiviranta H, Virtanen HE, Sundqvist E, Tuomisto JT, Tuomisto J, et al. Flame retardants in placenta and breast milk and cryptorchidism in newborn boys. Environ Health Perspect. 2007;115(10):1519–26.
52. Shen H, Main KM, Andersson AM, Damgaard IN, Virtanen HE, Skakkebaek NE, et al. Concentrations of persistent organochlorine compounds in human milk and placenta are higher in Denmark than in Finland. Hum Reprod. 2008;23(1):201–10.
53. Brucker-Davis F, Wagner-Mahler K, Delattre I, Ducot B, Ferrari P, Bongain A, et al. Cryptorchidism at birth in Nice area (France) is associated with higher prenatal exposure to PCBs and DDE, as assessed by colostrum concentrations. Hum Reprod. 2008;23(8):1708–18.
54. Palmer JR, Herbst AL, Noller KL, Boggs DA, Troisi R, Titus-Ernstoff L, et al. Urogenital abnormalities in men exposed to diethylstilbestrol in utero: a cohort study. Environ Health. 2009;8:37.
55. Palmer JR, Wise LA, Robboy SJ, Titus-Ernstoff L, Noller KL, Herbst AL, et al. Hypospadias in sons of women exposed to diethylstilbestrol in utero. Epidemiology. 2005;16(4):583–6.
56. Christiansen S, Scholze M, Dalgaard M, Vinggaard AM, Axelstad M, Kortenkamp A, et al. Synergistic disruption of external male sex organ development by a mixture of four antiandrogens. Environ Health Perspect. 2009;117(12):1839–46.

Additional References

European science foundation. Male reproductive health. http://www.esf.org/publications/science-policy-briefings.html
Fernández MF, Rivas A, Olea-Serrano F, Cerrillo I, Molina-Molina JM, Araque P, et al. Assessment of total effective xenoestrogen burden in adipose tissue and identification of chemicals responsible for the combined estrogenic effect. Anal Bioanal Chem. 2004;379(1):163–70.
Main KM, Skakkebaek NE, Virtanen HE, Toppari J. Genital anomalies in boys and the environment. Best Pract Res Clin Endocrinol Metab. 2010;24(2):279–89. Review.
Paulozzi LJ, Erickson JD, Jackson RJ. Hypospadias trends in two American surveillance systems. Pediatrics. 1997;100:831–4.
Robaire B, Hales BF. Mechanism of action of cyclophosphamide as a malemediated developmental toxicant. Adv Exp Med Biol. 2003;518:169–80.
Weidner IS, Møller H, Jensen TK, Skakkebæk NE. Risk factors for cryptorchidism and hypospadias. J Urol. 1999;161(5):1606–9.

Chapter 9
EDC Exposures and the Development of Reproductive and Nonreproductive Behaviors

Craige C. Wrenn, Ashwini Mallappa, and Amy B. Wisniewski

Abstract An endocrine-disrupting compound (EDC) is "an exogenous agent that interferes with synthesis, secretion, transport, metabolism, binding action, or elimination of natural blood-borne hormones that are present in the body and responsible for homeostasis, reproduction, and developmental processes Diamanti-Kandarakis (Endocr Rev 30:293–342, 2009)." These compounds exert their actions through traditional nuclear receptors as well as nontraditional membrane-bound receptors to affect neurotransmitter and enzymatic pathways involved in steroid synthesis and metabolism. EDCs can be naturally occurring or synthetic. A great deal of research has considered the impact of EDCs on hormone-sensitive anatomic development and physiologic function. Less work has been conducted on the possible influences of EDCs on hormone-sensitive behaviors.

Keywords Aggression • Androgen • Anxiety • Copulation • Estrogen • Learning • Memory

Endocrine-Disrupting Compounds (EDCS)

An endocrine-disrupting compound (EDC) is "an exogenous agent that interferes with synthesis, secretion, transport, metabolism, binding action, or elimination of natural blood-borne hormones that are present in the body and responsible for

---

C.C. Wrenn (✉)
Department of Pharmaceutical, Biomedical and Administrative Sciences,
College of Pharmacy and Health Sciences, Drake University,
Des Moines, IA 50311, USA
e-mail: craige.wrenn@drake.edu

A. Mallappa • A.B. Wisniewski
Division of Pediatric Endocrinology, University of Oklahoma Health Sciences Center,
Oklahoma City, OK 73104, USA

E. Diamanti-Kandarakis and A.C. Gore (eds.), *Endocrine Disruptors and Puberty*,
Contemporary Endocrinology, DOI 10.1007/978-1-60761-561-3_9,
© Springer Science+Business Media, LLC 2012

homeostasis, reproduction, and developmental processes [1]." These compounds exert their actions through traditional nuclear receptors as well as nontraditional membrane-bound receptors to affect neurotransmitter and enzymatic pathways involved in steroid synthesis and metabolism. EDCs can be naturally occurring or synthetic. A great deal of research has considered the impact of EDCs on hormone-sensitive anatomic development and physiologic function. Less work has been conducted on the possible influences of EDCs on hormone-sensitive behaviors.

Organizing and Activating Behaviors

When behaviors are sexually dimorphic, it is likely that sex steroid hormones play a role in their development and maintenance. Generally speaking, early exposure to gonadal hormones is needed to organize or establish neural networks that support behaviors. However, later exposure (typically around the time of puberty) is needed to then activate the expression of these male- or female-typical networks [2]. Thus, pre- or perinatal exposure to EDCs is expected to have an organizing effect, while adolescent or adult exposure should exert an activating impact, on sexually dimorphic behaviors. Below we present evidence for both organizational and activational influences of a variety of EDCs on male reproductive behaviors, as well as on non-reproductive behaviors that differ between males and females.

EDCs Studied in the Context of Sexually Dimorphic Behaviors

Several EDCs have been identified and can generally be categorized as exerting actions that are estrogenic, antiestrogenic, androgenic, or antiandrogenic. In the context of behavioral studies, most EDCs investigated exert estrogenic and/or anti-androgenic effects. For example, ethinyl estradiol (EE2) is a main ingredient of oral contraceptives and hormone replacement therapies used by women affected by pre-mature ovarian failure or menopause. EE2 is excreted in urine and thus found in water supplies at concentrations sufficient to alter physiologic function [3,4]. Bisphenol A (BPA) is an estrogenic/antiandrogenic chemical necessary to manufacture plastics and epoxy resins [5,6]. Methoxychlor (MXC) is a widely used insecticide in the Unites States, and, similar to BPA, MXC exerts both estrogenic and antiandrogenic activity [7]. Vinclozolin (VIN) is a fungicide that has antiandrogenic activity [8,9], and 2,3,7,8-tetrachlorodibenzodioxin (TCDD) is a herbicide that exerts both estrogenic and antiestrogenic actions [10,11]. Phthalates are used to manufacture plastics and personal-care products and have been shown to exhibit antiandrogenic actions [12]. Finally, genistein is a plant-derived EDC found in soybeans that exerts both estrogenic and antiestrogenic activity [13,14].

Reproductive Behaviors

Reproductive behaviors include any activities that bring together a sperm and an egg. Reproductive behaviors are evident with the onset and progression of puberty and can be divided into two types of activities – appetitive and consummatory. Appetitive behaviors include any actions that involve attracting, or competing for, a sexual partner. Consummatory behaviors are those that constitute the act of copulation with a sexual partner. As both types of behaviors emerge at puberty, the importance of endogenous sex steroids for their development and maintenance is clear [2]. Thus, appetitive and consummatory behaviors are both likely candidates to be influenced by the actions of EDCs.

Appetitive Behaviors. Attracting and competing for mates can be influenced by EDCs that exert estrogenic, antiestrogenic, androgenic, or antiandrogenic influences. Sand gobies (*Pomatoschistus minutus*) are small fish that spawn with multiple partners over the course of 1 year. In general, females of this species prefer to mate with large males. Male sand gobies attract females by fanning their pectoral fins, building nests, and caring for females' eggs deposited in those nests.

Female sand gobies prefer to mate with males not treated with the estrogenic EDC EE2 compared to males exposed to this EDC [15]. Furthermore, females do not exhibit the usual mating preference for larger animals when males are exposed to EE2 [16]. Although not known at this time, perhaps this decreased preference is due to slower pectoral fin fanning [17] or the impaired ability to compete for nests following EE2 treatment of male sand gobies [15]. In fathead minnows (*Pimephales promelas*), exposure to EE2 decreases the ability of males to successfully compete for, and defend, mating territories [18]. Thus, evidence for associations between EE2 exposure and impaired appetitive behaviors, particularly behaviors that are needed to compete for females, is evident in several species of fish.

Appetitive behaviors are also adversely influenced by EDCs in rodent species. Male Sprague Dawley (*Rattus norvegicus*) rats exposed to low doses of bisphenol-A (BPA) during gestation show more defensive interactions with competitors compared to controls [19], and mice exposed to the estrogenic EDC methoxychlor during gestation are less likely to approach or show interest in sexually receptive females [20]. Remarkably, EDCs not only negatively impact appetitive behaviors across multiple species, but they can do so across multiple generations within a species. For example, F_3 generation female rats that descended from F_0 females exposed to the antiandrogenic fungicide, vinclozolin, prefer to mate with F_3 males that have no history of EDC exposure [21].

Estrogenic EDCs such as EE2 and methoxychlor hinder the ability of males to both attract and compete for potential mates. Antiandrogenic EDCs such as vinclozolin impair the ability of males to attract females. While many studies of EDCs focus on abnormal male reproductive structure and function, fewer consider the impact of these compounds on behaviors that are necessary for mating to actually occur. When appetitive behaviors are investigated in the context of EDC exposure, impairment in the ability of males to attract or compete for females is frequently observed.

Consummatory Behaviors. EDCs adversely impact the ability of males to copulate in birds, rodents, and humans. Male Japanese quail (*Coturnix japonica*) exposed to methoxychlor during early incubation attempt fewer copulations with females compared to their unexposed counterparts [20]. In contrast, male mice exposed to methoxychlor postnatally exhibit more frequent intromissions and less time to ejaculate than controls [20]. Long Evans (*Rattus norvegicus*) rats exposed to the dioxin 2,3,7,8-Tetrachlorodibenzodioxin (TCDD) prenatally also exhibit more frequent intromissions with receptive females; however, they show an increased latency to ejaculate [22]. TCDD exerts both estrogenic and antiestrogenic actions, potentially contributing to the ambiguous outcomes observed with male copulatory behavior. Long Evans rats exposed to diets rich in the phytoestrogen genistein, during both gestation and lactation, are less likely to mount, intromit, and ejaculate with receptive females compared to unexposed males [23]. Males treated with vinclozolin prenatally also fail to intromit and ejaculate when exposed to receptive females [24]. Sprague Dawley rats treated with the estrogenic di(2-ethylhexyl) phthalate (DEHP) pre- and postnatally exhibit decreased mounts, intromissions, and ejaculations with receptive females compared to controls [25]. Finally, men exposed to BPA in the workplace report a fourfold increased risk for erectile dysfunction and a sevenfold increased risk for ejaculatory problems compared to men not routinely exposed [26].

In contrast to male appetitive behaviors that show consistent impairment following both pre- and postnatal exposure to estrogenic and antiandrogenic EDCs, the impact of EDCs on male consummatory behaviors is more varied. In both birds and humans, exposure to estrogenic EDCs is consistently associated with decreased male copulatory behaviors. In rodents, exposure to estrogenic EDCs results in both decreased and increased male copulatory behaviors. It is not clear at this time why consummatory behaviors respond to EDCs in a more variable manner than their appetitive counterparts.

Nonreproductive Behaviors

Numerous nonreproductive behaviors are sexually dimorphic including spatial learning and memory [27–30], anxiety-like behavior [31–33], and aggression [34]. As expected, each of these behaviors can be altered by manipulating sex steroid exposure during early development and at puberty. Hence, similar to reproductive behaviors, the potential of EDCs to impact spatial learning and memory, anxiety-like behavior, and aggression is high.

Spatial Learning and Memory. Spatial learning and memory is an attractive paradigm for exploring the endocrine disruption of sexually dimorphic, nonreproductive behaviors because male-typical patterns of spatial learning and memory result from

9 EDC Exposures and the Development of Reproductive... 245

the organizational effects of gonadal hormones on brain development [35,36]. Furthermore, investigators have shown that manipulation of gonadal hormones during development results in morphological changes in hippocampal neurons, and these structural changes correlate with spatial learning performance [37–39].

Lund et al. tested the hypothesis that exposure to dietary phytoestrogens alters spatial learning in memory in rats [40,41]. The design of these studies was to feed female rats either a phytoestrogen-free diet or one containing phytoestrogens (Phyto-600). These females were mated, and their pups were then fed the same diet so that EDC exposure in the phytoestrogen group was "lifelong." In adulthood, these offspring were tested on a radial maze that consists of a central hub with eight arms radiating out from the hub. In this task, food rewards are located at the end of these arms, and the animals are required to learn which arms consistently contain a reward from trial to trial (reference memory) and which food rewards have been taken within a trial (working memory). Males with "lifelong" dietary exposure to phytoestrogens were slower to acquire criterion performance in the radial maze than males on a diet free of phytoestrogens [41].

In a similar experiment, after acquisition of the learning task was complete, males were either switched to a phytoestrogen-free diet, or they remained on the phytoestrogen-containing diet for 25 days. These rats were then tested for 15 days on a task in which only four of the eight arms contained a food reward. Entries into never baited arms were recorded as reference memory errors; re-entries into arms within a trial were recorded as working memory errors. Males that remained on the phytoestrogen diet made more reference memory errors and fewer correct choices than males switched to the phytoestrogen-free diet [41].

The above-mentioned studies show that phytoestrogen exposure impairs male-typical patterns of spatial learning, but they leave questions unanswered regarding when the exposure must occur to exert these deleterious consequences. For example, are the effects of phytoestrogens on spatial cognition due to organizational or activational mechanisms of action? In other words, do phytoestrogens interfere with the development of the hippocampus resulting in impaired spatial learning later in adulthood (an organizational effect) or do they impair spatial learning by disrupting neural circuits during learning trials (an activational effect)? The switching of diets from phytoestrogen-containing to phytoestrogen-free food suggests an activational mechanism; however, continuous exposure to phytoestrogens from conception onward cannot discriminate between an organizational or activational mechanism of action.

A recent study in our laboratory addressed this issue by limiting exposure to genistein, the most common phytoestrogen found in soy [42,43], to either the gestational or lactational periods. Pregnant dams were fed chow containing 5 mg/kg of genistein only during gestation, only during lactation, or during both developmental periods [44]. Pregnant dams fed genistein-free food were used as the control group. At weaning, all males were switched to genistein-free food and tested for spatial learning as adults.

To test spatial learning in these males, we used the Morris water maze which requires the animals to learn the location of a hidden platform in a pool of water.

Latency to find the platform over several days of training decreases as the subjects learn the spatial location. We observed that male rats exposed to genistein during *both* gestation and lactation had higher latencies to find the platform. Graphically, this difference was seen as a shift to the right in the learning curve in the genistein-exposed rats. Male rats that were exposed to genistein only during gestation or only during lactation did not differ from controls in their latencies to find the platform [44]. These water maze data clearly show that genistein exerts an organizational effect on spatial learning and memory in males. An interesting aspect of these data is that no impairment was seen when genistein exposure was limited to either gestation or lactation suggesting that organizational effects during one time period can be mitigated by lack of exposure at other points in development.

While it is clear that developmental exposure to phytoestrogens has a negative impact on spatial learning and memory in male rats, many questions remain and should be the focus of future studies. Given that manipulation of gonadal hormones during early development produces morphological changes in hippocampal neurons and that these changes correlate with spatial learning ability [37–39], it seems obvious that the effect of phytoestrogens on hippocampal development should be examined to determine if they correlate with impairments in spatial learning and memory. Also, the mechanism through which phytoestrogens alter behavior and presumably the hippocampus should be identified. For example, genistein is known to bind to both types of estrogen receptors, ERα and ERβ, and to act as either an agonist [45] or antagonist [46] depending on the physiological context of the exposure. Other possible mechanistic explanations include effects on the expression of androgen receptors [47,48], androgen metabolism [49], and tyrosine kinase inhibition [50].

Phytoestrogens and Anxiety-Like Behavior. In addition to spatial learning and memory, anxiety-like behavior is sexually dimorphic in rats. In the elevated plus maze, a validated behavioral test of anxiety-like behavior in rodents [51,52], males exhibit more anxiety-like behavior than females [31–33]. Similar to spatial cognition, this sexual dimorphism is sensitive to manipulations of gonadal hormones during specific points in development [31,32].

As to the question of whether exposure to EDCs such as phytoestrogens alter anxiety-like behavior in males, some of the first studies were conducted by Lund and Lephart using the same diet that they used in their spatial cognition studies described above [53]. As before, the male rats had "lifelong" exposure to either phytoestrogen-containing or phytoestrogen-free food. The animals were tested in adulthood in the elevated plus maze in order to assess anxiety-like behavior. This apparatus consists of two runway arms of equal length that intersect at a right angle such that the shape of a "plus sign" is formed. This configuration provides four runways that animals can explore during a brief test session. The baseline behavior of rats is to prefer to explore the closed arms, where hiding from predators and other dangers is easier, and to avoid the presumably aversive open arms. This preference is interpreted by behavioral pharmacologists as an "anxiety-like" behavior because administration of antianxiety medications will decrease the time rats spend in the closed arms (for a full discussion of the validity of this paradigm see [52]). Thus, entries or time spent in the open arms are used as measures of whether an experimental manipulation increases anxiety-like behavior (anxiogenic) or decreases anxiety-like behavior (anxiolytic). When Lund and Lephart compared time spent in the open arms between male rats fed phytoestrogens throughout life versus those never exposed to EDCs, they found that males on the phytoestrogen diet spent significantly more time on the open arms, illustrating an anxiolytic effect of the treatment.

A study by Hartley et al. [54] addressed whether phytoestrogens alter anxiety-like behavior through activational mechanisms by randomly allocating adult male rats to a phytoestrogen-free diet or a diet containing 150 µg/g of the phytoestrogens genistein and daidzein. Males remained on these diets for 18 days and were then tested in two tests of anxiety-like behavior: the social interaction test [55] and elevated plus maze. In the social interaction test, the amount of time that the male rats spent in engaging in various social behaviors (sniffing, following, grooming, boxing, and wrestling) with another rat was measured. Rats that were on the phytoestrogen diet spent significantly less time engaging in social interactions which is interpreted as an anxiogenic effect. In the elevated plus maze, the phytoestrogen-fed rats had significantly fewer entries into the open arms, another anxiogenic effect. These behavioral findings were bolstered by measurements of plasma corticosterone concentration that showed that the phytoestrogen-fed rats had increased levels of this stress hormone in response to handling.

Patisaul and colleagues examined the question of whether phytoestrogens affect anxiety-like behavior using both dietary exposure and systemic injection. In the first of these studies [56], these investigators prepared rodent diet that was mixed with commercially available supplements containing genistein, daidzein, and other isoflavones. The resulting diet had an isoflavone content of 95 mg/kg of feed. Male rats in this study were maintained on the diet or on a soy-free diet for 1 week prior to testing in the elevated plus maze in order to test for an activational effect on anxiety-like behavior. Males that had been eating the phytoestrogen-containing diet spent significantly less time in the open arms, again suggestive of an anxiogenic effect.

More recent studies by Patisaul et al. [57] have focused specifically on the compound equol, which is a metabolite of the soy phytoestrogen daidzein [58]. Equol, at a dose of 10 mg/kg, was injected into male neonates daily from PND 0 to 4 [57]. These animals were then tested for anxiety-like behavior around 2 months of age (PND 56–61) in an elevated plus maze. During all phases of their experiment, the rats ate only phytoestrogen-free food so that any alterations in anxiety-like behavior could be attributed to organizational actions of the EDC. The results of this study showed an anxiogenic effect of equol in which male rats that were injected with equol as neonates had significantly fewer open-arm entries and significantly less open-arm time relative to vehicle-injected controls.

In a third set of studies, Patisaul's group tested whether equol had later, activating effects on anxiety-like behavior. In one experiment, these investigators performed three daily subcutaneous injections of 3 mg/kg of equol in 3-month-old male rats and then tested the animals in a light–dark box. This apparatus consists of a two-chambered box, one that has transparent or white walls (the "light" chamber) and a second that has black walls (the "dark" chamber). Based on a similar ethological reasoning that supports the elevated plus maze, rats' preference for the dark chamber is interpreted as an anxiety-like behavior, and anxiolytic drugs increase time spent in the light chamber [52]. Equol exposure did not influence the amount of time spent in the light chamber nor transitions into the light chamber. In another experiment in the same set of studies, the authors injected a higher dose of equol (20 mg/kg) daily for 3 days in adult male rats. No effect on open-arm entries or open-arm time was found.

Currently available findings regarding phytoestrogen exposure and anxiety-like behavior are discrepant as there are data in support of anxiolytic, anxiogenic, and null effects. However, the referenced studies are also quite varied in their design. For example, the only study that used prenatal exposure to EDCs showed an anxiolytic effect [53], while postnatal or adult treatments were anxiogenic [54,56,58]. Thus, it seems reasonable to hypothesize that prenatal exposure may have an organizational effect on neural systems that result in the inhibition of anxiety-like behavior in adulthood, while postnatal exposures exert an opposite effect. Formally testing this hypothesis will require an experimental design that limits exposure to critical time periods similar to what was done in the learning and memory studies described above [44].

Phytoestrogens and Aggression. Aggression is more commonly displayed by males of many species and is influenced by exposure to sex steroid hormones [59,60]. As such, aggression is another nonreproductive, sexually dimorphic behavior that is likely to be sensitive to EDC exposure.

Wisniewski and colleagues used C57BL/6 mice to examine whether the phytoestrogen genistein could exert organizational effects on aggressive behavior in this animal model by limiting exposure to the gestational and lactational periods [61]. Similar to the spatial learning studies described earlier, the design was to allow dams to eat food containing 0, 5, or 300 mg/kg of genistein. At weaning, male offspring were placed on a phytoestrogen-free diet and were tested for aggressive behavior at 70 days of age using a resident-intruder paradigm. In this behavioral test, an intruder male is placed in the subject's (resident) home cage, and offensive (attacks, boxing, tail rattles, and dominant posture) and defensive (retreats, freezing, and subordinate posture) behaviors are recorded. These investigators found no effect of genistein exposure during development on offensive behaviors, but a significant increase in defensive behavior in mice exposed to the 5 mg/kg concentration of food was observed. This effect was nonmonotonic as the higher concentration of 300 mg/kg did not alter behavior. These data were interpreted as showing that developmental exposure to genistein causes a decrease in male-typical aggression in adulthood.

Counter to the results of Wisniewski et al., male CD-1 (*Mus musculus*) mice exposed to BPA during gestation exhibit increased aggression toward unexposed animals at the beginning of sexual maturity, but not later in adulthood [62]. Adult male Syrian hamsters (*Mesocricetus auratus*) fed either a phytoestrogen-containing or phytoestrogen-free diet for 4 weeks engage in significantly more attacks on conspecifics, and this behavioral finding correlated with elevated serum testosterone [63].

The most recent study regarding the effects of phytoestrogens on aggressive behavior was conducted by Patisaul and Bateman [58]. Like the study by Wisniewski

9 EDC Exposures and the Development of Reproductive...

et al., this work tested an organizational hypothesis by limiting phytoestrogen exposure to an early developmental period. Equol (10 mg/kg) was injected subcutaneously in male neonatal rats once a day from PND 0 to PND 4. In adulthood (PND 121), aggression was assayed using the resident-intruder paradigm. The results showed that equol significantly increased the number of aggressive bouts compared to vehicle-injected controls. Counter to the data in mice from Wisniewski et al., these data were consistent with an organizational effect of the phytoestrogen in which aggressive behavior is increased.

As is the case with the data concerning anxiety-like behavior, the aggression findings are discrepant, but the discrepancies are very likely due to differences in very important experimental parameters. These parameters include species, specific phytoestrogen administered, route of administration, and timing of exposure. Future studies should focus on systematically varying these parameters in order to establish whether phytoestrogens alter male-typical aggressive behavior and whether any alterations are due to organizational or activational mechanisms.

Conclusions

In the current chapter, we have reviewed studies reporting effects of EDCs on behavior in males that emerge at puberty. One important point that can be made from these studies is that the effects of EDCs on male-typical behaviors are not limited to a specific domain but include reproductive behavior (both appetitive and consummatory), spatial cognition, anxiety-like behavior, and aggression. In each case, EDCs alter behaviors that are organized by and sensitive to gonadal hormones. We focused on male behaviors for the current chapter, but we do not wish the reader to assume that females are not sensitive to the deleterious influences of EDCs on behavior [64–66].

Hormones not only are important regulators of development and physiology, but they also play important organizational and activational roles in behavior. For this reason, EDCs that affect hormone function are expected to have significant behavioral impacts. Future animal studies should be designed in a way that can identify critical periods of EDC exposure and discern between organizational and activational effects of this compounds. These animal studies will be important in helping us understand the risks that EDCs found in the environment pose to human health and behavior.

References

1. Diamanti-Kandarakis E, Bourguignon JP, Giudice LC, Hauser R, Prins GS, Soto AM, Zoeller T, Gore AC. Endocrine-disrupting chemicals: an Endocrine Society Scientific Statement. Endocr Rev. 2009;30:293–342.
2. Nelson RJ, editor. An introduction to behavioral endocrinology. 3rd ed. Sunderland: Sinauer Associates; 2005.

3. Purdom CE, Hardiman PA, Bye VVJ, Eno NC, Tyler CR, Sumpter JP. Estrogenic effects of effluent from sewage treatment works. Chem Ecol. 1994;8:275–85.
4. Bell AM. Effects of an endocrine disruptor on courtship and aggressive behavior of male three-spined stickleback, *Gasterosteus aculeatus*. Anim Behav. 2001;62:775–80.
5. Krishnan AV, Stathis P, Permuth SF, Tokes L, Feldman D. Bisphenol-A: an estrogenic substance is released from polycarbonate flasks during autoclaving. Endocrinology. 1993;132: 2279–86.
6. Lee HJ, Chattopadhyay S, Gong EY, Ahn RS, Lee K. Antiandrogenic effects of bisphenol A and nonylphenol on the function of androgen receptor. Toxicol Sci. 2003;75:40–6.
7. Maness SC, McDonnell DP, Gaido KW. Inhibition of androgen receptor-dependent 9 transcriptional activity by DDT isomers and methoxychlor in HepG2 human hepatoma cells. Toxicol Appl Pharmacol. 1998;151:135–42.
8. Gray Jr LE, Ostby J, Monosson E, Kelce WR. Environmental antiandrogens: low doses of the fungicide vinclozolin alter sexual differentiation of the male rat. Toxicol Ind Health. 1999; 15:48–64.
9. McGary S, Henry PF, Ottinger MA. Impact of vinclozolin on reproductive behavior and endocrinology in Japanese quail (*Coturnix coturnix japonica*). Environ Toxicol Chem. 2001;20: 2487–93.
10. Mably TA, Moore RW, Peterson RE. In utero and lactational exposure of male rats to 2,3,7,8-tetrachlorodibenzo-p-dioxin: effects on androgenic status. Toxicol Appl Pharmacol. 1992;114:97–107.
11. Mably TA, Moore RW, Goy RW, Peterson RE. In utero and lactational exposure of male rats to 2,3,7,8-tetrachlorodibenzo-p-dioxin: Effects on sexual behavior and the regulation of luteinizing hormone secretion in adulthood. Toxicol Appl Pharmacol. 1992;114:108–17.
12. Lehmann KP, Phillips S, Sar M, Foster PM, Gaido KW. Dose-dependent alterations in gene expression and testosterone synthesis in the fetal testes of male rats exposed to di (n-butyl) phthalate. Toxicol Sci. 2004;81:60–8.
13. Morito K, Hirose T, Kinjo J, Hirakawa T, Okawa M, Nohara T, Ogawa S, Inoue S, Muramatsu M, Masamune Y. Interaction of phytoestrogens with estrogen receptors alpha and beta. Biol Pharm Bull. 2001;24:351–6.
14. Setchell K. Phytoestrogens: the biochemistry, physiology, and implications for human health of soy isoflavones. Am J Clin Nutr. 1998;68:1333S–46S.
15. Saaristo M, Craft JA, Lehtonen KK, Lindström K. Sand goby (*Pomatoschistus minutus*) males exposed to an endocrine disrupting chemical fail in nest and mate competition. Horm Behav. 2009;56:315–21.
16. Saaristo M, Craft JA, Lehtonen KK, Björk H, Lindström K. Disruption of sexual selection in sand gobies (Pomatoschistus minutus) by 17α-ethinyl estradiol, an endocrine disruptor. Horm Behav. 2009;55:530–7.
17. Saaristo M, Craft JA, Lehtonen KK, Lindström K. An endocrine disrupting chemical changes courtship and parental care in the sand goby. Aquat Toxicol. 2010;97:285–92.
18. Majewski AR, Blanchfield PJ, Palace VP, Wautier K. Waterborne 17α-ethynylestradiol affects aggressive behaviour of male fathead minnows (*Pimephales promelas*) under artificial spawning conditions. Water Qual Res J Canada. 2002;37:697–710.
19. Farabollini F, Porrini S, Della Seta D, Bianchi F, Dessi-Fulgheri F. Effects of perinatal exposure to bisphenol A on sociosexual behavior of female and male rats. Environ Health Perspect. 2002;110:409–13.
20. Eroschenko VP, Amstislavsky SY, Schwabel H, Ingermann RL. Altered behaviors in male mice, male quail, and salamander larvae following early exposures to the estrogenic pesticide methoxychlor. Neurotoxicol Teratol. 2002;24:29–36.
21. Crews D, Gore AC, Hsu TS, Dangleben NL, Spinetta M, Schallert T, Anway MD, Skinner MK. Transgenerational epigenetic imprints on mate preference. PNAS. 2007;104:5942–6.
22. Gray LE, Kelce WR, Monosson E, Ostby JS, Birnbaum LS. Exposure to TCDD during development permanently alters reproductive function in male Long Evans rats and hamsters:

reduced ejaculated and epididymal sperm numbers and sex accessory gland weights in offspring with normal androgenic status. Toxicol Appl Pharmacol. 1995;131:108–18.

23. Wisniewski AB, Klein SL, Lakshmanan Y, Gearhart JP. Exposure to genistein during gestation and lactation demasculinizes the reproductive system in rats. J Urol. 2003;169:1582–6.

24. Gray LE, Ostby JS, Kelce WR. Developmental effects of an environmental antiandrogen: the fungicide vinclozolin alters sex differentiation of the male rat. Toxicol Appl Pharmacol. 1994;129:46–52.

25. Moore RW, Rudy TA, Lin TM, Ko K, Peterson RE. Abnormalities of sexual development in male rats with in utero and lactational exposure to the antiandrogen plasticizer di(2-ethylhexyl) phthalate. Environ Health Perspect. 2001;109:229–37.

26. Li D, Zhou Z, Qing D, He Y, Wu T, Miao M, Wang J, Weng X, Ferber JR, Herrinton LJ, Zhu Q, Gao E, Checkoway H, Yuan W. Occupational exposure to bisphenol-A (BPA) and the risk of self-reported male sexual dysfunction. Hum Reprod. 2010;25:519–27.

27. Joseph R. Effects of rearing and sex on maze learning and competitive exploration in rats. J Psychol. 1979;101:37–43.

28. Beatty WW. Hormonal organization of sex differences in play fighting and spatial behavior. Prog Brain Res. 1984;61:320–4.

29. Jonasson Z. Meta-analysis of sex differences in rodent models of learning and memory: a review of behavioral and biological data. Neurosci Biobehav Rev. 2005;28(8):811–25.

30. Blokland A, Rutten K, Prickaerts J. Analysis of spatial orientation strategies of male and female Wistar rats in a Morris water escape task. Behav Brain Res. 2006;171:216–24.

31. Leret ML, Molina-Holgado F, Gonzalez MI. The effect of perinatal exposure to estrogens on the sexually dimorphic response to novelty. Physiol Behav. 1994;55:371–3.

32. Lucion AB, Charchat H, Pereira GAM, Rasia-Filho AA. Influence of early postnatal gonadal hormones on anxiety in adult male rats. Physiol Behav. 1996;60:1419–23.

33. Aguilar R, Gil L, Gray JA, et al. Fearfulness and sex in F2 Roman rats: males display more fear though both sexes share the same fearfulness traits. Physiol Behav. 2003;78:723–32.

34. Edwards DA. Mice: fighting by neonatally androgenized females. Science. 1968;161:1027–8.

35. Joseph R, Hess S, Birecree E. Effects of hormone manipulations and exploration on sex differences in maze learning. Behav Biol. 1978;24:364–77.

36. Williams CL, Barnett AM, Meck WH. Organizational effects of early gonadal secretions on sexual differentiation in spatial memory. Behav Neurosci. 1990;104:84–97.

37. Isgor C, Sengelaub DR. Effects of neonatal gonadal steroids on adult CA3 pyramidal neuron dendritic morphology and spatial memory in rats. J Neurobiol. 2003;55(2):179–90.

38. Isgor C, Sengelaub DR. Prenatal gonadal steroids affect adult spatial behavior, CA1 and CA3 pyramidal cell morphology in rats. Horm Behav. 1998;34(2):183–98.

39. Roof RL, Havens MD. Testosterone improves maze performance and induces development of a male hippocampus in females. Brain Res. 1992;572(1–2):310–3.

40. Lund TD, Lephart ED. Manipulation of prenatal hormones and dietary phytoestrogens during adulthood alter the sexually dimorphic expression of visual spatial memory. BMC Neurosci. 2001;2:21.

41. Lund TD, West TW, Tian LY, et al. Visual spatial memory is enhanced in female rats (but inhibited in males) by dietary soy phytoestrogens. BMC Neurosci. 2001;2:20.

42. Setchell KDR, Zimmer-Nechemias L, Cai J, Heubi JE. Exposure of infants to phyto-oestrogens from soy-based infant formula. Lancet. 1997;350(9070):23.

43. Franke A, Custer L. Daidzein and genistein concentrations in human milk after soy consumption. Clin Chem. 1996;42(6):955–64.

44. Ball ER, Caniglia MK, Wilcox JL, et al. Effects of genistein in the maternal diet on reproductive development and spatial learning in male rats. Horm Behav. 2010;57(3):313–22.

45. Kuiper GGJM, Lemmen JG, Carlsson B, et al. Interaction of estrogenic chemicals and phytoestrogens with estrogen receptor β. Endocrinology. 1998;139(10):4252–63.

46. Patisaul HB, Dindo M, Whitten PL, Young LJ. Soy isoflavone supplements antagonize reproductive behavior and estrogen receptor α- and β-dependent gene expression in the brain. Endocrinology. 2001;142(7):2946–52.

47. Fritz WA, Wang J, Eltoum I-E, Lamartiniere CA. Dietary genistein down-regulates androgen and estrogen receptor expression in the rat prostate. Mol Cell Endocrinol. 2002;186(1): 89–99.
48. Fritz WA, Cotroneo MS, Wang J, Eltoum I-E, Lamartiniere CA. Dietary diethylstilbestrol but not genistein adversely affects rat testicular development. J Nutr. 2003;133(7):2287–93.
49. Weber KS, Jacobson NA, Setchell KD, Lephart ED. Brain aromatase and 5-alpha reductase, regulatory behaviors and testosterone levels in adult rats on phytoestrogen diets. Proc Soc Exp Biol Med. 1999;221(2):131–5.
50. O'Dell TJ, Kandel ER, Grant SGN. Long-term potentiation in the hippocampus is blocked by tyrosine kinase inhibitors. Nature. 1991;353:558–60.
51. Pellow S, Chopin P, File SE, Briley M. Validation of open: closed arm entries in an elevated plus-maze as a measure of anxiety in the rat. J Neurosci Methods. 1985;14(3):149–67.
52. Cryan JF, Holmes A. The ascent of mouse: advances in modelling human depression and anxiety. Nat Rev Drug Discov. 2005;4:775–90.
53. Lund TD, Lephart ED. Dietary soy phytoestogens produce anxiolytic effects in the elevated plus-maze. Brain Res. 2001;913:180–4.
54. Hartley DE, Edwards JE, Spiller CE, et al. The soya isoflavone content of rat diet can increase anxiety and stress hormone release in the male rat. Psychopharmacology. 2003;167:46–53.
55. File SE, Seth P. A review of 25 years of the social interaction test. Eur J Pharmacol. 2003; 463(1–3):35–53.
56. Patisaul HB, Blum A, Luskin JR, Wilson ME. Dietary soy supplements produce opposite effects on anxiety in intact male and female rats in the elevated plus maze. Behav Neurosci. 2005;119:587–94.
57. Patisaul HB, Bateman HL. Neonatal exposure to endocrine active compounds or an ERβ agonist increases adult anxiety and aggression in gonadally intact male rats. Horm Behav. 2008;53(4):580–8.
58. Setchell KDR, Faughnan MS, Avades T, et al. Comparing the pharmacokinetics of daidzein and genistein with the use of 13 C-labeled tracers in premenopausal women. Am J Clin Nutr. 2003;77(2):411–9.
59. Trainor BC, Kyomen HH, Marler CA. Estrogenic encounters: how interactions between aromatase and the environment modulate aggression. Front Neuroendocrinol. 2006;27(2):170–9.
60. Compaan JC, Wozniak A, De Ruiter AJH, Koolhaas JM, Hutchison JB. Aromatase activity in the preoptic area differs between aggressive and nonaggressive male house mice. Brain Res Bull. 1994;35(1):1–7.
61. Wisniewski AB, Cernetich A, Gearhart JP, Klein SL. Perinatal exposure to genistein alters reproductive development and aggressive behavior in male mice. Physiol Behav. 2005; 84(2):327–34.
62. Kawai K, Nozaki T, Nishikata H, Aou S, Masato T, Kubo C. Aggressive behavior and serum testosterone concentration during the maturation process of male mice: the effects of fetal exposure to bisphenol A. Environ Health Perspect. 2003;111:175–8.
63. Moore TO, Karom M, O'Farrell L. The neurobehavioral effects of phytoestrogens in male Syrian hamsters. Brain Res. 2004;1016(1):102–10.
64. Razzoli M, Valsecchi P, Palanza P. Chronic exposure to low doses bisphenol A interferes with pair-bonding and exploration in female Mongolian gerbils. Brain Res Bull. 2005;65:249–54.
65. Porrini S, Belloni V, Della Seta D, Farabollini F, Giannelli G, Dessi-Fulgheri F. Early exposure to a low dose of bisphenol A affects socio-sexual behavior of juvenile female rats. Brain Res Bull. 2005;65:261–6.
66. Steinberg RM, Juenger TE, Gore AC. The effects of prenatal PCBs on adult female paced mating reproductive behaviors in rats. Horm Behav. 2007;51:364.

Part III
Developmental Exposure to Endocrine Disruptors and Metabolic Disorders

Chapter 10
Adipocytes as Target Cells for Endocrine Disruption

Amanda Janesick and Bruce Blumberg

Abstract Throughout the first two decades of human development, growth of adipose tissue primarily results from an increase in adipocyte number (hyperplasia). Once established, the number of fat cells remains relatively constant, and adipose tissue grows larger primarily by filling the resident cells with more fat (hypertrophy). Obese children exhibit an augmented rate of increase in adipocyte number, and correspondingly, obese adults possess more fat cells. Hence, childhood and adolescence appear to be critical periods in establishing the number of fat cells. Endocrine-disrupting chemicals (EDCs) can predispose a child to obesity by influencing all aspects of adipose tissue growth, starting from multipotent stromal cells (MSCs) and ending with mature adipocytes. EDC exposure can increase the number of preadipocytes, enhance the differentiation of preadipocytes into adipocytes, and augment the uptake of fat into existing adipocytes. Unlike genetic mechanisms, which require a mutation event, EDCs have the capacity to quantitatively alter gene expression by modulating cellular-signaling pathways and by introducing epigenetic changes that also alter gene expression. Therefore, EDC exposure could foster a swift change in the metabolic profile of a population, which might provide at least a partial explanation for the rapid rise in obesity. This review focuses on the developmental origins of the adipocyte and its connection to early-onset obesity with the aim of providing a foundation for formulating hypotheses regarding how EDCs can interfere with adipogenesis and contribute to the obesity epidemic.

A. Janesick (✉)
Department of Developmental and Cell Biology, 2011 Biological Sciences 3,
University of California, Irvine, CA, USA
e-mail: ajanesic@uci.edu

B. Blumberg
Department of Developmental and Cell Biology, 2011 Biological Sciences 3,
University of California, Irvine, CA, USA

Department of Pharmaceutical Sciences, University of California, Irvine, CA, USA
e-mail: Blumberg@uci.edu

E. Diamanti-Kandarakis and A.C. Gore (eds.), *Endocrine Disruptors and Puberty*,
Contemporary Endocrinology, DOI 10.1007/978-1-60761-561-3_10,
© Springer Science+Business Media, LLC 2012

Keywords Adipocyte • Adipogenesis • Endocrine-disrupting chemical (EDC) • Endocrine disruption • Epigenetic reprogramming • Epigenetics • Multipotent stromal cell (MSC) • Nuclear receptor • Obesity • Obesogen • PPAR gamma • Stem cell • Transgenerational effects

Introduction

On April 20, 2010, a group of retired military leaders reported that 27% of all Americans aged 17–24 were too overweight to join the military [1]. Although current recruitment goals are satisfied, the concern is that this trajectory will continue until only a small proportion of America's youth will be qualified to serve [2]. This is supported by a recent study projecting that by 2030, 86% of Americans will be overweight, contributing to upwards of 900 billion dollars yearly in additional health care costs [3]. Notwithstanding the national security implications, the major concern with childhood obesity is its strong link with and ability to predict adult obesity and metabolic disease [4–7]. Indeed, a recent longitudinal study showed that increased waist diameter and body mass index (BMI) in children and adolescents increased the risk for abdominal obesity, insulin resistance, thrombotic disorder, elevated low-density lipoproteins ("bad" cholesterol), and systemic inflammation, later in life [8, 9]. These are the signatures of metabolic syndrome, and while adults might rise above this outcome, prevention at earlier stages in life would be more effective.

Throughout the first two decades of human development, growth of adipose tissue primarily results from an increase in adipocyte number (hyperplasia) [10, 11]. After this, the number of fat cells remains relatively constant, and adipose tissue grows larger primarily by filling the resident cells with more fat (hypertrophy) [11]. Obese children exhibit an augmented rate of increase in adipocyte number, and correspondingly, obese adults possess more fat cells [10, 11]. Hence, researchers have hypothesized that childhood and adolescent periods are critical in establishing the number of fat cells [12, 13]. The inference is that a child who is overnourished will produce an excess of new fat cells and, in essence, be "stuck" with these cells later in life. In part, this explains why the military leaders suggest school nutrition intervention, and why celebrities like Chef Jamie Oliver champion the eradication of processed foods from school lunch programs [14]. While these are wise choices that will ultimately be beneficial, unfortunately, mounting evidence supports the idea that obesity is established before puberty [15] and even before birth [16]. Genetics is a popular mechanism to explain such a deterministic phenomenon, except that it cannot account for the rapid rise in obesity.

Endocrine-disrupting chemicals (EDCs) have the strong potential to predispose a child to obesity by influencing all aspects of adipose tissue growth, starting from multipotent stromal cells (MSCs) and ending with mature adipocytes [17]. Certain EDCs can create an immutable adipocyte landscape characterized by increased proliferation and differentiation of adipose progenitors, coupled with an inherent alteration in the MSC compartment that biases stem cells toward the adipose lineage [18]. Unlike genetic mechanisms, which require a mutation event, EDCs have the capacity to quantitatively alter gene expression by modulating cellular-signaling pathways and by introducing epigenetic changes that also alter gene expression. Epigenetic marks in chromatin are programmed during times of developmental plasticity (in the womb or during childhood and adolescence) [19]. Therefore, EDC exposure could foster a swift change in the metabolic profile of a population, which might provide at least a partial explanation for the rapid rise in obesity. This also vindicates the common experience that adipose tissue is often unyielding to shrinkage through improved diet and exercise. Many of the early events in adipogenesis remain unclear, thus, the processes targeted by EDCs may not be fully understood until fundamental mechanisms of adipose development have been resolved. This review focuses on the developmental origins of the adipocyte and its connection to early-onset obesity with the aim of providing a foundation for formulating hypotheses regarding how EDCs can interfere with adipogenesis and contribute to the obesity epidemic.

The Morphology of Adipose Tissue in Early-Onset Obesity

Humans are the fattest of all mammals at birth, typically possessing ~ 15% body fat, of which most is white adipose tissue (WAT) [20]. This is predicted to be the result of high encephalization [21] because our large brains consume an enormous amount of energy. During infancy, body fat composition nearly doubles (to 28%), but then declines during early childhood [22]. After about 8 years of age and throughout adolescence, adipose tissue grows, although the body distribution becomes sexually dimorphic. Males decrease their fat composition as a percentage of total body mass, whereas females increase their fat composition [23]. Under normal circumstances, body fat functions in the proper timing of puberty, again with a sexual dimorphism. A high-percent body fat delays puberty in males, but initiates early menarche in females [24]. Recent research, using 14C labeling, highlighted the childhood and adolescent periods as the main time of adipose hyperplastic growth [11]. In early adulthood, the total number of fat cells stabilizes; the number only increases when existing cells have reached full capacity through hypertrophic growth [11].

While more sophisticated, the 14C-labeling studies actually confirmed long-standing results. Nearly 40 years ago, obesity, particularly early-onset obesity, was linked to hypercellularity of adipose tissue [25–27]. Ten years later, researchers charted the steady increase in adipocytes during the first two decades of life and showed that obese children possessed more fat cells which multiplied at a faster rate than non-obese children [10]. Further studies demonstrated that once established, a person's adipocyte population is resistant to changes in number. For example, obese adolescent girls who were put on a strict diet and advised to engage in physical activity surprisingly showed increased hyperplastic adipocyte growth after 1.5 years, compared to the reference group [28]. Those girls that were most resistant to the treatment had the greatest increase in adipocyte number. This is supported by the 14C-labeled adipose studies, mentioned above, which also showed that the number of adipocytes remains relatively constant throughout adulthood, is higher in obese individuals, and cannot readily be reduced even after bariatric surgery (although adipocyte volume decreases) [11].

The body's resistance to changes in adipocyte number correlates well with the observation that the body resists major alterations in weight gain or loss. In the 1930s and 1940s, the general belief was that obesity was induced by a lack of will-power [29], perhaps due to some hypothalamic disruption that influenced satiety [30]. However, when obese patients were forced to follow a severe liquid diet, they reached a more normal weight, but this was, by no means, a homeostatic state [31–33]. They exhibited signs of starvation, and a preoccupation with the craving for food, stronger than addiction, an urge akin to an extreme thirst for water or the desire to breathe. In the opposite scenario, lean volunteers from the Vermont State Prison were asked to eat 2–3 times more than usual [34]. The authors of the study stated, "achieving a serious gain in weight cannot be undertaken as a secondary occupation." Although the prisoners gained weight at different rates, they were all able to return to their former weight. Importantly, the prisoners' weight gain was hypertrophic in nature, whereas the starving obese subjects' weight loss was hypotrophic. In each case, the individuals were struggling against their basic biology, their metabolic set point, defined by the number of cells in adipose depots.

Of course, this is not to say that adipocytes are static entities, or that only a fixed number of cells are allocated to each person. Although adipocytes are postmitotic, they do regenerate about once every 10 years [11] through the processes of apoptosis, autophagy, dedifferentiation, or necrosis, and subsequent renewal [35–37]. The adipocyte pool size can fluctuate; a high-fat diet, for example, will encourage hyperplasic growth of subcutaneous fat depots in adult mice [38]. Moreover, when cells have reached their lipid capacity, they will generate paracrine signals that result in the generation of more fat cells in rodents [39–41]. Not all fat tissue is equally capable of hyperplastic growth; in particular, visceral abdominal fat (VAT) does not readily increase its cell number [38, 42]. A very popular hypothesis is that many pathological consequences of obesity are due to excessive adipocyte hypertrophy (as a result of nutrient excess, or an inherent imbalance in fatty acid synthesis and oxidation [43]) coupled with an impaired ability to compensate with adipose tissue expansion [44–48]. The result is adiposopathy [49], which is often characterized by lipotoxicity, where lipid essentially "leaks" into other tissues generating proinflammatory signals and oxidative stress, leading to chronic hyperglycemia and increased circulation of triglycerides [50, 51]. One reason that thiazolidinedione antidiabetic medications are effective is because they protect against this ectopic fat "leakage" by encouraging the birth of new adipocytes in depots other than the VAT, thereby preventing adipocytes from bursting or leaking into the surrounding tissue [52].

Notwithstanding these results, increased adipocyte number, rather than increased size, is the morphology commonly shared among obese individuals who developed the disease early or during adolescence. Furthermore, adipose depot hypercellularity is also the morphology associated with resistance to weight loss. This review

10 Adipocytes as Target Cells for Endocrine Disruption

highlights programming events early in life that augment the proliferation and differentiation capacity of adipose precursors resulting in an increased number of fat cells. This is important because the obesity epidemic is now recognized as a crisis trending toward the very young [53]. Hence, there is a strong impetus to understand the mechanisms underlying early fat cell development, and how EDCs can influence the differentiation of preadipocytes to adipocytes and the commitment of stem cells to preadipocytes.

Endocrine Disruption During the Differentiation Phase

In the field of adipose biology, a heightened number of fat cells is simply an end point, and it becomes important to investigate the mechanistic basis. The mature adipocyte is generated from a white adipocyte precursor (often called a "preadipocyte") which is committed to the adipocyte fate and cannot differentiate into any other lineage (like bone, cartilage, muscle, or even brown fat) [54–57]. This precursor is derived from multipotent stromal cells (MSCs) found in almost all fetal and adult tissues [58], but most commonly cultured from adipose tissue or bone marrow. Most evidence supports the theory that MSCs are the progeny of perivascular cells that surround blood vessel walls [59, 60]. In support of this hypothesis, it was recently shown that the white adipose precursor exists within the adipose vascular network [61]. This is not surprising because adipose is a highly vascularized tissue and antiangiogenic agents reduce adipose mass [62, 63]. The differentiation of white adipocyte precursors into mature adipocytes (adipogenesis) is well characterized in the literature. The central regulator in this process is the peroxisome proliferator–activated receptor gamma (PPARγ), which associates with the retinoid X receptor (RXR) and binds DNA targets as a heterodimer [64]. Figure 10.1 depicts the many of the known events in adipocyte differentiation, focusing on their origin from multipotent precursors, the other main pathways that these precursors can differentiate along, commitment to preadipocytes, differentiation, and apoptosis.

PPARγ is first induced at the transcriptional level by CCAAT/enhancer-binding protein (C/EBP) β and δ [65, 66], and then engages in a feed-forward loop with C/EBPα, amplifying the adipogenic signal [67]. However, C/EBPα also induces Sirtuin-1 [68] which curbs adipogenesis via inhibition of PPARγ target genes [69]. The induction of adipogenesis is typically initiated in cell culture by differentiating agents [70] such as insulin, glucocorticoids, and methylisobutylxanthine, which act through the PI3K/AKT (phosphoinositide 3-kinase/AKT), glucocorticoid receptor, and cAMP protein kinase pathways, respectively. This induction cocktail primarily functions to increase the expression level of C/EBPα or PPARγ, not to activate PPARγ via the production of ligand. An exception to this is the insulin-induced transcription factor, sterol regulatory element–binding protein 1c (SREBP1c), which synthesizes fatty acids that can bind PPARγ [71]. Indeed, the fatty acid derivative, 15-deoxy-Δ12,14-prostaglandin J2 (15d-PGJ2), is the strongest candidate for an endogenous ligand for PPARγ [72, 73], although 15d-PGJ2 primarily functions

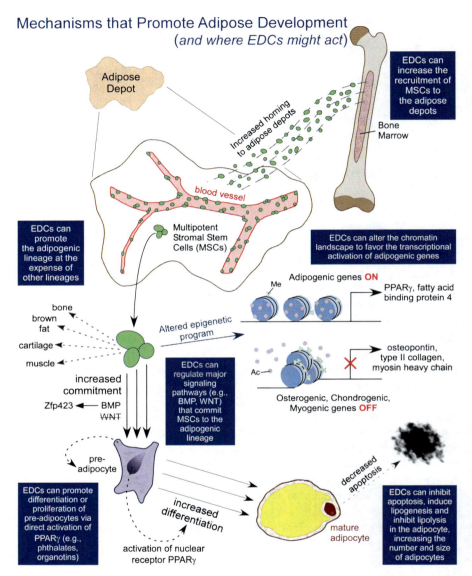

Fig. 10.1 EDCs have the potential to affect multiple aspects of adipocyte development. MSCs reside within the vasculature of adipose depots; however, EDCs could increase homing of circulating MSCs to adipose tissues. EDCs could also alter the epigenetic program of MSCs by turning on the expression of adipogenic genes and turning off osterogenic, chondrogenic, and myogenic genes. These changes, in addition to the possible involvement of EDCs in the modulation of cellular-signaling pathways (BMP, WNT), increase the commitment of MSCs to preadipocytes at the expense of other lineages. EDCs also play a direct role in the differentiation phase by binding PPARγ, promoting the proliferation and differentiation of preadipocytes. Finally, EDCs could exert effects on the mature adipocyte by inhibiting apoptosis, or enhancing the storage of fat, stimulating hypertrophic growth

in angiogenesis [74]. It was recently suggested that PPARγ may function in adipogenesis without requiring the ability to be activated by ligand [75]. However, the ligand-binding domain itself is required. When the ligand-binding domain of PPARγ was mutated, the receptor was unresponsive to known agonists, but the ability of preadipocytes to differentiate into adipocytes was unaffected in cell culture [75]. In contrast, deletion of the activation function 2 region of the ligand-binding domain rendered the receptor unable to support adipogenesis [75].

Several EDCs affect PPARγ activity during adipogenesis. The most well-known synthetic agonists of PPARγ are the thiazolidinedione class of antidiabetic agents including rosiglitazone (ROSI) and pioglitazone (PIO) [76]. In addition to increasing insulin sensitivity, these drugs encourage new fat-cell growth and relieve inflammatory stress from hypertrophic cells, but promote hyperplastic obesity. Environmental chemicals such as certain phthalates [77–79] and organotins [80–82] are agonistic ligands for PPARγ. Perfluoroalkyl acids either activate PPARγ weakly [83] or not at all, [84] despite activating PPARα or PPARδ. In cell culture models, phthalates and organotins have the expected effect of promoting preadipocyte differentiation [78, 80]. Importantly, prenatal exposure to tributyltin (TBT) in mice caused substantial storage of triglycerides in newborn tissues that normally have little to no fat at all [80]. The experiments did not distinguish between increased lipid accumulation in existing cells and an increased number of fat cells. However, given the fact that TBT exposure in *Xenopus laevis* tadpoles caused the entire testes to essentially turn into fat, it was concluded that TBT primarily functions to promote new adipocyte development at the expense of other cell types. Further studies (outlined in the next section) were undertaken to understand how TBT biases progenitor cells toward the adipocyte lineage.

To date, TBT is the only EDC known to cause in utero effects on adipocytes via PPARγ. Prenatal exposure to phthalates has not yet been linked to adiposity later in life; although high levels of urinary phthalate metabolites were positively correlated with waist diameter in men [85]. Other EDCs also promote adipogenesis, but do not act through PPARγ. Coplanar PCBs (e.g., PCB-77) bind the aryl hydrocarbon receptor in adipocytes and increase adipogenesis [86]. Bisphenol A (BPA) and BPA-related chemicals including alkylphenols stimulate adipogenesis in cell culture [87]. There are a variety of other nuclear receptors and their cofactors that are activated during adipogenesis [88, 89] including those known to be involved in energy metabolism such as the liver X receptor (LXR), the glucocorticoid receptor (GR) and the thyroid receptor (TR), and those more well known in other non-metabolic pathways such as nuclear receptor–related 1 (NURR-1) and the germ cell nuclear factor (GCNF) [88]. Any of these receptors could potentially be targeted by chemicals. Furthermore, RXR is also upregulated during adipogenesis and is activated by TBT [80, 90]. RXR is an obligate heterodimeric partner for PPARγ, as well as TR, NURR-1, LXR, and PPAR (among many other nuclear receptors). RXR itself can be activated in a subset of these heterodimers.

Despite the fact that many EDCs accumulate in adipose tissue and contribute to local effects, most studies have explored the broader metabolic consequences of EDCs, and then addressed the secondary effects on the adipocyte. For example, polychlorinated biphenyls (PCBs) and polybrominated diphenyl ethers (PBDEs) reduce thyroid function [91] possibly by competing for thyroid transport proteins [92]. High levels of maternal PCBs and PBDEs are correlated with reduced total and free T_4 levels in infant cord blood [93]. Thyroid hormone stimulates lipolysis in adipocytes by downregulating phosphodiesterase activity which normally functions to inhibit catecholamine-induced lipolysis [94, 95]. Thyroid hormone also down-regulates SREBP1c, resulting in the inhibition of lipogenesis [96]. The inference is that exposure to EDCs like PCBs and PBDEs will reduce thyroid hormone levels and cause a concomitant increase in lipid accumulation in adipocytes. Another large class of EDCs that globally affect energy metabolism is the synthetic estrogens, like diethylstilbestrol (DES) and bisphenol A (BPA) [97]. In adults, estrogens protect against adiposity through exergonic, energy-consuming reactions in glycolysis, fatty acid oxidation, and electron transport [97]. At the level of the adipocyte, estrogen receptor beta (ERβ) blocks the transcriptional activity of PPARγ and, hence, is antiadipogenic but prodiabetogenic [98]. In contrast to their effects in adults, low doses of estrogens given pre- or perinatally strikingly promote obesity in exposed animals [99]. Similarly, nicotine, which promotes weight loss in adults, increases adipose hypertrophy in young rats [100]. The overall conclusion from these studies is that EDCs can impact the differentiation of adipocytes in a variety of ways.

Endocrine Disruption During the Commitment Phase

Recent research has been directed toward understanding the commitment phase of adipocyte development, that is, how MSCs become preadipocytes. Recently, progenitor cells (which later matured into adipocytes) were purified and found to have the following phenotype: lin⁻, CD29⁺, CD34⁺, Sca-1⁺, CD24⁺ [101]. These cells could generate an entire adipose depot in lipodystrophic mice, suggesting that they are bona fide adipocyte precursors [101]. What remains largely unknown is the transcriptional program that turns stem cells into preadipocytes. Bone morphogenic protein (BMP) signaling is not only important in skeletal development but appears to be a vital component in stem cell commitment to the adipocyte lineage [102]. Recently, Zfp423, a downstream target of BMP signaling, was found to be upregulated in fibroblast clones with high adipogenic potential compared with cells having a low adipogenic potential [103]. However, Zfp423 expression remained unchanged during the preadipocyte to adipocyte transition, suggesting that it is involved in the commitment of stem cells to the preadipocyte lineage, but perhaps not in preadipocyte maintenance [103]. In addition to active BMP signaling, repression of noncanonical WNT5a signaling is required for MSCs to evade the osteogenic lineage and proceed toward the adipogenic lineage [104]. This aligns well with the observation that preadipocytes do not differentiate in the presence of WNT signaling [105].

There are very few studies that investigate how EDCs could bias the MSC population toward the adipogenic linage. In cell culture, organophosphates and

4-tertoctylphenol thwarted the bone differentiation capacity of MSCs [106]. When MSCs were cultured from adipose tissue, then treated with TBT or ROSI plus induction cocktail, their proliferation capacity decreased, and up to 60–80% of the cells differentiated into mature fat cells [18]. Cells treated with induction cocktail, but without TBT or ROSI, maintained higher proliferative levels, and only 25–40% of cells became adipocytes [18]. Interestingly, this effect was also seen after in utero exposure. Pregnant dams were treated with a single dose of TBT or ROSI, and the MSCs (cultured from the adipose tissue of embryos) were already predisposed to become fat cells, even without further treatment with TBT or ROSI [18]. In addition, the MSCs harvested from the TBT- or ROSI-exposed pups were preprogrammed to prefer the adipogenic fate because, when induced with bone differentiating cocktail, many cells still differentiated into adipocytes [18]. This suggests that the MSC population in prenatally TBT- or ROSI-treated animals is enriched in adipocyte precursors at the expense of osteoblasts [18]. This mutually exclusivity ability of a subset of MSCs to differentiate into bone or fat is why osteoporosis has been called "obesity of bone" [107, 108]. MSCs cultured from postmenopausal women with low bone density accumulate twofold more lipid and twofold less type I collagen (part of the bone extracellular matrix) compared to women with healthy bones [109]. Not surprisingly, diabetes medications that activate PPARγ increase the risk for bone fractures [110]. The results with TBT suggest that obesogen exposure may have a similar effect on osteoporosis.

The most recent evidence supports the idea that fat cells are regenerated from an existing population of MSCs that are found in the vasculature of adipose depots [61, 101]. However, it is also possible that circulating MSCs derived from bone marrow can be recruited to the depots. The relative contribution of resident and circulating MSCs to the obesogenic phenotype of TBT- or ROSI-treated animals is currently unknown. Most studies of MSC migration have been devoted to how MSCs escape from bone marrow and flood an injury site [111, 112]. MSCs are predicted to move by chemotaxis, losing adherence to their origin, rolling along blood vessels, and moving through the extracellular matrix to the tissue that is injured [113]. By a similar mechanism, these cells could migrate to adipose tissue upon receiving an appropriate signal. This idea was recently tested by transplanting GFP-labeled bone marrow–derived MSCs into irradiated wild-type mice, and determining whether these cells could populate fat pads in response to adipogenic signals [114]. ROSI or a high-fat diet increased migration of circulating bone marrow cells to omental or dorsal intrascapular fat depots [114]. However, this view has been challenged in a subsequent paper, which states that bone marrow–derived circulating progenitor cells fail to differentiate into adipocytes [115]. Whether or not MSCs can be recruited to adipose depots from bone marrow is probably a moot point since it now appears that MSCs are pericytes that are found in close proximity with vasculature. Whether they are solely localized in the bone marrow and adipose vasculature [61, 101] or instead are found throughout the body [59] is currently controversial. More studies will be required to determine where adipogenic MSCs are located, whether or not they migrate to adipose depots in response to dietary stimuli, and what effects EDCs such as TBT and phthalates have on these processes.

Epigenetic Modifications During Puberty

Although ordinary genetic variability can account for why some people have the propensity to become obese, the rapid increase in obesity argues against a genetic explanation. Epigenetic phenomena occur much faster than gradual genetic mutation and can easily become established in a population within a single generation [116]. For this reason, epigenetics is a more likely explanation for the "epidemic" of obesity. Moreover, epigenetics can also underlie the rapid metabolic adaptations that occur in the womb under dietary stress, and perhaps during other developmental time windows, such as adolescence. While EDCs such as TBT can act on adipose tissue directly (e.g., by binding to PPARγ and inducing adipocyte differentiation), they can also target adipose development via more subtle, epigenetic modifications. EDC exposure has been linked with alterations in the expression of proteins that remodel the chromatin landscape, such as DNA methyltransferases, histone acetyltransferases, deacetylases, and methyltransferases. For example, in the uteri of animals that were prenatally exposed to DES, there was a significant increase in the mRNA expression of DNA methyltransferase 1 (Dnmt1) and DNA methyltransferase 3b (Dnmt3b) [117]. In another example, exposure to the commonly used fungicide vinclozolin caused a decrease in the expression of Dnmt1, Dnmt3a, Dnmt3L, and euchromatic histone methyltransferase (Ehmt1) in the testes of male rats, and some of these genes remained inhibited in subsequent generations [118].

Since EDCs typically have targeted effects on a particular metabolic pathway, the question of specificity arises. It was shown that Dnmt1 and Dnmt3 can be recruited to individual genes simply by being escorted by transcription factors to specific sequences of DNA, suggesting that the specificity of the EDC effect is conferred by the active transcriptional programs in individual cells [119, 120]. Of course, identification of bona fide DNA sequences (often CpG islands) modulated by EDCs is a challenge and requires a genomic approach. Over 6,000 genes have been predicted to affect body mass, and there are ten times more genes that foster an increase in weight, rather than a decrease [121]. It is likely that epigenetic changes related to adipogenesis and obesity originate within the stem cell compartment. The reason that diet and exercise primarily alter adipocyte volume rather than adipocyte number is because the stem cells are already biased, through changes in gene regulation, to replenish the adipocyte pool to its "set point," if for any reason the number of adipocytes declines. In support of this model, MSCs from mice exposed to TBT in utero exhibited alterations in the methylation status of the CpG islands of adipogenic genes such as AP2 and PPARγ which led to an increase in the number of preadipocytes at birth and an increased propensity to differentiate into adipocytes upon stimulation [18].

Even more profound is the possibility that epigenetic changes caused by environmental exposures are accumulated or inherited across generations. Vinclozolin, for example, is linked to infertility throughout multiple generations, [122] due to a "memory" maintained within the male germ cells [123]. While there are currently no data regarding transgenerational effects of obesogens, it is reasonable to hypothesize that obesogenic compounds such as TBT, BPA, or phthalates could also influence the propensity to be obese for some generations after the initial exposure. An example in humans serves to illustrate this point. In the Overkalix region of Sweden, it was demonstrated that food availability during the prepubescent period affected the longevity and mortality from cardiovascular disease of a boy's grandchildren. A single winter of overeating could lead to a 6-year decrease in longevity of a boy's grandsons, but not granddaughters [124]. While this is a nutritional study, it is quite possible that chemical exposure during development could have similar transgenerational effects. The human prepubescent period is especially susceptible to epigenetic changes because the testes or ovaries are developing, and the primordial germ cells incorporate sex-specific imprinting patterns in mice [125]. One hypothesis proposed to explain the Overkalix phenomenon is that stress caused by improper nutrition affects downstream proteins involved in imprinting [126]. One such protein candidate is BORIS (Brother of the Regulator of Imprinted Sites) which is expressed in the male testes only in germ cells undergoing genome remethylation [127]. During the prepubescent period, an exogenous chemical could regulate BORIS and permanently impact the methylation status of DNA. Hence, while it is debatable whether adolescence is the most critical period in establishing obesity, it is likely to be a period when epigenetic changes can be "locked in" for the future.

Conclusion

A somewhat vestigial function of the adipocyte, especially in Western societies, is its incredibly efficient storage capacity for fat. Hence, a resounding presumption is that unhealthy eating and a convenience food-driven society are enabling agents that contribute to self-induced corpulence. However, babies, children, and adolescents seem destined, almost programmed, to keep their "baby fat" throughout life. We have argued that the morphology of adipose tissue associated with early-onset obesity provides a basis for why weight loss in obese people (the state of being "reduced obese") is not often a homeostatic state of normalcy and weight gain is observed in more than 90% of cases. We argued that EDCs could exacerbate this problem by increasing the differentiation capacity of preadipocytes, or biasing the MSC pool toward the adipocyte lineage. Moreover, we discussed how EDCs have the potential to affect epigenetic reprogramming events that might cause irreversible modifications that create transgenerational inheritance of obesity.

The concept of early disruption in adipocyte development explains why each person seems to possess a metabolic set point, which is regulated like a thermostat. The thermostat will resist change in temperature, no matter if a window is opened or a hot oven is turned on. If the set point is altered, the new temperature will be maintained. Early nutrition and chemical exposure could alter an individual's set point, making the fight against weight gain that much more difficult. Obesity is difficult to reverse once established; therefore, it makes sense to shift the focus toward preventative measures. On May 11, 2010, First Lady Michelle Obama launched a campaign called Let's Move! As a result of strong communication between science and politics, Mrs. Obama drew attention to EDCs, calling for increased research in this area, reformulation of plastics, and screening for chemicals that are obesogenic [128]. With national attention to this problem, the hope is that the contribution of EDCs to the obesity epidemic will finally be recognized so that appropriate action can be taken to reduce exposure at critical periods during life.

Acknowledgments Work in the authors' laboratory was supported by a grant from the NIH R01 ES015849. A.J. is a predoctoral trainee of NSF IGERT DGE 0549479.

References

1. Christeson W, Taggart AD, Messner-Zidell S. Too fat to fight. Washington DC: Mission: Readiness; 2010.
2. Kiernan M, Eismeier T. 9 million young adults are too overweight to join the military. New report shows: Mission: Readiness; 2010.
3. Wang Y, Beydoun MA, Liang L, Caballero B, Kumanyika SK. Will all Americans become overweight or obese? Estimating the progression and cost of the US obesity epidemic. Obesity. 2008;16(10):2323–30.
4. Guo SS, Roche AF, Chumlea WC, Gardner JD, Siervogel RM. The predictive value of childhood body mass index values for overweight at age 35 y. Am J Clin Nutr. 1994;59(4):810–9.
5. Must A, Jacques PF, Dallal GE, Bajema CJ, Dietz WH. Long-term morbidity and mortality of overweight adolescents. A follow-up of the Harvard Growth Study of 1922 to 1935. N Engl J Med. 1992;327(19):1350–5.
6. Whitaker RC, Wright JA, Pepe MS, Seidel KD, Dietz WH. Predicting obesity in young adulthood from childhood and parental obesity. N Engl J Med. 1997;337(13):869–73.
7. Guo SS, Wu W, Chumlea WC, Roche AF. Predicting overweight and obesity in adulthood from body mass index values in childhood and adolescence. Am J Clin Nutr. 2002;76(3):653–8.
8. Grundy SM, Brewer Jr HB, Cleeman JI, Smith Jr SC, Lenfant C. Definition of metabolic syndrome: report of the national heart, lung, and blood institute/American Heart Association conference on scientific issues related to definition. Circulation. 2004;109(3):433–8.
9. Sun SS, Liang R, Huang TT, et al. Childhood obesity predicts adult metabolic syndrome: the Fels longitudinal study. J Pediatr. 2008;152(2):191–200.
10. Knittle JL, Timmers K, Ginsberg-Fellner F, Brown RE, Katz DP. The growth of adipose tissue in children and adolescents. Cross-sectional and longitudinal studies of adipose cell number and size. J Clin Invest. 1979;63(2):239–46.
11. Spalding KL, Arner E, Westermark PO, et al. Dynamics of fat cell turnover in humans. Nature. 2008;453(7196):783–7.
12. Dietz WH. Critical periods in childhood for the development of obesity. Am J Clin Nutr. 1994;59(5):955–9.
13. Dietz WH. Periods of risk in childhood for the development of adult obesity–what do we need to learn? J Nutr. 1997;127(9):1884S–6S.
14. Wishnow J. Jamie Oliver's TED prize wish: teach every child about food. Long Beach. 2010.
15. Wardle J, Brodersen NH, Cole TJ, Jarvis MJ, Boniface DR. Development of adiposity in adolescence: five year longitudinal study of an ethnically and socioeconomically diverse sample of young people in Britain. BMJ. 2006;332(7550):1130–5.

16. Gluckman PD, Hanson MA, Beedle AS, Raubenheimer D. Fetal and neonatal pathways to obesity. Front Horm Res. 2008;36:61–72.
17. Janesick A, Blumberg B. The role of environmental obesogens in the obesity epidemic. In: Lustig RH, editor. Obesity before birth, vol. 30. New York: Springer; 2011. p. 383–399.
18. Kirchner S, Kieu T, Chow C, Casey S, Blumberg B. Prenatal exposure to the environmental obesogen tributyltin predisposes multipotent stem cells to become adipocytes. Mol Endocrinol. 2010;24(3):526–39.
19. Hanson MA, Gluckman PD. Developmental origins of health and disease: new insights. Basic Clin Pharmacol Toxicol. 2008;102(2):90–3.
20. Kuzawa CW. Adipose tissue in human infancy and childhood: an evolutionary perspective. Am J Phys Anthropol. 1998;Suppl 27:177–209. Volume 107, Issue Supplement 27 (Supplement: Yearbook of Physical Anthropology).
21. Foley RA, Lee PC. Ecology and energetics of encephalization in hominid evolution. Philos Trans R Soc Lond B Biol Sci. 1991;334(1270):223–31; discussion 232.
22. Hager A, Sjostrm L, Arvidsson B, Bjorntorp P, Smith U. Body fat and adipose tissue cellularity in infants: a longitudinal study. Metabolism. 1977;26(6):607–14.
23. Rogol AD, Roemmich JN, Clark PA. Growth at puberty. J Adolesc Health. 2002;31(6 Suppl):192–200.
24. Lee JM, Kaciroti N, Appugliese D, Corwyn RF, Bradley RH, Lumeng JC. Body mass index and timing of pubertal initiation in boys. Arch Pediatr Adolesc Med. 2010;164(2):139–44.
25. Hirsch J, Knittle JL. Cellularity of obese and nonobese human adipose tissue. Fed Proc. 1970;29(4):1516–21.
26. Knittle JL. Obesity in childhood: a problem in adipose tissue cellular development. J Pediatr. 1972;81(6):1048–59.
27. Salans LB, Cushman SW, Weismann RE. Studies of human adipose tissue. Adipose cell size and number in nonobese and obese patients. J Clin Invest. 1973;52(4):929–41.
28. Hager A, Sjorstrom L, Arvidsson B, Bjorntorp P, Smith U. Adipose tissue cellularity in obese school girls before and after dietary treatment. Am J Clin Nutr. 1978;31(1):68–75.
29. Newburgh LH, Johnston MW. The nature of obesity. J Clin Invest. 1930;8(2):197–213.
30. Hetherington AW, Ranson SW. The spontaneous activity and food intake of rats with hypothalamic lesions. Am J Physiol. 1942;136(4):609–17.
31. Glucksman ML, Hirsch J. The response of obese patients to weight reduction I. A clinical evaluation of behavior. Psychosom Med. 1968;30(1):1–11.
32. Glucksman ML, Hirsch J. The response of obese patients to weight reduction III. The perception of body size. Psychosom Med. 1969;31(1):1–7.
33. Glucksman ML, Hirsch J, McCully RS, Barron BA, Knittle JL. The response of obese patients to weight reduction II. A quantitative evaluation of behavior. Psychosom Med. 1968;30(4): 359–73.
34. Sims EA, Horton ES. Endocrine and metabolic adaptation to obesity and starvation. Am J Clin Nutr. 1968;21(12):1455–70.
35. Gesta S, Tseng YH, Kahn CR. Developmental origin of fat: tracking obesity to its source. Cell. 2007;131(2):242–56.
36. Prins JB, O'Rahilly S. Regulation of adipose cell number in man. Clin Sci (Lond). 1997; 92(1):3–11.
37. Singh R, Xiang Y, Wang Y, et al. Autophagy regulates adipose mass and differentiation in mice. J Clin Invest. 2009;119(11):3329–39.
38. Joe AW, Yi L, Even Y, Vogl AW, Rossi FM. Depot-specific differences in adipogenic progenitor abundance and proliferative response to high-fat diet. Stem Cells. 2009;27(10):2563–70.
39. Lau DC, Shillabeer G, Wong KL, Tough SC, Russell JC. Influence of paracrine factors on preadipocyte replication and differentiation. Int J Obes. 1990;14(Suppl 3):193–201.
40. Marques BG, Hausman DB, Martin RJ. Association of fat cell size and paracrine growth factors in development of hyperplastic obesity. Am J Physiol. 1998;275(6 Pt 2):R1898–908.
41. Shillabeer G, Forden JM, Lau DC. Induction of preadipocyte differentiation by mature fat cells in the rat. J Clin Invest. 1989;84(2):381–7.

42. Adams M, Montague CT, Prins JB, et al. Activators of peroxisome proliferator-activated receptor gamma have depot-specific effects on human preadipocyte differentiation. J Clin Invest. 1997;100(12):3149–53.
43. Lelliott C, Vidal-Puig AJ. Lipotoxicity, an imbalance between lipogenesis de novo and fatty acid oxidation. Int J Obes Relat Metab Disord. 2004;28(Suppl 4):S22–8.
44. Heilbronn L, Smith SR, Ravussin E. Failure of fat cell proliferation, mitochondrial function and fat oxidation results in ectopic fat storage, insulin resistance and type II diabetes mellitus. Int J Obes Relat Metab Disord. 2004;28(Suppl 4):S12–21.
45. Kim JY, van de Wall E, Laplante M, et al. Obesity-associated improvements in metabolic profile through expansion of adipose tissue. J Clin Invest. 2007;117(9):2621–37.
46. Virtue S, Vidal-Puig A. It's not how fat you are, it's what you do with it that counts. PLoS Biol. 2008;6(9):e237.
47. Gustafson B, Gogg S, Hedjazifar S, Jenndahl L, Hammarstedt A, Smith U. Inflammation and impaired adipogenesis in hypertrophic obesity in man. Am J Physiol Endocrinol Metab. 2009 American Journal of Physiology, vol. 297, no. 5, pp. E999–E1003.
48. Weyer C, Foley JE, Bogardus C, Tataranni PA, Pratley RE. Enlarged subcutaneous abdominal adipocyte size, but not obesity itself, predicts type II diabetes independent of insulin resistance. Diabetologia. 2000;43(12):1498–506.
49. Bays HE, Gonzalez-Campoy JM, Henry RR, et al. Is adiposopathy (sick fat) an endocrine disease? Int J Clin Pract. 2008;62(10):1474–83.
50. de Ferranti S, Mozaffarian D. The perfect storm: obesity, adipocyte dysfunction, and metabolic consequences. Clin Chem. 2008;54(6):945–55.
51. Gregor MG, Hotamisligil GS. Adipocyte stress: the endoplasmic reticulum and metabolic disease. J Lipid Res. 2007;48:1905–14.
52. Medina-Gomez G, Gray SL, Yetukuri L, et al. PPAR gamma 2 prevents lipotoxicity by controlling adipose tissue expandability and peripheral lipid metabolism. PLoS Genet. 2007; 3(4):e64.
53. Lee JM, Pilli S, Gebremariam A, et al. Getting heavier, younger: trajectories of obesity over the life course. Int J Obes (Lond). 2010;34(4):614–23.
54. Park KW, Halperin DS, Tontonoz P. Before they were fat: adipocyte progenitors. Cell Metab. 2008;8(6):454–7.
55. Seale P, Bjork B, Yang W, et al. PRDM16 controls a brown fat/skeletal muscle switch. Nature. 2008;454(7207):961–7.
56. Timmons JA, Wennmalm K, Larsson O, et al. Myogenic gene expression signature establishes that brown and white adipocytes originate from distinct cell lineages. Proc Natl Acad Sci USA. 2007;104(11):4401–6.
57. Cornelius P, MacDougald OA, Lane MD. Regulation of adipocyte development. Annu Rev Nutr. 1994;14:99–129.
58. da Silva Meirelles L, Chagastelles PC, Nardi NB. Mesenchymal stem cells reside in virtually all post-natal organs and tissues. J Cell Sci. 2006;119(Pt 11):2204–13.
59. Crisan M, Yap S, Casteilla L, et al. A perivascular origin for mesenchymal stem cells in multiple human organs. Cell Stem Cell. 2008;3(3):301–13.
60. da Silva ML, Caplan AI, Nardi NB. In search of the in vivo identity of mesenchymal stem cells. Stem Cells. 2008;26(9):2287–99.
61. Tang W, Zeve D, Suh JM, et al. White fat progenitor cells reside in the adipose vasculature. Science. 2008;322(5901):583–6.
62. Rupnick MA, Panigrahy D, Zhang CY, et al. Adipose tissue mass can be regulated through the vasculature. Proc Natl Acad Sci USA. 2002;99(16):10730–5.
63. Kahn CR. Medicine. Can we nip obesity in its vascular bud? Science. 2008;322(5901): 542–3.
64. Tontonoz P, Hu E, Spiegelman BM. Stimulation of adipogenesis in fibroblasts by PPAR gamma 2, a lipid-activated transcription factor. Cell. 1994;79(7):1147–56.
65. Wu Z, Bucher NL, Farmer SR. Induction of peroxisome proliferator-activated receptor gamma during the conversion of 3 T3 fibroblasts into adipocytes is mediated by C/EBPbeta, C/EBPdelta, and glucocorticoids. Mol Cell Biol. 1996;16(8):4128–36.

10 Adipocytes as Target Cells for Endocrine Disruption

66. Shao D, Lazar MA. Peroxisome proliferator activated receptor gamma, CCAAT/enhancer-binding protein alpha, and cell cycle status regulate the commitment to adipocyte differentiation. J Biol Chem. 1997;272(34):21473–8.
67. Rosen ED, Sarraf P, Troy AE, et al. PPAR gamma is required for the differentiation of adipose tissue in vivo and in vitro. Mol Cell. 1999;4(4):611–7.
68. Jin Q, Zhang F, Yan T, et al. C/EBPalpha regulates SIRT1 expression during adipogenesis. Cell Res. 2010;20(4):470–9.
69. Picard F, Kurtev M, Chung N, et al. Sirt1 promotes fat mobilization in white adipocytes by repressing PPAR-gamma. Nature. 2004;429(6993):771–6.
70. Student AK, Hsu RY, Lane MD. Induction of fatty acid synthetase synthesis in differentiating 3 T3-L1 preadipocytes. J Biol Chem. 1980;255(10):4745–50.
71. Kim JB, Wright HM, Wright M, Spiegelman BM. ADD1/SREBP1 activates PPARgamma through the production of endogenous ligand. Proc Natl Acad Sci USA. 1998;95(8): 4333–7.
72. Kliewer SA, Lenhard JM, Willson TM, Patel I, Morris DC, Lehmann JM. A prostaglandin J2 metabolite binds peroxisome proliferator-activated receptor gamma and promotes adipocyte differentiation. Cell. 1995;83(5):813–9.
73. Forman BM, Tontonoz P, Chen J, Brun RP, Spiegelman BM, Evans RM. 15-deoxy-delta 12, 14-prostaglandin J2 is a ligand for the adipocyte determination factor PPAR gamma. Cell. 1995;83(5):803–12.
74. Kim EH, Surh YJ. The role of 15-deoxy-delta(12,14)-prostaglandin J(2), an endogenous ligand of peroxisome proliferator-activated receptor gamma, in tumor angiogenesis. Biochem Pharmacol. 2008;76(11):1544–53.
75. Walkey CJ, Spiegelman BM. A functional peroxisome proliferator-activated receptor-gamma ligand-binding domain is not required for adipogenesis. J Biol Chem. 2008;283(36): 24290–4.
76. Lehmann JM, Moore LB, Smith-Oliver TA, Wilkison WO, Willson TM, Kliewer SA. An antidiabetic thiazolidinedione is a high affinity ligand for peroxisome proliferator-activated receptor gamma (PPAR gamma). J Biol Chem. 1995;270(22):12953–6.
77. Bility MT, Thompson JT, McKee RH, et al. Activation of mouse and human peroxisome proliferator-activated receptors (PPARs) by phthalate monoesters. Toxicol Sci. 2004;82(1): 170–82.
78. Hurst CH, Waxman DJ. Activation of PPARalpha and PPARgamma by environmental phthalate monoesters. Toxicol Sci. 2003;74(2):297–308.
79. Feige JN, Gelman L, Rossi D, et al. The endocrine disruptor monoethyl-hexyl-phthalate is a selective peroxisome proliferator-activated receptor gamma modulator that promotes adipogenesis. J Biol Chem. 2007;282(26):19152–66.
80. Grun F, Watanabe H, Zamanian Z, et al. Endocrine-disrupting organotin compounds are potent inducers of adipogenesis in vertebrates. Mol Endocrinol. 2006;20(9):2141–55.
81. Kanayama T, Kobayashi N, Mamiya S, Nakanishi T, Nishikawa J. Organotin compounds promote adipocyte differentiation as agonists of the peroxisome proliferator-activated receptor gamma/retinoid X receptor pathway. Mol Pharmacol. 2005;67(3):766–74.
82. Hiromori Y, Nishikawa J, Yoshida I, Nagase H, Nakanishi T. Structure-dependent activation of peroxisome proliferator-activated receptor (PPAR) gamma by organotin compounds. Chem Biol Interact. 2009;180(2):238–44.
83. Vanden Heuvel JP, Thompson JT, Frame SR, Gillies PJ. Differential activation of nuclear receptors by perfluorinated fatty acid analogs and natural fatty acids: a comparison of human, mouse, and rat peroxisome proliferator-activated receptor-alpha, -beta, and -gamma, liver X receptor-beta, and retinoid X receptor-alpha. Toxicol Sci. 2006;92(2):476–89.
84. Takacs ML, Abbott BD. Activation of mouse and human peroxisome proliferator-activated receptors (alpha, beta/delta, gamma) by perfluorooctanoic acid and perfluorooctane sulfonate. Toxicol Sci. 2007;95(1):108–17.
85. Stahlhut RW, van Wijngaarden E, Dye TD, Cook S, Swan SH. Concentrations of urinary phthalate metabolites are associated with increased waist circumference and insulin resistance in adult U.S. males. Environ Health Perspect. 2007;115(6):876–82.

86. Arsenescu V, Arsenescu RI, King V, Swanson H, Cassis LA. Polychlorinated biphenyl-77 induces adipocyte differentiation and proinflammatory adipokines and promotes obesity and atherosclerosis. Environ Health Perspect. 2008;116(6):761–8.
87. Masuno H, Iwanami J, Kidani T, Sakayama K, Honda K. Bisphenol a accelerates terminal differentiation of 3 T3-L1 cells into adipocytes through the phosphatidylinositol 3-kinase pathway. Toxicol Sci. 2005;84(2):319–27.
88. Fu M, Sun T, Bookout AL, et al. A nuclear receptor atlas: 3 T3-L1 adipogenesis. Mol Endocrinol. 2005;19(10):2437–50.
89. Rosen ED, MacDougald OA. Adipocyte differentiation from the inside out. Nat Rev Mol Cell Biol. 2006;7(12):885–96.
90. le Maire A, Grimaldi M, Roecklin D, et al. Activation of RXR-PPAR heterodimers by organotin environmental endocrine disruptors. EMBO Rep. 2009;10(4):367–73.
91. Hallgren S, Sinjari T, Hakansson H, Darnerud PO. Effects of polybrominated diphenyl ethers (PBDEs) and polychlorinated biphenyls (PCBs) on thyroid hormone and vitamin A levels in rats and mice. Arch Toxicol. 2001;75(4):200–8.
92. Cheek AO, Kow K, Chen J, McLachlan JA. Potential mechanisms of thyroid disruption in humans: interaction of organochlorine compounds with thyroid receptor, transthyretin, and thyroid-binding globulin. Environ Health Perspect. 1999;107(4):273–8.
93. Herbstman JB, Sjodin A, Apelberg BJ, et al. Birth delivery mode modifies the associations between prenatal polychlorinated biphenyl (PCB) and polybrominated diphenyl ether (PBDE) and neonatal thyroid hormone levels. Environ Health Perspect. 2008;116(10):1376–82.
94. Smith CJ, Vasta V, Degerman E, Belfrage P, Manganiello VC. Hormone-sensitive cyclic GMP-inhibited cyclic AMP phosphodiesterase in rat adipocytes. Regulation of insulin- and cAMP-dependent activation by phosphorylation. J Biol Chem. 1991;266(20):13385–90.
95. Van Inwegen RG, Robison GA, Thompson WJ. Cyclic nucleotide phosphodiesterases and thyroid hormones. J Biol Chem. 1975;250(7):2452–6.
96. Viguerie N, Millet L, Avizou S, Vidal H, Larrouy D, Langin D. Regulation of human adipocyte gene expression by thyroid hormone. J Clin Endocrinol Metab. 2002;87(2):630–4.
97. Chen JQ, Brown TR, Russo J. Regulation of energy metabolism pathways by estrogens and estrogenic chemicals and potential implications in obesity associated with increased exposure to endocrine disruptors. Biochim Biophys Acta. 2009;1793(7):1128–43.
98. Foryst-Ludwig A, Clemenz M, Hohmann S, et al. Metabolic actions of estrogen receptor beta (ERbeta) are mediated by a negative cross-talk with PPARgamma. PLoS Genet. 2008;4(6):e1000108.
99. Newbold RR, Padilla-Banks E, Jefferson WN. Environmental estrogens and obesity. Mol Cell Endocrinol. 2009;304(1–2):84–9.
100. Somm E, Schwitzgebel VM, Vauthay DM, et al. Prenatal nicotine exposure alters early pancreatic islet and adipose tissue development with consequences on the control of body weight and glucose metabolism later in life. Endocrinology. 2008;149(12):6289–99.
101. Rodeheffer MS, Birsoy K, Friedman JM. Identification of white adipocyte progenitor cells in vivo. Cell. 2008;135(2):240–9.
102. Tang QQ, Otto TC, Lane MD. Commitment of C3H10T1/2 pluripotent stem cells to the adipocyte lineage. Proc Natl Acad Sci USA. 2004;101(26):9607–11.
103. Gupta RK, Arany Z, Seale P, et al. Transcriptional control of preadipocyte determination by Zfp423. Nature. 2010;464(7288):619–23.
104. Bilkovski R, Schulte DM, Oberhauser F, et al. Role of WNT-5a in the determination of human mesenchymal stem cells into preadipocytes. J Biol Chem. 2010;285(9):6170–8.
105. Ross SE, Hemati N, Longo KA, et al. Inhibition of adipogenesis by Wnt signaling. Science. 2000;289(5481):950–3.
106. Hoogduijn MJ, Rakonczay Z, Genever PG. The effects of anticholinergic insecticides on human mesenchymal stem cells. Toxicol Sci. 2006;94(2):342–50.
107. Rosen CJ, Bouxsein ML. Mechanisms of disease: is osteoporosis the obesity of bone? Nat Clin Pract Rheumatol. 2006;2(1):35–43.

10 Adipocytes as Target Cells for Endocrine Disruption

108. Verma S, Rajaratnam JH, Denton J, Hoyland JA, Byers RJ. Adipocytic proportion of bone marrow is inversely related to bone formation in osteoporosis. J Clin Pathol. 2002;55(9): 693–8.
109. Rodriguez JP, Montecinos L, Rios S, Reyes P, Martinez J. Mesenchymal stem cells from osteoporotic patients produce a type I collagen-deficient extracellular matrix favoring adipogenic differentiation. J Cell Biochem. 2000;79(4):557–65.
110. Habib ZA, Havstad SL, Wells K, Divine G, Pladevall M, Williams LK. Thiazolidinedione use and the longitudinal risk of fractures in patients with type 2 diabetes mellitus. J Clin Endocrinol Metab. 2010;95(2):592–600.
111. Hofstetter CP, Schwarz EJ, Hess D, et al. Marrow stromal cells form guiding strands in the injured spinal cord and promote recovery. Proc Natl Acad Sci USA. 2002;99(4):2199–204.
112. Orlic D, Kajstura J, Chimenti S, et al. Bone marrow cells regenerate infarcted myocardium. Nature. 2001;410(6829):701–5.
113. Liu ZJ, Zhuge Y, Velazquez OC. Trafficking and differentiation of mesenchymal stem cells. J Cell Biochem. 2009;106(6):984–91.
114. Crossno Jr JT, Majka SM, Grazia T, Gill RG, Klemm DJ. Rosiglitazone promotes development of a novel adipocyte population from bone marrow-derived circulating progenitor cells. J Clin Invest. 2006;116(12):3220–8.
115. Koh YJ, Kang S, Lee HJ, et al. Bone marrow-derived circulating progenitor cells fail to transdifferentiate into adipocytes in adult adipose tissues in mice. J Clin Invest. 2007;117(12): 3684–95.
116. Gluckman PD, Hanson MA, Spencer HG. Predictive adaptive responses and human evolution. Trends Ecol Evol. 2005;20(10):527–33.
117. Bromer JG, Wu J, Zhou Y, Taylor HS. Hypermethylation of homeobox A10 by in utero diethylstilbestrol exposure: an epigenetic mechanism for altered developmental programming. Endocrinology. 2009;150(7):3376–82.
118. Anway MD, Rekow SS, Skinner MK. Transgenerational epigenetic programming of the embryonic testis transcriptome. Genomics. 2008;91(1):30–40.
119. Robertson KD, Ait-Si-Ali S, Yokochi T, Wade PA, Jones PL, Wolffe AP. DNMT1 forms a complex with Rb, E2F1 and HDAC1 and represses transcription from E2F-responsive promoters. Nat Genet. 2000;25(3):338–42.
120. Burgers WA, Fuks F, Kouzarides T. DNA methyltransferases get connected to chromatin. Trends Genet. 2002;18(6):275–7.
121. Reed DR, Lawler MP, Tordoff MG. Reduced body weight is a common effect of gene knockout in mice. BMC Genet. 2008;9:4.
122. Anway MD, Cupp AS, Uzumcu M, Skinner MK. Epigenetic transgenerational actions of endocrine disruptors and male fertility. Science. 2005;308(5727):1466–9.
123. Anway MD, Skinner MK. Epigenetic transgenerational actions of endocrine disruptors. Endocrinology. 2006;147(6 Suppl):S43–9.
124. Kaati G, Bygren LO, Pembrey M, Sjostrom M. Transgenerational response to nutrition, early life circumstances and longevity. Eur J Hum Genet. 2007;15(7):784–90.
125. Hajkova P, Erhardt S, Lane N, et al. Epigenetic reprogramming in mouse primordial germ cells. Mech Dev. 2002;117(1–2):15–23.
126. Pembrey ME. Time to take epigenetic inheritance seriously. Eur J Hum Genet. 2002;10(11): 669–71.
127. Loukinov DI, Pugacheva E, Vatolin S, et al. BORIS, a novel male germ-line-specific protein associated with epigenetic reprogramming events, shares the same 11-zinc-finger domain with CTCF, the insulator protein involved in reading imprinting marks in the soma. Proc Natl Acad Sci USA. 2002;99(10):6806–11.
128. Obama M, Barnes M. Solving the problem of childhood obesity within a generation. Washington DC; 2010 http://www.letsmove.gov/sites/letsmove.gov/files/TaskForce_on_ Childhood_Obesity_May2010_FullReport.pdf.

Chapter 11
Altered Glucose Homeostasis Resulting from Developmental Exposures to Endocrine Disruptors

Alan Schneyer and Melissa Brown

Abstract The recent increase in obesity and diabetes incidence is too rapid to be entirely accounted for by dietary and exercise components alone, raising the possibility that environmental factors also contribute to this epidemic. It has been proposed that exposure to industrial chemicals possessing endocrine-disrupting activity causes changes in metabolic homeostatic systems during embryonic development which could lead to development of adult obesity. This chapter reviews evidence for the effects of potential endocrine disruptors on homeostatic systems regulating insulin secretion and glucose uptake. The effects of endocrine disruptors on adipocyte differentiation and glucose uptake are also considered. Alterations in these pathways could lead to enhanced weight gain and eventually, diabetes in adults.

Keywords Adipocyte • EDC • Endocrine-disrupting chemicals • Glucose homeostasis • Insulin • Obesity

Introduction

Diabetes affects nearly 24 million Americans, while obesity affects 33% of adults in the USA, a figure that has doubled over the past four decades [1]. Of even greater concern is the threefold increase in obesity among children less than 18 years old [1], suggesting that adult incidence of obesity will increase further as these children age. Since obesity is associated with serious comorbidities including type 2 diabetes (T2D), cardiovascular disease, and hypertension, the potential impact of these trends on healthcare costs in the USA is enormous. This increase in obesity is too rapid to be accounted for entirely by changes in lifestyle and food consumption, suggesting

---

A. Schneyer (✉) • M. Brown
Pioneer Valley Life Science Institute, University of Massachusetts
Amherst, Springfield, MA, USA
e-mail: alan.schneyer@bhs.org

E. Diamanti-Kandarakis and A.C. Gore (eds.), *Endocrine Disruptors and Puberty*,
Contemporary Endocrinology, DOI 10.1007/978-1-60761-561-3_11,
© Springer Science+Business Media, LLC 2012

that environmental influences may be contributing to the increased obesity prevalence. Over the same time period that includes the dramatic rise in obesity, the number and prevalence of industrial and natural compounds that interfere with hormonal systems, referred to as EDCs, have also dramatically increased in human tissues and fluids. This observation led to the hypothesis that exposure to industrial chemicals possessing EDC activity causes changes in metabolic homeostatic systems during development which lead to adult obesity [2]. This hypothesis is supported by recent studies demonstrating that chronic exposure to the nearly ubiquitous EDC, Bisphenol A (BPA), that is widely used for production of plastic food and drink containers or linings is linked to T2D, cardiovascular disease, and liver abnormalities in humans [3]. Interestingly, BPA is detectable in greater than 95% of the US population [4], and detection of substantial BPA in human serum, follicular fluid, cord blood, and amniotic fluid [5] demonstrates that BPA crosses the human placenta where it could alter fetal development. In fact, administration of BPA to pregnant rodents at doses approximating those found in humans caused increased weight in adult offspring of treated dams [6, 7], potentially linking fetal/neonatal BPA exposure to development of obesity, T2D, and cardiovascular disease in adults. Therefore, these animal studies support epidemiological analyses that fetal exposure to EDCs modifies developing endocrine organs and homeostatic mechanisms that alter glucose and lipid homeostasis leading to obesity, insulin resistance, and/or T2D in adults. In this chapter, we will review data mainly from animal models that support this hypothesis. Given the documented exposure of pregnant human females to EDCs, including BPA [4], and the realization that the developing fetus and neonate are especially sensitive to EDC exposure [8] along with the more recent identification of epigenetic changes induced during fetal development by EDCs, it seems likely that much of EDC-influenced adult disease has fetal origins. Therefore, this chapter will focus primarily on fetal exposures leading to alterations in weight or glucose homeostasis in adults.

Direct Actions of EDCs on Islet Function and Glucose Homeostasis

The role of estrogen signaling in regulating glucose and lipid homeostasis through both ER-α and -β receptors has been recognized for some time [9], including modulation of insulin sensitivity and pancreatic insulin content [10]. Therefore, it is not surprising that EDCs with estrogenic potential can also influence insulin sensitivity and glucose homeostasis. Estradiol binds to a membrane-localized estrogen receptor in β-cells, which induces cGMP production, protein kinase G activation, and closing ATP-dependent potassium channels, leading to increased insulin secretion [11]. Using primary β-cells in islets isolated from 10–12-week-old OF1 male mice, 1 nM BPA or diethylstilbestrol (DES) was found to have half the potency of estradiol in accelerating the frequency of glucose-induced calcium spikes [12]. Longer (24 h) exposure to 10 or 100 ng/ml BPA or 0.1–100 ng/ml nonylphenol was found to enhance insulin secretion from cultured rat islets incubated in 16.7 mM glucose,

11 Altered Glucose Homeostasis Resulting from Developmental Exposures... 275

but no effect of BPA or nonylphenol was observed from acute (1 h) exposure [13]. These results suggest that estrogenic EDCs might mimic estradiol's effects on islets in some circumstances and thus, influence glucose homeostasis in vivo.

In support of this concept, 10 µg/kg estradiol was found to decrease blood glucose levels in male mice after administration of a corn oil carbohydrate load, an effect that was dose responsive over a range of 1–100 µg/kg [14]. Similarly, administration of 10 or 100 µg/kg BPA was slightly more effective at reducing serum glucose levels than estradiol, although the lowest 1 µg/kg dose of BPA was ineffective. This enhanced glucose clearance was mediated by increased plasma insulin assessed 30 min after injection of 10 µg/kg estradiol or BPA. When exposed to BPA or estradiol for 2 days, pancreatic insulin content was significantly increased, suggesting that longer exposure enhances insulin production and/or secretion from β-cells. However, by 4 days of treatment with either estradiol or BPA, these mice became hyperinsulinemic and insulin resistant [14], suggesting that chronic exposure to estrogenic EDCs like BPA could lead to reduced insulin sensitivity with potential loss of glucose control and hyperglycemia. Thus, estrogenic EDCs mimic estradiol's ability to enhance β-cell function, leading to enhanced glucose tolerance under acute exposure conditions, but detrimental actions on insulin sensitivity under chronic exposures.

Pancreatic islets also contain α-cells that secrete glucagon necessary for inducing gluconeogenesis and glycogenolysis during fasting periods to maintain blood glucose. BPA, DES, and estradiol (all 10^{-9} M) suppressed calcium spikes regulating glucagon secretion induced by low glucose, suggesting that in addition to altering glucose homeostasis mediated by insulin, BPA and DES can also alter glucose concentrations by modulating glucagon secretion [15].

Genistein, a naturally occurring flavonoid in legumes, can also be classified as an EDC. Genistein has weak estrogenic effects and has been implicated as a soy component that can moderate hyperglycemia through reduction of blood glucose levels in diabetic animals [16]. At least some of this effect is mediated by direct stimulation of insulin secretion from β-cells in mouse islets [17], similar to that observed with BPA and estradiol (see above). Genistein also induced a significant increase in proliferation of human β-cells [18]. Moreover, administration of 0.25 g/kg genistein in feed (producing a serum concentration of 0.59 µM) significantly reduced serum glucose in mice previously made hyperglycemic by streptozotocin (STZ) injection and produced higher insulin levels [18]. In fact, genistein increased β-cell mass in STZ-treated mice through stimulation of ß-cell proliferation [18]. These observations indicate that some EDCs, including genistein, can have beneficial effects on diabetic or prediabetic animals and may be a valuable natural treatment for diabetes.

Fetal EDC Exposure and Weight Gain

One of the earliest EDCs to be studied in detail is DES, a compound that was once widely prescribed to pregnant women to prevent miscarriage. Administration of low-dose (1 µg/kg/day) DES to neonatal female rats did not alter body weight during treatment but induced increased body weight as adults [19]. This effect was not seen in males. A higher DES dose (1 mg/kg/day) caused a significant decrease in body weight during treatment which disappeared at puberty and then produced an increase in body weight as adults. This increased body weight was shown to be due to increased body fat [20]. Interestingly, by 18 months of age, this difference in body weight was no longer evident, apparently masked by increased variability between individuals [21]. Individual fat-pad weights were investigated at 6–8 months of age when the DES-treated mice were still significantly heavier, and all visceral fat pads (inguinal, parametrial, gonadal, and retroperitoneal) were increased relative to untreated mice, but no changes in brown fat were observed [22]. At both 2 and 6 months of age, DES-treated females had significantly elevated leptin, adiponectin, and triglyceride levels, and at 6 months, these mice were hyperinsulinemic, suggesting that the increased body weight and adiposity was accompanied by alterations in glucose and lipid homeostasis [22]. No significant differences in activity or food intake were identified in this study [22] although altered glucose homeostasis was observed in some animals. Thus, DES exposure early in neonatal life may alter the adipocyte development program leading to increased adiposity in adults.

Other estrogenic EDCs have been observed to alter body weight. Neonatal genistein exposure (50 mg/kg/day) at a dose equally estrogenic to low-dose (1 μg/kg/day) DES caused a significant 20% increase in female body weight at 4 months of age [23]. BPA, originally designed as an estrogenic compound but now thought to act also through nonclassical estrogen signaling pathways, also induced increased weight after fetal exposure (2.4 μg/kg administered to pregnant female mice) [24]. In this study, pups were delivered by cesarean section to control for intrauterine location and sex steroid exposure during development. Although all pups had equal weights at birth, the BPA-treated pups were significantly heavier by weaning, and this effect was observed in both males and females. In support of these observations, BPA was administered in drinking water to pregnant and lactating female CD-1 mice at two doses (1 μg/ml or 10 μg/ml) [6]. Exposure was continued by giving pups access to this BPA-containing water supply after weaning up to day 31, which may model bottle-feeding of human infants using BPA-containing plastic bottles. Both male and female pups were significantly heavier at postnatal day 31 and had significantly increased abdominal adipose tissue weights. In females, serum cholesterol was significantly elevated in BPA-treated mice, whereas in males, serum triglycerides and nonesterified fatty acids were significantly elevated. Gestational exposure of rats to low or high dose (0.1 or 1.2 mg/kg/day) of BPA also resulted in significantly increased body weights in both male and female offspring as pups and adults [7]. In another study, pregnant females were administered 0.5 or 10 mg/kg/day of genistein, resveratrol, zearalenone, and BPA or 0.5 or 10 μg/kg/day of DES [25]. Mice were examined at 4, 8, 12, and 16 weeks of age. Although differences were not significant until 16 weeks of age, all treated mice were heavier than untreated controls. These animal studies in which neonatal exposure to estrogenic EDCs induces increased adult body weight and altered lipid homeostasis suggest that at least some adult obesity seen in human populations could be due to, or exacerbated by, developmental exposures to low doses of EDCs.

11 Altered Glucose Homeostasis Resulting from Developmental Exposures...

That these EDC effects are likely to be mediated, at least in part, via estrogen signaling is supported by comparing low versus high phytoestrogen diets [26]. Paradoxically, switching pregnant mice to a low phytoestrogen diet resulted in increased endogenous estradiol production which had an estrogenizing effect on fetuses. In both males and females, body weights were reduced at birth, but by adulthood, they became obese and had significantly elevated leptin concentrations and increased gonadal fat-pad weights. At 90 days of age, males in this study were relatively glucose intolerant, being slower to clear the glucose challenge than untreated controls. These results demonstrate that even small increases in endogenous estrogen signaling during gestation result in increased body weight, altered fat deposition, and decreased glucose tolerance as adults.

Not all studies of fetal EDC exposure find weight gain as an outcome. This may be due to different doses, administration routes, exposure windows, examination ages, or strains or species of rodents studied. In one recent study, pregnant female CD-1 mice were fed chow with 4 μg/kg DES (1 μg/kg/day assuming 5 g eaten/day), 1 μg/kg BPA (0.25/kg/day eaten), or no supplement [27]. Weight at weaning of BPA-treated pups was significantly elevated in both males and females compared to untreated pups, but no effect of DES was observed. Males exposed to BPA had more body fat at 7 weeks of age compared to DES-treated mice, but neither group was different from untreated controls. In addition, there were no differences in body weight in weeks 3–9, that is, after weaning for males or females. No effects on glucose tolerance were observed for males, but females treated developmentally with DES were relatively glucose intolerant compared to untreated controls or to BPA-treated females. On a high-fat diet, females treated with BPA or DES had lower body fat compared to untreated controls, while in males, only the DES group had lower body weights. These results are not consistent with studies showing that DES, BPA, or other estrogenic compounds cause increased weight in adults as reviewed above. Differences among studies include diets, routes of administration, dose calculations, and other parameters so that direct comparisons are challenging. Differences between studies could indicate that dose, timing, and/or route of EDC administration are critical parameters that need to be controlled to more directly compare the effects of EDCs in different animal models. More importantly, it is critical to determine which of these paradigms is most relevant for modeling human exposures to assess their relative importance as an inducer of obesity.

Direct Actions of EDCs on Adipocyte Number and Function

The complex feedback loops regulating glucose and fat homeostasis suggest that the effects of fetal EDC exposure leading to increased body weight and fat content could be direct on adipocytes or indirect and act via modulating energy consumption versus expenditure. Several studies examine direct actions of EDCs on adipocytes or adipocyte precursors. Treatment of C3H10T1/2 multipotent precursor cells with 4-tert-octylphenol inhibited their differentiation into osteoblasts and upregulated peroxisome proliferator-activated receptor γ (PPARγ) [28]. When exposed to treatments that promote maturation of preadipocytes to adipocytes (insulin, dexamethasone, and 1-methyl-3-isobutylxanthine (MIX)), triglyceride and adiponectin content, both markers of adipocyte differentiation, were increased. These findings suggest that exposure to 4-tert-octylphenol could directly increase the number of adipocytes and thus, potentially, accumulation of fat. Organotins are another class of EDCs that include tributyltin chloride which was first used as antifouling agent in paints and persists in fish and shellfish [29]. Tributyltin binds directly to both PPARγ and retinoid X receptor (RXR) nuclear receptors and can induce adipogenesis in vitro and in vivo in a similar manner to other known PPAR agonists [30]. Similarly, tributyltin sensitized both human and mouse multipotent stromal cells derived from white adipose tissue to undergo adipogenesis, including increasing lipid content and adipogenic gene expression [31]. Moreover, stromal stem cells taken from mice exposed to tributyltin in utero exhibited an increased adipogenic capacity, reduced osteoblast potential, and enhanced lipid accumulation after induction. Stromal stem cells from treated mice also had higher PPARγ target gene expression, including increased Fabp4 (fatty acid-binding protein 4), as well as hypomethylation of the Fabp4 promoter. These results support the concept that EDCs can have direct effects on adipogenesis. Moreover, neonatal EDC exposure alters the adipogenic program toward increased adipocyte differentiation that could explain the increased relative fat content of mice treated in utero with EDCs noted above.

BPA treatment also alters adipocytes. In 3T3-F442A adipocytes, BPA enhanced basal and insulin-stimulated glucose uptake and enhanced expression of glucose transporter 4 (GLUT4) protein [32]. Therefore, BPA exposure might lead to increased adiposity via enhanced glucose uptake which is then stored as lipid. In another model of adipocyte differentiation, treatment of differentiated 3T3-L1 preadipocytes with nonylphenol or BPA stimulated lipid accumulation in a dose-dependent manner [33]. These treatments also upregulated expression of genes involved in lipid metabolism, including hormone-sensitive lipase, phospholipase A2, and phospholipase C, as well as CD-36, an LDL receptor [33]. These studies suggest that both nonylphenol and BPA can induce uptake and/or synthesis of triglycerides in adipocytes which would be expected to increase fat mass. Further support for this concept stems from the observation that BPA treatment of both visceral and subcutaneous human adipocytes suppressed adiponectin secretion as well as or better than equimolar estradiol treatments [34]. Since adiponectin protects against insulin resistance and inflammation, it could be hypothesized that BPA's actions on insulin sensitivity may be mediated, at least in part, via this mechanism.

Summary

It is now clear from epidemiological studies that the presence of EDCs in human fluids is associated with a number of adult metabolic diseases including diabetes [3] and insulin resistance [35]. While these associations raise concern that at least a portion of the increase in obesity and diabetes might be related to such exposures, they do not address mechanistic questions. Animal studies, as summarized above, lend support to the concept that these EDCs can directly alter critical cells and systems involved in glucose and lipid homeostasis and at least in animal models, induce weight gain and/or glucose intolerance in adults after gestational and/or neonatal exposures. The challenge for environmental scientists is to find ways to evaluate the effects of EDCs in human tissues and identify the mechanisms involved, which will then provide markers for these actions that can be explored directly in exposed humans. Another challenge is to design more integrative approaches to study the effects of EDCs on metabolism since adult metabolic disease often results from alterations or imbalances in highly integrated control systems with many levels of feedback regulation. Moreover, we now know from genetic studies that portions of the human population carry genetic polymorphisms that predispose to metabolic disease, so EDCs may function as an environmental modulator acting on those genetically prone to disease. Thus, there is a need for studies exploring interactions between EDC exposure and genetic variability to determine if specific gene changes make carriers more susceptible to EDC exposure. Many of these advances are on the horizon, and more definitive answers about the risks of EDCs will be forthcoming in the near future.

References

1. Ogden CL, Carroll MD, Curtin LR, McDowell MA, Tabak CJ, Flegal KM. Prevalence of overweight and obesity in the united states, 1999–2004. JAMA. 2006;295(13):1549–55.
2. Baillie-Hamilton PF. Chemical toxins: a hypothesis to explain the global obesity epidemic. J Altern Complement Med. 2002;8(2):185–92.
3. Lang IA, Galloway TS, Scarlett A, et al. Association of urinary bisphenol A concentration with medical disorders and laboratory abnormalities in adults. JAMA. 2008;300(11):1303–10.
4. Vandenberg LN, Hauser R, Marcus M, Olea N, Welshons WV. Human exposure to bisphenol A (BPA). Reprod Toxicol. 2007;24(2):139–77.
5. Ikezuki Y, Tsutsumi O, Takai Y, Kamei Y, Taketani Y. Determination of bisphenol A concentrations in human biological fluids reveals significant early prenatal exposure. Hum Reprod. 2002;17(11):2839–41.
6. Miyawaki J, Sakayama K, Kato H, Yamamoto H, Masuno H. Perinatal and postnatal exposure to bisphenol a increases adipose tissue mass and serum cholesterol level in mice. J Atheroscler Thromb. 2007;14(5):245–52.
7. Rubin BS, Murray MK, Damassa DA, King JC, Soto AM. Perinatal exposure to low doses of bisphenol A affects body weight, patterns of estrous cyclicity, and plasma LH levels. Environ Health Perspect. 2001;109(7):675–80.
8. Bern H. The fragile fetus. In: Colborn T, Clement T, editors. Chemically-induced alterations in sexual and functional development: the wildlife/human connection. Princeton: Princeton Scientific Publishing; 1992.
9. Ropero AB, Alonso-Magdalena P, Quesada I, Nadal A. The role of estrogen receptors in the control of energy and glucose homeostasis. Steroids. 2008;73(9–10):874–9.
10. Alonso-Magdalena P, Ropero AB, Carrera MP, et al. Pancreatic insulin content regulation by the estrogen receptor ER alpha. PLoS One. 2008;3(4):e2069.

11. Nadal A, Rovira JM, Laribi O, et al. Rapid insulinotropic effect of 17beta-estradiol via a plasma membrane receptor. FASEB J. 1998;12(13):1341–8.
12. Nadal A, Ropero AB, Laribi O, Maillet M, Fuentes E, Soria B. Nongenomic actions of estrogens and xenoestrogens by binding at a plasma membrane receptor unrelated to estrogen receptor alpha and estrogen receptor beta. Proc Natl Acad Sci USA. 2000;97(21):11603–8.
13. Adachi T, Yasuda K, Mori C, et al. Promoting insulin secretion in pancreatic islets by means of bisphenol A and nonylphenol via intracellular estrogen receptors. Food Chem Toxicol. 2005;43(5):713–9.
14. Alonso-Magdalena P, Morimoto S, Ripoll C, Fuentes E, Nadal A. The estrogenic effect of bisphenol A disrupts pancreatic beta-cell function in vivo and induces insulin resistance. Environ Health Perspect. 2006;114(1):106–12.
15. Alonso-Magdalena P, Laribi O, Ropero AB, et al. Low doses of bisphenol A and diethylstilbestrol impair Ca2+ signals in pancreatic alpha-cells through a nonclassical membrane estrogen receptor within intact islets of Langerhans. Environ Health Perspect. 2005;113(8):969–77.
16. Ae PS, Choi MS, Cho SY, et al. Genistein and daidzein modulate hepatic glucose and lipid regulating enzyme activities in C57BL/KsJ-db/db mice. Life Sci. 2006;79(12):1207–13.
17. Liu D, Zhen W, Yang Z, Carter JD, Si H, Reynolds KA. Genistein acutely stimulates insulin secretion in pancreatic beta-cells through a cAMP-dependent protein kinase pathway. Diabetes. 2006;55(4):1043–50.
18. Fu Z, Zhang W, Zhen W, et al. Genistein induces pancreatic beta-cell proliferation through activation of multiple signaling pathways and prevents insulin-deficient diabetes in mice. Endocrinology. 2010;151(7):3026–37.
19. Newbold RR, Padilla-Banks E, Jefferson WN, Heindel JJ. Effects of endocrine disruptors on obesity. Int J Androl. 2008;31(2):201–8.
20. Newbold RR, Padilla-Banks E, Jefferson WN. Environmental estrogens and obesity. Mol Cell Endocrinol. 2009;304(1–2):84–9.
21. Newbold RR, Jefferson WN, Padilla-Banks E. Long-term adverse effects of neonatal exposure to bisphenol A on the murine female reproductive tract. Reprod Toxicol. 2007;24(2):253–8.
22. Newbold RR, Padilla-Banks E, Snyder RJ, Phillips TM, Jefferson WN. Developmental exposure to endocrine disruptors and the obesity epidemic. Reprod Toxicol. 2007;23(3):290–6.
23. Newbold RR, Padilla-Banks E, Snyder RJ, Jefferson WN. Developmental exposure to estrogenic compounds and obesity. Birth Defects Res A Clin Mol Teratol. 2005;73(7):478–80.
24. Howdeshell KL, Hotchkiss AK, Thayer KA, Vandenbergh JG, vom Saal FS. Exposure to bisphenol A advances puberty. Nature. 1999;401(6755):763–4.
25. Nikaido Y, Yoshizawa K, Danbara N, et al. Effects of maternal xenoestrogen exposure on development of the reproductive tract and mammary gland in female CD-1 mouse offspring. Reprod Toxicol. 2004;18(6):803–11.
26. Ruhlen RL, Howdeshell KL, Mao J, et al. Low phytoestrogen levels in feed increase fetal serum estradiol resulting in the "fetal estrogenization syndrome" and obesity in CD-1 mice. Environ Health Perspect. 2008;116(3):322–8.
27. Ryan KK, Haller AM, Sorrell JE, Woods SC, Jandacek RJ, Seeley RJ. Perinatal exposure to bisphenol-a and the development of metabolic syndrome in CD-1 mice. Endocrinology. 2010;151(6):2603–12.
28. Miyawaki J, Kamei S, Sakayama K, Yamamoto H, Masuno H. 4-tert-octylphenol regulates the differentiation of C3H10T1/2 cells into osteoblast and adipocyte lineages. Toxicol Sci. 2008; 102(1):82–8.
29. Appel KE. Organotin compounds: toxicokinetic aspects. Drug Metab Rev. 2004;36(3–4): 763–86.
30. Grun F, Watanabe H, Zamanian Z, et al. Endocrine-disrupting organotin compounds are potent inducers of adipogenesis in vertebrates. Mol Endocrinol. 2006;20(9):2141–55.
31. Kirchner S, Kieu T, Chow C, Casey S, Blumberg B. Prenatal exposure to the environmental obesogen tributyltin predisposes multipotent stem cells to become adipocytes. Mol Endocrinol. 2010;24(3):526–39.

32. Sakurai K, Kawazuma M, Adachi T, et al. Bisphenol A affects glucose transport in mouse 3T3-F442A adipocytes. Br J Pharmacol. 2004;141(2):209–14.
33. Wada K, Sakamoto H, Nishikawa K, et al. Life style-related diseases of the digestive system: endocrine disruptors stimulate lipid accumulation in target cells related to metabolic syndrome. J Pharmacol Sci. 2007;105(2):133–7.
34. Hugo ER, Brandebourg TD, Woo JG, Loftus J, Alexander JW, Ben-Jonathan N. Bisphenol A at environmentally relevant doses inhibits adiponectin release from human adipose tissue explants and adipocytes. Environ Health Perspect. 2008;116(12):1642–7.
35. Lee DH, Lee IK, Jin SH, Steffes M, Jacobs Jr DR. Association between serum concentrations of persistent organic pollutants and insulin resistance among nondiabetic adults: results from the National Health and Nutrition Examination Survey 1999–2002. Diabetes Care. 2007; 30(3):622–8.

Chapter 12
Developmental Exposure to Endocrine Disrupting Chemicals: Is There a Connection with Birth and Childhood Weights?

Elizabeth E. Hatch, Jessica W. Nelson, Rebecca Troisi, and Linda Titus

Abstract Childhood obesity has increased dramatically over the last several decades in developed countries, and more recently in developing countries. Most research on causes of obesity has focused on various aspects of diet and lack of physical activity as the primary risk factors. Clearly, a balance between energy intake and energy expenditure is critical to maintain a healthy body weight, but other factors have also been linked to the obesity epidemic. One area of concern is increasing exposure to endocrine disrupting chemicals (EDCs), especially during the prenatal period. Animal studies have suggested that fetal exposure to certain EDCs may cause systemic alterations in aspects of 'fetal programming' related to adipocyte differentiation and function, appetite regulation, and other body systems involved in weight homeostasis. One plausible pathway between prenatal exposures to EDCs and obesity might be through effects on fetal growth, since both growth retardation and high birth weight have been associated with later obesity. Some studies have found associations between several classes of EDCs, including PCBs, organochlorine pesticides such as DDT and hexachlorobenzene (HCB), phenols,

---

E.E. Hatch (✉)
Department of Epidemiology, Boston University School
of Public Health, Boston, MA, USA
e-mail: eehatch@bu.edu

J.W. Nelson
Department of Environmental Health, Boston University
School of Public Health, Boston, MA, USA

R. Troisi
Epidemiology and Biostatatistics Program, Division of Cancer
Epidemiology and Genetics, National Cancer Institute, National Institutes
of Health, Department of Health and Human Services, Bethesda, MD, USA

L. Titus
Department of Community & Family Medicine and Pediatrics,
Dartmouth Medical School and Dartmouth-Hitchcock
Medical Center, Lebanon, NH, USA

E. Diamanti-Kandarakis and A.C. Gore (eds.), *Endocrine Disruptors and Puberty*,
Contemporary Endocrinology, DOI 10.1007/978-1-60761-561-3_12,
© Springer Science+Business Media, LLC 2012

and PFCs and fetal growth retardation. Although animal data have suggested that EDCs can affect offspring obesity, thus far, data from human epidemiological studies that have directly examined prenatal EDC exposure in relation to childhood growth are not sufficient to draw conclusions. Several methodologic challenges exist in conducting these studies, including timing of measurement of EDCs in gestation, accounting for potential confounders occurring during pregnancy and childhood, and accurately measuring adiposity.

Keywords Birth weight • Developmental origins of health and disease • Endocrine-disrupting chemicals • In utero exposures • Obesity

Introduction

Childhood overweight and obesity have increased dramatically over the last several decades in developed countries, and more recently in developing countries. In the United States, the prevalence of childhood obesity among 6–11 year olds increased from 4% in the early 1970s to 15.3% in 1999–2000, [1] although the increase appears to have leveled off more recently [2]. US data also suggest that children in recent birth cohorts are becoming heavier at younger ages [3]. Similar increases in the prevalence of childhood obesity have occurred in most European countries, although later and of a smaller magnitude than in the USA [4]. Because of obesity's association with many adverse health outcomes, [4] including higher mortality rates, [5] and of the stubborn persistence of childhood obesity into adulthood, [6] it is important to identify risk factors that might be amenable to prevention.

Most research on causes of obesity, both during childhood and adult life, has focused on various aspects of diet and lack of physical activity as the primary risk factors. Clearly, a balance between energy intake and energy expenditure is important in maintaining a healthy body weight, but some investigators have suggested that other factors may also play a causal role in the obesity epidemic [7–10]. One area of concern is increasing exposure to endocrine disrupting chemicals (EDCs), especially during the prenatal period. Animal studies have suggested that fetal exposure to certain EDCs may cause systemic alterations in aspects of "fetal programming" related to adipocyte differentiation and function, appetite regulation, and other body systems involved in weight homeostasis [7, 11, 12]. Several EDCs are suspected to be "obesogens," based primarily upon evidence from studies in animals. Chemicals on the list of suspects include diethylstilbestrol (DES), phthalates, bisphenol A (BPA), organotins, perfluorinated chemicals (PFCs), and persistent organochlorines such as dichlorodiphenyltrichloroethane (DDT) and polychlorinated biphenyls (PCBs).

One plausible pathway between prenatal exposures to EDCs and childhood obesity might be through effects on fetal growth or length of gestation. Although epidemiologic evidence is mixed, some studies have found associations between several classes of EDCs, including PCBs, organochlorine pesticides such as DDT and hexachlorobenzene (HCB), phenols, and PFCs and fetal growth restriction. Some studies suggest that there is a J-shaped relation between birth weight and subsequent obesity, with both low birth weight (especially combined with rapid catch-up growth) and high birth weight having an association with later obesity, independent of gestational length [9, 13–17]. In addition, there is mixed evidence that phthalates may be associated with length of gestation, although the connection between length of gestation and obesity has been infrequently studied.

In this chapter, we will briefly review the literature on the association between fetal growth, as represented by birth weight, and childhood or later obesity. We will discuss several examples of factors that appear to affect fetal growth, and thus may be part of the causal pathway to later obesity. We will review the limited epidemiologic literature that has directly evaluated in utero exposures to EDCs in relation to childhood obesity. Throughout the chapter, we will discuss potential biologic mechanisms that may operate during the prenatal period to affect later obesity risk. Finally, we will discuss methodological challenges of these studies, and some possible solutions.

Birth Weight and Subsequent Adiposity

Birth weight is determined by genetic and environmental factors and is often used as a proxy for attained fetal growth. Birth weight is highly correlated with length of gestation, and a more accurate assessment of fetal growth requires adjustment for gestational length. Instead of low birth weight, defined as <2,500 g, a more common measure of poor fetal growth is "small for gestational age" (SGA). SGA is usually defined as birth weight less than the tenth percentile for each specific gestational week, either using a standard reference population or the specific data under study [18]. (The term IUGR, intrauterine growth retardation, is also used.) Babies born small display a tendency to undergo rapid compensatory (catch-up) growth in infancy [19, 20]. On the other end of the growth spectrum, babies who are at or above the 90th percentile for their gestational age are considered "large for gestational age" (LGA). Infant macrosomia, variably defined as a birth weight ≥4,000, 4,200, or 4,500 g, is also sometimes used to describe babies that are born large [21]. Macrosomia is frequently a consequence of gestational diabetes. Numerous studies have evaluated the relationship between birth weight and childhood obesity, but it is important to recognize that birth weight is a mixture of the effects of fetal growth and the duration of gestation. Also, gestational length itself was subject to measurement error in older studies since women's menstrual cycle lengths are variable and the exact date of conception was rarely known. It may be more useful to evaluate growth potential, an infant's size relative to its genetic potential based on parental sizes, as not all babies born at the same gestational age had the same growth experience in utero.

Epidemiological evidence indicates the possibility of two separate pathways through which fetal growth affects susceptibility to later obesity. The first of these is the well-established positive association between birth weight and childhood offspring adiposity (comprehensively reviewed in [17, 22]), an association which appears to be independent of gestational age [22, 23]. For example, in 300,000 Danish children, increased birth weight was associated with the risk of overweight at ages 6 through 13 [24]. A positive association between birth weight and adiposity in adults has also been shown, including in a study of 4,300 Danish men [22, 25] and >160,000 American women [25]. Some investigators have suggested that the positive association between birth weight and later BMI is the result of a higher proportion of lean body mass [26].

The second possible pathway between fetal growth and obesity is based on findings from certain studies showing a J-shaped association between birth weight and later adiposity, i.e., that small babies are also at risk for developing obesity [9, 15, 17]. These findings are less consistent than those for the high birth weight-obesity association, though lower birth weight is an established risk factor for obesity-related diseases such as hypertension, cardiovascular disease, and insulin resistance [22, 27, 28]. In a British longitudinal cohort of children born in 1958 and followed to age 33 [13], the shape of the association between birth weight and subsequent BMI became more J-shaped as age increased. Babies born with low birth weight who underwent more rapid growth before age seven were at the greatest risk for being overweight as young adults [13].

Studies that examine birth weight in relation to measures of central (or "visceral") adiposity, as opposed to BMI, appear to have stronger findings [17]. Central adiposity is hypothesized to be more detrimental than overall or subcutaneous obesity as it is associated with an increased risk of metabolic syndrome, a constellation of conditions including abnormal lipid levels, elevated blood pressure, and hyperinsulinemia, which are themselves associated with diabetes and cardiovascular disease risk [17, 29]. After adjusting for attained BMI, a number of studies in children and adults have found an inverse association between birth weight and measures of central adiposity, including subscapular-to-triceps skinfold ratio and waist-to-hip ratio [28, 30, 31]. Other studies that used dual energy X-ray absorptiometry (DXA) to more precisely measure body composition also have found associations between lower birth weight and higher percent body fat and fat mass [32] and lower bone and muscle mass in adults [33]. In one study, SGA babies had greater visceral-to-subcutaneous fat ratios, measured by magnetic resonance imaging, at age 6 compared with babies born appropriate for gestational age, [34] despite having similar total fat mass and lean mass. Additionally, unlike with heavier babies, the risk of greater visceral adiposity appears most pronounced for SGA infants who undergo rapid catch-up growth [15].

Various hypotheses have been proposed to explain these associations between fetal growth, at both ends of the spectrum, and later adiposity. One, the "developmental origins of adult disease" postulates that there are certain critical windows during fetal development in which "programming" of different biological systems, including body weight regulation, occurs based on nutritional, hormonal, and other cues from the mother [7]. The related Barker hypothesis suggests that an undernourished fetal environment, for which lower birth weight may serve as a surrogate, leads to permanent changes in regulation of body weight aimed at conserving nutrients [35, 36]. In the case of mismatch, however, differences in the environment experienced in utero or in early childhood, and later in life, can lead to maladaptive responses. For example, when faced with the unrestricted diet of many developed countries, the individual whose physiology predicts a constrained environment may be more at risk for obesity and related metabolic disorders. Specific biological mechanisms could include effects on developing adipocytes, alterations in the neuroendocrine signaling pathways that regulate appetite, changes in structure and function of the pancreas, and regulation of other genes that control carbohydrate and lipid metabolism [9, 15, 17].

It is possible that the epidemiological evidence for the relationship between birth weight and later obesity is due to factors other than the fetal environment, including aspects of the postnatal environment or genetics. Also, the associations could be confounded by maternal BMI or gestational weight gain. For example, a study of African-American young adults found that birth weight was positively correlated with adiposity measured by skinfold thickness, but the association was fully attenuated when adjusted for child's sex, maternal prepregnancy BMI, and birth order [37]. Few studies have had the ability to comprehensively account for these factors. It is also difficult to separate the effects of birth weight and postnatal growth; some studies show evidence that growth in infancy may be more important than prenatal growth [38, 39]. These associations may also depend on the population studied either because of real population differences or the distribution of birth weight [40].

Gestational Diabetes and Offspring Body Measures

Much of the original evidence suggesting a role for fetal programming of offspring obesity was based on studies of maternal prepregnancy diabetes in relation to offspring macrosomia. The biological mechanism proposed to explain this association involves heightened fetal pancreatic insulin production in response to maternal hyperglycemia [41], although other transplacental substances [42], as well as genetic [43] and lifestyle factors [44, 45], may play a role.

Whether offspring overweight/obesity also is influenced by gestational diabetes, a milder (usually diet-controlled) but more common and usually temporary form of diabetes occurring during pregnancy is of interest. The issue has been addressed by several studies, but in many of these, the analysis either combined women with gestational diabetes with those who had insulin-dependent (or noninsulin dependent) diabetes mellitus before the pregnancy, or did not include comparisons with offspring of women without these conditions.

Results of the few studies that have compared obesity in the offspring of women with and without gestational diabetes are mixed. A review published in 2007 [46] identified four relevant reports. Of three retrospective cohort studies [41, 45, 47], one found no association with obesity [45], and two suggested a 40% increase in risk for obesity or overweight, although the findings were not statistically significant [41, 47]. In a prospective study, offspring of mothers with gestational diabetes who were large for gestational age had greater childhood BMI, and circumference and skinfold measures than control offspring (no gestational diabetes in the mother) or offspring of mothers with gestational diabetes who had normal birth weights [48]. The authors note these findings became more apparent between ages 4 and 7.

Results published subsequent to the 2007 review are not consistent regarding the influence of gestational diabetes on offspring body size. A prospective study in Denmark showed an increased risk of overweight (OR 1.8) and an even greater risk of metabolic syndrome (OR 4.1) in the young adult (ages 18–27) offspring of women with diet-treated gestational diabetes (defined by several indicators, including fasting glucose) compared to those of community controls after adjustment for covariates (maternal age at delivery, maternal prepregnancy BMI, ethnic origin, family occupational social class) [43]. The findings also suggested that the association with offspring metabolic syndrome was stronger for gestational diabetes than Type 1 diabetes in the mother (OR 2.6), whereas the association with overweight was somewhat stronger for offspring of mothers with Type 1 diabetes (OR 2.3). The study was limited by small numbers (fewer than 200 women with gestational diabetes) and incomplete follow-up of the offspring (56%) [43].

The largest study to date, a prospective follow-up study conducted in Germany, included 232 offspring of women with gestational diabetes, 757 offspring of women with Type 1 diabetes, and 431 offspring of mothers without any form of diabetes [49]. The investigators examined gestational diabetes, determined by oral glucose tolerance tests, in relation to the prevalence of offspring overweight (> 90th percentile of BMI) at ages 2, 8, and 11 and insulin resistance at ages 8 and 11. The results of this study showed a statistically significant increase in the prevalence of overweight from age 2 (17%) to 11 (31%) in the children of women with gestational diabetes, when compared to offspring of women without diabetes (11% at age 2 and 16% at age 11). In contrast to the study by Clausen et al. [43], no increase in the prevalence of overweight was seen in the offspring of women with Type 1 diabetes. Potential confounders of the relationship between gestational diabetes and offspring overweight, including maternal BMI, were not controlled in the analysis.

Another recent prospective study assessed gestational diabetes and glycosuria (the presence of glucose in the urine) in relation to offspring birth weight and body measures at ages 9–11, including BMI, waist circumference, and fat mass as measured by DXA [44]. In this study, the offspring of women diagnosed with either of these conditions had higher birth weight and an increased risk of macrosomia compared to offspring of women without gestational diabetes or glycosuria. The adjusted odds ratios for macrosomia were 5.50 (95% CI 1.18–10.30) in the offspring of women with gestational diabetes and 1.58 (95% CI 1.18–2.12) for the offspring of women with glycosuria, respectively. Anthropometric measures, available for nearly 7,000 pairs of mothers and offspring, suggested that offspring (ages 9–11) of mothers with gestational diabetes had an increased risk of overweight/obesity, but the association was not apparent after adjustment for maternal prepregnancy BMI.

The data addressing the association of gestational diabetes and birth weight and childhood adiposity are mixed. Results of studies that take into account the greater birth size associated with this condition seem to indicate that birth size is the more important predictor of childhood adiposity. However, the results to date have not adequately addressed the possibility of confounding by a shared maternal-offspring obesogenic environment.

> **Box 1** Birth Weight and Later Obesity
>
> - Most studies have found a positive association between birth weight and higher body mass index in childhood and adulthood, although some authors suggest this may be due to greater lean body mass, and incomplete control of confounding factors cannot be completely ruled out.
> - Gestational diabetes is associated with macrosomia and possibly with greater childhood BMI, but findings are mixed, and many studies do not control for important confounders.
> - Some studies suggest that fetal growth restriction and low birth weight are also associated with later obesity.
> - Poor fetal growth, especially combined with rapid early "catch-up" growth, is associated with central adiposity.

Examples of Fetal Growth Disruptors Which May Be Associated with Childhood/Adolescent Obesity

Maternal Smoking During Pregnancy and Offspring Body Mass

In developed countries, the most prevalent modifiable risk factor for low birth weight is maternal smoking during pregnancy, which reduces birth weight by an average of 150–300 g [50]. The association between maternal smoking and low birth weight is remarkably consistent across diverse, albeit westernized, study populations. Substantial evidence indicates that babies born to smoking mothers are at an increased risk for overweight and obesity, [46, 50, 51] which increases risk of adult obesity-related morbidity and mortality.

Several reviews and meta-analyses of studies investigating the association between exposure to maternal smoking and offspring obesity have been published in recent years. A systematic review covering eight studies, half of which were cohort in design, showed a relationship between maternal smoking and childhood overweight or obesity on the order of 50–100% elevation in risk [46]. Similar results were observed in the subset of individual studies that adjusted for potential confounders including parental obesity and socioeconomic status, maternal breastfeeding, and measures of childhood nutrition and activity.

A recent meta-analysis pooled the results of 14 studies conducted over a 40-year period (1966–2006), including some omitted from the earlier review, and representing nearly 85,000 children [51]. Offspring overweight and obesity, respectively, were defined as body mass index (BMI) at or above the 85th or 90th percentile and BMI at or above the 95th or 97th percentile. In all but two of the analyzed studies, both of which included assessments at older ages, outcomes were assessed between ages 2 and 8. Differences were noted between the characteristics of mothers who smoked and those who did not; those who did were more likely to have a lower income and to be less educated, heavier, and less likely to breastfeed; their children had more rapid weight gain in infancy and were less active than offspring of non-smokers. The results of the meta-analysis showed a modest association between maternal smoking during pregnancy and offspring overweight and obesity (OR 1.5). The association between maternal smoking and child overweight remained evident after adjusting for covariates, which generally included a measure of the mothers' weight and socioeconomic status. About half of the studies additionally adjusted for a measure of fetal growth/birth weight. Thus, the authors inferred that the association between intrauterine smoking exposure and childhood overweight is independent of fetal growth restriction.

A meta-analysis of 17 studies published through 2007, most of which were follow-up in design, and all but one of which assessed outcomes during childhood, also found a link between intrauterine exposure to maternal smoking and offspring overweight [50]. The OR of 1.64 was evident after adjustment for confounders, including maternal socioeconomic status, obesity, breastfeeding, and offspring birth weight. Evidence of an association was absent in only one of the 17 studies. Studies subsequent to the meta-analysis have also shown an association between maternal smoking and offspring obesity in childhood [52].

Studies assessing dose–response (e.g., cigarettes smoked per day) have demonstrated consistent findings of a relationship between maternal smoking during pregnancy and offspring overweight. A recent meta-analysis [51] found a dose–response in all seven studies that measured the amount of maternal smoking, including a study in which outcomes were assessed into adulthood [53]. In that study, 6,000 offspring born in 1958 in the UK were assessed at ages 7, 11, 16, 23, and 33; the prevalence of obesity at ages 16 and 33 increased with the amount of maternal smoking (measured as the number of cigarettes smoked per day after the fourth month of pregnancy) [53]. One of the studies provided additional data showing a more pronounced association with smoking throughout the pregnancy than in early pregnancy alone, but this was not universally observed [51].

Only a few studies have assessed maternal smoking during pregnancy and outcomes in children past the age of 10 years. The UK follow-up study cited above included assessments through age 33; the data suggested that the association between maternal smoking and offspring overweight/obesity became more pronounced with increasing age. At least one other study has suggested similar increases with age up until age 8 years [54]. In a recent study conducted in Canada, intrauterine smoking exposure was related to increased risk of subcutaneous and intra-abdominal fat among children in late but not early puberty [55]. Thus, if the association between intrauterine smoke exposure and increased body measures is causal, the effects potentially may be long lasting and increase with offspring age.

The association between maternal smoking and offspring overweight is unlikely to be explained by prepregnancy influences on the uterine environment or by postnatal exposure to the mother's smoking. The meta-analysis by Oken et al. [51] showed that the association between maternal smoking and offspring body mass was present when smoking mothers were compared to nonsmokers or to a comparison

> **Box 2** Maternal Smoking and Birth Weight and Offspring Obesity
>
> - Gestational exposure to maternal smoking is associated with intrauterine growth restriction as measured by decreased birth weight and other markers.
> - Mechanisms for this relationship have not been identified but may include nicotine effects on oxygenation and vascularization.
> - Gestational exposure to maternal smoking is associated with increased risk for overweight in offspring.
> - The association between maternal smoking and offspring overweight may be mediated by rapid postnatal growth (rather than intrauterine growth restriction per se.)

group of women who smoked, but not during the index pregnancy. A prospective study in Brisbane, Australia, found greater mean differences in BMI, and higher risk of overweight (OR 1.3) and obesity (OR 1.4) at age 14 in offspring who were exposed prenatally to maternal smoking, compared with offspring of nonsmokers, whereas body measures of the offspring of mothers who smoked before or after (but not during) pregnancy were similar to those of the offspring of nonsmoking mothers [56]. These results suggest that other factors, including pre- or postnatal maternal smoking, do not account for the association between intrauterine smoke exposure and later obesity. Additionally, based on simulations, publication bias is unlikely to account for the association [50, 51].

Several biological mechanisms have been posited to explain an influence of maternal smoking on fetal growth, and some of these may be involved in offspring overweight. They include an appetite-suppressing effect of on the mother during pregnancy, reduced oxygen uploading to the fetus from elevated carbon monoxide levels, nicotine-induced vasoconstriction, or cyanide compounds interfering with fetal oxidative metabolism [53]. Other mediators have been postulated, including but not limited to alterations in leptin levels and leptin receptor expression, and insulin or glucocorticoid dysregulation [46]. As noted above, however, recent evidence indicates that the association between low birth weight and later obesity could also be explained by rapid postnatal catch-up growth, which may mediate the relationship between intrauterine smoking exposure and offspring obesity.

Diethylstilbestrol (DES)

DES is often invoked as a model for potential effects of exposures to ubiquitous endocrine disrupting chemicals occurring at lower levels in the environment [57]. DES is a nonsteroidal estrogen first synthesized in 1938, and commonly prescribed to pregnant women during the 1940s through the 1960s. Originally developed to alleviate menopausal symptoms, DES was prescribed for use in pregnancy to bolster the decline in pregnancy hormones thought to be related to miscarriage and preterm birth [58]. Although several clinical trials found that DES was not efficacious [59, 60], and might even be harmful [59], it was heavily promoted by pharmaceutical companies and researchers and its use continued until 1971 when it was linked to a rare vaginal cancer among female offspring [61]. Subsequently, effects of in utero and, more recently, transgenerational exposure [62–65] have been studied extensively in humans and animal models. Results from animal models and human epidemiologic studies have been remarkably consistent in finding teratogenic, carcinogenic, and adverse reproductive effects [66–69]. Such studies suggest the involvement of epigenetic changes (altered gene function) that arise from hormonal stimulation occurring during a critical window of prenatal development [70–72]. To date, the genes implicated in epigenetic alteration include the hormone-responsive genes lactoferrin [73] and epidermal growth factor; [72, 74, 75] the Hox [76, 77] and Wnt genes [78] involved with embryonic reproductive tract development; and c-fos [74, 79] and c-jun [79], proto-oncogenes involved in hormone regulation. Thus far, most of the focus in DES research has been on mechanisms related to cancer and possible transmission of risk to the third generation; whether the epigenetic changes associated with DES exposure might alter the risk of obesity in humans is uncertain. One study of DES-exposed mice showed changes in genes involved in fat cell distribution but not in genes involved in adipocyte differentiation [80].

A relation between prenatal DES exposure and obesity in humans has not been reported, but a possible mechanism by which DES might affect later obesity is through effects on fetal growth. Growth restriction in mice following prenatal DES exposure has been described, [68, 81] although relatively few animal studies have evaluated growth restriction as a specific endpoint, because outcomes later in life have been of greater interest. Brackbill and Berendes [82] reanalyzed data from a human clinical trial of DES conducted during the 1950s and reported an approximately twofold increase in the risk of low birth weight in the DES-exposed group. A recent extension of this finding from a large cohort of DES exposed and unexposed daughters followed by the National Cancer Institute found lower mean birth weight, a 1.6-fold increase in SGA, and a two- to threefold increase in preterm birth among the prenatally exposed, compared to unexposed, daughters [83]. Higher cumulative dose of DES and exposure earlier in gestation was related to preterm birth, but did not appear to affect SGA or birth weight, compared with lower doses or first administration later in pregnancy.

Interest in the possible obesogenic effects of prenatal DES exposure followed a recent report of increased body weights and adiposity among mice exposed to DES during the perinatal period [80]. Newbold et al. treated groups of 4–8 mice with low (1 µg/kg/day) and high doses (1,000 kg/day) of DES or vehicle during neonatal days 1–5, which corresponds to third trimester exposure in humans. At the low dose, DES did not affect body weight during treatment, but when mice were measured postnatally at 4–6 months of age, there was a significant increase in body weight compared to controls among females but not males. At the high dose, DES

Box 3 Diethylstilbestrol, Fetal Growth, and Obesity

- Limited evidence from animal and human studies suggests that in utero DES exposure may result in fetal growth restriction.
- Results from two animal studies on the effect of perinatal DES exposure in relation to offspring obesity are conflicting, although both found that female DES-exposed mice were more glucose intolerant than controls.
- Possible mechanisms for obesogenic effects are speculative, but might operate through effects on fetal growth, or epigenetic changes related to metabolism and adipocyte biology.

resulted in growth restriction during treatment. Using mouse densitometry to measure outcomes, few differences were found in the high dose group at 2 months of age, but at 6 months of age, DES mice were significantly heavier (45 ± 2.5 vs. 37.9 ± 2.5 g) and had higher estimated fat weight (18.4 ± 1.8 vs. 11.5 ± 1.3 g), higher percentage of fat mass (40.1% vs. 29.9%), and lower percentage of lean mass (59.2 vs. 70.1%). DES mice also had larger fat pads, particularly inguinal (0.207 ± 0.041 vs. 0.108 ± 0.015 g) and retroperitoneal (0.555 ± 0.080 vs. 0.291 ± 0.029 g); there were no significant differences in the weight of brown fat. Animals exposed at the high dose also had significantly increased levels of leptin, adiponectin, glucose, and interleukin-6, (8 animals per group were measured) at 6 months of age (16 animals per group were measured). Insulin levels in the exposed mice were elevated at 6 months of age, but were significantly lower at 2 months of age, whereas triglycerides were elevated at 2 months but not affected at 6 months of age. There was some suggestion that decreased physical activity and increased food intake might partially explain the higher body weights of the DES exposed [11]. However, when DES-exposed and unexposed animals had access to a running wheel, they actually had lower body weights than controls [84]. It was unclear whether body weight differences persisted over the whole life span; the authors mentioned that the differences were not statistically significant at 18 months due to increased variability among the groups [80]. In a more recent study, Ryan et al. evaluated effects of DES and BPA on offspring obesity also found that female mice exposed to DES perinatally were more glucose intolerant than controls, but found no evidence for effects on body weight or fat composition [85] using a dose of 1 µg/kg/day to correspond to the low dose in the Newbold study [80]. The protocols differed in timing of exposure; in the Ryan study, [85] exposure was through maternal diet prior to breeding and continuing until postnatal day 21, whereas in the Newbold study, newborn mice were exposed to DES during neonatal day 1 through 5 [80]. These differences in protocol may explain the different findings in the two studies. Human data on body mass index and adiposity following prenatal DES exposure have not been reported, although early results suggested that DES-exposed daughters might have a higher risk of anorexia [86].

Endocrine Disrupting Chemicals in the Environment

Several classes of chemicals with endocrine disrupting potential have been studied in relation to fetal growth and/or length of gestation in humans. However, only a limited number of epidemiologic studies have evaluated possible effects of exposure to EDCs during the prenatal period in relation to growth during childhood. We will provide a brief overview of selected studies that evaluated fetal growth and gestational length in relation to EDCs, followed by a discussion of studies that have looked at childhood growth in greater depth. We will cover several classes of EDCs including persistent organochlorine chemicals (focusing on PCBs and DDT/DDE), phthalates, phenols, and perfluorinated chemicals (PFCs).

Persistent Organochlorines: PCBs

PCBs were banned in the USA in the late 1970s but are highly persistent and are still detected worldwide in people and the environment. Numerous studies have evaluated the association between PCBs and fetal growth and/or length of gestation, and there have been several reviews of the topic [87–89]. A smaller number of studies have also evaluated PCBs in relation to growth in childhood (Table 12.1). Accidental contamination of food sources with high levels of PCBs resulted in large fetal growth deficits and some evidence for reduced growth in childhood [90, 91], but findings for lower levels of exposure have been inconsistent. The specific mechanisms through which PCBs may impact fetal growth and/or later obesity are not known, but several possibilities have been suggested. Certain PCB congeners have been shown to be estrogenic, and others antiestrogenic [92]; estrogen levels in pregnancy have been positively associated with higher birth weight in humans [93], and animal studies have demonstrated that exposure to estrogenic compounds during development increases the number of adipocytes and alters subsequent functioning of these cells [7]. In addition, studies have found that certain PCB congeners affect thyroid hormones, and neonatal thyroid status is involved in long-term metabolic programming [94].

Several studies which measured circulating concentrations of PCBs before or during pregnancy have reported reductions in birth weight among the more highly exposed participants, ranging from -126 g decrement for a tenfold increase in PCB level in one study to –480 g for the lowest to highest tertile in another study [95–98]. In the only study to our knowledge that enrolled women prior to pregnancy and evaluated preconception levels of PCBs in relation to birth weight, Murphy et al. found a large reduction in mean birth weight related to antiestrogenic PCB congeners [98]. Other studies have found effects on birth weight in males but not females [99–101] and stronger effects in smokers, [102] whereas others have had largely null results [103–105].

Most studies have controlled for numerous potential confounders including maternal characteristics (e.g., age, prepregnancy BMI, height, education, smoking),

Table 12.1 Low-level exposure to EDCs during pregnancy in relation to growth in childhood: summary of pregnancy cohort studies

Citation	Study population, birth year(s), sample size	Exposure	Covariates	Outcome, age when assessed	Results
Jacobson 1990 [110]	Children of Michigan women who ate sportsfish, and who were potentially exposed to PCBs and PBBs via farm products 1980–1981	Cord blood, breast milk, and child age 4 serum levels of total PCBs; not lipid adjusted	25 potential confounders in 5 domains: SES, other demographic, perinatal risk, other environmental chemicals, situational variables	Weight, height, and head circumference at age 4 collected by study staff	Children with cord blood PCBs >5.0 ng/ml were on average 1.8 kg lighter than children with levels <1.5 ng/ml (not adjusted for height). No associations seen with height or head circumference
					Breast milk and child serum levels associated with reduced activity levels
	n = 323				Results not affected by control for PBBs, DDT, lead
Patandin 1998 [95]	Offspring of pregnant women in the Netherlands 1990–1992 n = 205	Late pregnancy serum and cord blood levels of PCBs [118, 138, 153, 180]; not lipid adjusted	SES, maternal health, gestational age, parity, gender, alcohol and smoking	Weight, length, and head circumference at 10 days, 3, 7, 18, and 42 months measured by study personnel; measures converted to SDS using healthy Dutch children as reference	PCBs had negative effect on birth weight and growth rate until 3 months of age. No effects from 3–42 months

(continued)

Table 12.1 (continued)

Citation	Study population, birth year(s), sample size	Exposure	Covariates	Outcome, age when assessed	Results
Gladen, 2000 [113]	Children from the North Carolina Infant Feeding Study 1978–1982 n=594	Indices of transplacental and breast milk PCB and DDE exposure based on levels in breast milk, maternal blood, cord blood, placenta	Child age, maternal (usually prepregnancy) weight, race, breastfeeding	Weight and height at age 14, self-reported by child or parents with instructions on measurement	PCBs positively associated with weight adjusted for height; differences only significant when restricted to white girls. DDE positively associated with height and weight adjusted for height in males. Associations less obvious with lactational exposure
Verhulst 2008 [111]	Offspring of women selected from maternity wards in Belgium 2002–2004 n=138	Cord blood levels of PCBs [118, 138, 153, 170, 180], DDE, HCB, dioxin-like compounds (assessed through dioxin-responsive chemical-activated luciferase expression assay); lipid adjusted	Parental health status, household income, smoking (ever and during pregnancy), age, family composition, height and weight; child diet, health status, breastfeeding	Height and weight collected by questionnaires to parents every 6 months from 12 to 36 months old; BMI SDS calculated.	Associations between DDE and BMI SDS modified by smoking; smoking enhanced the positive relationship between DDE and BMI at 3 years. PCBs associated with higher BMI SDS at ages 1–3. No association seen with PCB-118, HCB, or dioxin-like compounds

| Mendez 2011 [128] | Environment and Childhood Project, Catalonia, Spain 2004–2006 n=657 mothers n=518 infancy growth measures n=502, BMI at 14 months | Maternal first trimester serum levels of DDE, HCB, βHCH, PCBs [118, 138, 153, 180] | Maternal and paternal age, education, BMI, occupation, country of origin, pregnancy weight gain, pregnancy tobacco use | Infant weight and length measured at birth and 14 months. Repeated measures of infant growth from birth to 6 months abstracted from medical charts | DDE level positively associated with rapid infant weight gain between birth and 6 months. DDE level also associated with higher BMI z-score at 14 months. Effects seen in normal weight women only. PCBs, HCB, and βHCH not independently associated with growth |
| Cupul-uicab 2010 [130] | Male infants of mothers from Chiapas, Mexico 2002–2003 n=788 | DDE and DDT measured in maternal serum within 1 day of delivery; lipid adjusted | Child age, maternal smoking in pregnancy, hospital of recruitment, rural vs. urban residence, gestational age at birth, maternal education, height and prepregnancy BMI | Height and BMI assessed at multiple home visits by trained personnel, median age 18 months; SDS, rate of change in growth measured by interaction terms | No association seen between DDE or DDT and either height or BMI with height |

(continued)

Table 12.1 (continued)

Citation	Study population, birth year(s), sample size	Exposure	Covariates	Outcome, age when assessed	Results
Smink 2008 [112]	Children from the Asthma Multicenter Infants Cohort Study in Menorca, Spain 1997–1998 n=405	Cord blood measurements of HCB, PCBs (7 congeners), DDE, DDT; not lipid adjusted	Maternal age, education, SES, parity, 1st trimester smoking, alcohol, prepregnancy weight, diet during pregnancy, and height; infant weight and length, breastfeeding; simultaneous adjustment for other pollutants	Weight and height measured by trained personnel at 6.5 years; overweight and "at risk" defined based on BMI using 85% of WHO standards; BMI z-scores calculated	Compared to those in the low-exposure group, children moderately and highly exposed to HCB had a 1.7-fold increased risk of being overweight, and 2.0-fold increased risk of being obese Sub-analysis in normal weight women found slightly smaller effects Results for DDT, DDE, and PCBs not reported, but adjustment for them in HCB models did not change results
Karmaus 2009 [127]	Adult daughters of Michigan fish-eaters cohort 1973–1991 n=176	Maternal serum PCB and DDE at time of pregnancy, extrapolated using multiple serum measurements; not lipid adjusted	Maternal age, height, BMI; offspring age, birth weight, breastfeeding, parity	Weight and height assessed at age 20–50: in 2001–2002, height and weight collected by phone interview; in 2006–2007, BMI measured by study personnel	DDE positively associated with BMI and weight. Compared to lowest quintile of exposure, BMI was 1.65 times higher for 2nd quintile, and 2.88 times higher for quintiles 3–5 combined No association seen with PCBs

Study/reference	Population	Exposure measure	Covariates adjusted	Outcome measure	Results
Gladen 2004 [129]	Adolescent male offspring from Philadelphia subset of Collaborative Perinatal Project 1961–1965 n = 304	Maternal third trimester serum levels of ppDDE, ppDDT, opDDT; lipid adjusted	Age, maternal height and BMI, breastfeeding, smoking in pregnancy, birth order, race, maternal age at birth, maternal age at menarche, SES	Height, weight, skinfold thickness, skeletal age, collected by trained personnel at multiple visits, up to age 20	No association seen between any measure of DDT and the outcomes
Michels Blanck 2002 [116]	Daughters of women in the Michigan PBB cohort, a group highly exposed to PBBs and PCBs due to an industrial accident During or after 1973–1974 (when accident occurred) n = 308	Estimated prenatal PBB and PCB exposure using maternal serum levels at enrollment and decay models; not lipid adjusted	Daughters' age, physical activity, income, maternal education, maternal alcohol and smoking, breastfeeding, maternal age, and height or BMI	Height and weight at age 5–24, self-reported for older girls, measured with help from mother for younger girls	PCB exposure above the median was associated with lower weight adjusted for height (−11 lbs) in daughters. A slight increase in weight for moderate exposure to PBBs was seen, but none with higher exposure
Ribas-Fito 2006 [132]	Subsample of offspring from the multicenter US Collaborative Perinatal Project 1959–1966 n = 1712	Third trimester maternal serum levels of DDE, DDT, and PCBs; models adjusted for lipids	Study center, age, gender, race, SES, maternal smoking during pregnancy, maternal age, lipid concentrations, maternal height, parity, and prepregnancy BMI	Height and weight at ages 1, 4, and 7 years measured by study staff	DDE associated with reduced height at all three ages. The association was seen most strongly in African-American children, who were also most highly exposed. Relationships were not seen between DDT and PCBs and height

(continued)

Table 12.1 (continued)

Citation	Study population, birth year(s), sample size	Exposure	Covariates	Outcome, age when assessed	Results
Rylander 2007 [115]	Low- and normal-birth-weight children of a cohort of Swedish fishermen's wives from the west and east coast; the latter consumes more contaminated fish 1973–1991 n=468	Geographic residence used as proxy for PCB exposure; a subset of women had serum PCB 153 measured long after birth (lipid adjusted)	Maternal age, height, smoking, education, parity, infant sex	Weight and height at 4 and 7 years old, reported by mothers and child health centers and school health services	East coast children had slightly lower weight and height than west coast children at both ages A similar pattern with weight was seen with PCB 153 These differences were observed primarily in normal birth weight children, and not low birth weight
Jusko 2006 [133]	Subsample of offspring from the Child Health and Development Study cohort conducted in the San Francisco Bay Area, USA 1964–1967 n=399	Maternal serum levels of DDT and DDE taken in the second or third trimester; lipid adjusted	Demographic factors, height, prepregnancy BMI, parity, prenatal care, smoking, alcohol, and medication use in pregnancy, chronic hypertension, preeclampsia, diabetes, paternal height, occupation, education, and infant sex, gestational age at blood draw, PCB levels	Weight, height, head circumference, and other anthropometric measures collected at age 5 by pediatrician; also used race- and sex-standardized weight and height z-scores	Head circumferences were slightly larger in kids in the 75th percentile of DDE exposure compared to the 25th percentile. Mean differences in z-scores for height and weight were positive for all DDT measures, but not statistically significant

Study	Population	Exposure	Covariates	Outcome measures	Results
Lamb 2006 [114]	African-American children in the Columbia-Presbyterian cohort of the National Collaborative Perinatal Project 1959–1962 n=150	Maternal serum concentrations of PCBs collected in the third trimester; models adjusted for lipids	DDE, gestational age at birth, mother's height, prepregnancy BMI, smoking and age at birth, parity, breastfeeding and SES	Height and weight collected by study staff at birth, 4, 7, and 17 years of age	PCBs, especially *ortho*-substituted PCBs, were associated with reduced weight in girls up to 17 years of age, but not boys. The strongest associations were seen at 4 and 7 years old
Hertz-Picciotto 2005 [99]	Subsample of offspring from the Child Health and Development Study cohort conducted in the San Francisco Bay Area, USA 1964–1967 n=399	Maternal serum levels of PCBs collected in the second or third trimester; lipid adjusted	Demographic factors, height, prepregnancy BMI, parity, prenatal care, smoking, alcohol, and medication use in pregnancy, chronic hypertension, preeclampsia, diabetes, paternal height, occupation, education, and infant sex, gestational age at blood draw, DDT levels	Weight, height, head circumference, and other anthropometric measures collected at age 5 by pediatrician. Used weight and height z-scores	Comparing those in the 90th vs. 10th percentiles of PCB exposure, PCBs were positively associated with height and weight z-scores and sitting height in 5-year-old girls. Similar but smaller associations for height and weight z-scores were seen in boys, as well as a suggestive increase in biacromial distance

gestational age at birth, and infant sex, but to our knowledge, none have adjusted for gestational weight gain which is positively associated with birth weight and later obesity [106, 107]. A few studies have adjusted for additional factors such as inter-pregnancy interval, and maternal health and medication use. Residual confounding, particularly poorly measured data on maternal smoking, is an important concern in these studies. Maternal smoking is consistently related to lower birth weight and may also be related to PCB exposure [108]. Alcohol consumption has also been associated with higher levels of PCBs [109]. Thus, poor control of potential con-founders such as maternal smoking, alcohol use, and factors related to family socio-economic status (SES) may have affected the results of these studies.

Fewer studies have evaluated the effect of prenatal PCB exposure on postnatal growth (Table 12.1). Studies in pregnant women exposed to high levels of PCBs through accidental food contamination suggest that childhood growth is reduced [90, 91], but evidence for an association between prenatal exposure to chronic low levels of PCBs and childhood growth is mixed (Table 12.1). A study of 4-year-old offspring from a cohort in Michigan exposed to presumably high levels of PCBs from diet found that weight was reduced by an average of 1.8 kg in children who had the highest levels of total PCBs in cord blood, compared to those with the low-est levels [110]. Interestingly, activity levels in these children were also significantly reduced [110]. In a Dutch birth cohort study of 205 infants, PCBs were associated with lower birth weights and reduced growth rates up to 3 months of age, but not with growth measured from 3 to 42 months of age [95]. In contrast, a Belgian study of 138 infants with repeated growth measures between the age of 1 and 3 found a positive association between prenatal exposure to PCBs and BMI and an inverse association with length [111]. PCBs in serum of pregnant women from the San Francisco Bay area were evaluated with regard to growth of their children at age 5. The authors reported that "prenatal PCBs were associated with greater growth in 5-year-old girls, with no apparent effect on 5-year-old boys"; however, detailed data were not given [99]. A study of 405 children in Spain measured prenatal exposures to several types of chemicals in relation to growth at age 6.5; results for PCBs were not directly reported, but evaluation of PCB level simultaneous with other chemi-cals did not affect the results for other chemicals, suggesting that there was not an important effect of PCBs [112]. Gladen et al. evaluated growth around the time of puberty in a cohort of 594 children from a birth cohort study conducted in North Carolina; there was some indication of a positive association between estimated transplacental exposure PCBs and body weight, but only among girls [113]. A study based on a small subset of 150 offspring from the Collaborative Perinatal Project classified PCB congeners in terms of their degree of *ortho*-substitution, which deter-mines the rigidity of their molecular structure and whether or not they may have dioxin-like biologic activity [114]. Higher levels of ortho-substituted PCBs were associated with reductions in weight measured several times between ages 1 and 17, in girls but not boys. The results for boys were not as clear, but there were sugges-tive patterns of increased weight and height with higher levels of most of the conge-ners, although the results were not statistically significant. Studies which have used geographic location as a proxy for PCB exposure [115] or extrapolated pregnancy exposures from maternal serum collected primarily after pregnancy [116] have also reported inverse associations between PCB level and growth, while a study that evaluated lactational exposure to PCBs found no association with growth in infancy [117]. A recent study of serum levels of PCBs and dioxin among Russian boys measured at age 8–9 found that those exposed to higher levels of both classes of chemicals had deficits in BMI and height when they were measured 3 years later, around the time of puberty [118].

In summary, results of studies of PCBs in relation to growth in childhood are inconsistent, and results may differ by gender. These studies suffer from all the same potential biases (measurement error and confounding) mentioned above for studies on fetal growth. In addition, results from the various studies are difficult to compare as they have focused on growth at different time points from infancy to adulthood. Also, few studies have attempted to analyze PCBs by grouping them based on their suspected biologic effects, and thus may be limited by an additional type of measurement error.

Persistent Organochlorines: DDT, DDE, HCB

The persistent organochlorine pesticides most frequently studied in relation to fetal growth and/or growth in childhood include DDT, its major metabolite, DDE, and hexachlorobenzene (HCB). Like the PCBs, these compounds are lipophilic, bioaccumulative, and, although their use has been banned in most developed countries, are still found worldwide in both people and the environment [119]. DDE may impact fetal growth and programming of body weight through its estrogenicity [80]. In addition to being associated with a decrease in thyroid levels [120], HCB has been found to interfere with the glucocorticoid signaling pathway, a system closely involved in adipogenesis [121, 122].

The association between DDT and/or DDE and fetal growth as well as length of gestation has been evaluated in numerous studies, but findings are mixed. A recent review article concluded that the "evidence is suggestive of an association but is limited" due to potential problems with bias in some of the studies [89]. A total of 15 studies conducted up to 2005 were reviewed by Farhang [123]. Overall, the findings point to more consistent effects for DDT and/or DDE on preterm birth than on fetal growth restriction. Farhang [123] found some evidence for an increased risk of preterm birth for DDE, but not DDT, in a case–control study of male offspring nested within the California Child Health and Development Study. There was no association between either DDT or DDE and fetal growth retardation. One of the largest and best conducted studies found increased odds for both outcomes (a 3.1-fold increased odds for preterm birth and 2.6 for SGA) [124], but more recently published studies show little association between DDT and DDE and either birth weight or length of gestation [102, 125].

A few studies have evaluated HCB levels in relation to birth weight or gestational length with mixed findings. Sagiv et al. found no effects for cord blood levels of HCB in relation to pregnancy outcomes [102]. Eggesbo et al. found a small reduction in birth weight in relation to HCB levels in breast milk in a Norwegian cohort [126], while Fenster et al. noted slight reductions in gestational length in a California study [125].

A total of nine studies have evaluated the relationship between prenatal DDT and/or DDE exposure in relation to growth in childhood (height, weight, and/or body mass index) with mixed findings (Table 12.1). Gladen et al. found that boys around the time of puberty (aged 10–14) who had been exposed to the highest level of DDE (4+ ppm) were on average 6.9 kg heavier than those exposed to the lowest levels (<=1 ppm), but no effects were seen in girls [113]. In a study of a cohort exposed to high levels of organochlorines through consumption of Lake Michigan fish, higher DDE levels during pregnancy (extrapolated from measurements taken from the mothers after pregnancy) were associated higher BMI in adult female offspring; male offspring were not studied [127]. A recent study in Belgium reported that DDE was positively associated with BMI in early childhood; the effect was most evident among smokers and almost negligible among nonsmokers [111]. A recent study of 657 mothers and infants in Spain found that level of pre-natal DDE was associated with rapid weight gain in infancy and with elevations in BMI z-scores at 14 months among normal weight mothers but not overweight or obese mothers. Effects were nonlinear, with declines in rapid growth among infants whose mothers were in the top 3% of DDE exposure. The relative risk of rapid early growth by quartiles of exposure was similar across the second to fourth quartiles (RR from 2.42 to 2.47) among the normal weight mothers [128]. In con-trast, several other studies found no effect of pregnancy levels of DDE or DDT on BMI or weight gain in offspring [129, 130]. Two studies have evaluated the rela-tionship between lactational exposure to DDT and DDE and growth during the first year of life and neither found effects on growth according to levels of DDT or DDE [117, 131].

Studies of DDE in relation to childhood height also show mixed results. One study of 1,712 US children born between 1959 and 1966 found a reduction in height at 1, 4, and 7 years of age in the highest category of DDE exposure ($\geq$60 µg/l) com-pared to the lowest category (<15 µg/l) [132]. There was some indication that the effect differed by both race and gender, with the strongest effects reported for African-American females. The Gladen study found increased height among boys with higher compared with lower prenatal DDE exposures but found no consistent pattern in girls measured around the time of puberty [113]. In contrast, three studies found no association between estimates of prenatal exposure to DDE and height [111, 128, 129].

Two studies have evaluated the effects of prenatal exposure to HCB, which was banned for use as a pesticide in most countries during the 1970s. However, because it is highly persistent and a by-product of the manufacture of other organochlorines, exposure to HCB still occurs, although at much lower levels. One study found no effect of HCB on growth measured in very early childhood [111]. In contrast, a study of body measures of 400 6-year-old children whose mothers had been enrolled during pregnancy in Menorca, Spain [112],showed a twofold higher risk of childhood obesity among children with the highest quartile of HCB measured in cord blood compared with the lowest quartile. This study, along with the recent study by Mendez [128], appears to be one of the most methodologically sound studies conducted thus far on the effect of prenatal EDC exposure and offspring obesity. Numerous confounding factors were considered, sub-analyses were conducted among normal weight women to evaluate whether maternal BMI was affecting the results, HCB was measured in cord blood, and results were adjusted for levels of PCBs and DDT.

Emerging Chemicals of Concern: Phthalates, PFCs, BPA, Organotins

After decades of focus on the potential health effects of persistent, bioaccumulative chemicals, attention has turned more recently to less persistent chemicals to which humans are widely exposed from a variety of consumer products such as plastics, cosmetics, and building materials. In this section, we will focus on two classes of endocrine disruptors – phthalates and PFCs – that have been studied in relation to fetal growth and for which there are also some suspicions of effects on obesity and related metabolic conditions. In addition, we will touch on possible effects of bisphenol A and organotins for which there is a wealth of experimental literature, but little evidence for obesogenic effects based on human studies.

Phthalates are a ubiquitous class of chemicals used widely in consumer products. Some of the heavier-weight phthalates such as DEHP are used as plasticizers, while lower molecular weight phthalates such as DEP are incorporated into cosmetics, fragrances, and used to coat certain medications. Animal studies have consistently demonstrated that certain phthalates have relatively potent antiandrogenic effects [134], although evidence is more limited for this effect in humans [135]. While the relationship between androgen exposure and fetal programming of body weight is less clear, androgen levels have been found to be associated with birth weight in humans [136] and postnatal growth in animals [137]. Some phthalates are also agonists of peroxisome proliferator–activated receptors (PPAR-gamma), [138] receptors which play a critical role in adipogenesis and may have lasting impacts on adipocyte differentiation [139]. These chemicals may also have effects on thyroid hormone levels [140, 141].

Several studies have evaluated phthalates in relation to gestational age at birth, with equivocal findings. Three studies reported that higher levels of phthalates were related to shorter gestational length [142–144], whereas two found that phthalates appeared to lengthen gestation [16, 145]. Fewer studies have evaluated phthalates in relation to birth weight; Wolff et al. reported no effect on birth weight [16], while Zhang suggested that there was a higher risk of low birth weight with higher phthalate exposures [146]. To our knowledge there have been no longitudinal studies which have evaluated phthalate exposure during pregnancy in relation to the development of obesity among offspring, although there have been some cross-sectional studies of phthalates and BMI in children [147] and adults [148, 149].

Polyfluoroalkyl chemicals (PFCs) are widely used in commercial and industrial applications as surfactants, paper and textile coatings, and in food packaging due to their water and oil repelling properties [150]. Biomonitoring studies have documented widespread human exposure to PFCs [150]; major sources of human exposure are not definitive, but may include diet (either directly from food or migration from food packaging), drinking water, and house dust (reviewed in [151]). PFCs bind to PPAR-alpha, a receptor closely involved in lipid metabolism, as well as to PPAR-gamma [152]. In mice, prenatal exposure to a low dose of PFOA resulted in increased leptin and insulin levels at mid-life [153]. There have been no prospective cohort studies to evaluate pregnancy levels of PFCs and effects on obesity in offspring, and findings have been mixed from cross-sectional studies assessing the relationship between PFCs and body weight in adults [154–155].

Four studies have suggested that PFCs may reduce fetal growth, although results have been somewhat inconsistent. In a study of 1,400 women and their infants randomly selected from the Danish National Birth Cohort, head circumference, abdominal circumference, birth length, and placental weight were all reduced in infants who had been exposed to higher levels of perfluorooctanoic acid (PFOA), one of the most prevalent types of PFCs. A cross-sectional study of 239 births evaluated cord serum levels of PFCs in relation to birth size and gestational length [157]. Both perfluorooctane sulfonate (PFOS) and PFOA were associated with decrements in birth weight, ponderal index, and head circumference but not with gestational length after adjustment for lipids and other potential confounders. Two other studies suggested that PFOS, but not PFOA, was associated with reduced birth weight [158, 159].

To our knowledge, only one study has investigated the association between prenatal exposure to PFCs and postnatal growth [160]. Using data from the Danish National Birth Cohort, Andersen et al. found that infants with higher prenatal exposure to PFOS and PFOA tended to have lower weights and body mass index at 5 and 12 months of age, after adjusting for birth weight. When the data were stratified by gender, effects were much more apparent among boys and there was little evidence for an association in girls.

There is a wealth of animal and laboratory data to suggest that BPA and other phenols and organotins have properties that promote obesity in prenatally exposed animals. In certain animal studies, the estrogenic chemical BPA has been found to result in obesity after in utero exposure [7, 161], though another study did not discern these effects once animals reached adulthood [85]. BPA has also been shown in vitro to inhibit adiponectin release from adipose tissue [162]. Organotins are potent PPAR-gamma agonists, and they also bind to RXR, the heterodimer partner of PPAR-gamma [94]. To our knowledge, no human studies have evaluated the association between obesity and BPA, other phenols, or organotins in offspring exposed during pregnancy. One study has evaluated pregnancy exposure to phenols in relation to birth weight [16]. Wolff et al. found that higher levels of 2,5 dichlorophenol were associated with lower birth weight in boys, and exposure to benzophenone (BP3), a common ingredient in UV filters, was related to lower birth weight in girls, but higher birth weight in boys.

12 Developmental Exposure to Endocrine Disrupting Chemicals... 307

> **Box 4** Associations Between EDCs and Fetal and Childhood Growth
>
> - High level (accidental) exposure to PCBs may reduce birth weight and childhood growth, but evidence is inconsistent for an influence of lower level (environmental) exposures.
> - Most studies have not shown an association between prenatal DDT/DDE exposure and childhood body measures.
> - HCBs have not been frequently studied, and results so far are inconsistent for a relationship with offspring body measures.
> - Evidence is inconsistent for an association between phthalates and birth weight and gestational length.
> - Certain PFCs may be associated with lower fetal growth and birth size; one study also linked PFCs with reduced growth among infant boys.
> - Although animal and laboratory studies suggest BPA may be obesogenic, human studies are sparse.

Methodologic Challenges in Studying EDCs and Offspring Obesity

Studying effects of intrauterine exposures on outcomes beyond the immediate neonatal period is fraught with methodological challenges including the usual problems of bias and confounding that must always be considered in epidemiologic studies. Also problematic is that the actual intrauterine exposure to the fetus is rarely known and is often inferred through the mother's exposure, which may or may not be a valid proxy. Moreover, the exposure may be indirect, and not mediated through the maternal environment, for example, as with paternal smoking or occupational exposure. Some exposures may exert effects on the gametes prior to conception, [163] or they may be more accurately measured during the preconception time period [98, 164]. In addition, timing of exposure during the prenatal period is likely to be critical, but may not be precisely known, for the particular outcome of interest. Finally, exposures occurring postnatally may be correlated or interact with those occurring prenatally; hence, a life course approach is warranted [165]. For example, in studying childhood obesity, we must attempt to separate the effects of in utero from early life exposures, including catch-up growth and lactational exposures. The in utero versus early life contribution of many factors, such as parental smoking, dietary patterns, and other SES-related factors, may be difficult to parse. These problems may be particularly intractable in studies of prenatal exposures to environmental contaminants.

Other challenges relate to the difficulties of exposure measurement. Often, because of scarce study resources, biospecimens are collected only once during pregnancy, and timing of exposure becomes key. This is somewhat less crucial for lipophilic chemicals that have long half-lives, but many of the "emerging" chemicals suspected to be obesogenic have short half-lives. In the case of offspring overweight and obesity, the critical time period is unknown and may differ for different biological mechanisms. Epigenetic mechanisms may operate throughout pregnancy but at discrete points of time [166]. Measuring exposures at the wrong time period is a type of misclassification bias that is most likely nondifferential. With dichotomous exposures, one would expect such misclassification to attenuate effects of the exposure on the outcome. In contrast, with chemical exposures, measurements are usually continuous or may be categorized into multiple levels to evaluate potential dose response. The effect of nondifferential misclassification in this setting is difficult to predict and may not result in attenuated associations [167].

There also exist questions about how chemical biomonitoring measurements should be modeled with regard to lipid or creatinine adjustment [147, 168–170]. Most persistent chemicals accumulate in fat, and it is customary to analyze serum concentrations adjusted for lipid levels. Similarly, nonpersistent chemicals are often analyzed with some type of adjustment for urinary creatinine, a by-product of muscle activity and a measure of urinary dilution [171, 172]. However, because serum lipids and creatinine may be on the pathway to associations with body size and fat distribution, the best approach for modeling these variables is not clear.

Additional questions exist about whether maternal adiposity may be related to concentrations of EDCs through elevated lipids and fat reserve [173]. Also, plasma volume expansion in pregnancy could affect analyte concentrations and is related to maternal weight [174].

An additional source of bias related to measurement error, but perhaps mainly a problem of confounding, is due to the fact that humans are exposed simultaneously to low levels of a multitude of chemicals [175] which may have different mechanisms of action and health effects. Exposure to these chemicals may be correlated, and thus, it may be difficult to determine the true chemical of interest. In addition, chemicals could interact, and analyzing them singly could result in an underestimate or overestimate of effects [176]. Also, measuring exposure to complex mixtures, thus mimicking environmental exposure, may be more relevant and more important than measuring exposure to individual EDCs [177, 178], yet epidemiologists have only rarely attempted to analyze these combinations [179], even within the same class of chemicals [135]. Many of the studies of EDCs and fetal or childhood growth reviewed above did not control simultaneously for exposure to multiple chemicals, much less attempt to analyze complex mixtures of chemicals.

Yet another challenge is related to the routes of exposure for EDCs. Human exposure to many EDCs is thought to be primarily through foods via bioaccumulation up the food chain, or through packaging materials that result in migration of EDCs into foods (e.g., BPA in cans, PFCs in fast-food containers). How can we separate the effects of EDCs from the effects of certain dietary exposures/patterns during pregnancy? For example, mothers who eat more canned foods or drink more soft drinks during pregnancy may have higher exposures to BPA or PFCs, but they may also have greater gestational weight gain leading to higher birth weight and offspring obesity, or they may have unhealthy dietary patterns that could be associated directly or indirectly with childhood obesity. Thus, careful control for maternal dietary factors in pregnancy is important in studies of in utero exposure to EDCs and childhood obesity. Inert components of medications provide another source of high-dose exposure to some EDCs, particularly phthalates [180, 181]. Here, the same

Box 5 Methodologic Challenges in Epidemiologic Studies of In Utero EDC Exposure in Relation to Childhood Obesity

- Difficult to separate contributions of genetics and prepregnancy (maternal and paternal), in utero, lactational, and childhood exposures.
- Confounding by lifestyle and SES-related variables is likely but difficult to account for.
- Exposure misclassification, such as measurement during a noncritical time period, will often reduce likelihood of finding an association.
- Simultaneous exposure to mixtures of EDCs may result in antagonistic or synergistic effects; mixture effects are rarely considered in epidemiologic studies.
- Use of body mass index may result in misclassification of adiposity, the true outcome of interest, leading to potential underestimation of effects.

caution would apply as in dietary exposures: if phthalates appear to be associated with childhood obesity, are the effects due to phthalate exposure, due to medications consumed by the mother during pregnancy, or due to the health condition that led the mother to take medications in the first place ("confounding by indication")?

Misclassification or misspecification of the outcome of interest may be problematic in studies of prenatal risk factors for childhood obesity. BMI is known to be a poor proxy for adiposity in adulthood [182] and may also be prone to misclassification in childhood [183]. Such misclassification is unlikely to be related to exposure measurements but could obscure relationships between EDCs and the outcome of interest, adiposity. For example, animals exposed to tributyltin in utero had greater amounts of adipose tissue, but no increase in body weight, compared with unexposed animals [184]. If obesogens have the same effect in humans, it would be impossible to discern by assessing BMI alone. Yet, most studies use BMI because it is relatively easy to measure compared with other methods such as bioelectrical impedance, DXA scans, MRI, or measurements of waist circumference or skinfold thicknesses. Another advantage of BMI compared with more precise measures of adiposity is that it may be routinely collected in schools, medical records, or registry data, making special follow-up measurements of children in a birth cohort unnecessary.

If BMI is used to measure obesity, there are questions about how it should be statistically modeled, particularly in regard to childhood overweight. Most studies evaluate measures on a continuous scale such as weight (controlling for length), BMI, or a standardized BMI measure such as BMI standard deviation scores (SDS) or z-scores, which may make it difficult to decipher nonlinear effects. On the other hand, there may be loss of information when dichotomous outcomes are used, such as childhood "at risk of overweight" (BMI>=85%) and overweight (BMI>=95%) which are based on comparisons to standard growth curves for each age [2]. Another recent approach is the use of quantile regression, which allows an assessment of nonlinear effects and also provides more information than a dichotomous outcome measure of obesity using BMI cutpoints [185, 186].

Recommendations for Future Directions

As alluded to above, separating contributions of family/SES risk factors, paternal exposures, and timing of exposure to the fetus and child (preconception, in utero, lactational, childhood) is essential but difficult to accomplish in practice. Davey Smith [187] has outlined several approaches for optimizing methods to reach valid interpretations:

1. Compare the strength of the association between the maternal exposure (a direct in utero exposure) and offspring health outcome to that of the association between the paternal exposure and offspring outcome. If the association with the maternal exposure (e.g., alcohol consumption) is stronger than that with the paternal exposure, the evidence is consistent with a causal relationship between the maternal, in utero, exposure and the offspring outcome. When the associations with maternal and paternal exposures are attenuated after mutual adjustment, the involvement of other lifestyle confounders is suspected. However, analytic findings should be interpreted carefully, as there may be settings in which the father's exposure influences the intrauterine environment, in which case mutual adjustment might underestimate the exposure's influence. For example, if secondhand smoke from the father influences the intrauterine environment, adjustment for paternal smoking might result in an underestimate of the influence of cigarette smoke on offspring outcomes.
2. Under some circumstances (e.g., medication use), it may be possible to distinguish intrauterine exposures from pre- and postpregnancy exposures, but this is not commonplace because "lifestyle" exposures occurring during pregnancy also generally occur before and after pregnancy as well. If some women stop the exposure (e.g., smoking) during pregnancy, but resume afterward, a comparison can be made among that group, women exposed continuously and an unexposed group. Rarely, an exposure may occur only during pregnancy (e.g., diethylstilbestrol), facilitating the identification of intrauterine effects. Exposures may also occur during a brief period in history (e.g., high levels of radiation), allowing comparison of women contemporaneously exposed during or after pregnancy. Many, if not most, exposures that disrupt the intrauterine environment may also contaminate breast milk, in which case, it is important to know whether the child was breast-fed. Lipophilic chemical exposures during lactation may be much greater than in utero exposures [164]; it is also possible that the route of exposure invokes different risks. Finally, some intrauterine exposures (e.g., alcohol use) may indirectly influence the postnatal environment. For example, mothers who drink during pregnancy may be more likely to drink heavily postpartum which may adversely affect their ability to care for the child.
3. Studies of exposed versus unexposed siblings also reduce the likelihood of confounding, assuming the mother's age is taken into account and environmental/ lifestyle confounders remain stable over time. This scenario is unlikely to occur when studying low-level exposures to environmental chemicals, but might be

possible in the setting of exposures that occurred during one pregnancy but not another.

In addition, improving exposure assessment may help clarify associations between EDCs and subsequent obesity. Collecting multiple measurements, preferably before and during pregnancy, and measuring chemicals in cord blood might reduce misclassification. Measuring exposures during the critical windows for the development of obesity is also extremely important. Most studies have evaluated exposures in maternal urine or serum, but other biological matrices such as hair, nails, meconium, and amniotic fluid should be considered depending on the chemical of interest [188]. Developing techniques to evaluate exposures to mixtures of chemicals that have similar mechanisms of action, as advocated by Kortenkamp [177], is very important as the field moves forward.

It is also essential that future studies of EDCs and obesity collect extensive information on potential confounding factors, especially maternal smoking history, maternal and paternal BMI, gestational weight gain, and other lifestyle or SES-related factors. Information on diet would also be useful to sort out exposure pathways. Finally, effects may differ according to offspring gender; thus, this should continue to be evaluated in future studies.

Finally, the outcome of interest – obesity – must be accurately assessed. More precise measures of adiposity, such as skinfold thickness, DXA, or MRI, may help determine whether there are associations between EDCs and the development of excess adipose tissue in humans, and whether there are differences by fat distribution or tissue type (i.e., central vs. subcutaneous, lean mass vs. fat mass). Additionally, to investigate the possible pathways to obesity, it would be useful to evaluate fetal and childhood growth patterns in the same study, and to collect information on other potential markers of metabolic disturbances, such as leptin, adiponectin, glucose, and insulin levels. Postnatal behavioral characteristics of the children, including diet and physical activity levels, should also be assessed.

Summary/Conclusions

Birth weight appears to be associated with growth in childhood, and there may be different effects/pathways at the two ends of the birth weight spectrum. This suggests that postnatal growth may be partially programmed by the in utero experience. The strongest data thus far comes from studies of maternal smoking and offspring obesity; however, the mechanisms for this association are not fully known. Human studies are sparse for most chemicals. The limited data suggest that high (accidental) exposure to PCBs may reduce fetal growth and birth weight, but evidence for associations with low-level (environmental) exposure is mixed. Some well-conducted studies suggest that DDT/DDE is associated with fetal growth restriction and shortened gestation, but findings overall are inconsistent. Recent studies of

PFCs suggest small effects on fetal growth. Although animal data have suggested that EDCs can affect offspring obesity, thus far, data from human epidemiological studies are not sufficient to draw conclusions. Several methodologic challenges exist in conducting these studies, including timing of measurement of EDCs in gestation, accounting for potential confounders occurring during pregnancy and childhood, and accurately measuring adiposity.

References

1. Ogden CL, Carroll MD, Flegal KM. Epidemiologic trends in overweight and obesity. Endocrinol Metab Clin North Am. 2003;32(4):741–60. vii.
2. Ogden CL, Carroll MD, Curtin LR, Lamb MM, Flegal KM. Prevalence of high body mass index in US children and adolescents, 2007–2008. JAMA. 2010;303(3):242–9.
3. Lee JM, Pilli S, Gebremariam A, Keirns CC, Davis MM, Vijan S, Freed GL, Herman WH, Gurney JG. Getting heavier, younger: trajectories of obesity over the life course. Int J Obes (Lond). 2010;34(4):614–23.
4. Ebbeling CB, Pawlak DB, Ludwig DS. Childhood obesity: public-health crisis, common sense cure. Lancet. 2002;360(9331):473–82.
5. Franks PW, Hanson RL, Knowler WC, Sievers ML, Bennett PH, Looker HC. Childhood obesity, other cardiovascular risk factors, and premature death. N Engl J Med. 2010;362(6): 485–93.
6. Must A, Strauss RS. Risks and consequences of childhood and adolescent obesity. Int J Obes Relat Metab Disord. 1999;23(Suppl 2):S2–S11.
7. Heindel JJ, vom Saal FS. Role of nutrition and environmental endocrine disrupting chemicals during the perinatal period on the aetiology of obesity. Mol Cell Endocrinol. 2009;304(1–2): 90–6.
8. Keith SW, Redden DT, Katzmarzyk PT, Boggiano MM, Hanlon EC, Benca RM, Ruden D, Pietrobelli A, Barger JL, Fontaine KR, Wang C, Aronne LJ, Wright SM, Baskin M, Dhurandhar NV, Lijoi MC, Grilo CM, DeLuca M, Westfall AO, Allison DB. Putative contributors to the secular increase in obesity: exploring the roads less traveled. Int J Obes (Lond). 2006;30(11):1585–94.
9. McAllister EJ, Dhurandhar NV, Keith SW, Aronne LJ, Barger J, Baskin M, Benca RM, Biggio J, Boggiano MM, Eisenmann JC, Elobeid M, Fontaine KR, Gluckman P, Hanlon EC, Katzmarzyk P, Pietrobelli A, Redden DT, Ruden DM, Wang C, Waterland RA, Wright SM, Allison DB. Ten putative contributors to the obesity epidemic. Crit Rev Food Sci Nutr. 2009; 49(10):868–913.
10. Newbold RR, Padilla-Banks E, Jefferson WN, Heindel JJ. Effects of endocrine disruptors on obesity. Int J Androl. 2008;31(2):201–8.
11. Newbold RR, Padilla-Banks E, Snyder RJ, Phillips TM, Jefferson WN. Developmental exposure to endocrine disruptors and the obesity epidemic. Reprod Toxicol. 2007;23(3):290–6.
12. Heindel JJ. Animal models for probing the developmental basis of disease and dysfunction paradigm. Basic Clin Pharmacol Toxicol. 2008;102(2):76–81.
13. Parsons TJ, Power C, Manor O. Fetal and early life growth and body mass index from birth to early adulthood in 1958 British cohort: longitudinal study. BMJ. 2001;323(7325):1331–5.
14. Ong KK. Size at birth, postnatal growth and risk of obesity. Horm Res. 2006;65(Suppl 3): 65–9.
15. McMillen IC, Rattanatray L, Duffield JA, Morrison JL, MacLaughlin SM, Gentili S, Muhlhausler BS. The early origins of later obesity: pathways and mechanisms. Adv Exp Med Biol. 2009;646:71–81.

16. Wolff MS, Engel SM, Berkowitz GS, Ye X, Silva MJ, Zhu C, Wetmur J, Calafat AM. Prenatal phenol and phthalate exposures and birth outcomes. Environ Health Perspect. 2008;116(8): 1092–7.
17. Oken E, Gillman MW. Fetal origins of obesity. Obes Res. 2003;11(4):496–506.
18. Wilcox AJ. Birth weight and fetal growth. Fertility and pregnancy: an epidemiologic perspective. New York: Oxford University Press; 2010.
19. Chakraborty S, Joseph DV, Bankart MJ, Petersen SA, Wailoo MP. Fetal growth restriction: relation to growth and obesity at the age of 9 years. Arch Dis Child Fetal Neonatal Ed. 2007;92(6):F479–83.
20. Ong KK, Ahmed ML, Emmett PM, Preece MA, Dunger DB. Association between postnatal catch-up growth and obesity in childhood: prospective cohort study. BMJ. 2000;320(7240): 967–71.
21. Henriksen T. The macrosomic fetus: a challenge in current obstetrics. Acta Obstet Gynecol Scand. 2008;87(2):134–45.
22. Sorensen HT, Sabroe S, Rothman KJ, Gillman M, Fischer P, Sorensen TI. Relation between weight and length at birth and body mass index in young adulthood: cohort study. BMJ. 1997;315(7116):1137.
23. O'Callaghan MJ, Williams GM, Andersen MJ, Bor W, Najman JM. Prediction of obesity in children at 5 years: a cohort study. J Paediatr Child Health. 1997;33(4):311–6.
24. Rugholm S, Baker JL, Olsen LW, Schack-Nielsen L, Bua J, Sorensen TI. Stability of the association between birth weight and childhood overweight during the development of the obesity epidemic. Obes Res. 2005;13(12):2187–94.
25. Curhan GC, Chertow GM, Willett WC, Spiegelman D, Colditz GA, Manson JE, Speizer FE, Stampfer MJ. Birth weight and adult hypertension and obesity in women. Circulation. 1996;94(6):1310–5.
26. Singhal A, Wells J, Cole TJ, Fewtrell M, Lucas A. Programming of lean body mass: a link between birth weight, obesity, and cardiovascular disease? Am J Clin Nutr. 2003;77(3):726–30.
27. Curhan GC, Willett WC, Rimm EB, Spiegelman D, Ascherio AL, Stampfer MJ. Birth weight and adult hypertension, diabetes mellitus, and obesity in US men. Circulation. 1996;94(12): 3246–50.
28. Byberg L, McKeigue PM, Zethelius B, Lithell HO. Birth weight and the insulin resistance syndrome: association of low birth weight with truncal obesity and raised plasminogen activator inhibitor-1 but not with abdominal obesity or plasma lipid disturbances. Diabetologia. 2000;43(1):54–60.
29. Fall CHD. Developmental origins of cardiovascular disease, type 2 diabetes, and obesity in humans. In: Wintour EM, Owens J, editors. Early life origins of health and disease. Georgetown: Springer; 2006.
30. Barker M, Robinson S, Osmond C, Barker DJ. Birth weight and body fat distribution in adolescent girls. Arch Dis Child. 1997;77(5):381–3.
31. Law CM, Barker DJ, Osmond C, Fall CH, Simmonds SJ. Early growth and abdominal fatness in adult life. J Epidemiol Community Health. 1992;46(3):184–6.
32. Kensara OA, Wootton SA, Phillips DI, Patel M, Jackson AA, Elia M. Fetal programming of body composition: relation between birth weight and body composition measured with dual-energy X-ray absorptiometry and anthropometric methods in older Englishmen. Am J Clin Nutr. 2005;82(5):980–7.
33. Gale CR, Martyn CN, Kellingray S, Eastell R, Cooper C. Intrauterine programming of adult body composition. J Clin Endocrinol Metab. 2001;86(1):267–72.
34. Ibanez L, Lopez-Bermejo A, Suarez L, Marcos MV, Diaz M, de Zegher F. Visceral adiposity without overweight in children born small for gestational age. J Clin Endocrinol Metab. 2008;93(6):2079–83.
35. Barker DJ. Maternal nutrition, fetal nutrition, and disease in later life. Nutrition. 1997; 13(9):807–13.
36. Gluckman PD, Hanson MA, Cooper C, Thornburg KL. Effect of in utero and early-life conditions on adult health and disease. N Engl J Med. 2008;359(1):61–73.

37. Stettler N, Tershakovec AM, Zemel BS, Leonard MB, Boston RC, Katz SH, Stallings VA. Early risk factors for increased adiposity: a cohort study of African American subjects followed from birth to young adulthood. Am J Clin Nutr. 2000;72(2):378–83.
38. Monteiro PO, Victora CG. Rapid growth in infancy and childhood and obesity in later life–a systematic review. Obes Rev. 2005;6(2):143–54.
39. Eriksson M, Tynelius P, Rasmussen F. Associations of birthweight and infant growth with body composition at age 15–the COMPASS study. Paediatr Perinat Epidemiol. 2008;22(4):379–88.
40. Victora CG, Adair L, Fall C, Hallal PC, Martorell R, Richter L, Sachdev HS. Maternal and child undernutrition: consequences for adult health and human capital. Lancet. 2008;371(9609):340–57.
41. Gillman MW, Rifas-Shiman S, Berkey CS, Field AE, Colditz GA. Maternal gestational diabetes, birth weight, and adolescent obesity. Pediatrics. 2003;111(3):e221–e6.
42. Vohr BR, Boney CM. Gestational diabetes: the forerunner for the development of maternal and childhood obesity and metabolic syndrome? J Matern Fetal Neonatal Med. 2008;21(3):149–57.
43. Clausen TD, Mathiesen ER, Hansen T, Pedersen O, Jensen DM, Lauenborg J, Schmidt L, Damm P. Overweight and the metabolic syndrome in adult offspring of women with diet-treated gestational diabetes mellitus or type 1 diabetes. J Clin Endocrinol Metab. 2009;94(7):2464–70.
44. Lawlor DA, Fraser A, Lindsay RS, Ness A, Dabelea D, Catalano P, Davey Smith G, Sattar N, Nelson SM. Association of existing diabetes, gestational diabetes and glycosuria in pregnancy with macrosomia and offspring body mass index, waist and fat mass in later childhood: findings from a prospective pregnancy cohort. Diabetologia. 2010;53(1):89–97.
45. Whitaker RC, Pepe MS, Seidel KD, Wright JA, Knopp RH. Gestational diabetes and the risk of offspring obesity. Pediatrics. 1998;101(2):E9.
46. Huang JS, Lee TA, Lu MC. Prenatal programming of childhood overweight and obesity. Matern Child Health J. 2007;11(5):461–73.
47. Malee MP, Verma A, Messerlian G, Tucker R, Vohr BR. Association between maternal and child leptin levels 9 years after pregnancy complicated by gestational diabetes. Horm Metab Res. 2002;34(4):212–6.
48. Vohr BR, McGarvey ST, Tucker R. Effects of maternal gestational diabetes on offspring adiposity at 4–7 years of age. Diabetes Care. 1999;22(8):1284–91.
49. Boerschmann H, Pfluger M, Henneberger L, Ziegler AG, Hummel S. Prevalence and predictors of overweight and insulin resistance in offspring of mothers with gestational diabetes mellitus. Diabetes Care. 2010;33(8):1845–9.
50. Ino T. Maternal smoking during pregnancy and offspring obesity: meta-analysis. Pediatr Int. 2010;52(1):94–9.
51. Oken E, Levitan EB, Gillman MW. Maternal smoking during pregnancy and child overweight: systematic review and meta-analysis. Int J Obes (Lond). 2008;32(2):201–10.
52. Nagel G, Wabitsch M, Galm C, Berg S, Brandstetter S, Fritz M, Klenk J, Peter R, Prokopchuk D, Steiner R, Stroth S, Wartha O, Weiland SK, Steinacker J. Determinants of obesity in the Ulm Research on Metabolism, Exercise and Lifestyle in Children (URMEL-ICE). Eur J Pediatr. 2009;168(10):1259–67.
53. Power C, Jefferis BJ. Fetal environment and subsequent obesity: a study of maternal smoking. Int J Epidemiol. 2002;31(2):413–9.
54. Chen A, Pennell ML, Klebanoff MA, Rogan WJ, Longnecker MP. Maternal smoking during pregnancy in relation to child overweight: follow-up to age 8 years. Int J Epidemiol. 2006; 35(1):121–30.
55. Syme C, Abrahamowicz M, Mahboubi A, Leonard GT, Perron M, Richer L, Veillette S, Gaudet D, Paus T, Pausova Z. Prenatal exposure to maternal cigarette smoking and accumulation of intra-abdominal fat during adolescence. Obesity (Silver Spring). 2010;18(5):1021–5.
56. Al Mamun A, Lawlor DA, Alati R, O'Callaghan MJ, Williams GM, Najman JM. Does maternal smoking during pregnancy have a direct effect on future offspring obesity? Evidence from a prospective birth cohort study. Am J Epidemiol. 2006;164(4):317–25.
57. Environment CoHAAit. Hormonally active agents in the environment. Washington DC: National Academy Press; 1999.

12 Developmental Exposure to Endocrine Disrupting Chemicals... 315

58. Smith O. Diethylstilbesterol in the prevention and treatment of complications of pregnancy. Am J Obstet Gynecol. 1948;56:821–34.
59. Dieckmann WJ, Davis ME, Rynkiewicz LM, Pottinger RE. Does the administration of diethylstilbestrol during pregnancy have therapeutic value? Am J Obstet Gynecol. 1953;66(5):1062–81.
60. Ferguson JH. The importance of controls in a clinical experiment; stilbestrol therapy in pregnancy. Obstet Gynecol. 1954;3(4):452–7.
61. Herbst AL, Ulfelder H, Poskanzer DC. Adenocarcinoma of the vagina. Association of maternal stilbestrol therapy with tumor appearance in young women. N Engl J Med. 1971;284(15):878–81.
62. Titus-Ernstoff L, Troisi R, Hatch EE, Hyer M, Wise LA, Palmer JR, Kaufman R, Adam E, Noller K, Herbst AL, Strohsnitter W, Cole BF, Hartge P, Hoover RN. Offspring of women exposed in utero to diethylstilbestrol (DES): a preliminary report of benign and malignant pathology in the third generation. Epidemiology. 2008;19(2):251–7.
63. Titus-Ernstoff L, Troisi R, Hatch EE, Wise LA, Palmer J, Hyer M, Kaufman R, Adam E, Strohsnitter W, Noller K, Herbst AL, Gibson-Chambers J, Hartge P, Hoover RN. Menstrual and reproductive characteristics of women whose mothers were exposed in utero to diethylstilbestrol (DES). Int J Epidemiol. 2006;35(4):862–8.
64. Walker BE. Tumors of female offspring of mice exposed prenatally to diethylstilbestrol. J Natl Cancer Inst. 1984;73(1):133–40.
65. Walker BE. Tumors in female offspring of control and diethylstilbestrol-exposed mice fed high-fat diets. J Natl Cancer Inst. 1990;82(1):50–4.
66. Newbold RR. Lessons learned from perinatal exposure to diethylstilbestrol. Toxicol Appl Pharmacol. 2004;199(2):142–50.
67. McLachlan JA, Newbold RR, Bullock BC. Long-term effects on the female mouse genital tract associated with prenatal exposure to diethylstilbestrol. Cancer Res. 1980;40(11):3988–99.
68. McLachlan JA, Newbold RR, Shah HC, Hogan MD, Dixon RL. Reduced fertility in female mice exposed transplacentally to diethylstilbestrol (DES). Fertil Steril. 1982;38(3):364–71.
69. Newbold RR, McLachlan JA. Vaginal adenosis and adenocarcinoma in mice exposed prenatally or neonatally to diethylstilbestrol. Cancer Res. 1982;42(5):2003–11.
70. Newbold RR, Padilla-Banks E, Jefferson WN. Adverse effects of the model environmental estrogen diethylstilbestrol are transmitted to subsequent generations. Endocrinology. 2006;147(6 Suppl):S11–7.
71. McLachlan JA, Burow M, Chiang TC, Li SF. Gene imprinting in developmental toxicology: a possible interface between physiology and pathology. Toxicol Lett. 2001;120(1–3):161–4.
72. Nelson KG, Sakai Y, Eitzman B, Steed T, McLachlan J. Exposure to diethylstilbestrol during a critical developmental period of the mouse reproductive tract leads to persistent induction of two estrogen-regulated genes. Cell Growth Differ. 1994;5(6):595–606.
73. Li S, Washburn KA, Moore R, Uno T, Teng C, Newbold RR, McLachlan JA, Negishi M. Developmental exposure to diethylstilbestrol elicits demethylation of estrogen-responsive lactoferrin gene in mouse uterus. Cancer Res. 1997;57(19):4356–9.
74. Falck L, Forsberg JG. Immunohistochemical studies on the expression and estrogen dependency of EGF and its receptor and C-fos proto-oncogene in the uterus and vagina of normal and neonatally estrogen-treated mice. Anat Rec. 1996;245(3):459–71.
75. Okada A, Sato T, Ohta Y, Buchanan DL, Iguchi T. Effect of diethylstilbestrol on cell proliferation and expression of epidermal growth factor in the developing female rat reproductive tract. J Endocrinol. 2001;170(3):539–54.
76. Block K, Kardana A, Igarashi P, Taylor HS. In utero diethylstilbestrol (DES) exposure alters Hox gene expression in the developing mullerian system. FASEB J. 2000;14(9):1101–8.
77. Ma L, Benson GV, Lim H, Dey SK, Maas RL. Abdominal B (AbdB) Hoxa genes: regulation in adult uterus by estrogen and progesterone and repression in mullerian duct by the synthetic estrogen diethylstilbestrol (DES). Dev Biol. 1998;197(2):141–54.
78. Mericskay M, Carta L, Sassoon D. Diethylstilbestrol exposure in utero: a paradigm for mechanisms leading to adult disease. Birth Defects Res A Clin Mol Teratol. 2005;73(3):133–5.

79. Yamashita S, Takayanagi A, Shimizu N. Effects of neonatal diethylstilbestrol exposure on c-fos and c-jun protooncogene expression in the mouse uterus. Histol Histopathol. 2001;16(1): 131–40.
80. Newbold RR, Padilla-Banks E, Jefferson WN. Environmental estrogens and obesity. Mol Cell Endocrinol. 2009;304(1–2):84–9.
81. Wardell RE, Seegmiller RE, Bradshaw WS. Induction of prenatal toxicity in the rat by diethylstilbestrol, zeranol, 3,4,3'',4'',-tetrachlorobiphenyl, cadmium, and lead. Teratology. 1982; 26(3):229–37.
82. Brackbill Y, Berendes HW. Dangers of diethylstilboestrol: review of a 1953 paper. Lancet. 1978;2(8088):520.
83. Hatch EE, Troisi R, Wise LA, Titus-Ernstoff L, Hyer M, Palmer JR, Strohsnitter WC, Robboy SJ, Anderson D, Kaufman R, Adam E, Hoover RN. Preterm birth, fetal growth, and age at menarche among women exposed prenatally to diethylstilbestrol (DES). Reprod Toxicol. 2011;31:151–7.
84. Newbold RR, Padilla-Banks E, Snyder RJ, Jefferson WN. Perinatal exposure to environmental estrogens and the development of obesity. Mol Nutr Food Res. 2007;51(7):912–7.
85. Ryan KK, Haller AM, Sorrell JE, Woods SC, Jandacek RJ, Seeley RJ. Perinatal exposure to bisphenol-a and the development of metabolic syndrome in CD-1 mice. Endocrinology. 2010;151(6):2603–12.
86. Gustavson CR, Gustavson JC, Noller KL, O'Brien PC, Melton LJ, Pumariega AJ, Kaufman RH, Colton T. Increased risk of profound weight loss among women exposed to diethylstilbestrol in utero. Behav Neural Biol. 1991;55(3):307–12.
87. Longnecker MP, Rogan WJ, Lucier G. The human health effects of DDT (dichlorodiphenyltrichloroethane) and PCBS (polychlorinated biphenyls) and an overview of organochlorines in public health. Annu Rev Public Health. 1997;18:211–44.
88. Stillerman KP, Mattison DR, Giudice LC, Woodruff TJ. Environmental exposures and adverse pregnancy outcomes: a review of the science. Reprod Sci. 2008;15(7):631–50.
89. Wigle DT, Arbuckle TE, Turner MC, Berube A, Yang Q, Liu S. Krewski D. Epidemiologic evidence of relationships between reproductive and child health outcomes and environmental chemical contaminants. J Toxicol Environ Health B Crit Rev. 2008;11(5–6):373–517.
90. Guo YL, Lambert GH, Hsu CC. Growth abnormalities in the population exposed in utero and early postnatally to polychlorinated biphenyls and dibenzofurans. Environ Health Perspect. 1995;103(Suppl 6):117–22.
91. Rogan WJ, Gladen BC, Hung KL, Koong SL, Shih LY, Taylor JS, Wu YC, Yang D, Ragan NB, Hsu CC. Congenital poisoning by polychlorinated biphenyls and their contaminants in Taiwan. Science. 1988;241(4863):334–6.
92. Cooke PS, Sato T, Buchanan DL. Disruption of steroid hormone signaling by PCBs. In: Robertson LR, Hansen LG, editors. PCBs: recent advances in environmental toxicology and health effects. Lexington: University Press of Kentucky; 2001.
93. Nagata C, Iwasa S, Shiraki M, Shimizu H. Estrogen and alpha-fetoprotein levels in maternal and umbilical cord blood samples in relation to birth weight. Cancer Epidemiol Biomarkers Prev. 2006;15(8):1469–72.
94. Grun F, Blumberg B. Endocrine disrupters as obesogens. Mol Cell Endocrinol. 2009; 304(1–2):19–29.
95. Patandin S, Koopman-Esseboom C, de Ridder MA, Weisglas-Kuperus N, Sauer PJ. Effects of environmental exposure to polychlorinated biphenyls and dioxins on birth size and growth in Dutch children. Pediatr Res. 1998;44(4):538–45.
96. Halldorsson TI, Meltzer HM, Thorsdottir I, Knudsen V, Olsen SF. Is high consumption of fatty fish during pregnancy a risk factor for fetal growth retardation? A study of 44,824 Danish pregnant women. Am J Epidemiol. 2007;166(6):687–96.
97. Konishi K, Sasaki S, Kato S, Ban S, Washino N, Kajiwara J, Todaka T, Hirakawa H, Hori T, Yasutake D, Kishi R. Prenatal exposure to PCDDs/PCDFs and dioxin-like PCBs in relation to birth weight. Environ Res. 2009;109(7):906–13.

12 Developmental Exposure to Endocrine Disrupting Chemicals...

98. Murphy LE, Gollenberg AL, Buck Louis GM, Kostyniak PJ, Sundaram R. Maternal serum preconception polychlorinated biphenyl concentrations and infant birth weight. Environ Health Perspect. 2010;118(2):297–302.

99. Hertz-Picciotto I, Charles MJ, James RA, Keller JA, Willman E, Teplin S. In utero polychlorinated biphenyl exposures in relation to fetal and early childhood growth. Epidemiology. 2005;16(5):648–56.

100. Rylander L, Stromberg U, Hagmar L. Dietary intake of fish contaminated with persistent organochlorine compounds in relation to low birthweight. Scand J Work Environ Health. 1996;22(4):260–6.

101. Sonneborn D, Park HY, Petrik J, Kocan A, Palkovicova L, Trnovec T, Nguyen D, Hertz-Picciotto I. Prenatal polychlorinated biphenyl exposures in eastern Slovakia modify effects of social factors on birthweight. Paediatr Perinat Epidemiol. 2008;22(3):202–13.

102. Sagiv SK, Tolbert PE, Altshul LM, Korrick SA. Organochlorine exposures during pregnancy and infant size at birth. Epidemiology. 2007;18(1):120–9.

103. Grandjean P, Bjerve KS, Weihe P, Steuerwald U. Birthweight in a fishing community: significance of essential fatty acids and marine food contaminants. Int J Epidemiol. 2001;30(6):1272–8.

104. Longnecker MP, Klebanoff MA, Brock JW, Guo X. Maternal levels of polychlorinated biphenyls in relation to preterm and small-for-gestational-age birth. Epidemiology. 2005;16(5):641–7.

105. Weisskopf MG, Anderson HA, Hanrahan LP, Kanarek MS, Falk CM, Steenport DM, Draheim LA. Maternal exposure to Great Lakes sport-caught fish and dichlorodiphenyl dichloroethylene, but not polychlorinated biphenyls, is associated with reduced birth weight. Environ Res. 2005;97(2):149–62.

106. Schack-Nielsen L, Michaelsen KF, Gamborg M, Mortensen EL, Sorensen TI. Gestational weight gain in relation to offspring body mass index and obesity from infancy through adulthood. Int J Obes (Lond). 2010;34(1):67–74.

107. Wrotniak BH, Shults J, Butts S, Stettler N. Gestational weight gain and risk of overweight in the offspring at age 7 y in a multicenter, multiethnic cohort study. Am J Clin Nutr. 2008;87(6):1818–24.

108. Lackmann GM, Angerer J, Tollner U. Parental smoking and neonatal serum levels of polychlorinated biphenyls and hexachlorobenzene. Pediatr Res. 2000;47(5):598–601.

109. Rogan WJ, Gladen BC, McKinney JD, Carreras N, Hardy P, Thullen J, Tingelstad J, Tully M. Polychlorinated biphenyls (PCBs) and dichlorodiphenyl dichloroethene (DDE) in human milk: effects of maternal factors and previous lactation. Am J Public Health. 1986;76(2):172–7.

110. Jacobson JL, Jacobson SW, Humphrey HE. Effects of exposure to PCBs and related compounds on growth and activity in children. Neurotoxicol Teratol. 1990;12(4):319–26.

111. Verhulst SL, Nelen V, Hond ED, Koppen G, Beunckens C, Vael C, Schoeters G, Desager K. Intrauterine exposure to environmental pollutants and body mass index during the first 3 years of life. Environ Health Perspect. 2009;117(1):122–6.

112. Smink A, Ribas-Fito N, Garcia R, Torrent M, Mendez MA, Grimalt JO, Sunyer J. Exposure to hexachlorobenzene during pregnancy increases the risk of overweight in children aged 6 years. Acta Paediatr. 2008;97(10):1465–9.

113. Gladen BC, Ragan NB, Rogan WJ. Pubertal growth and development and prenatal and lactational exposure to polychlorinated biphenyls and dichlorodiphenyl dichloroethene. J Pediatr. 2000;136(4):490–6.

114. Lamb MR, Taylor S, Liu X, Wolff MS, Borrell L, Matte TD, Susser ES, Factor-Litvak P. Prenatal exposure to polychlorinated biphenyls and postnatal growth: a structural analysis. Environ Health Perspect. 2006;114(5):779–85.

115. Rylander L, Stromberg U, Hagmar L. Weight and height at 4 and 7 years of age in children born to mothers with a high intake of fish contaminated with persistent organochlorine pollutants. Chemosphere. 2007;67(3):498–504.

116. Blanck HM, Marcus M, Rubin C, Tolbert PE, Hertzberg VS, Henderson AK, Zhang RH. Growth in girls exposed in utero and postnatally to polybrominated biphenyls and polychlorinated biphenyls. Epidemiology. 2002;13(2):205–10.

117. Pan IJ, Daniels JL, Herring AH, Rogan WJ, Siega-Riz AM, Goldman BD, Sjodin A. Lactational exposure to polychlorinated biphenyls, dichlorodiphenyltrichloroethane, and dichlorodiphenyldichloroethylene and infant growth: an analysis of the pregnancy, infection, and nutrition babies study. Paediatr Perinat Epidemiol. 2010;24(3):262–71.

118. Burns JS, Williams PL, Sergeyev O, Korrick S, Lee MM, Revich B, Altshul L, Del Prato JT, Humblet O, Patterson Jr DG, Turner WE, Needham LL, Starovoytov M, Hauser R. Serum dioxins and polychlorinated biphenyls are associated with growth among Russian boys. Pediatrics. 2011;127(1):e59–e68.

119. Longnecker MP. Invited commentary: why DDT matters now. Am J Epidemiol. 2005;162(8): 726–8.

120. Langer P. Persistent organochlorinated pollutants (PCB, DDE, HCB, dioxins, furans) and the thyroid–review 2008. Endocr Regul. 2008;42(2–3):79–104.

121. Lelli SM, Ceballos NR, Mazzetti MB, Aldonatti CA, Martin S, de Viale LC. Hexachlorobenzene as hormonal disruptor–studies about glucocorticoids: their hepatic receptors, adrenal synthesis and plasma levels in relation to impaired gluconeogenesis. Biochem Pharmacol. 2007; 73(6):873–9.

122. Sargis RM, Johnson DN, Choudhury RA, Brady MJ. Environmental endocrine disruptors promote adipogenesis in the 3 T3-L1 cell line through glucocorticoid receptor activation. Obesity (Silver Spring). 2010;18(7):1283–8.

123. Farhang L, Weintraub JM, Petreas M, Eskenazi B, Bhatia R. Association of DDT and DDE with birth weight and length of gestation in the child health and development studies, 1959–1967. Am J Epidemiol. 2005;162(8):717–25.

124. Longnecker MP, Klebanoff MA, Zhou H, Brock JW. Association between maternal serum concentration of the DDT metabolite DDE and preterm and small-for-gestational-age babies at birth. Lancet. 2001;358(9276):110–4.

125. Fenster L, Eskenazi B, Anderson M, Bradman A, Harley K, Hernandez H, Hubbard A, Barr DB. Association of in utero organochlorine pesticide exposure and fetal growth and length of gestation in an agricultural population. Environ Health Perspect. 2006;114(4):597–602.

126. Eggesbo M, Stigum H, Longnecker MP, Polder A, Aldrin M, Basso O, Thomsen C, Skaare JU, Becher G, Magnus P. Levels of hexachlorobenzene (HCB) in breast milk in relation to birth weight in a Norwegian cohort. Environ Res. 2009;109(5):559–66.

127. Karmaus W, Osuch JR, Eneli I, Mudd LM, Zhang J, Mikucki D, Haan P, Davis S. Maternal levels of dichlorodiphenyl-dichloroethylene (DDE) may increase weight and body mass index in adult female offspring. Occup Environ Med. 2009;66(3):143–9.

128. Mendez MA, Garcia-Esteban R, Guxens M, Vrijheid M, Kogevinas M, Goni F, Fochs S, Sunyer J. Prenatal organochlorine compound exposure, rapid weight gain, and overweight in infancy. Environ Health Perspect;119(2):272–8.

129. Gladen BC, Klebanoff MA, Hediger ML, Katz SH, Barr DB, Davis MD, Longnecker MP. Prenatal DDT exposure in relation to anthropometric and pubertal measures in adolescent males. Environ Health Perspect. 2004;112(17):1761–7.

130. Cupul-Uicab LA, Hernandez-Avila M, Terrazas-Medina EA, Pennell ML, Longnecker MP. Prenatal exposure to the major DDT metabolite 1,1-dichloro-2,2-bis(p-chlorophenyl)ethylene (DDE) and growth in boys from Mexico. Environ Res. 2010;110(6):595–603.

131. Rogan WJ, Gladen BC, McKinney JD, Carreras N, Hardy P, Thullen J, Tingelstad J, Tully M. Polychlorinated biphenyls (PCBs) and dichlorodiphenyl dichloroethene (DDE) in human milk: effects on growth, morbidity, and duration of lactation. Am J Public Health. 1987;77(10): 1294–7.

132. Ribas-Fito N, Gladen BC, Brock JW, Klebanoff MA, Longnecker MP. Prenatal exposure to 1,1-dichloro-2,2-bis (p-chlorophenyl)ethylene (p, p'-DDE) in relation to child growth. Int J Epidemiol. 2006;35(4):853–8.

133. Jusko TA, Koepsell TD, Baker RJ, Greenfield TA, Willman EJ, Charles MJ, Teplin SW, Checkoway H, Hertz-Picciotto I. Maternal DDT exposures in relation to fetal and 5-year growth. Epidemiology. 2006;17(6):692–700.

134. Hauser R, Calafat AM. Phthalates and human health. Occup Environ Med. 2005;62(11):806–18.

135. Swan SH, Main KM, Liu F, Stewart SL, Kruse RL, Calafat AM, Mao CS, Redmon JB, Ternand CL, Sullivan S, Teague JL. Decrease in anogenital distance among male infants with prenatal phthalate exposure. Environ Health Perspect. 2005;113(8):1056–61.
136. Carlsen SM, Jacobsen G, Romundstad P. Maternal testosterone levels during pregnancy are associated with offspring size at birth. Eur J Endocrinol. 2006;155(2):365–70.
137. Smith AS, Birnie AK, French JA. Maternal androgen levels during pregnancy are associated with early-life growth in Geoffroy's marmosets, Callithrix geoffroyi. Gen Comp Endocrinol. 2010;166(2):307–13.
138. Hurst CH, Waxman DJ. Activation of PPARalpha and PPARgamma by environmental phthalate monoesters. Toxicol Sci. 2003;74(2):297–308.
139. Grun F, Blumberg B. Perturbed nuclear receptor signaling by environmental obesogens as emerging factors in the obesity crisis. Rev Endocr Metab Disord. 2007;8(2):161–71.
140. Boas M, Frederiksen H, Feldt-Rasmussen U, Skakkebaek NE, Hegedus L, Hilsted L, Juul A, Main KM. Childhood exposure to phthalates – associations with thyroid function, insulin-like growth factor I (IGF-I) and growth. Environ Health Perspect. 2010;118:1458–64.
141. Meeker JD. Human epidemiologic studies of exposure to endocrine-disrupting chemicals and altered hormone levels. In: Shaw I, editor. Endocrine-disrupting chemicals in food. Boca Raton: CRC press; 2009.
142. Meeker JD, Hu H, Cantonwine DE, Lamadrid-Figueroa H, Calafat AM, Ettinger AS, Hernandez-Avila M, Loch-Caruso R, Tellez-Rojo MM. Urinary phthalate metabolites in relation to preterm birth in Mexico city. Environ Health Perspect. 2009;117(10):1587–92.
143. Whyatt RM, Adibi JJ, Calafat AM, Camann DE, Rauh V, Bhat HK, Perera FP, Andrews H, Just AC, Hoepner L, Tang D, Hauser R. Prenatal di(2-ethylhexyl)phthalate exposure and length of gestation among an inner-city cohort. Pediatrics. 2009;124(6):e1213–e20.
144. Latini G, De Felice C, Presta G, Del Vecchio A, Paris I, Ruggieri F, Mazzeo P. In utero exposure to di-(2-ethylhexyl)phthalate and duration of human pregnancy. Environ Health Perspect. 2003;111(14):1783–5.
145. Adibi JJ, Hauser R, Williams PL, Whyatt RM, Calafat AM, Nelson H, Herrick R, Swan SH. Maternal urinary metabolites of Di-(2-Ethylhexyl) phthalate in relation to the timing of labor in a US multicenter pregnancy cohort study. Am J Epidemiol. 2009;169(8):1015–24.
146. Zhang Y, Lin L, Cao Y, Chen B, Zheng L, Ge RS. Phthalate levels and low birth weight: a nested case-control study of Chinese newborns. J Pediatr. 2009;155(4):500–4.
147. Wolff MS, Teitelbaum SL, Windham G, Pinney SM, Britton JA, Chelimo C, Godbold J, Biro F, Kushi LH, Pfeiffer CM, Calafat AM. Pilot study of urinary biomarkers of phytoestrogens, phthalates, and phenols in girls. Environ Health Perspect. 2007;115(1):116–21.
148. Hatch EE, Nelson JW, Stahlhut RW, Webster TF. Association of endocrine disruptors and obesity: perspectives from epidemiological studies. Int J Androl. 2010;33(2):324–32.
149. Stahlhut RW, van Wijngaarden E, Dye TD, Cook S, Swan SH. Concentrations of urinary phthalate metabolites are associated with increased waist circumference and insulin resistance in adult U.S. males. Environ Health Perspect. 2007;115(6):876–82.
150. Calafat AM, Wong LY, Kuklenyik Z, Reidy JA, Needham LL. Polyfluoroalkyl chemicals in the U.S. population: data from the National Health and Nutrition Examination Survey (NHANES) 2003–2004 and comparisons with NHANES 1999–2000. Environ Health Perspect. 2007;115(11):1596–602.
151. Lau C, Anitole K, Hodes C, Lai D, Pfahles-Hutchens A, Seed J. Perfluoroalkyl acids: a review of monitoring and toxicological findings. Toxicol Sci. 2007;99(2):366–94.
152. Vanden Heuvel JP, Thompson JT, Frame SR, Gillies PJ. Differential activation of nuclear receptors by perfluorinated fatty acid analogs and natural fatty acids: a comparison of human, mouse, and rat peroxisome proliferator-activated receptor-alpha, -beta, and -gamma, liver X receptor-beta, and retinoid X receptor-alpha. Toxicol Sci. 2006;92(2):476–89.
153. Hines EP, White SS, Stanko JP, Gibbs-Flournoy EA, Lau C, Fenton SE. Phenotypic dichotomy following developmental exposure to perfluorooctanoic acid (PFOA) in female CD-1 mice: Low doses induce elevated serum leptin and insulin, and overweight in mid-life. Mol Cell Endocrinol. 2009;304(1–2):97–105.

154. Nelson JW, Hatch EE, Webster TF. Exposure to polyfluoroalkyl chemicals and cholesterol, body weight, and insulin resistance in the general U.S. population. Environ Health Perspect. 2010;118(2):197–202.
155. Fei C, McLaughlin JK, Tarone RE, Olsen J. Fetal growth indicators and perfluorinated chemicals: a study in the Danish national birth cohort. Am J Epidemiol. 2008;168(1):66–72.
156. Olsen GW, Gilliland FD, Burlew MM, Burris JM, Mandel JS, Mandel JH. An epidemiologic investigation of reproductive hormones in men with occupational exposure to perfluorooctanoic acid. J Occup Environ Med. 1998;40(7):614–22.
157. Apelberg BJ, Witter FR, Herbstman JB, Calafat AM, Halden RU, Needham LL, Goldman LR. Cord serum concentrations of perfluorooctane sulfonate (PFOS) and perfluorooctanoate (PFOA) in relation to weight and size at birth. Environ Health Perspect. 2007;115(11):1670–6.
158. Stein CR, Savitz DA, Dougan M. Serum levels of perfluorooctanoic acid and perfluorooctane sulfonate and pregnancy outcome. Am J Epidemiol. 2009;170(7):837–46.
159. Washino N, Saijo Y, Sasaki S, Kato S, Ban S, Konishi K, Ito R, Nakata A, Iwasaki Y, Saito K, Nakazawa H, Kishi R. Correlations between prenatal exposure to perfluorinated chemicals and reduced fetal growth. Environ Health Perspect. 2009;117(4):660–7.
160. Andersen CS, Fei C, Gamborg M, Nohr EA, Sorensen TI, Olsen J. Prenatal exposures to perfluorinated chemicals and anthropometric measures in infancy. Am J Epidemiol. 2010;172(11): 1230–7.
161. Rubin BS, Soto AM. Bisphenol A: perinatal exposure and body weight. Mol Cell Endocrinol. 2009;304(1–2):55–62.
162. Hugo ER, Brandebourg TD, Woo JG, Loftus J, Alexander JW, Ben-Jonathan N. Bisphenol A at environmentally relevant doses inhibits adiponectin release from human adipose tissue explants and adipocytes. Environ Health Perspect. 2008;116(12):1642–7.
163. Louis GM, Cooney MA, Lynch CD, Handal A. Periconception window: advising the pregnancy-planning couple. Fertil Steril. 2008;89(2 Suppl):e119–e21.
164. Bloom MS, Buck-Louis GM, Schisterman EF, Kostyniak PJ, Vena JE. Changes in maternal serum chlorinated pesticide concentrations across critical windows of human reproduction and development. Environ Res. 2009;109(1):93–100.
165. Gillman MW. A life-course approach to obesity. In: Kuh D, Ben-Shlomo Y, editors. A life course approach to chronic disease epidemiology. Oxford: Oxford University Press; 2004. p. 189–217.
166. Lillycrop KA, Burdge GC. Epigenetic changes in early life and future risk of obesity. Int J Obes (Lond). 2010 Jun 15. [Epub ahead of print].
167. Rothman KJ, Greenland S, Lash TL. Validity in epidemiologic studies. In: Rothman KJ, Greenland S, Lash TL, editors. Modern epidemiology. 3rd ed. Philadelphia: Lippincott, Williams and Wilkins; 2008.
168. Longnecker MP. Pharmacokinetic variability and the miracle of modern analytical chemistry. Epidemiology. 2006;17(4):350–1.
169. Webster TF. Pharmacokinetics of POPs: simple models with different implications for half-lives and steady state levels. Organohalogen Comp. 2006;68:344–7.
170. Wolff MS, Anderson HA, Britton JA, Rothman N. Pharmacokinetic variability and modern epidemiology–the example of dichlorodiphenyltrichloroethane, body mass index, and birth cohort. Cancer Epidemiol Biomarkers Prev. 2007;16(10):1925–30.
171. Gerchman F, Tong J, Utzschneider KM, Zraika S, Udayasankar J, McNeely MJ, Carr DB, Leonetti DL, Young BA, de Boer IH, Boyko EJ, Fujimoto WY, Kahn SE. Body mass index is associated with increased creatinine clearance by a mechanism independent of body fat distribution. J Clin Endocrinol Metab. 2009;94(10):3781–8.
172. Hauser R, Meeker JD, Park S, Silva MJ, Calafat AM. Temporal variability of urinary phthalate metabolite levels in men of reproductive age. Environ Health Perspect. 2004;112(17):1734–40.
173. Schildkraut JM, Demark-Wahnefried W, DeVoto E, Hughes C, Laseter JL, Newman B. Environmental contaminants and body fat distribution. Cancer Epidemiol Biomarkers Prev. 1999;8(2):179–83.

12 Developmental Exposure to Endocrine Disrupting Chemicals... 321

174. Faupel-Badger JM, Hsieh CC, Troisi R, Lagiou P, Potischman N. Plasma volume expansion in pregnancy: implications for biomarkers in population studies. Cancer Epidemiol Biomarkers Prev. 2007;16(9):1720–3.

175. Centers for Disease Control and Prevention (CDC) NCfHSN. Third national report on human exposure to environmental chemicals. Atlanta: CDC; 2005.

176. Longnecker MP, Ryan JJ, Gladen BC, Schecter AJ. Correlations among human plasma levels of dioxin-like compounds and polychlorinated biphenyls (PCBs) and implications for epidemiologic studies. Arch Environ Health. 2000;55(3):195–200.

177. Kortenkamp A. Ten years of mixing cocktails: a review of combination effects of endocrine-disrupting chemicals. Environ Health Perspect. 2007;115(Suppl 1):98–105.

178. Kortenkamp A. Low dose mixture effects of endocrine disrupters: implications for risk assessment and epidemiology. Int J Androl. 2008;31(2):233–40.

179. Ibarluzea Jm J, Fernandez MF, Santa-Marina L, Olea-Serrano MF, Rivas AM, Aurrekoetxea JJ, Exposito J, Lorenzo M, Torne P, Villalobos M, Pedraza V, Sasco AJ, Olea N. Breast cancer risk and the combined effect of environmental estrogens. Cancer Causes Control. 2004;15(6): 591–600.

180. Hauser R, Duty S, Godfrey-Bailey L, Calafat AM. Medications as a source of human exposure to phthalates. Environ Health Perspect. 2004;112(6):751–3.

181. Hernandez-Diaz S, Mitchell AA, Kelley KE, Calafat AM, Hauser R. Medications as a potential source of exposure to phthalates in the U.S. population. Environ Health Perspect. 2009; 117(2):185–9.

182. Rothman KJ. BMI-related errors in the measurement of obesity. Int J Obes (Lond). 2008; 32(Suppl 3):S56–9.

183. Wang MC, Bachrach LK. Validity of the body mass index as an indicator of adiposity in an ethnically diverse population of youths. Am J Hum Biol. 1996;8:641–51.

184. Grun F, Watanabe H, Zamanian Z, Maeda L, Arima K, Cubacha R, Gardiner DM, Kanno J, Iguchi T, Blumberg B. Endocrine-disrupting organotin compounds are potent inducers of adipogenesis in vertebrates. Mol Endocrinol. 2006;20(9):2141–55.

185. Gillman MW, Kleinman K. Antecedents of obesity - analysis, interpretation, and use of longitudinal data. Am J Epidemiol. 2007;166(1):14–6. author reply 17–8.

186. Terry MB, Wei Y, Esserman D. Maternal, birth, and early-life influences on adult body size in women. Am J Epidemiol. 2007;166(1):5–13.

187. Smith GD. Assessing intrauterine influences on offspring health outcomes: can epidemiological studies yield robust findings? Basic Clin Pharmacol Toxicol. 2008;102(2):245–56.

188. Jones R, Golding J. Choosing the types of biological sample to collect in longitudinal birth cohort studies. Paediatr Perinat Epidemiol. 2009;23(Suppl 1):103–13.

Part IV
EDCs and Human Health

Chapter 13
The Impact of Endocrine Disruptors on Female Pubertal Timing*

Jean-Pierre Bourguignon and Anne-Simone Parent

Abstract Secular changes in pubertal timing and particular forms of sexual precocity after migration to new geographical areas whose levels of environment differ from those in the natal country suggest that endocrine-disrupting compounds (EDCs) may be involved. Studies in humans based on measurement of some EDC in relation to pubertal timing are inconclusive for several possible reasons including predominant genetic determinism and effects of complex EDC mixtures. Based on animal models, some possible neuroendocrine mechanisms are discussed as well as the interrelation between the homeostasis of reproduction and energy balance.

Keywords Adiponectin • DDE • Development • DDT • Energy balance • GnRH • Homeostasis • Intrauterine growth retardation • Leptin • Menarche • Nutrition • Puberty • Reproduction

Pubertal Timing as an Endpoint for Endocrine Disruption

Female puberty is the life period when pituitary–ovarian maturation leads to a series of physical changes: initially, onset of breast development and increased height and weight gain; ultimately, first menses (menarche) followed, after months or years, by regular (ovulatory) cycling. A central event in the onset of puberty is an increase in

---

*This work was supported by grants from the European Commission (EDEN project, contract QLRT-2001-00269), the Fonds National de la Recherche Scientifique (FRS-FNRS 3.4.573.05 F and 3.4.567.09 F), the Faculty of Medicine at the University of Liège (Léon Frédéricq Foundation), and the Belgian Study Group for Paediatric Endocrinology.

J.-P. Bourguignon (✉) • A.-S. Parent
Developmental Neuroendocrinology Unit, GIGA Neurosciences,
University of Liège, Liège, Belgium

Department of Pediatrics, CHU ND des Bruyères, Chênée, Belgium
e-mail: jpbourguignon@ulg.ac.be; asparent@ulg.ac.be

E. Diamanti-Kandarakis and A.C. Gore (eds.), *Endocrine Disruptors and Puberty*,
Contemporary Endocrinology, DOI 10.1007/978-1-60761-561-3_13,
© Springer Science+Business Media, LLC 2012

frequency and amplitude of gonadotropin-releasing hormone (GnRH) secretion in the hypothalamus. This event is controlled by redundant inhibitory and excitatory mechanisms that are respectively inactivated and activated at the onset of puberty [1]. For initial pubertal signs (breast development) or subsequent events (menarche), it is generally agreed that timing can vary physiologically within a 5-year period. These variations are predominantly determined by genetic factors, while environmental factors play a comparatively minor role [2]. A striking reflection of environmental effects on pubertal timing arose through the secular advance in menarcheal age. Since this was observed between the mid-nineteenth and the mid-twentieth centuries both in USA and Western Europe and more recently in developing countries [2], improvements in health and nutritional status with industrialization were thought to be an explanation. The end or slowdown of this process between 1960 and 2000 [2] was consistent with that explanation. Two large American studies published around the year 2000 [3, 4] and, very recently, findings obtained in Denmark and Belgium [5, 6] provided evidence of earlier onset of puberty. However, onset of breast development, an initial sign, was more affected than age at menarche, a relatively final pubertal sign [3–6]. Moreover, the age distribution of pubertal signs showed recently skewing toward earlier ages for initial signs [6, 7] and toward later ages for final signs [6]. The latter finding is consistent with a French study showing that the secular trend toward earlier menarcheal age was associated with a trend toward later occurrence of regular (ovulatory) cycling [8]. Because those changes in pubertal timing were concomitant with the epidemic of obesity in USA, the pathophysiological involvement of fat mass, possibly through leptin [3, 4, 9], was hypothesized in that country. However, the early breast development reported recently in Denmark was not significantly associated with changes in adiposity [5]. Thus, the possible involvement of other factors such as endocrine-disrupting chemicals (EDCs) was postulated to account for the current secular changes in pubertal timing [10], though no direct evidence substantiates that hypothesis so far.

Suggestive evidence of environmental effects on female pubertal timing also came from studies in children migrating for international adoption. As cohorts, they appeared to mature earlier than children in the foster countries and in the countries of origin [2, 11]. Also, sexual precocity requiring treatment was seen in those migrating children much more frequently than in others [12, 13]. Increased serum levels of DDE, a persisting derivative of the estrogenic insecticide DDT, were found among migrating children. We therefore hypothesized that early exposure to this EDC and subsequent withdrawal due to migration could account for a neuroendocrine mechanism of secondary central precocious puberty in those girls [2, 12]. Using an immature female rat model exposed during early postnatal life to DDT for 5 days, we have been able to induce sexual precocity with involvement of hypothalamic mechanisms [14, 15]. The involvement of DDT in the pathogenesis of sexual precocity in migrating girls could be further substantiated by the observation of lower serum DDE or DDT levels in the girls with normal or late pubertal timing than in precocious girls. Direct evidence, however, is lacking and difficult to obtain since those girls have no reason to attend the clinic and have a blood sample drawn. Moreover, many other EDCs could have affected the migrating children in early

13 The Impact of Endocrine Disruptors on Female Pubertal Timing

life, whereas they cannot be identified several years later. Other factors including recovery from earlier nutritional as well as psychosocial deprivation could play some role as well [16]. One further example suggesting EDC-related development of sexual precocity inferred from geographical clusters of children has been reported in Italy, where the mycotoxin zearalenone has been incriminated in increased frequency of sexual precocity [17].

Association Between Exposure to Endocrine Disruptors and Variations in Human Female Pubertal Timing

Linking exposure to particular EDCs and physiological or abnormal variations in pubertal timing is a challenge for several reasons [18, 19]. Humans as well as animals are likely exposed simultaneously to a variety of EDCs acting as mixtures that change in composition and dose with time. Since compounds mixed at concentrations that are inactive when used as single EDCs have been shown to become active when used as mixtures [20], the significance of studies on single-EDC effects is limited, and addition of the effects of several individual EDCs cannot predict the effect of mixtures. The study of mixtures, however, is complex and laborious. So far, studies on EDC effects on pubertal timing have been performed with single classes of EDCs only. When exposure is assessed at the time of pubertal development or disorders, the identified EDCs could be different from those which were acting during fetal/perinatal life, a most critical time for EDC effects on pubertal timing [20, 21]. Keeping in mind that the relevance of the studied relationship between EDCs and pubertal disorders is limited due to the above reasons; the available information mainly based on the study of single EDCs is summarized in Tables 13.1 (breast development) and 13.2 (menarche). Prenatal and postnatal exposures are separated whenever possible.

Links between EDCs and puberty have been evaluated in some experimental studies. After exposure to the insecticide DDT or its derivative DDE, discrepant data were obtained since early onset of breast development [12] and early menarche [22, 23] were observed in some studies, whereas others found female pubertal timing to be within normal age limits [24–26]. In the female monkey, delayed nipple growth and short follicular phase were seen after exposure to the pesticide methoxychlor [27]. The timing of breast development was not affected after exposure to polybrominated biphenyls (PBBs) or polychlorinated biphenyls (PCBs) [24, 26, 28, 29]. Early menarche, however, was reported after exposure to PBBs [28] or PCBs [25], though several studies found no alteration of menarcheal age in association with PCBs [22, 24, 29, 30]. Dioxins differentially affected timing of breast development that was delayed [29, 31] and timing of menarche that was normal [29, 31, 32]. Early breast development was reported in association with postnatal exposure to phthalates [33, 34]. Phytoestrogens were associated with delayed timing of breast development [26, 34] but no effects on menarcheal age [35]. The discrepant findings for breast development and menarche emphasize the importance of studying

Table 13.1 Variations in timing of onset of breast development in relation to pre- and/or postnatal exposure to endocrine disrupters

Pubertal timing	Early			Normal			Delayed		
Exposure	Prenatal	Postnatal	Unspecified	Prenatal	Postnatal	Unspecified	Prenatal	Postnatal	Unspecified
DDE (+ DDT)			Krstevska-Konstantinova et al., [12]		Wolff et al., [25]	Gladen et al., [23]			
Methoxychlor									
PBBs						Blanck et al., [27]			
PCBs					Den Hond et al., [28]; Wolff etal., [25]	Gladen et al., [23]			
Dioxins							Leijs et al., [30]	Den Hond et al., [28]	
Phthalates		Colon et al., [32]; Wolff et al., [33]							
Phytoestrogens soy formula								Wolff et al., [25, 33]	

DDE dichlorodiphenylchloroethane, *DDT* dichlorodiphenyltrichloroethane, *PBB* polybrominated biphenyl, *PCB* polychlorinated biphenyl, *B*2, *B*3 Tanner's stages 2 and 3 of breast development (Modified from Ref. [19])

Table 13.2 Variations in timing of menarche in relation with pre- and/or postnatal exposure to endocrine disrupters

Pubertal timing	Early			Normal			Delayed		
Exposure	Prenatal	Postnatal	Unspecified	Prenatal	Postnatal	Unspecified	Prenatal	Postnatal	Unspecified
DDE (+ DDT)	Vasiliu et al., [21]	Ouyang et al., [22]			Denham et al., [24]	Gladen et al., [23]			
Methoxychlor								(Monkey) Golub et al., [26]	
PBBs			Blanck et al., [27]						
PCBs		Denham et al., [24]		Vasiliu et al., [29]; Yang et al., [30]	Den Hond et al., [28]	Gladen et al., [23]			
Dioxins				Leijs et al., [30] Warner et al., [31]	Den Hond et al., [28]				
Phthalates									
Phytoestrogens soy formula					Strom et al., [34]				

DDE dichlorodiphenylchloroethane, *DDT* dichlorodiphenyltrichloroethane, *PBB* polybrominated biphenyl, *PCB* polychlorinated biphenyl, *B*2, *B*3 Tanner's stages 2 and 3 of breast development (modified from Ref. [19])

different pubertal signs, possibly involving different mechanisms at several times throughout the pubertal process. Overall, those studies did not enable a differentiation of effects depending on prenatal or postnatal period of exposure to the EDCs.

Possible Mechanisms of EDC Effects on Female Sexual Maturation

Overall, most EDC interactions with sex steroid effects are translated into either estrogen agonist action or androgen antagonism with the ratio of estrogen/androgen actions as an ultimate determinant [36]. In concordance with this concept, premature breast development can occur after exposure to phthalates that are considered to act primarily as androgen antagonists [33]. Because sex steroids can act at the different levels of the hypothalamic–pituitary–gonadal system (Table 13.3), the imbalance between estrogen and androgen effects caused by EDCs potentially affects all those levels. Direct disruption of the peripheral female reproductive system involves predominantly effects mimicking estrogens [19]. Such direct peripheral effects of EDCs could influence the circulating levels of endogenous peripheral hormones. Then, changes in neuroendocrine and pituitary function could result indirectly from altered peripheral feedback. Alternatively or additionally, changes in hypothalamic and pituitary function could result directly from EDC neuroendocrine effects as illustrated by alteration of the preovulatory gonadotropin surge following exposure to sex steroids during fetal or perinatal life [37]. Female pubertal development can involve either hypothalamic–

Table 13.3 Some possible mechanisms of EDC maturational effects on the hypothalamic–pituitary–gonadal system in female individuals

Level possibly targeted by EDCs	Developmental effects of increased estrogen/androgen balance	Mechanisms
Central nervous system: suprahypothalamic afferents	Structural changes?	Primary central/ neuroendocrine or secondary to altered feedback effects of gonadal hormones
Hypothalamus: GnRH neurons and surrounding neuronal– glial system	Facilitation (or inhibition) of pulsatile GnRH secretion	
	Female more sensitive than the male?	
	Alteration of sexually dimorphic control of ovulation	
Pituitary gland: gonadotrophic cells	Early pubertal stimulation or increased prepubertal inhibition (negative feedback)	Response to neuroendocrine effects or peripheral feedback
Gonads: sex steroid production/ effects and gametogenesis	Alteration of folliculogenesis	Primary peripheral or secondary to altered neuroendocrine control
Peripheral tissues: Sex steroid effects	Early/increased stimulation of estrogen-sensitive tissues (breast, uterus)	

pituitary maturation that will secondarily increase sex steroid secretion or direct peripheral interaction in the tissues targeted by sex steroids or both mechanisms.

Due to obvious limits in assessment of neuroendocrine function in the clinical setting, the use of experimental models is required to tackle neuroendocrine effects of EDCs. We exposed female rats early and transiently to DDT in order to mimic the conditions thought to account for sexual precocity in girls migrating for international adoption [2, 12]. Because fetal or early postnatal exposure to testosterone or estradiol would masculinize the CNS and alter the neuroendocrine mechanism of estradiol stimulation of GnRH secretion [38], the animals were exposed to DDT starting postnatal day 6 until day 10. Ex vivo study of GnRH release from hypothalamic explants showed premature acceleration of pulsatile GnRH secretion, early vaginal opening, and early first estrus [14]. Further evidence of possible hypothalamic–pituitary effects of DDT in vivo was obtained through a premature developmental reduction in LH response [14]. Our interpretation of sexual precocity after migration [2, 14] is that during exposure to DDT, estrogenic effects can account for both peripheral and central (neuroendocrine) stimulation. However, due to concomitant negative feedback inhibition at the pituitary level, the central effects are not translated into gonadotropin stimulation of the ovaries until the pituitary inhibition disappears following migration in a DDT-free environment. In a study of internationally adopted girls aged 5–8 years, Teilmann and coworkers [39] reported that serum FSH and estradiol levels were already elevated in several girls before they eventually showed clinical evidence of sexual precocity, confirming early pituitary–ovarian activity after migration. The above mechanism is comparable to that operating in other conditions with peripheral precocious puberty (e.g., congenital adrenal hyperplasia, adrenal or gonadal tumors) followed by secondary central precocious puberty after the peripheral disorder is cured by medical or surgical treatment [2]. Consistent with this concept, central precocity should not be manifested in conditions of persisting exposure to DDT. We attempted to demonstrate that continued administration of DDT to the female rats was not associated with evidence of precocious central maturation, but we failed due to toxic effects including malnourishment and growth failure [14].

The effects of EDCs on sexual maturation in laboratory rodents were reviewed previously [40]. Using triphenyltin in the female rat [41] and phthalates in the male [42], peripheral effects can be biphasic, low or high doses causing early or delayed puberty, respectively. In the few studies where hypothalamic–pituitary function was assessed, LH secretion was reduced together with early vaginal opening after DES [43] or irregular cycling after BPA [44]. Still, the possibly coexisting central and peripheral effects remain a matter of confusion, and the early developmental reduction in LH secretory response to GnRH in the rat [14] is difficult to separate from negative feedback inhibition. As another example, BPA administration to neonatal female rats for 4 days resulted in early vaginal opening and acyclicity but no change in sexual receptivity and FOS induction in GnRH neurons after steroid priming [45]. This suggests predominant peripheral effects in those conditions and highlights the importance of simultaneous study of central and peripheral effects.

Further elucidation of EDC involvement in the neuroendocrine mechanism of female sexual precocity may come from recent progress in identification of estrogen-responsive cells – neurons and/or glial cells through expression of ER alpha and/or ER beta and nongenomic effects and estrogen-responsive genes. Based on murine and ovine animal studies on the sex steroid feedback control of the ovulatory cycle, it appears that GnRH neurons are unlikely to respond directly to stimulatory effects of estrogens that are mediated through afferent neurons [46–48]. Among them, the kisspeptin neurons appear to play an important role as estrogen-sensitive final gate to the GnRH neurons [48, 49] with a gender dimorphism that could account for the female predominance of sexual precocity. This mechanistic hypothesis is consistent with preliminary data showing higher peripheral kisspeptin levels in patients with precocious puberty than in controls [50]. Not only estrogens or estrogenic compounds of peripheral or environmental origin could influence that system but also locally originating estrogens through brain aromatase [51]. Nongenomic interactions of estrogens with the GnRH neurons were reported besides the effects mediated through kisspeptin neurons in the monkey [52]. Recent studies indicate that kisspeptin expression or kisspeptin fiber density in the rodent hypothalamus is reduced after neonatal exposure to BPA [53] or genistein [54], respectively. Though these data suggest negative feedback effects of those EDCs, further studies with emphasis on developmental changes are needed to substantiate or invalidate the involvement of kisspeptin in possible stimulatory effects of EDCs on pubertal timing.

Female Sexual Precocity: An Early Homeostatic Disturbance Involving the Control of Reproduction and Energy Balance?

The hypothalamic–neuroendocrine system is a common integrative site for regulation of homeostasis including reproduction and energy balance. That regulation likely involves, during fetal and neonatal life, programming mechanisms that account for the "developmental origin of health and diseases" [55]. Peripheral messengers from the gonads (e.g., sex steroids) and from adipose tissue (e.g., leptin, adiponectin) can also play important regulatory roles in linking energy balance and reproduction throughout life. In Table 13.4 are proposed some of the connections between homeostasis of reproduction and homeostasis of energy balance, throughout life, as discussed in two recent reviews [20, 21]. Feto-maternal malnourishment or overfeeding with fat as well as fetal exposure to sex steroids or EDCs such as DES and BPA were shown to possibly result in low birth weight, early puberty, ovulatory disorders, obesity in adulthood, and metabolic syndrome [55–58]. Some of those effects are sexually dimorphic since DES causes a metabolic syndrome in female but not in male mice [56]. Both sex steroids and leptin are involved in early organizational effects in the hypothalamus that could affect, respectively, the central control of reproduction [59] and energy balance [60]. There are several lines of experimental and clinical evidence of cross talk between the systems controlling reproduction and

13 The Impact of Endocrine Disruptors on Female Pubertal Timing

Table 13.4 Some characteristics shared in disorders of homeostasis of reproduction and energy balance

Environmental stressors in fetal life	Endocrine-disrupting chemicals (EDCs)	Under- / Over-feeding	Examples of interaction
Neuroendocrine endpoint	Estrogen/androgen balance	Energy balance	DES-induced late obesity and metabolic syndrome
Ante-/neonatal programming	Estrogen-mediated brain masculinization	Leptin-mediated brain organization	Leptin permissive for puberty and reproduction
Neonatal manifestations	Disorders of sex differentiation	Low birth weight	Combined anomalies
Late manifestations	Disorders of puberty and reproduction	Obesity and metabolic syndrome	Combined anomalies

energy balance. During postnatal life, leptin is an important link between energy balance and reproduction through facilitatory effects on GnRH secretion [2, 61]. Clinically, cross talk is supported by the recent finding that rapid weight gain in infancy between birth and 9 months predicts increased adiposity at 10 years and early menarcheal age [62]. Ibanez et al. [63] reported that intrauterine growth retardation (IUGR) was associated with increased risk of premature pubarche, hyperinsulinism, ovarian hyperandrogenism, and polycystic ovary syndrome (PCOS). Of particular interest is the observation that IUGR is associated with increased visceral adiposity and reduced serum levels of adiponectin in childhood [64]. The latter effect parallels BPA-induced reduction of adiponectin production by adipocytes [65, 66]. All together, those findings suggest that both the reproductive and energy balance systems share a possible fetal/early postnatal determinism of disorders such as sexual precocity under the influence of early nutritional conditions and/or early exposure to EDCs, with mechanistic cross talk between the two systems.

References

1. Bourguignon JP. Control of the onset of puberty. In: Pescovitz OH, Eugster E, editors. Pediatric endocrinology: mechanisms, manifestations and management. Philadelphia: Lippincott Williams & Wilkins; 2004. p. 285–98.
2. Parent AS, Teilmann G, Juul A, Skakkebaek NE, Toppari J, Bourguignon JP. The timing of normal puberty and the age limits of sexual precocity: variations around the world, secular trends, and changes after migration. Endocr Rev. 2003;24:668–93.
3. Herman-Giddens ME, Slora EJ, Wasserman RC, Bourdony CJ, Bhapkar MV, Koch GG, Hasemeier CM. Secondary sexual characteristics and menses in young girls seen in office practice: a study from the Pediatric Research in Office Settings Network. Pediatrics. 1997;99:505–12.
4. Lee PA, Guo SS, Kulin HE. Age of puberty: data from the United States of America. APMIS. 2001;109:81–8.

5. Aksglaede L, Sørensen K, Petersen JH, Skakkebæk NE, Juul A. Recent decline in age at breast development: the Copenhagen Puberty Study. Pediatrics. 2009;123:e932–9.
6. Roelants M, Hauspie R, Hoppenbrouwers K. References for growth and pubertal development from birth to 21 years in Flanders, Belgium. Ann Hum Biol. 2009;36:680–94.
7. Papadimitriou A, Pantsiotou S, Douros K, Papadimitriou DT, Nicolaidou P, Fretzayas A. Timing of pubertal onset in girls: evidence for non-gaussian distribution. J Clin Endocrinol Metab. 2008;93:4422–5.
8. Clavel-Chapelon F. E3N-EPIC (European prospective investigation into cancer) Group Evolution of age at menarche and at onset of regular cycling in a large cohort of French women. Hum Reprod. 2002;17:228–32.
9. Himes JH. Examining the evidence for recent secular changes in the timing of puberty in US children in light of increases in the prevalence of obesity. Mol Cell Endocrinol. 2006;254–255: 13–21.
10. Teilmann G, Juul A, Skakkebaek NE, Toppari J. Putative effects of endocrine disrupters on pubertal development in the human. Best Pract Res Clin Endocrinol Metab. 2002;16:105–21.
11. Proos LA, Hofvander Y, Tuvemo T. Menarcheal age and growth pattern of Indian girls adopted in Sweden. I. Menarcheal age. Acta Paediatr Scand. 1991;80:852–8.
12. Krstevska-Konstantinova M, Charlier C, Craen M, Du Caju M, Heinrichs C, de Beaufort C, Plomteux G, Bourguignon JP. Sexual precocity after immigration from developing countries to Belgium: evidence of previous exposure to organochlorine pesticides. Hum Reprod. 2001;16:1020–6.
13. Teilmann G, Pedersen CB, Skakkebæk NE, Jensen TK. Increased risk of precocious puberty in internationally adopted children in Denmark. Pediatrics. 2006;118:e391–9.
14. Rasier G, Parent AS, Gérard A, Lebrethon MC, Bourguignon JP. Early maturation of gonado-tropin-releasing hormone secretion and sexual precocity after exposure of infantile female rats to estradiol or dichlorodiphenyltrichloroethane. Biol Reprod. 2007;77:734–42.
15. Rasier G, Parent AS, Gérard A, Denooz R, Lebrethon MC, Charlier C, Bourguignon JP. Mechanisms of interaction of endocrine disrupting chemicals with glutamate-evoked secretion of gonadotropin-releasing hormone. Toxicol Sci. 2008;102:33–41.
16. Dominé F, Parent AS, Rasier G, Lebrethon MC, Bourguignon JP. Assessment and mechanism of variations in pubertal timing in internationally adopted children: a developmental hypothe-sis. Eur J Endocrinol. 2006;155:S17–25.
17. Massart F, Saggese G. Oestrogenic mycotoxin exposures and precocious pubertal develop-ment. Int J Androl. 2010;33:369–76.
18. Buck Louis GM, Gray Jr LE, Michele Marcus M, Ojeda S, Pescovitz OA, Feldman Witchel S, Sippell W, Abbott DA, Soto A, Tyl RW, Bourguignon JP, Skakkebaek NE, Swan SH, Golub MS, Wabitsch M, Toppari J, Euling SY. Environmental factors and puberty timing: expert panel research needs. Pediatrics. 2008;121(suppl3):S192–207.
19. Diamanti-Kandarakis E, Bourguignon JP, Giudice LC, Hauser R, Prins GS, Soto AM, Zoeller RT, Gore AC. Endocrine-disrupting chemicals: an Endocrine Society scientific statement. Endocr Rev. 2009;30:293–342.
20. Bourguignon JP, Rasier G, Lebrethon MC, Gérard A, Naveau E, Parent AS. Neuroendocrine disruption of pubertal timing and interactions between homeostasis of reproduction and energy balance. Mol Cell Endocrinol. 2010;324:110–20.
21. Bourguignon JP, Parent AS. Early homeostatic disturbances of human growth and maturation by endocrine disrupters. Curr Opin Pediatr. 2010;22:470–7.
22. Vasiliu O, Muttineni J, Karmaus W. In utero exposure to organochlorines and age at menarche. Hum Reprod. 2004;19:1506–12.
23. Ouyang F, Perry MJ, Venners SA, Chen C, Wang B, Yang G, Fang Z, Zang T, Wang L, Xu X, Wang X. Serum DDT, age at menarche, and abnormal menstrual cycle length. Occup Environ Med. 2005;62:878–84.
24. Gladen BC, Ragan B, Rogan WJ. Pubertal growth and development and prenatal and lacta-tional exposure to polychlorinated biphenyls and dichlorodiphenyl dichloroethene. J Pediatr. 2000;136:490–6.

25. Denham M, Schell LM, Deane G, et al. Relationship of lead, mercury, mirex, dichlorodiphenyldichloroethylene, hexachlorobenzene, and polychlorinated biphenyls to timing of menarche among Akwesasne Mohawk girls. Pediatrics. 2005;115:e127–34.
26. Wolff MS, Britton JA, Boguski L, et al. Environmental exposures and puberty in inner-city girls. Environ Res. 2008;107:393–400.
27. Golub MS, Hogrefe CE, Germann SL, Lasley BL, Natarajan K, Tarantal AF. Effects of exogenous estrogenic agents on pubertal growth and reproductive system maturation in female Rhesus monkeys. Toxicol Sci. 2003;74:103–13.
28. Blanck HM, Marcus M, Tolbert PE, et al. Age at menarche and tanner stage in girls exposed in utero and postnatally to polybrominated biphenyl. Epidemiology. 2000;11:641–7.
29. Den Hond E, Roels HA, Hoppenbrouwers K, et al. Sexual maturation in relation to polychlorinated aromatic hydrocarbons: Sharpe and Skakkebaek's hypothesis revisited. Environ Health Perspect. 2002;110:771–6.
30. Yang CY, Yu ML, Guo HR, Lai TJ, Hsu CC, Lambert G, Guo YL. The endocrine and reproductive function of the female Yucheng adolescents prenatally exposed to PCBs/PCDFs. Chemosphere. 2005;61:355–60.
31. Leijs MM, Koppe JG, Olie K, et al. Delayed initiation of breast development in girls with higher prenatal dioxin exposure; a longitudinal cohort study. Chemosphere. 2008;73:999–1004.
32. Warner M, Samuels S, Mocarelli P, et al. Serum dioxin concentrations and age at menarche. Environ Health Perspect. 2004;112:1289–92.
33. Colon I, Caro D, Bourdony CJ, Rosario O. Identification of phthalate esters in the serum of young Puerto Rican girls with premature breast development. Environ Health Perspect. 2000;108: 895–900.
34. Wolff MS, Teitelbaum SL, Pinney SM, Windham G, Liao L, Biro F, Kushi LH, Erdmann C, Hiatt RA, Rybak ME, Calafat AML and the Breast Cancer and Environment Research Centers. Investigation of relationships between urinary biomarkers of phytoestrogens, phthalates, and phenols and pubertal stages in girls. Environ Health Perspect. 2010;118:1039–46.
35. Strom BL, Schinnar R, Ekhard E, Ziegler EE, Kurt T, Barnhart MD, Mary D, Sammel ScD, George A, Macones MD, Virginia A, Stallings MD, Jean M, Drulis BA, Steven E, Nelson BA, Sandra A, Hanson BA. Exposure to soy-based formula in infancy and endocrinological and reproductive outcomes in young adulthood. JAMA. 2001;286:807–14.
36. Rivas A, Fisher J, McKinnell C, Atanassova N, Sharpe RM. Induction of reproductive tract developmental abnormalities in the male rat by lowering androgen production or action in combination with a low dose of diethylstilbestrol: evidence for importance of the androgen-estrogen balance. Endocrinology. 2002;143:4797–808.
37. Gorski RA. Influence of age on a perinatal administration of a low dose of androgen. Endocrinology. 1968;82:1001–4.
38. Matagne V, Rasier G, Lebrethon MC, Gérard A, Bourguignon JP. Estradiol stimulation of pulsatile GnRH secretion in vitro: correlation with perinatal exposure to sex steroids and induction of sexual precocity in vivo. Endocrinology. 2004;145:2775–83.
39. Teilmann G, Boas M, Petersen JH, Main KM, Gormsen M, Damgaard K, Brocks V, Skakkebæk NE, Jensen KT. Early pituitary-gonadal activation in 5–8 years old adopted girls before clinical signs of puberty: a study of 99 foreign adopted girls and 93 controls. J Clin Endocrinol Metab. 2007;92:2538–44.
40. Rasier G, Toppari J, Parent AS, Bourguignon JP. Female sexual maturation and reproduction after prepubertal exposure to estrogens and endocrine disrupting chemicals: a review of rodent and human data. Mol Cell Endocrinol. 2006;254–255:187–201.
41. Grote K, Andrade AJM, Wichert Grande S, Kuriyama SN, Talsness CE, Appel KE, Chahoud I. Effects of peripubertal exposure to triphenyltin on female sexual development of the rat. Toxicology. 2006;222:17–24.
42. Ge RS, Chen GR, Dong Q, Akingbemi B, Sottas CM, Santos M, Sealfon SC, Bernard DJ, Hardy MP. Biphasic effects of postnatal exposure to diethylhexylphthalate on the timing of puberty in male rats. J Androl. 2007;28:513–20.

43. Kubo K, Arai O, Omura M, Watanabe R, Ogata R, Aou S. Low-dose effects of bisphenol A on sexual differentiation of the brain and behavior in rats. Neurosci Res. 2003;45:345–56.
44. Rubin BL, Murray MK, Damassa DA, King JC, Soto AM. Perinatal exposure to low doses of bisphenol A affects body weight, patterns of estrous cyclicity and plasma LH levels. Environ Health Perspect. 2001;109:675–80.
45. Adewale HB, Jefferson WN, Newbold RR, Patisaul HB. Neonatal bisphenol-A exposure alters rat reproductive development and ovarian morphology without impairing activation of gonadotropin-releasing hormone neurons. Biol Reprod. 2009;81:690–9.
46. Herbison AE. Estrogen positive feedback to gonadotropin-releasing hormone (GnRH) neurons in the rodent: the case for the rostral periventricular area of the third ventricle (RP3V). Brain Res Rev. 2008;57:277–87.
47. Goodman RL, Jansen HT, Billings HJ, Coolen LM, Lehman MN. Neural systems mediating seasonal breeding in the ewe. J Neuroendocrinol. 2010;22:674–81.
48. Caraty A, Franceschini I, Hoffman GE. Kisspeptin and the preovulatory gonadotrophin-releasing hormone/luteinising hormone surge in the ewe: Basic aspects and potential applications in the control of ovulation. J Neuroendocrinol. 2010;22:710–5.
49. Kauffman AS. Gonadal and nongonadal regulation of sex differences in hypothalamic Kiss1 neurones. J Neuroendocrinol. 2010;22:682–91.
50. de Vries L, Shtaif B, Phillip M, Gat-Yablonski G. Kisspeptin serum levels in girls with central precocious puberty. Clin Endocrinol. 2009;71:524–8.
51. Cornil CA, Charlier TD. Rapid behavioural effects of oestrogens and fast regulation of their local synthesis by brain aromatase. J Neuroendocrinol. 2010;22:664–73.
52. Terasawa E, Kurian JR, Guerriero KA, Kenealy BP, Hutz ED, Keen KL. Recent discoveries on the control of gonadotrophin-releasing hormone neurones in nonhuman primates. J Neuroendocrinol. 2010;22:630–8.
53. Navarro VM, Sanchez-Garrido MA, Castellano JM, Roa J, García-Galiano D, Pineda R, Aguilar E, Pinilla L, Tena-Sempere M. Persistent impairment of hypothalamic KiSS-1 system after exposures to estrogenic compounds at critical periods of brain sex differentiation. Endocrinology. 2009;150:2359–67.
54. Bateman HL, Patisaul HB. Disrupted female reproductive physiology following neonatal exposure to phytoestrogens or estrogen specific ligands is associated with decreased GnRH activation and kisspeptin fiber density in the hypothalamus. Neurotoxicology. 2008;29:988–97.
55. Gluckman PD, Hanson MA. The developmental origins of the metabolic syndrome. Trends Endocrinol Metab. 2004;15:183–7.
56. Newbold RR, Padilla-Banks E, Jefferson WN. Environmental estrogens and obesity. Mol Cell Endocrinol. 2009;304:84–9.
57. Sloboda DM, Howie GJ, Pleasants A, Gluckman PD, Vickers MH. Pre- and postnatal nutritional histories influence reproductive maturation and ovarian function in the rat. PLoS One. 2009;4:e6744.
58. Heindel JJ, vom Saal FS. Role of nutrition and environmental endocrine disrupting chemicals during the perinatal period on the aetiology of obesity. Mol Cell Endocrinol. 2009;304:90–6.
59. McCarthy MM, Auger AP, Bale TL, De Vries GJ, Dunn GA, Forger NG, Murray EK, Nugent BM, Schwarz JM, Wilson ME. The epigenetics of sex differences in the brain. J Neurosci. 2009;29:12815–23.
60. Bouret SG. Role of early hormonal and nutritional experiences in shaping feeding behavior and hypothalamic development. J Nutr. 2010;140:653–7.
61. Lebrethon MC, Aganina A, Fournier M, Gérard A, Parent AS, Bourguignon JP. Effects of in vivo and in vitro administration of ghrelin, leptin and neuropeptide mediators on pulsatile gonadotrophin-releasing hormone secretion from male rat hypothalamus before and after puberty. J Neuroendocrinol. 2007;19:181–8.
62. Ong KK, Emmett P, Northstone K, Golding J, Rogers I, Ness AR, Wells JC, Dunger DB. Infancy weight gain predicts childhood body fat and age at menarche in girls. J Clin Endocrinol Metab. 2009;94:1527–32.

63. Ibáñez L, Francois I, Potau N, de Zegher F. Precocious pubarche, hyperinsulinism and ovarian hyperandrogenism in girls: relation to reduced fetal growth. J Clin Endocrinol Metab. 1998;83: 3558–662.
64. Ibáñez L, Lopez-Bermejo A, Díaz M, Suarez L, de Zegher F. Low-birth weight children develop lower sex hormone binding globulin and higher dehydroepiandrosterone sulfate levels and aggravate their visceral adiposity and hypoadiponectinemia between six and eight years of age. J Clin Endocrinol Metab. 2009;94:3696–9.
65. Somm E, Schwitzgebel VM, Toulotte A, Cederroth CR, Combescure C, Nef S, Aubert ML, Hüppi PS. Perinatal exposure to bisphenol A alters early adipogenesis in the rat. Environ Health Perspect. 2009;117:1549–55.
66. Hugo ER, Brandebourg TD, Woo JG, Loftus J, Alexander JW, Ben-Jonathan N. Bisphenol A at environmentally relevant doses inhibits adiponectin release from human adipose tissue explants and adipocytes. Environ Health Perspect. 2008;116:1642–7.

Chapter 14
The Influence of Endocrine Disruptors on Male Pubertal Timing

Xiufeng Wu, Ningning Zhang, and Mary M. Lee

Abstract Wildlife observations and toxicological studies in animals have revealed that endocrine disruptors have adverse effects on reproductive health, including altering the timing of pubertal development in both males and females. In contrast to the early puberty observed with EDC exposure in females, most antiandrogenic and estrogenic compounds cause a delay in male pubertal onset accompanied by dysfunction of testicular steroidogenic pathways and perturbation in secondary sexual maturation. Animal studies have identified critical windows of vulnerability for susceptibility that vary among compounds. For several estrogenic compounds such as DES and DDE, the juvenile and peripubertal windows are key exposure periods for the outcome of preputial separation, a hallmark of pubertal onset in male rodents. Gestational and neonatal exposures have been shown to have no impact on pubertal timing. Conversely, gestational exposures to PCBs, dioxins, and flutamide (an antiandrogen) delay pubertal onset. A small but growing number of epidemiologic studies have shown that EDCs can also alter pubertal timing and progression in humans. Recent studies demonstrate an association of lead, dioxins, endosulfan, and PCBs on delaying onset of puberty and age of attaining pubertal milestones and a weaker association of serum organochlorines with earlier puberty. Cumulatively, these studies have identified key factors that affect vulnerability and toxicity. Dose, duration, and timing of exposure, genetic susceptibility, and the endogenous endocrine milieu may all modulate the effects of exposures to EDCs. Human studies are further complicated by the long interval between exposure and outcome and the

---

X. Wu • N. Zhang • M.M. Lee (✉)
Department of Pediatrics, University of Massachusetts
Medical School, Worcester, MA, USA
e-mail: mary.lee@umassmemorial.org

E. Diamanti-Kandarakis and A.C. Gore (eds.), *Endocrine Disruptors and Puberty*,
Contemporary Endocrinology, DOI 10.1007/978-1-60761-561-3_14,
© Springer Science+Business Media, LLC 2012

340   X. Wu et al.

likelihood of mixed exposures which may have opposing, additive, or synergistic actions on the reproductive system. In conclusion, animal and human data confirm perturbations in pubertal onset and attainment of late pubertal stages with early-life exposures to EDCs.

Keywords Dioxins • Endocrine disrupting compounds • Genital staging • Male puberty • Preputial separation • Pubertal • Testicular volume • Testosterone

Background

As advances are made toward understanding the genetic control of puberty, the interplay of environment factors and key pubertal processes is also being elucidated. Improved public health measures and better nutrition are believed to be major contributors to the secular trend toward earlier age at menarche over the past century. Similar temporal shifts in recent decades to an earlier age of breast and pubic hair onset in US girls, however, have been attributed in part to environmental factors such as estrogenic endocrine disrupting compounds (EDCs) [1,2]. Some US studies report a comparable temporal trend in boys toward an earlier age of pubic hair development and onset of puberty, while others have not observed such changes in male pubertal timing [3]. Consequently, the potential impact of environmental influences on secular trends in male pubertal timing is uncertain, although a rise in the incidence of male genital and reproductive disorders heightens concern regarding the detrimental effects of hormonally active chemicals on reproductive health [4]. Observations of reproductive toxicity in wildlife, as well as insights from animal exposure models helped identify the vulnerability of the reproductive system and spurred investigations of parallel EDC toxicities in humans. This chapter summarizes data from animal toxicology studies and human exposures on the consequences of EDC exposures on male pubertal outcomes.

Factors such as pharmacogenetic variations in metabolism and elimination of chemical compounds, polymorphisms in receptor and signaling pathway genes, and robustness of the hypothalamic-pituitary-gonadal (HPG) axis all influence individual and species vulnerability to EDC toxicity. Animal studies have identified critical windows of susceptibility during development and age-specific threshold doses. In contrast to classic hormone–receptor pharmacokinetics, a nontraditional U-tonic dose–response curve is observed for certain EDCs with different toxicities at low versus high exposures. Consequently, exposures below the lowest observed adverse effect level have had unexpected and often delayed reproductive consequences [5]. Three major factors complicate efforts to investigate the toxicity of EDCs in human epidemiological studies or accidental exposures: (1) the long interval from exposure to pubertal onset in boys, (2) the difficulties in accurate determination of body burdens at different windows, and (3) the real-life occurrence of exposures to multiple compounds rather than a single chemical. In elucidating the consequences of

14 The Influence of Endocrine Disruptors on Male Pubertal Timing

endocrine disruptors on puberty, an additional challenge is teasing out the confounding effects of environmental exposures from the considerable influence of genetics, nutrition, and emotional and physical well-being on pubertal timing.

Pubertal Timing

Environmental EDCs are being explored as key contributors to the shift toward an earlier age of onset of breast development in recent decades [6]. Whether a similar trend in pubertal onset has occurred for boys remains unresolved, probably reflecting fewer studies and discrepancies among studies due to methodological differences in assessment of male puberty [3]. After an expert panel exhaustively reviewed the published literature, the majority of participants agreed that data from 1940 to 1994 were inconclusive at demonstrating secular changes in timing of male pubertal development [3]. In males, however, an increased incidence of hypospadias, cryptorchidism, and testicular germ cell cancers has been observed in a distinct geographic distribution, and sperm counts and fertility rates have declined worldwide [1,4]. To explain these observations, the testicular dysgenesis syndrome hypothesizes that these findings are caused by fetal exposure to EDCs that disrupt testicular development and perinatal CNS programming leading to long-term effects on male reproductive health [4]. Exposure to EDCs during key developmental windows is speculated to play a similar role in shifting the age of pubertal onset [2,7,8].

Pubertal onset is a sensitive biomarker of reproductive health that marks the transition from relative quiescence of the HPG axis to the sexually mature secretory profile necessary for adult reproductive competence. The GPR54/kisspeptin system is believed to integrate key neuronal and environmental/nutritional signals and serves as the upstream regulator of the HPG axis through convergence upon the hypothalamic gonadotropin-releasing hormone (GnRH) neurons. Familial patterns, genetics, nutrition, somatic growth, and overall health all have a major impact on pubertal timing. Physical signs of puberty are preceded by activation of the HPG axis, initially characterized by nocturnal gonadotropin pulses that stimulate early morning secretion of testosterone. In boys, an increase in testicular volume is the first sign of pubertal onset. In contrast, peak height velocity, often examined in epidemiological studies as an outcome measure of puberty, occurs at a late stage of sexual maturation and is therefore an imprecise parameter of pubertal onset. Pubic and axillary hair growth represent adrenal sexual maturation rather than central or gonadal maturation and is specific to humans, with no rodent correlates. Adrenarche, however, is usually closely temporally related to gonadal maturation; thus, discordance of adrenarche and gonadarche may be an indication of reproductive system perturbation. Pubertal milestones are not easily identified in boys; therefore, investigators have used surrogate measures such as voice change, peak height velocity, and attainment of adult height [9]. Preputial separation serves as a physical sign of pubertal onset in male rodents (see Table 14.1). Other indices of sexual maturation in rodents include testicular descent, spermatogenesis, body size, weight of reproductive organs, and testosterone production [10–13].

Table 14.1 Normal range of male pubertal milestones in rats and mice

	Rats			Mice
	Sprague–Dawley	Wistar	Long-Evans	CD-1
Age of PPS (PND)	39.1–44.4[a,b]	39.5[c]	39.5–40.4[d]	30.5 [e]
Testis descent (PND)		17.7[f]		21–25[g,] h
Body weight on the day of PPS (g)	227.0–254.6[b]	201.3[i]	178–195[d]	25.0 [e]
Increase in plasma T level (PND)	40–51[a]			28–45[j]

[a] [52]
[b] [13]
[c] [53]
[d] [22]
[e] [54]
[f] [55]
[g] [56]
[h] [57]
[i] [58]
[j] [12]
Abbreviation: *PND* Postnatal day

Environmental EDCs and Pubertal Development in Rodents

Animal studies support vulnerability of male pubertal-timing end points to EDCs, especially those that interfere with estrogen or androgen signaling pathways [14,15]. EDCs alter pubertal onset by both central and peripheral mechanisms. For example, EDC exposure during the perinatal period of central nervous system sexual differentiation can disrupt neuroendocrine organization and perturb negative feedback mechanisms of the HPG axis [16]. In vitro studies can assess the estrogenic or antiandrogenic potency and molecular mechanism of individual compounds. These in vitro findings, however, may not have anticipated in vivo effects as the endogenous endocrine milieu and genetic susceptibility temper outcomes. For example, estrogenic compounds can have opposite actions in females versus males due to the widely divergent concentrations of endogenous estrogens. In general, estrogenic compounds delay rather than advance puberty in the male. Moreover, a specific compound may elicit different effects on pubertal endpoints depending on dose and timing of exposure. For example, the fungicide, triphenyltin, delays vaginal opening (VO) at high doses, whereas it causes earlier VO when administered at low doses, and prenatal genistein exposure delays puberty in rats, while postnatal exposure causes earlier pubertal onset [14,15]. Adverse effects of EDCs on pubertal measures can therefore vary depending on the animal model, the category of the compound, the dose administered, and the dosing window (gestational, lactational, or juvenile exposure).

Estrogenic agents shown in animal models to affect male pubertal outcomes include diethylstilbestrol (DES), methoxychlor, bis (2-ethylhexyl) phthalate (DEHP), bisphenol A (BPA), styrene, and tributyltin (IV) chloride (see Table 14.2). DES, a nonsteroidal estrogen, is the quintessential endocrine disruptor that has been recognized as having reproductive toxicity in rats, mice, and humans. The effects of different exposure windows were assessed using a dose of 60 µg/L drinking water

14 The Influence of Endocrine Disruptors on Male Pubertal Timing

Table 14.2 EDCs causing delayed pubertal onset in male rats

EDC	Exposure window	Outcome measure	Strain	Reference
Aroclor 1221 (PCB mixture)	Maternal (GD 16 and 18)	Delay age of PPS	SD	[51]
Bisphenol A	Maternal or PND 21–91	Delay age of PPS	SD	[19]
DDE	Neonatal	Delay age of PPS	SD	[25]
DDE	PND 17–21	Delay age of PPS	SD	[13]
DE-71	PND 23–50	Delay age of PPS	W	[34]
Diethylstilbestrol	GD0-PND 21 & PND 21–100	Delay age of PPS	W	[11]
Diethylstilbestrol	PND 33–53	Delay age of PPS	SD	[17]
Diethylstilbestrol	PND 1–5 & PND 35–39	Delay age of PPS	SD	[13]
Methoxychlor	PND 21–45	Delay age of PPS	LE	[22]
Prochloraz	PND 23–43 & PND 23–50	Delay age of PPS	SD	[30, 59]
DEHP	PND 21–48	Advance age of PPS	LE	[21]
Vinclozolin	PND 17–21 or PND 23 to PND 55–57	Delay age of PPS and increase serum testosterone	SD	[13,26]
Vinclozolin + Iprodione	PND 23 to PND 55–57	Delay age of PPS	SD	[26]
Ethynylestradiol	PND 6–10	Delay age of PPS, increase body weight at PPS	SD	[13]
Flutamide	PND 35–39	Delay age of PPS, increase body weight at PPS	SD	[13]
Tamoxifen	PND 1–5	Delay age of PPS, increase body weight at PPS	SD	[13]
Benzophenone	PND 1[a]	Decrease juvenile steroidogenesis	SD	[25]
Bis(2-ethylhexyl) phthalate	PND 1[a]	Decrease juvenile steroidogenesis	SD	[25]
Methoxychlor	PND 1[a]	Decrease juvenile steroidogenesis	SD	[25]
Styrene	PND 1[a]	Decrease juvenile steroidogenesis	SD	[25]
Tributyltin(IV)chloride	PND 1[a]	Decrease juvenile steroidogenesis	SD	[25]
2, 4, 5-trichlorophe-noxyacetic acid	PND 1[a]	Decrease juvenile steroidogenesis	SD	[25]

SD Sprague–Dawley, *LE* Long-Evans rats, *W* Wistar, *GD* gestational days, *PND* postnatal day, *PPS* preputial separation

[a] One dose

(equivalent to ~6.5 µg/kg BW/day) which was the midpoint in the dosing range of prior DES endocrine toxicity studies [11]. Exposure of rats to DES 60 µg/L from conception to PND 21 and from PND 21–100 delays testicular descent by 3 and 2 days, and delays PPS by 5 and 2 days, respectively, while exposure from birth to PND 21 delays testicular descent by 2 days but does not affect PPS. Prenatal or neonatal exposures that ended by PND 10 had no effect on PPS. For the PND 21–100 exposure window, a decrease in the DES dose from 60 to 30 µg/L shifted the delay in PPS from 5 to 2 days while 10 µg/L had no effect [11]. The key finding in this study was that prenatal and early neonatal exposure to DES had no effect on male pubertal measures. Effects on testicular descent and PPS were only observed when DES was given beyond PND 10 or started at weaning.

Yoshimura and colleagues confirmed that *prenatal* exposure to DES, even at much higher doses of 100 and 300 µg/kg, had no effect on PPS [13]. However, *postnatal* exposure to DES at 100 or 300 µg/kg from PND 1 to 5 or from PND 35 to 39 delayed PPS by 4–5 days and 2–4 days, respectively, while the same doses at PND 17–21 had no effect [13]. Shin and colleagues examined the effects of gavage DES to juvenile rats at doses of 10, 20, and 40 µg/kg/day for 20 days from PND 33 to 53 on PPS and other indices of pubertal maturation [17]. Only the two higher DES doses of 20 and 40 µg/kg/day significantly delayed PPS by 4–5 days. Androgen-responsive tissue weight and hormone concentrations were measured on PND 54. DES at all doses caused a decrease in serum testosterone by three- to sevenfold and luteinizing hormone (LH) by three- to fourfold. DES caused a dose-dependent decline in the weight of the levator ani and bulbocavernosus muscles and the glans penis, but only the two higher doses were associated with a decreased weight of the testes, epididymides, ventral prostate, and seminal vesicles and a reduction in the number of spermatogenic cells [17].

DDE is a metabolite of the organochlorine pesticide DDT. Although it has been shown to have antiandrogenic, antiestrogenic, and estrogenic actions, depending on the model system, the vulnerable window for pubertal outcomes is similar to that of DES. In a comparison of various exposure periods in rats, neither prenatal nor neonatal exposure to DDE altered timing of PPS; only postnatal administration from PND 35 to 39 of DDE 100 mg/kg caused a significant delay in PPS by 4 days [13].

Ethinyl estradiol (EE) is a potent and highly specific estrogen. Comparison of estradiol administration at different postnatal windows (PND 1–5, 6–10, 11–15, 17–21, and 35–39) revealed that the pubertal outcome of PPS is most sensitive to exogenous EE at PND 6–10. At 10 µg/kg, PPS is delayed by more than 4 days, and body weight at PPS is increased by 17% [13]. Tamoxifen is a selective estrogen receptor modulator that has mixed agonist/antagonist properties depending on the cellular context. In contrast to estradiol, for tamoxifen, the newborn period is most vulnerable; 1 and 3 mg/kg on PND 1–5 delayed PPS by 6 and 12 days, respectively [13].

Bisphenol A (BPA) is a common component of polycarbonate plastics, epoxy linings of food cans, and dental sealants and composites. BPA dosing windows and concentrations remain controversial due to significant strain-specific differences in vulnerability and the complexity of establishing BPA-free experimental conditions. In rats administered BPA 100 mg/kg by gavage from PND 23–53, only 67% had

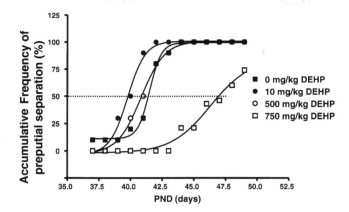

Fig. 14.1 Biphasic effect of di (2-ethylhexyl)phthalate (DEHP) exposures on preputial separation. Average age was calculated as the intercept at 50% accumulative frequency, shown as the dotted line (Reprinted with permission from Ge RS, Chen GR, Dong Q, Akingbemi B, Sottas CM, Santos M, Sealfon SC, Bernard DJ and Hardy MP. Journal of Andrology 2007; 28 (4): 513–520.)

reached PPS at PND 53 as compared to 100% of vehicle-treated rats [18]. Of those who had PPS, the average age was delayed by 1 day. In a three-generation dietary BPA exposure model, BPA in chow at 7,500 ppm, estimated to be equivalent to 500 mg/kg/day, delayed the age of PPS by 3–6 days in F1, F2, and F3 generations, and reduced the body weight at PPS, raising the concern of cofounding effects of body weight on PPS timing [19]. Exposure at lower doses had no effect. In a mouse study, maternal ingestion of BPA at a much lower dose of 2 ng/g from GD 11 to 17 caused a decrease in body weight and weight of seminal vesicles and epididymis in the adult [20].

Concerns have arisen recently about the endocrine disrupting properties of phthalate esters such as diethylhexylphthalate (DEHP), a plasticizer used commonly in certain plastic bottles, infant toys, and medical tubing. Exposure of Long-Evans rats from PND 21 to 48 caused a biphasic dose-dependent effect on PPS (Fig. 14.1) and testosterone production [21]. Daily gavage with 10 mg/kg DEHP advanced PPS by 2 days and raised serum testosterone concentrations, whereas 750 mg/kg delayed PPS by 5 days and lowered serum testosterone. Pituitary mRNA expression of LH and AR were unchanged suggesting that these were direct effects on steroidogenesis and not mediated centrally.

Methoxychlor is a synthetic organochlorine with weak estrogenic activity used as an agricultural pesticide. The effects of methoxychlor administered by daily gavage (25–200 mg/kg/day) from weaning (PND 21) to adulthood on endocrine function, PPS, and weight of androgen-responsive tissues were examined [22]. Doses of 100 and 200 mg/kg/day caused a delay in timing of PPS, downregulated basal and hCG-stimulated testosterone production, upregulated pituitary prolactin and FSH content, and decreased weight of androgen-dependent organs [22]. Surprisingly, cohousing conditions affected the timing of PPS. At a dose of 200 mg/kg/day, treated rats that were cohoused with a female had a delay in PPS of 13 days, whereas those cohoused with an untreated male had PPS delay of only 6 days [22].

4-Methylbenzylidene camphor (4-MBC) is an estrogenic chemical widely used in the cosmetic industry for UV protection. Lifelong exposure of rats to 4-MBC (starting preconception through adulthood) through chow at an average daily intake of 7, 24, and 47 mg/kg caused a dose-dependent delay in PPS of 1–3 days but did not alter body weight or the weight of the testis, seminal vesicle, or epididymis [23]. Gestational exposure to a maternal subcutaneous dose of 20 mg/kg/day 4-MBC increased serum LH at PND 30 [24]. Higher 4-MBC doses of 100 and 500 mg/kg/day decreased testicular weight, serum LH, GnRH, and glutamate of prepubertal offspring (PND 15), whereas it increased serum LH and FSH in peripubertal offspring (PND 30).

Octylphenol and nonylphenol are industrial additives used as nonionic surfactants in the production of plastics and lubricants. These alkylphenols have weak estrogenic activity and are found in wastewater as biodegradation products. Daily exposure to 100 mg/kg nonylphenol by gavage from PND 23 to 53 resulted in only 33% of rats reaching PPS by PND 53, compared to 100% of controls [18]. Of these, PPS was delayed by 1 day. The nonylphenol also caused lower seminal vesicle and coagulation gland weights, and abnormalities in spermatogenesis. To illustrate the unpredictability of exposures to mixtures of compounds, gavage of both nonylphenol and BPA resulted in amelioration of the effects of each individual compound with fewer animals having spermatogenic cycle defects and no decrease in seminal vesicle weight rather than the anticipated synergistic effects [18].

Kuwada and colleagues examined the consequences of a single subcutaneous injection of a number of compounds in the newborn rat pup on their testis weight and steroidogenic capacity at puberty [25]. In terms of testis weight, 4 µM tributyltin (IV) chloride, 2 mM styrene, and 0.8 or 4 mM bis (2-ethylhexyl) phthalate (DEHP) had the greatest inhibition at PND 21, but by PND 50, all testis weights were comparable. The neonatal dose of all estrogenic compounds tested (40 mM benzophenone; 0.8 mM or 4 mM DEHP, 2 mM styrene, 40 mM 2,4,5-trichlorophenoxyacetic acid, and 4 µM tributyltin (IV) chloride) reduced steroidogenesis in the PND 21 testis, but these adverse effects were mostly restored at PND 50 [25].

Antiandrogens that have been examined for effects on pubertal outcomes include p,p'-dichlorodiphenyldichloroethylene (DDE), vinclozolin, diethylhexylphthalate (DEHP), 2,4,5-trichlorophenoxyacetic acid, prochloraz, linuron, and polybrominated diphenyl ethers. Vinclozolin and iprodione (IPRO) are dicarboximide fungicides used in vineyards that act as androgen receptor (AR) antagonists. In Sprague–Dawley rats, while prenatal and neonatal exposure to vinclozolin did not affect timing of PPS, postnatal vinclozolin 100 mg/kg on PND 17–21 and 30 or 100 mg/kg at PND 35–39 caused a delay in PPS by 2–3 days, though the weight of androgen-responsive tissues was unchanged [13]. To explore the actions of two antiandrogenic fungicides in combination, rats were dosed with a range of vinclozolin concentrations (0, 10, 30, 60, and 100 mg/kg/day) with and without the addition of iprodione 50 mg/kg/day from PND 23 to 55–57 [26]. Vinclozolin alone caused a dose-dependent delay in PPS timing, and the addition of iprodione shifted this further by 2–3 days. Vinclozolin also reduced the weight of androgen-responsive tissues; iprodione further reduced the weight of glans penis, seminal vesicle, epididymis, and levator ani and bulbocavernosus muscle [13,26]. Vinclozolin increases testosterone production through central effects – stimulation of LH secretion due to its antiandrogenic properties and loss of negative feedback on the HPG axis [27]. However, iprodione, as a much weaker antiandrogen, had no effect on LH secretion and antagonized the vinclozolin-induced increase in serum testosterone [26]. This study illustrates that a mixture of two anti-androgens can have synergistic effects on androgen-mediated pubertal measures and antagonistic effects on endocrine function.

Flutamide, a nonsteroidal androgen antagonist, was administered to male Sprague–Dawley rats by oral gavage for 20 days from PND 33 at doses of 0, 1, 5, or 25 mg/kg/day. The two highest doses of 5 and 25 mg/kg/day significantly delayed PPS, increased body weight at time of PPS, and decreased weights of multiple androgen-responsive tissues [28]. In the higher flutamide treatment groups, serum testosterone was increased, probably due to indirect central effects although gonadotropins were not measured. A similar study design confirmed these central effects of flutamide: pubertal exposure to flutamide 10 mg/kg/day increased serum LH concentrations in conjunction with a dose-dependent reduction in the weight of the epididymis, seminal vesicle, and ventral prostate [29]. In contrast to DES and other estrogenic compounds, *gestational* exposure to flutamide 10 mg/kg delayed PPS by 7–8 days [13].

Prochloraz (PCZ), an imidazole fungicide, is an androgen receptor (AR) antagonist that inhibits androgen steroidogenesis. Experiments to assess the effects of pubertal PCZ exposure on reproductive measures in Sprague–Dawley rats found that PCZ inhibited testosterone production at much lower doses (15.6 mg/kg/day) than needed to delay PPS (125 mg/kg/day) or decrease reproductive organ weights (62.5–125 mg/kg/day) [30]. LH was unchanged or lower. The steroid hormone profile suggested that PCZ was inhibiting the CYP17 enzyme needed for androgen synthesis [30]. The finding that PPS timing was unchanged despite an eightfold reduction in serum testosterone indicated either that PPS timing correlates better with another androgen such as dihydrotestosterone or that the drop in testosterone was insufficient to affect androgen-responsive tissues and PPS.

Dioxin-like compounds comprise a group of organochlorine contaminants that have been found to have adverse effects on reproductive development. The toxicity of the aromatic hydrocarbons is mediated through the aryl hydrocarbon receptor and expressed in toxic equivalents relative to the reference compound, TCDD. In males, gestational exposure to PCB mixtures with dioxin activity and either gestational or lactational exposure to dioxins themselves have been associated with delayed PPS across species and strains. DE-71 is a commercial mixture of polybrominated diphenyl ethers (PBDEs) that is produced in high abundance as fire retardants for commercial products such as electronic equipment and textiles. PBDE concentration has been increasing in human milk and plasma samples [31–33]. Male Wistar rats gavaged daily with 0–60 mg/kg DE from PND 23–53 had a 2 day delay in PPS at the higher doses of 30 and 60 mg/kg [34]. Chronic gestational exposure to TCDD through maternal diet also caused a delay in PPS of 1.8 and 1.9 days at low and medium doses and of 4.4 days at the highest dose [35]. To develop an animal exposure model for assessing effects of gestational hydrocarbon exposure, Hamm and colleagues prepared a mixture of dioxins, furans, and non-ortho PCBs that reflected the relative ratio of the compounds contaminating food products [36]. Gavage of this mixture (0–1.0 mcg TEQs) to Long Evans rats at GD 15 caused a 1–2-day delay in PPS and lowered seminal vesicle weight. A similar study used a single injection of TCDD (0, 0.2 or 1 mg/kg/dose) at GD 15 to examine the effects of dioxins on the androgen steroidogenic pathway [37]. This study found that maternal injection of 1 mg/kg decreased 17-hydroxylase activity and reduced epididymal weights in male offspring at PND 30. At PND 45, although serum testosterone was lower, the expression of 3βHSD, 17βHSD, and 5α-reductase were increased, and 17-hydroxylase was unchanged from the control group. By day 60, there appeared to be no differences in serum androgens or steroidogenic enzyme expression in dioxin-treated offspring; thus, the effects of dioxin on steroidogenesis were transient. In C57Bl/6 mice, injection of TCDD 1 μg/kg at GD 15 decreased testis weights at PND 30 and 60 [38]. At PND 30, dioxin-treated offspring also had down-regulated androgen receptor expression, germ cell destruction, and decreased PCNA as a marker of proliferation. These changes were less noticeable by PND 60, again suggesting that the adverse effects of gestational dioxin exposure were greatest in early puberty and were attenuated later in sexual maturation.

Animal studies have also identified adverse neuroendocrine effects of various EDCs that can impact pubertal timing (discussed in Chap. 13). Other studies have demonstrated effects of EDCs on weight gain and metabolic control, both of which can impact reproductive function (discussed in Chap. 4). Insights from animal exposure models on direct actions of EDCs in conjunction with indirect effects mediated through body composition and weight gain confirm that pubertal-timing end points are sensitive to disruption by hormonally active compounds. These data suggest that perturbation of pubertal end points in humans may ensue from industrial, accidental, and low-level ubiquitous environmental exposures.

EDCs and Human Pubertal Development

Many chemical contaminants that are readily measurable in the general population have estrogenic or antiandrogenic effects, and therefore can affect pubertal processes. Despite efforts to limit hydrocarbon emissions, and bans on the manufacture and use of persistent organic pollutants such as DDT and PCBs, the general population is continuing to be exposed to these organic compounds, primarily through dietary ingestion of contaminated food sources from past use or use of new endocrine active chemicals in consumer products [5]. Investigation of pubertal outcomes in humans has focused on accidental exposures of communities from industrial accidents or children living in heavily contaminated neighborhoods (Table 14.3). Examining these highly exposed populations increases the likelihood of detecting detrimental consequences. Nevertheless, the data on pubertal onset is limited as

14 The Influence of Endocrine Disruptors on Male Pubertal Timing 349

Table 14.3 EDCs and pubertal development in boys

Exposure	Exposure window	N	Measure	Outcome
Endosulfan[a]	Pubertal	207	Genital stage	Delay
Lead[b]	Pubertal	489	Testicular volume Genital stage	Delay
PCB[c]	Pubertal	80	Genital stage 5	Delay
Dioxin-like cpds[c]	Pubertal	80	Genital stage 5	No effect
PCB, HCB, DDE[d]	Pubertal	887	Staging on school exams	Early
DDE[e], PCBs[e]	Prenatal + lactational	278	Self-reported staging	No effect
PCB[f]	Prenatal	196	Genital stage	No effect
DHEP[g]	Neonatal	13	genital stage	No effect

[a][45]
[b][44]
[c][46]
[d][47]
[e][42]
[f][41]
[g][40]

human epidemiologic studies of EDC exposures have often focused on mature sexual function and fertility as more easily assessed outcome measures.

Epidemiologic studies of pubertal-timing end points are generally small studies that are limited by self-reported pubertal staging, incomplete exposure assessment, or use of indirect or late pubertal biomarkers. Because some of the animal data suggest that adverse effects seen at earlier stages of pubertal development are attenuated at later stages, these latter studies may miss disruption of early pubertal events. Risk assessments are also complicated by the long latency period between exposure and pubertal onset and the probability that individuals are exposed to a mixture of compounds rather than a single agent. The subtle physical signs of pubertal onset may not be evident unless a detailed physical examination is performed by an experienced clinician, which is often not feasible in an epidemiologic study [9]. While a number of clinical parameters are used as surrogates for pubertal timing, events that occur late in sexual maturation, such as menarche or peak height velocity in boys, may reflect the tempo of pubertal progression rather than the timing of pubertal onset. While the studies are not designed or powered to define causation conclusively, they demonstrate biologically plausible associations of early-life EDC exposures with altered timing of puberty that parallel the evidence from animal studies. In general, EDC exposure is associated with earlier onset of puberty or pubertal milestones in girls, while delayed puberty is the predominant finding in males.

Several studies have been unable to identify consistent associations of environmental exposures with pubertal-timing end points (see Table 14.3). One of these examined exposure to neonatal phthalates, plasticizers used in medical and household plastics with antiandrogenic and weak estrogenic activity in vitro. These high production chemicals were added to polyvinyl chloride products for flexibility and durability and also used in cosmetics, solvents, and detergents, thus are ubiquitous contaminants in food, air, and soil [39]. Nineteen adolescents (14–15 years of age) presumed to have high phthalate exposures during extracorporeal membrane oxygenation for cardiovascular surgery as infants were evaluated for their growth and pubertal maturation [40]. The 13 boys in the group were in Tanner stages 2–5, and their hormonal values were consistent with their sexual maturity rating; thus, the investigators concluded that no adverse effects on sexual maturation were observed. This small study lacked exposure measures and had no control group; therefore, the role of phthalates in pubertal timing remains uncertain.

Several other studies have failed to find associations of organochlorine exposures with pubertal development. The Faroese birth cohort study examined the correlation of cord blood concentrations of PCBs with pubertal stages and testicular volume in 196 boys examined at ages 13.3–14.2 [41]. While there was a weak association of cord PCB concentrations with serum hormone values, overall, the investigators concluded that they found no clear associations of prenatal PCB exposures with pubertal development. This study, however, examined spermaturia as one of the outcome measures which is a late pubertal milestone that occurs in midpuberty. Moreover at the time of the study, the boys were already at an age when the majority of boys have progressed beyond stage 2, making it difficult to assess pubertal onset. Surprisingly, despite evaluation at an age when absence of puberty is considered clinically delayed, almost 16% of the boys were prepubertal in this cohort. The North Carolina Infant Feeding Study estimated perinatal exposures by measuring PCBs and DDE in breast milk, maternal blood, cord blood, and placenta and incorporating breastfeeding history into the model [42]. Among 138 boys who returned for the pubertal follow-up studies, the age at attainment of self-reported pubertal stages was not associated with measures of prenatal or lactational PCB or DDE exposure. A potential limitation of this study design is the reliance on self-reported pubertal staging, which is less accurate than physical examination and may therefore be less sensitive for detecting associations. This study was also unable to assess pubertal onset as 73% of the boys were already at genital stage 3 or higher at entry.

Animal data suggest that exposures to EDCs in boys are more likely to cause delayed pubertal onset. While few in number, human studies have primarily reported an association of environmental exposures with delay in onset or a later age of attainment of surrogate indices of puberty. In a cross-sectional analysis of baseline data, Russian boys with blood lead levels ≥ 5 µg/dL had a 43% reduced odds of having entered puberty (defined by physician measured testicular volume of >3 mL) compared to those with lead levels <5 µg/dL [43]. Follow-up analysis of this longitudinal cohort confirmed that within currently acceptable lead ranges, boys with lead levels ≥ 5 µg/dL had pubertal onset 6–8 months later than those with lead levels <5 µg/dL [44]. These studies confirm that even relatively low blood lead levels within CDC-defined acceptable ranges are associated with later pubertal onset in boys.

The effects of endosulfan on male pubertal development were examined in 117 boys residing in a village below cashew plantations which had been sprayed regularly with endosulfan for >20 year [45]. The village obtained water from streams originating at the cashew plantations. Ninety boys from a village about 20 km away were studied as the control group. Sexual maturity rating was assessed by examination, and hormone and endosulfan values were measured. Endosulfan was detected

14 The Influence of Endocrine Disruptors on Male Pubertal Timing

in 78% of the study boys at a total mean value of 7.47 ppb compared to 29% of the controls at a mean value of 1.37 ppb. Sexual maturity rating was positively related to age and endosulfan exposure, and serum testosterone and LH were negatively related. This study showed that endosulfan exposure was associated with a delay in pubertal maturation and decreased testosterone and LH secretion.

To assess the impact of PCBs and dioxin-like compounds on pubertal development, sexual maturity stage, testicular volume, and reproductive hormone concentrations were compared in 15–19-year-old Flemish adolescents (n = 80) residing in two industrial suburbs, one heavily polluted by two waste incinerators and the other by a smelter, and those living in a rural nonindustrialized area [46]. Three PCB congeners were measured to estimate PCB exposures, and dioxin bioactivity was measured in an AHR-luciferase cellular assay (CALUX). Interestingly, serum concentrations of PCBs were higher only in the suburb with the waste incinerators, whereas dioxin-like concentrations measured by CALUX were similar among the boys in any of the three communities. Sixty-two percent of the boys with the higher concentrations of PCB congeners had reached genital stage 5 versus 92% and 100% in the other two communities. Although unrelated to exposure measures, boys residing in the industrial suburbs had smaller testicular volumes than those in the rural area. Hormone concentrations among the boys were similar in all three areas. One key finding of this study was that a doubling of the concentration of PCB congener 138 increased the odds by 3.5-fold of being in genital stage 3 or 4 rather than 5, showing an association of PCB exposure with pubertal delay.

In a more recent cross-sectional Flemish study, 887 14–15-year-old boys had measurement of serum organochlorines and determination of pubertal status from their last annual school exam [47]. The median stages of this cohort were G4 and P4, as the investigators were interested in assessing late stages of pubertal maturation. Serum concentrations of hexachlorobenzene, p,p'-DDE, and PCBs among this cohort were associated with earlier sexual maturation (increased odds of genital stage 3). These investigators also noted that higher levels of serum HCB and blood lead were associated with a lower and a higher, respectively, risk of gynecomastia. These findings are consistent with preliminary data from an ongoing longitudinal cohort study that found increased odds of having entered puberty (defined as testis volume >3 ml) with increasing PCB exposures [60]. This prospective cohort study of 489 boys residing in Chapaevsk was enrolled at ages 8 and 9 and is being followed with annual physical examinations and biennial hormonal measurements, enabling pubertal onset and progression to be assessed in relation to measured peripubertal organochlorine and heavy metal concentrations [44]. At enrollment, the mean total serum TEQs in these boys was 21.1 pg/g lipid (range 4.0–174.7), severalfold higher than children in other European countries. Boys in the higher quartiles of TCDD and PCDD had later onset compared to those in the lowest quartile, whereas there was a trend toward earlier onset at higher quartiles of PCBs. Among this cohort, an increase from <1.3 to 4.0+ pg/g lipid of serum TCDD was associated with a 5½-month later attainment of pubertal onset. These preliminary data suggest an association of peripubertal dioxin concentrations with a delayed onset of puberty defined by testicular volume or genital staging [60].

The increased incidence of detrimental biomarkers of male reproductive health have raised concerns regarding the impact of chronic, low-level environmental exposures to hormonally active chemicals on reproductive health [1,2,7,8,48,49]. Data from animal studies find clear associations of EDCs with alterations in pubertal development and adverse reproductive outcomes. Human data, while less conclusive, also demonstrate effects of exposures on the timing of pubertal onset and a delay in attainment of pubertal milestones. These actions of EDCs on pubertal timing and tempo suggest that environmental factors play a role in the secular trends in pubertal timing that have been reported. While an increasing body of evidence supports the concern that EDCs are affecting pubertal maturation, yet unanswered is whether this confers an unrecognized health risk to future reproductive or metabolic health in the adult. Children with early puberty are at risk for accelerated skeletal maturation and short adult height, psychosocial problems such as poor self-image and substance abuse, eating and affective disorders, and risk-taking behavior [50]. Delayed pubertal maturation has also been associated with concurrent psychological stresses, eating disorders, and victimization and later development of affective disorders and osteopenia. Early pubertal onset has also been associated with an increased risk for breast, prostate, and testicular cancers. Premature adrenarche is associated with an increased risk for metabolic syndrome, obesity, type 2 diabetes mellitus, and cardiovascular disease. Small shifts in pubertal timing may have larger public health implications as a population shift in pubertal timing can impact growth patterns and confer health risks for later disease. Therefore, understanding modifiable determinants of altered pubertal timing may identify exposures that would translate to early-life and lifelong preventive measures that might protect against later adverse health effects.

Key Points

- Endocrine disruptors are either synthetic chemicals, byproducts of industrial processes, or natural plant compounds that can interfere with any aspect of hormonal signaling.
- Environmental exposure to EDCs is ubiquitous, and detectable levels of organic pollutants are measurable in all populations studied.
- The atypical U-tonic dose–response curve for some of these compounds suggests that adverse reproductive health effects may ensue at low-level environmental exposures.
- Assessment of reproductive toxicity in humans is complicated by the long latency period from the critical exposure window to the onset of secondary sexual maturation.
- The actions of EDCs on pubertal onset and progression are primarily caused by their estrogenic or antiandrogenic properties and most commonly associated with pubertal delay in males.
- Animal and human data demonstrate that exposures to EDCs have pronounced effects on pubertal timing which is modified by timing (window of exposure) and dosing.

References

1. Acerini CL, Hughes IA. Endocrine disrupting chemicals: a new and emerging public health problem? Arch Dis Child. 2006;91:633–41.
2. Euling SY, Selevan SG, Pescovitz OH, Skakkebaek NE. Role of environmental factors in the timing of puberty. Pediatrics. 2008;121(Suppl 3):S167–71.
3. Euling SY, Herman-Giddens ME, Lee PA, et al. Examination of US puberty-timing data from 1940 to 1994 for secular trends: panel findings. Pediatrics. 2008;121(Suppl 3):S172–91.
4. Sharpe RM, Skakkebaek NE. Are oestrogens involved in the falling sperm counts and disorders of the male reproductive tract? Lancet. 1993;341:1392–5.
5. Diamanti-Kandarakis E, Bourguignon JP, Giudice LC, et al. Endocrine-disrupting chemicals: an Endocrine Society scientific statement. Endocr Rev. 2009;30:293–342.
6. Biro FM, Galvez MP, Greenspan LC, et al. Pubertal assessment method and baseline characteristics in a mixed longitudinal study of girls. Pediatrics. 2010;126:e583–90.
7. Jacobson-Dickman E, Lee MM. The influence of endocrine disruptors on pubertal timing. Curr Opin Endocrinol Diabetes Obes. 2009;16:25–30.
8. Schoeters G, Den Hond E, Dhooge W, et al. Endocrine disruptors and abnormalities of pubertal development. Basic Clin Pharmacol Toxicol. 2008;102:168–75.
9. Rockett JC, Lynch CD, Buck GM. Biomarkers for assessing reproductive development and health: part 1–pubertal development. Environ Health Perspect. 2004;112:105–12.
10. Ge RS, Hardy MP. Variation in the end products of androgen biosynthesis and metabolism during postnatal differentiation of rat Leydig cells. Endocrinology. 1998;139:3787–95.
11. Odum J, Lefevre PA, Tinwell H, et al. Comparison of the developmental and reproductive toxicity of diethylstilbestrol administered to rats in utero, lactationally, preweaning, or postweaning. Toxicol Sci. 2002;68:147–63.
12. Wu X, Arumugam R, Zhang N, Lee MM. Androgen profiles during pubertal Leydig cell development in mice. Reproduction. 2010;140:113–21.
13. Yoshimura S, Yamaguchi H, Konno K, et al. Observation of preputial separation is a useful tool for evaluating endocrine active chemicals. J Toxicol Pathol. 2005;18:141–57.
14. Dickerson SM, Gore AC. Estrogenic environmental endocrine-disrupting chemical effects on reproductive neuroendocrine function and dysfunction across the life cycle. Rev Endocr Metab Disord. 2007;8:143–59.
15. Rasier G, Toppari J, Parent AS, Bourguignon JP. Female sexual maturation and reproduction after prepubertal exposure to estrogens and endocrine disrupting chemicals: a review of rodent and human data. Mol Cell Endocrinol. 2006;254–255:187–201.
16. Navarro VM, Sanchez-Garrido MA, Castellano JM, et al. Persistent impairment of hypothalamic KiSS-1 system after exposures to estrogenic compounds at critical periods of brain sex differentiation. Endocrinology. 2009;150:2359–67.
17. Shin JH, Kim TS, Kang IH, et al. Effects of postnatal administration of diethylstilbestrol on puberty and thyroid function in male rats. J Reprod Dev. 2009;55:461–6.
18. Tan BL, Kassim NM, Mohd MA. Assessment of pubertal development in juvenile male rats after sub-acute exposure to bisphenol A and nonylphenol. Toxicol Lett. 2003;143:261–70.
19. Tyl RW, Myers CB, Marr MC, et al. Three-generation reproductive toxicity study of dietary bisphenol A in CD Sprague-Dawley rats. Toxicol Sci. 2002;68:121–46.
20. vom Saal FS, Cooke PS, Buchanan DL, et al. A physiologically based approach to the study of bisphenol A and other estrogenic chemicals on the size of reproductive organs, daily sperm production, and behavior. Toxicol Ind Health. 1998;14:239–60.
21. Ge RS, Chen GR, Dong Q, et al. Biphasic effects of postnatal exposure to diethylhexylphthalate on the timing of puberty in male rats. J Androl. 2007;28:513–20.
22. Gray Jr LE, Ostby J, Ferrell J, et al. A dose-response analysis of methoxychlor-induced alterations of reproductive development and function in the rat. Fundam Appl Toxicol. 1989;12:92–108.

23. Durrer S, Ehnes C, Fuetsch M, et al. Estrogen sensitivity of target genes and expression of nuclear receptor co-regulators in rat prostate after pre- and postnatal exposure to the ultraviolet filter 4-methylbenzylidene camphor. Environ Health Perspect. 2007;115(Suppl 1):42–50.
24. Carou ME, Szwarcfarb B, Deguiz ML, et al. Impact of 4-methylbenzylidene-camphor (4-MBC) during embryonic and fetal development in the neuroendocrine regulation of testicular axis in prepubertal and peripubertal male rats. Exp Clin Endocrinol Diabetes. 2009;117:449–54.
25. Kuwada M, Kawashima R, Nakamura K, et al. Neonatal exposure to endocrine disruptors suppresses juvenile testis weight and steroidogenesis but spermatogenesis is considerably restored during puberty. Biochem Biophys Res Commun. 2002;295:193–7.
26. Blystone CR, Lambright CS, Cardon MC, et al. Cumulative and antagonistic effects of a mixture of the antiandrogens vinclozolin and iprodione in the pubertal male rat. Toxicol Sci. 2009; 111:179–88.
27. Monosson E, Kelce WR, Lambright C, et al. Peripubertal exposure to the antiandrogenic fungicide, vinclozolin, delays puberty, inhibits the development of androgen-dependent tissues, and alters androgen receptor function in the male rat. Toxicol Ind Health. 1999;15:65–79.
28. Shin JH, Kim HS, Moon HJ, et al. Effects of flutamide on puberty in male rats: an evaluation of the protocol for the assessment of pubertal development and thyroid function. J Toxicol Environ Health A. 2002;65:433–45.
29. Yamada T, Kunimatsu T, Sako H, et al. Comparative evaluation of a 5-day Hershberger assay utilizing mature male rats and a pubertal male assay for detection of flutamide's antiandrogenic activity. Toxicol Sci. 2000;53:289–96.
30. Blystone CR, Furr J, Lambright CS, et al. Prochloraz inhibits testosterone production at dosages below those that affect androgen-dependent organ weights or the onset of puberty in the male Sprague Dawley rat. Toxicol Sci. 2007;97:65–74.
31. de Boer J, Wester PG, Klamer HJ, et al. Do flame retardants threaten ocean life? Nature. 1998;394:28–9.
32. Ohta S, Ishizuka D, Nishimura H, et al. Comparison of polybrominated diphenyl ethers in fish, vegetables, and meats and levels in human milk of nursing women in Japan. Chemosphere. 2002;46:689–96.
33. Sjodin A, Hagmar L, Klasson-Wehler E, et al. Flame retardant exposure: polybrominated diphenyl ethers in blood from Swedish workers. Environ Health Perspect. 1999;107:643–8.
34. Stoker TE, Laws SC, Crofton KM, et al. Assessment of DE-71, a commercial polybrominated diphenyl ether (PBDE) mixture, in the EDSP male and female pubertal protocols. Toxicol Sci. 2004;78:144–55.
35. Bell DR, Clode S, Fan MQ, et al. Toxicity of 2,3,7,8-tetrachlorodibenzo-p-dioxin in the developing male Wistar(Han) rat. II: chronic dosing causes developmental delay. Toxicol Sci. 2007; 99:224–33.
36. Hamm JT, Chen CY, Birnbaum LS. A mixture of dioxins, furans, and non-ortho PCBs based upon consensus toxic equivalency factors produces dioxin-like reproductive effects. Toxicol Sci. 2003;74:182–91.
37. Cooke GM, Price CA, Oko RJ. Effects of in utero and lactational exposure to 2,3,7,8-tetrachlorodibenzo-p-dioxin (TCDD) on serum androgens and steroidogenic enzyme activities in the male rat reproductive tract. J Steroid Biochem Mol Biol. 1998;67:347–54.
38. Jin MH, Ko HK, Hong CH, Han SW. In utero exposure to 2,3,7,8-Tetrachlorodibenzo-p-Dioxin affects the development of reproductive system in mouse. Yonsei Med J. 2008;49:843–50.
39. Matsumoto M, Hirata-Koizumi M, Ema M. Potential adverse effects of phthalic acid esters on human health: a review of recent studies on reproduction. Regul Toxicol Pharmacol. 2008;50: 37–49.
40. Rais-Bahrami K, Nunez S, Revenis ME, et al. Follow-up study of adolescents exposed to di(2-ethylhexyl) phthalate (DEHP) as neonates on extracorporeal membrane oxygenation (ECMO) support. Environ Health Perspect. 2004;112:1339–40.
41. Mol NM, Sorensen N, Weihe P, et al. Spermaturia and serum hormone concentrations at the age of puberty in boys prenatally exposed to polychlorinated biphenyls. Eur J Endocrinol. 2002;146:357–63.

14 The Influence of Endocrine Disruptors on Male Pubertal Timing 355

42. Gladen BC, Ragan NB, Rogan WJ. Pubertal growth and development and prenatal and lactational exposure to polychlorinated biphenyls and dichlorodiphenyl dichloroethene. J Pediatr. 2000;136:490–6.

43. Hauser R, Sergeyev O, Korrick S, et al. Association of blood lead levels with onset of puberty in Russian boys. Environ Health Perspect. 2008;116:976–80.

44. Williams PL, Sergeyev O, Lee MM, et al. Blood Lead Levels and Delayed Onset of Puberty in a Longitudinal Study of Russian Boys. Pediatrics. 2010;125(5):e1088–96.

45. Saiyed H, Dewan A, Bhatnagar V, et al. Effect of endosulfan on male reproductive development. Environ Health Perspect. 2003;111:1958–62.

46. Den Hond E, Roels HA, Hoppenbrouwers K, et al. Sexual maturation in relation to polychlorinated aromatic hydrocarbons: Sharpe and Skakkebaek's hypothesis revisited. Environ Health Perspect. 2002;110:771–6.

47. Den Hond E, Dhooge W, Bruckers L, et al. Internal exposure to pollutants and sexual maturation in Flemish adolescents. J Expo Sci Environ Epidemiol. 2010;21(3):224–33.

48. Mouritsen A, Aksglaede L, Sorensen K, et al. Hypothesis: exposure to endocrine-disrupting chemicals may interfere with timing of puberty. Int J Androl. 2010;33:346–59.

49. Toppari J, Juul A. Trends in puberty timing in humans and environmental modifiers. Mol Cell Endocrinol. 2010;324:39–44.

50. Golub MS, Collman GW, Foster PM, et al. Public health implications of altered puberty timing. Pediatrics. 2008;121(Suppl 3):S218–30.

51. Dickerson SM, Cunningham SL, Patisaul HB, et al. Endocrine disruption of brain sexual differentiation by developmental PCB exposure. Endocrinology. 2011;152(2):581–94.

52. Korenbrot CC, Huhtaniemi IT, Weiner RI. Preputial separation as an external sign of pubertal development in the male rat. Biol Reprod. 1977;17(2):298–303.

53. Gaytan F, Bellido C, Aguilar R, et al. Balano-preputial separation as an external sign of puberty in the rat: correlation with histologic testicular data. Andrologia. 1988;20(5):450–3.

54. Lau C, Thibodeaux JR, Hanson RG, et al. Effects of perfluorooctanoic acid exposure during pregnancy in the mouse. Toxicol Sci. 2006;90(2):510–8.

55. Dalsenter PR, de Araujo SL, de Assis HC, et al. Pre and postnatal exposure to endosulfan in Wistar rats. Hum Exp Toxicol. 2003;22(4):171–5.

56. O'Shaughnessy PJ, Sheffield JW. Effect of temperature and the role of testicular descent on post-natal testicular androgen production in the mouse. J Reprod Fertil. 1991;91(1):357–64.

57. Klonisch T, Fowler PA, Hombach-Klonisch S. Molecular and genetic regulation of testis descent and external genitalia development. Dev Biol. 2004;270(1):1–18.

58. Stoker TE, Laws SC, Guidici DL, et al. The effect of atrazine on puberty in male wistar rats: an evaluation in the protocol for the assessment of pubertal development and thyroid function. Toxicol Sci. 2000;58(1):50–9.

59. Noriega NC, Ostby J, Lambright C, et al. Late gestational exposure to the fungicide prochloraz delays the onset of parturition and causes reproductive malformations in male but not female rat offspring. Biol Reprod. 2005;72(6):1324–35.

60. Korrick SA, Lee MM, Williams PL, et al. Dioxin exposure and age of pubertal onset among Russian boys. Environ Health Perspect. 2011;119:1339–44.

Chapter 15
Secular Trends in Pubertal Timing: A Role for Environmental Chemical Exposure?

Vincent F. Garry and Peter Truran

Abstract Puberty is a time of great developmental change in human physiology leading to psychosocial and physical maturation. In this process, the interplay of pituitary, adrenal, thyroid and sex hormones is complex and critical for developmental success. It is fertile soil for the disruption of the process by environmental chemicals.

In this chapter we discuss the normal physiologic events of puberty and how they may be altered by broad based population exposure to environmental chemicals. We examine the properties of selected endocrine disrupting chemicals (EDCs) in terms of extent and level of exposure in the human population and their potential to produce effects as individual chemicals. For some of these chemicals and chemical classes including Bisphenol A, Dioxins, and Phthalates human exposure is global, every day and almost unavoidable. Finally, we will offer different views on the possible cumulative effects of these chemicals on puberty.

Keywords Endocrine disruptors • Endocrinology • Environmental phenols • Menarche • Puberty • Tanner stages

Puberty

Background and Introduction

Puberty is a sign of sexual maturation, an event governed by culture and time of transition to adulthood [1, 2]. It is a complex series of developmental changes in physique and psyche.

V.F. Garry (✉) • P. Truran
Minneapolis, MN, USA
e-mail: garry001@umn.edu

E. Diamanti-Kandarakis and A.C. Gore (eds.), *Endocrine Disruptors and Puberty*,   357
Contemporary Endocrinology, DOI 10.1007/978-1-60761-561-3_15,
© Springer Science+Business Media, LLC 2012

Table 15.1 Tanner stages of sexual maturity

	Boys		Girls	
Stage	Genital development	Pubic hair growth	Breast development	Pubic hair growth
1	Prepubertal; no change in size or proportion of testes, scrotum, and penis from early childhood	Prepubertal; no pubic hair	Prepubertal; nipple elevation only	Prepubertal; no pubic hair
2	Enlargement of scrotum and testes, reddening and change in texture in skin of scrotum, and little or no penis enlargement	Sparse growth of hair at base of penis	Small raised breast bud	Sparse growth of hair along labia
3	Increase first in length then width of penis and growth of testes and scrotum	Darkening, coarsening and curling, increase in amount	General enlargement and raising of breast and areola	Pigmentation, coarsening and curling, with an increase in amount
4	Enlargement of penis with growth in breadth and development of glands, further growth of testes and scrotum, and darkening of scrotal skin	Hair resembles adult type but not spread to medial thighs	Further enlargement with projection of areola and nipple as secondary mound	Hair resembles adult type but not spread to medial thighs
5	Adult size and shape genitalia	Adult type and quantity, spread to medial thighs	Mature adult contour, with areola in same contour as breast, and only nipple projecting	Adult type and quantity, spread to medial thighs

[a] Table adapted from source data: Tanner JM, Growth at Adolescence. Oxford. Blackwell Scientific Publications, 1962

In the late 1960s and early 1970s, Tanner and Marshall put forward, through a series of publications, standard clinical measures to monitor the onset, progress, and duration of puberty [3, 4]. The end result was a set of norms that have proved useful over the years to identify overall trends in the timing of the physical changes of puberty (Table 15.1). Tanner staging provides a first step in the clinical investigation of early (precocious) and late-onset puberty. Data derived from these measures of puberty together with the timing (age) of menarche provide observational biomarkers that enable the detection of aberrations in the timing of events suggestive of endocrine disruptive effects in human populations (Fig. 15.1).

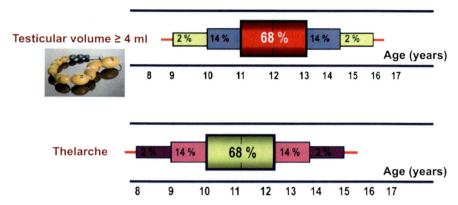

Fig. 15.1 Onset of pubertal development in human males and females. "Reprinted from Molecular and Cellular Endocrinology, 324, Martos-Moreno GA, Chowen JA, Argente J, Metabolic signals in human puberty: effects of over and undernutrition, 70–81, Copyright (2010), with permission from Elsevier

In the sections that follow, we will examine the framework of puberty including, briefly, the physiology and endocrinology of the process and the known factors that modify the onset of puberty. The effects of nutritional status, genetic and racial differences, and neurobehavioral stressors will be considered. Based on limited human data, the potential role of endocrine disruption in current secular trends in the onset of puberty will be examined.

Physiology and Endocrinology

During this developmental transition, physical growth including muscle development, shift in the distribution of body fat, development of secondary sexual characteristics, bone growth, and maturation occur at an accelerated pace. At the root of these changes, there is an uptick of metabolism mediated by heightened hormonal activity of the hypothalamic-pituitary-adrenal (HPA) and hypothalamic-pituitary-gonadal (HPG) axes. The key event in this endocrinologic process is the heightened pulsatile secretion of gonadotropin-releasing hormone (GnRH). The initiation, triggering, and orchestration of the activation of GnRH-secreting neurons of the hypothalamus require kisspeptin neuropeptide and its receptor (Gpr54) for GnRH function [5]. Other stimulatory factors include leptin [6], glutamate, norepinephrine, and growth factors [5, 7]. Sustained pulsatile release of GnRH from the hypothalamus leads to secretion of the pituitary gonadotropins, luteinizing hormone and follicle-stimulating hormone, into the circulation. In turn, the ovaries and testes undergo growth and produce the reproductive hormones estradiol and testosterone.

The end results are the body changes associated with puberty (Tanner stages), including development of secondary sex characteristics, menses, growth spurt, increased body mass, voice changes, and psychosocial development.

Psychosocial and Neurobehavioral Aspects

Psychosocial factors including stress, nutritional status, and disease [8] are known to influence the timing of puberty. For example, one study examined the impact of a number of stressors on the timing of puberty (overweight, behavior problems, family conflict, and father absence). In girls who had two risk factors, menarche occurred on average 2 months earlier. In those girls who had all four risk factors, menarche occurred 8 months earlier [9]. Further, early maturing girls developed behavioral problems more frequently than their peers who developed on time [10]. Seemingly, these same stressors have less-pronounced effects on boys [11, 12]. The role of gonadal steroid hormones in the organization of the developing brain, and thereby direct adult hormone responsiveness to dictate sex-specific behavior, is likely to have a role in these stress responses [13, 14]. Similarly, the onset of adrenal androgen production, i.e., "adrenal puberty" prior to reactivation of the hypothalamic GnRH pulsatile secretion, is thought by some to carry with it the development of different behavioral responses [15].

Nutrition, Genetics, Ethnicity, and Age at Menarche

Under- and overnutrition are considered important factors in the age of onset of menarche. Beginning in the 1940s, the secular trend for the onset of menarche has shown a gradual decrease in the age of menarche in the USA and Western Europe [16, 17]. Similar trends were recorded for India [18], China [19, 20], Thailand [21], South Korea [22, 23], Japan [24], Gambia [25], South Africa [26], Egypt [27], Cameroon [28], and Nigeria [29]. Improved nutrition is often cited for observed reduction in the age of onset of menarche. Oddly, black girls in the USA (National Health and Nutrition Examination Survey 1988–1994; NHANES III) were noted to have an earlier onset of puberty [30]. The reason(s) for this disparity as noted earlier in the PROS study need to be further defined [31].

There are few studies examining secular trend in the onset of puberty in boys. Review of available studies encompassing the years from 1940 to 1994 suggests that current data for boys in the USA are insufficient to evaluate secular trends [32]. More recent data from Denmark indicate that age of attainment of puberty may be decreasing in boys [33].

Correlations with increased body mass index [30, 33], obesity, and increased meat protein intake have been reported as plausible contributing causes for early onset of puberty. In all, the data imply that the metabolic signals for the onset of puberty have been altered to achieve energy homeostasis [34]. Recent works in the study of the regulation of the progress of puberty demonstrate that the adipocyte-derived hormone leptin and the intestinal tract–derived hormone ghrelin may play a critical role in the onset of puberty. These studies also indicate that these peripheral hormones have a major role in the regulation of energy metabolism [7]. As a consequence of the effects of leptin, ghrelin, and other peripheral hormones, weight gain, growth spurt, and earlier sexual maturity are heightened with increased energy efficiency during puberty. Although obesity is a contributing factor in the early onset of puberty, nutritional status seems inadequate to explain the worldwide shift in the age at menarche (AAM),

Genetics and Ethnicity

The timing of puberty has a genetic component; children whose parents underwent early/late puberty are more likely to experience similar pubertal timing. The genes involved in the timing of the onset of puberty, in HPG regulation, and in steroidogenesis are complex and are likely to be multifactorial [35]. Interruption or disturbance by congenital anomaly (e.g., hypothalamic hamartoma) of the organization of these networks can produce drastic changes in the onset of puberty [36, 37]. Genome-wide studies show that parts of the variation of age at menarche are associated with at least two genetic loci, 6q21 and 9Q31.23 [36] in individuals of European descent. Other genes and gene products based on analysis of patients with hypogonadotropic hypogonadism have been implicated as well [5].

Ethnic difference in AAM have been reported for southern Europeans (Mediterranean Basin) compared to northern Europeans [16]; and black girls compared to white girls in the USA and Denmark [30, 38–40]. In general, black girls undergo puberty earlier than Caucasians. The significance of these differences is not clear.

Consideration of Endocrine Disruption and Puberty

It is fairly obvious that substances that have hormone-like activity are likely to affect puberty. In the same context, nutrition, exercise, and genetic heritage all contribute to the timing of the onset of puberty.

At issue is the extent of exposure to endocrine-disrupting agents and the level of exposure from in utero through pubescence. The Fourth National Report on Human Exposure to Environmental Chemicals [41] from the Centers for Disease Control and Prevention (CDC) provides some insights on the magnitude and prevalence of endocrine-disruptor exposure in the general population. Samples of blood, serum, and urine were obtained from a representative subpopulation (N approximately 2,500) selected at random from CDC-NHANES I through III in the USA. Included in the subsample were children aged 6 years and above. Among the chemicals studied were volatile disinfection by-products primarily derived from chlorinated drinking water and used in recreational swimming areas, environmental phenols including Bisphenol A, herbicides including Atrazine, organochlorine insecticides including DDT and its metabolite DDE, organophosphates including chlorpyrifos and its metabolites, metals including arsenic and lead, perfluorochemicals including PFOA, phthalates and their metabolites, phytoestrogens including genistein, polychlorinated biphenyls, dioxin-like chemicals, polycyclic aromatic hydrocarbons including Naphthalene, and solvents including benzene. Nearly all the chemical and chemical groups cited have endocrine-disrupting properties [16, 42–45]. In the paragraphs that follow, we will present for your review illustrations of the available human data (biomarker and epidemiologic studies) regarding endocrine disruption and onset of puberty. Data tables from the Fourth National Report of the CDC [41] will be used to introduce each chemical toxicant to give the reader an estimate of the pervasiveness of human population exposure to endocrine disruptors. The intent here is not to be comprehensive but rather to examine the properties of selected EDCs in terms of extent and level of exposure in the human population and their potential to produce effects on puberty as individual chemicals. Finally, we will offer different views on the possible cumulative effects of these chemicals on puberty.

Arsenic

Although arsenic is a well-known carcinogen, other noncancer adverse health effects can be significant, including adverse pregnancy outcomes [46] and developmental delays [47]. Sources of exposure for the general population include diet, e.g., shellfish and groundwater or wells [48]. In the USA, the geometric mean concentration of arsenic in urine is 8.30 ug/l (Cl 7.19–9.57 ug/l). Mexican Americans (9.29 ug/l) and non-Hispanic blacks (11.6 ug/l) have somewhat higher levels of arsenic in urine compared to non-Hispanic whites (7.12 ug/l). By contrast, the median concentration of arsenic in urine of subjects exposed to arsenic in groundwater in the West Bengal region of India is 387 ug/l [49]. In a relatively small study from the same region, where there was arsenic exposure through well water, the mean age of menarche was 12.5 years; in nonexposed villagers, menarche occurred at age 11.7 years. The differences were significant ($p < 0.05$). Similar data were reported in an animal study; 1 month's delay in the onset of puberty was noted in female rats [50]. In vitro studies in mammalian cell lines show significant dose-related effects on steroid receptor gene expression. Lower doses yielded enhanced expression, while higher doses inhibited expression with somewhat similar effects on thyroid and retinoid receptor expression [51].

Lead

The major environmental exposures to lead (current and historic) are lead-based paints, leaded gasoline, lead soldering of food cans, and lead plumbing still in use in older homes. Based on NHANES I, II, and III, the geometric mean levels for all ages of blood lead varied little (1.66, 1.45, and 1.43 ug/dl) from 1999 to 2004. For children aged 1–5 years, geometric mean blood lead levels were somewhat higher (2.33, 1.70, and 1.77 ug/dl). Using data derived from NHANES III for girls from ages 8–18 years [52], blood lead levels of 3 ug/dl or higher were associated with decreased height regardless of race. More importantly, significant delays in puberty (breast and pubic hair development) were noted for African-American girls and for Mexican-American girls but not white girls. Similarly, in studies of a South African (Johannesburg) girls birth cohort [53], puberty was significantly delayed in those girls who had blood lead levels ≥5 ug/dl. In a study of Egyptian boys and girls, children with blood lead levels >10 ug/dl showed puberty delay [54]. Russian boys with blood lead levels ≥5 ug/dl demonstrated a 5–8-month delay in puberty onset [55].

Altogether, the data from these studies are consistent; elevated blood lead levels are associated with pubertal delay. Further, animal studies confirm delayed timing of puberty. Suppression of insulin-like growth factor, luteinizing hormone, and estradiol secretions are likely mechanism(s) for pubertal delay [56, 57].

Environmental Phenols

Bisphenol A (BPA), Octylphenol, and Nonylphenol

Bisphenol A (BPA) is a phenolic chemical commonly used in the manufacture of hard plastics such as dishes, toys, bottles, storage containers, protective linings of some canned foods, epoxy paints, and some dental composites. BPA is one of the highest volume chemicals produced worldwide. Leaching from industrial sites and from use is not uncommon. Urinary, circulating, and tissue biomonitoring studies indicate widespread exposure to BPA [58]. Data from the Fourth National Report show the following: The geometric mean concentration for BPA in urine for all ages (N=2517) was 2.64 ug/l. For children aged 6–11 years, 3.55 ug/l, and teenagers 12–19 years, 3.74 ug/l; the urinary concentration was somewhat higher. It is generally agreed (CASCADE network) that BPA has weak estrogenic and antiandrogenic properties [59]. Beyond this point, there appears to be little agreement. Some epidemiologic data suggests a relationship with early onset of puberty countered by other studies that suggest higher BPA concentrations were associated with delayed puberty [60]. Risk assessments from various sources reviewed by Beronius et al. [61] provide a well-founded analysis of the conflicting views regarding the overall hazard presented by BPA.

Much like BPA, alkylphenols such as *tertiary octylphenol* (OP) and *nonylphenol* (NP) are xenoestrogens [62, 63]. These surfactant chemicals are components of paints and detergents and are used as adjuvants to augment the efficacy of some herbicides and other pesticides. These chemicals are not as widely dispersed, and therefore, urine concentration (geometric mean for children age 6–11 of approximately 0.357 ug/l) is relatively low [41]. There is some epidemiologic evidence that NP is a more potent xenoestrogen than BPA. In a study of pubertal school girls and boys, early onset of puberty was correlated with the level of NP in urine. The geometric mean concentration of NP was 1.27 ug/l; the highest concentration being 178 ug/l. Direct comparison of NP with OP suggest that OP may be more potent than NP in the early initiation of puberty [62].

Phytoestrogens

Phytoestrogens are plant-derived polycyclic phenols with weak estrogenic activity when consumed in dietary form. In nonsoy-based diets, most come from grains (lignans). Soy and soy- based diets are high in isoflavones, with daidzein and genistein being the predominant isoflavones in soy products. These dietary sources can yield estrogenic compounds at levels 100–1,000 times greater concentration than endogenous estrogens (Table 15.2) [41]. Surprisingly, there is little current evidence to indicate that phytoestrogens can affect the timing of puberty in humans. One study of the DONALD participants (Dortmund Nutritional and Anthropometric Longitudinally Designed) demonstrated that girls with the highest dietary isoflavone diet had the greatest puberty delay. Study participants were characterized by high socioeconomic status compared with the general German population [64]. In a multiethnic longitudinal study of girls from New York City, Northern California, and Cincinnati, Ohio [65], examination of urinary levels of phthalates, phytoestrogens, and phenols indicated that higher urinary concentrations of daidzein, an isoflavone, was associated with a modest delay in breast development. These two recent human studies are consistent with data from animal studies [66] indicating a delay in timing of puberty by high-level phytoestrogen exposure. The dietary sources noted in the human studies represent a daily exposure scenario [65] beginning in utero and continuing through childhood to puberty and through adulthood.

Table 15.2 Phytoestrogens and their geometric mean concentration in urine (ug/l)[a]

	NHANES I–III		
Phytoestrogen	I	II	III
Daidzein	75	52	67
Enterodiol	27	36	40
Enterolactone	239	259	298
Equol	8	9	8
Genistein	24	33	31

[a] Data taken from Fourth National Report [41]. For the general population of the USA, the cumulative load of phytoestrogen exposure from dietary sources is over 300 ug/l

Phthalates

Phthalates are a group of synthetic chemicals commonly added to plastics to impart flexibility and resilience to polyvinyl chloride products. They are widely used in the home and in the workplace. The breadth of use extends from perfumes and detergents to plastic toys and medical devices. There are many members of this chemical family [41]. Out of that group, two members of the phthalate family of chemicals will be considered based on reported peer-reviewed human studies, di-2-ethylhexyl phthalate (DEHP), and one of its major metabolites, MEHP, and diethyl phthalate (DEP), and one of its major metabolites, MEP.

MEP (monoethyl phthalate), a major urinary metabolite of DEP, is a solvent used in many consumer products including fragrances. The reported geometric mean urinary concentration in the general US population varies from 179 to 193 ug/ml according to NHANES I through III [41].

MEHP (mono-2-ethylhexyl phthalate) is primarily used to produce flexibility in plastics. MEHP is formed by hydrolysis of DEHP and then absorbed from the gastrointestinal tract. It is one of the 30 metabolites of DEHP excreted in the urine. The geometric mean concentration varies from 2.3 to 4.3 ug/ml.

To a greater or lesser degree, these metabolites (MEP and MEHP) have been associated with early or premature breast development. Colon et al. [67] found significantly higher levels of phthalates in serum in patients with premature breasts compared to normal controls. Plasma DEHP and MEHP levels were significantly higher in boys with gynecomastia [68] than in unaffected control subjects. Similarly, girls with premature thelarche (breast development) showed significantly higher urinary levels of monomethyl phthalate (MMP) and MEHP [69]. Spot urine collected in a cross-sectional study of girls with precocious puberty and control subjects [70] did not show significant difference in phthalate levels. In this study population, MEHP was not detected in the urine of almost half the study subjects. Animal studies suggest that some phthalates have antiandrogenic properties [41]. Some others may have weak estrogen activity at least in vitro [70]. These properties may contribute to the effects of phthalates on pubertal timing.

Perfluorochemicals Perfluorooctanoic Acid (PFOA) and Perfluorooctane Sulfonate (PFOS)

Members of the perfluorochemical (PFCs) chemical family in polymeric form have applications in waterproofing, protective coatings, nonstick coatings for cookware, electric wire insulation polishes, flame retardants, food packaging, and other uses. Due to their widespread use, chemical stability, persistence in the environment, and bioaccumulation in humans, animals and wildlife, PFCs have been a subject of concern to scientists and citizens alike [71–73]. In 2000, the primary global manufacturer began to phase out a major PFC after a reported end product perfluorooctane sulfonate (PFOS) was found to bioaccumulate in humans and wildlife [74, 75]. In the subsequent 5-year period following phase out, there was a decline in blood plasma concentration of PFOS in specimens obtained from US regional blood banks [74]. The reported geometric mean concentration (PFOS) post phase out was 14.5 ng/ml. These data differ somewhat from serum concentrations reported [41] in a population-based national sample (geometric mean 20.7 ug/l) obtained in the third NHANES study survey (years 2003–2004). It appears that continuing sources of human exposure to PFOS remain.

PFOA and PFOS are commonly detected in human serum, more so in the US population than in the developing countries of Asia and South America [41]. The overall half-life of PFOA in human serum is about 3.5 years and for PFOS about 4.5 years. Concentrations in males tend to be higher than in females [75]. Curiously, since PFOA and PFOS are both lipo- and hydrophobic, rather than bioaccumulating in fat, PFOA and PFOS bind to serum protein [76].

In regard to puberty and puberty onset, there are a number of epidemiologic studies each having a different approach and outcome. As part of a breast cancer study [77], a correlation between girls arriving at Tanner stage IIB (breast development) and the concentration of PFOA (median concentration detected 6.4 ng/ml) was identified. In a cohort study of British mothers and their daughters, PFCs in serum samples obtained in pregnancy were correlated with age of onset of menarche of their female offspring. In this report, the authors state no correlation of PFCs with onset of the daughters' menarche. However, authors mention that serum concentrations of carboxylates (including PFOA and other perfluorinated carboxylates) were associated with increased odds of earlier menarche [78]. The median concentration of PFOS (19.8 ng/ml) and PFOA (3.7 ng/ml) were in the same range as the compiled results [77] from NHANES (I thru III) studies dealing with adult women (weighted mean of PFOA is 3.77 ng/ml and PFOS is 19.14 ug/ml).

A cross-sectional study [79] was done on children who were residents of a community living near a fluoropolymer manufacturing plant where drinking water was contaminated with PFOA. Following parental consent, serum specimens were obtained from children ages 8–18 years. Measurements of PFOA and PFOS were conducted and correlated with sex hormone measurements. Self-reported questionnaire data was used to assess age at menarche. No comparable questionnaire information was obtained for boys. Total testosterone level greater than 50 ug/dl was set as the cutoff point to determine onset of puberty in boys. The median PFOA and PFOS serum concentration in the children was 28.2 and 20.0 ng/ml, respectively. The median serum concentrations of PFOA (4.2 ng/ml) and PFOS (17.5 ng/ml) determined in NHANES survey collected in the same time frame (2005–2006) for PFOA may be significantly different.

For boys, increasing PFOS was associated with delay of 190 days in the onset of puberty. For girls, higher concentrations of PFOA or PFOS were associated with reduced odds of entry into menarche (130 and 138 days delay, respectively). Several animal studies suggest that higher PFOA concentrations are associated with lower testosterone levels in male rats [80] and higher progesterone levels in female mice [81]. While initial animal studies suggested that activation of peroxisome proliferation might play a major role in the endocrine disrupting effects of PFOA, more recent work point to specific effects on steroid metabolism [82]. Another work points to disruption of thyroid hormone action [76].

Organochlorines: DDT, PCBs and Dioxins

Much has been written about the endocrine disrupting properties of the organochlorine class of chemical toxicants and the onset of puberty. Exposure of the general population of the USA and other developed countries tends to be limited as a result of governmental regulatory actions [41].

Serum levels of 2,3,7,8-Tetrachlorodibenzo-p-dioxin (TCDD) detectable in the parts per trillion range, given these limits of detection, was detectable only at the 90th percentile and higher at levels (10% US population sample) not exceeding eight parts per trillion [41]. Sources of general population exposure tend to be limited to burning wood products [41]. Exposures to some dioxin-like and nondioxin-like PCB congeners are more frequent in the general population (e.g., PCB 126 [3,3',4,4,5-Pentachlorobiphenyl]; geometric mean 22.7 parts per trillion [NHANES II]). Since exposure to the organochlorine congeners is due to a mixture of compounds, levels of exposure are measured in terms of toxic equivalents (TEQ) of the mixture according to the cumulative potency of the individual congeners (WHO 2005).

In recent investigations [82, 83], prepubertal boys (age 8–9 years) residing in Chapaevsk, Russia, with high levels of dioxins and dioxin-like PCB congeners were studied. Local environmental contamination was due to by-products of the chlorinated chemical manufacturing process and waste incineration. These Russian boys had three times higher serum levels (median TEQ 21.1 pg/g lipid) than similar-aged European children. Delayed puberty onset was noted and consistent with earlier animal studies [83].

In a similar vein, levels of DDE (pp-'Dichlorodiphenyldichlorethene), a metabolite of DDT and indicator of past exposure to this organochlorine pesticide, are in the parts per billion range (geometric mean concentration 238–295 ppb). Only 5% of the US general population show demonstrable levels (19–28 ppb) of DDT, the parent chemical. Though DDT use has been banned in the US since 1972, the pesticide continues to be used for malarial control in some developing countries. In this connection, in a small sample of girls with early puberty residing in Belgium, all 15 immigrant girls in the sample had detectable DDE in serum (median concentration 1.04 ng/ml). Thirteen out of 15 nonimmigrant girls had no detectable DDE in serum [84].

Female offspring from the Michigan angler cohort dating to the 1970s, with in utero DDE exposures of 15 ug/l, was associated with an early age of menarche in female progeny [85]. In a study from China, the age at menarche was earlier as the concentration of DDT increased. Altogether, these earlier studies suggest that DDT exposure can give rise to an earlier age of menarche. Later studies from the USA [86, 87], in both rural and urban minority girls in a more contemporary setting, showed no such effect on puberty.

Speculations on the Early Onset of Puberty

The available human data regarding puberty are fairly consistent for an association between late onset of puberty and a single toxicant exposure. The relationship between a toxicant exposure and the early onset of puberty seems less certain in humans, although animal data for early puberty induction by environmental endocrine disruptors is much stronger. In humans, it is not possible to directly attribute the worldwide secular trend toward early onset of menarche/puberty from the 1940s to the present day to environmental exposures. However, it needs to be considered that early puberty is associated with increased BMI, obesity, and type 2 diabetes, and recent research supports links between endocrine disruptors and the metabolic syndrome [88–91]. In one view, cumulative exposure to a sea of estrogens and antiandrogens may promote early onset of puberty [42, 92]. A second view suggests that EDCs alter energy balance (metabolic endocrine disruption) promoting obesity and early onset of puberty [45, 93]. A third more radical view suggests that cumulative exposure of the general population to EDCs leads to an uptick of energy metabolism and that this continuous process leads to or is part of an epigenetic alteration of the onset of puberty [94–97]. Clearly, much remains to be done. The prevention of long-term health consequences of early puberty, with or without associated early obesity, needs to be addressed. We believe that pervasive exposure to endocrine-disrupting chemicals is likely to play a role in this epidemic.

References

1. Stubbs ML. Cultural perceptions and practices around menarche and adolescent menstruation in the United States. Ann N Y Acad Sci. 2008;1135:58–66.
2. Weisfield GE. Adolescence. In: Ember CR, Ember M, editors. The encyclopedia of sex and gender: men and women in the World's cultures, vol. 1. New York: Springer; 2003. p. 42–53.
3. Marshall WA, Tanner JM. Variations in the pattern of pubertal changes in boys. Arch Dis Child. 1970;45:13–23.
4. Marshall WA, Tanner JM. Variations in pattern of pubertal changes in girls. Arch Dis Child. 1969;44:291–303.
5. DiVall SA, Radovick S. Pubertal development and menarche. Ann N Y Acad Sci. 2008; 1135:19–28.
6. Rogol AD. Sex steroids, growth hormone, leptin and the pubertal growth spurt. Endocr Dev. 2010;17:77–85.
7. Roa J, García-Galiano D, Castellano JM, Gaytan F, Pinilla L, Tena-Sempere M. Metabolic control of puberty onset: new players, new mechanisms. Mol Cell Endocrinol. 2010;324:87–94.
8. Brougham MF, Kelnar CJ, Wallace WH. The late endocrine effects of childhood cancer treatment. Pediatr Rehabil. 2002;5:191–201.
9. Moffitt TE, Caspi A, Belsky J, Silva PA. Childhood experience and the onset of menarche: a test of a sociobiological theory. Child Dev. 1992;63:47–58.
10. Short MB, Rosenthal SL. Psychosocial development and puberty. Ann N Y Acad Sci. 2008;1135:36–42.
11. Walvoord EC. The timing of puberty: is it changing? Does it matter? J Adolesc Health. 2010; 47:433–9.

15 Secular Trends in Pubertal Timing... 369

12. Oldehinkel AJ, Verhulst FC, Ormel J. Mental health problems during puberty: tanner stage-related differences in specific symptoms. The TRAILS study. J Adolesc. 2011;34(1):73–85.
13. McCarthy MM. How it's made: organisational effects of hormones on the developing brain. J Neuroendocrinol. 2010;22:736–42.
14. Biegon A et al. Unique distribution of aromatase in the human brain: in vivo studies with PET and [N-methyl-11 C]vorozole. Synapse. 2010;64:801–7.
15. Del Giudice M. Sex, attachment, and the development of reproductive strategies. Behav Brain Sci. 2009;32:1–67.
16. Karapanou O, Papadimitriou A. Determinants of menarche. Reprod Biol Endocrinol. 2010; 8:115–21.
17. Parent A-S, Teilmann G, Juul A, Skakkeb NE, Toppari J, Bourguignon J-P. The timing of normal puberty and the age limits of sexual precocity: variations around the world, secular trends, and changes after migration. Endocr Rev. 2003;24:668–93.
18. Khanna G, Kapoor S. Secular trend in stature and age at menarche among Punjabi Aroras residing in New Delhi, India. Coll Antropol. 2004;28:571–5.
19. Graham MJ, Larsen U, Xu X. Secular trend in age at menarche in China: a case study of two rural counties in Anhui Province. J Biosoc Sci. 1999;31:257–67.
20. Huen KF, Leung SS, Lau JT, Cheung AY, Leung NK, Chiu MC. Secular trend in the sexual maturation of southern Chinese girls. Acta Paediatr. 1997;86:1121–4.
21. Mahachoklertwattana P et al. Earlier onset of pubertal maturation in Thai girls. J Med Assoc Thai. 2002;85:1127–34.
22. Hwang JY, Shin C, Frongillo EA, Shin KR, Jo I. "Secular trend in age at menarche for South Korean women born between, and 1986: the Ansan Study. Ann Hum Biol. 1920;30(2003): 434–42.
23. Cho GJ et al. Age at menarche in a Korean population: secular trends and influencing factors. Eur J Pediatr. 2010;169:89–94.
24. Tsuzaki S, Matsuo N, Ogata T, Osano M. Lack of linkage between height and weight and age at menarche during the secular shift in growth of Japanese children. Ann Hum Biol. 1989;16: 429–36.
25. Prentice S, Fulford AJ, Jarjou LM, Goldberg GR, Prentice A. Evidence for a downward secular trend in age of menarche in a rural Gambian population. Ann Hum Biol. 2010;37:717–21.
26. Jones LL, Griffiths PL, Norris SA, Pettifor JM, Cameron N. Age at menarche and the evidence for a positive secular trend in urban South Africa. Am J Hum Biol. 2009;21:130–2.
27. Hosny LA et al. Assessment of pubertal development in Egyptian girls. J Pediatr Endocrinol Metab. 2005;18:577–84.
28. Pasquet P, Biyong AM, Rikong-Adie H, Befidi-Mengue R, Garba MT, Froment A. Age at menarche and urbanization in Cameroon: current status and secular trends. Ann Hum Biol. 1999;26:89–97.
29. Odujinrin OM, Ekunwe EO. Epidemiologic survey of menstrual patterns amongst adolescents in Nigeria. West Afr J Med. 1991;10:244–9.
30. Anderson SE, Must A. Interpreting the continued decline in the average age at menarche: results from two nationally representative surveys of U.S. girls studied 10 years apart. J Pediatr. 2005;147:753–60.
31. Kaplowitz PB, Slora EJ, Wasserman RC, Pedlow SE, Herman-Giddens ME. Earlier onset of puberty in girls: relation to increased body mass index and race. Pediatrics. 2001;108:347–53.
32. Euling SY et al. Examination of US puberty-timing data from 1940 to 1994 for secular trends: panel findings. Pediatrics. 2008;121 Suppl 3:172–91.
33. Sørensen K, Aksglaede L, Petersen JH, Juul A. Recent changes in pubertal timing in healthy Danish boys: associations with body mass index. J Clin Endocrinol Metab. 2010;95:263–70.
34. Martos-Moreno GA, Chowen JA, Argente J. Metabolic signals in human puberty: effects of over and undernutrition. Mol Cell Endocrinol. 2010;324:70–81.
35. Ojeda SR et al. Gene networks and the neuroendocrine regulation of puberty. Mol Cell Endocrinol. 2010;324:3–11.

36. Gajdos ZK, Henderson KD, Hirschhorn JN, Palmert MR. Genetic determinants of pubertal timing in the general population. Mol Cell Endocrinol. 2010;324:21–9.
37. Parent AS et al. Early onset of puberty: tracking genetic and environmental factors. Horm Res. 2005;64:41–7.
38. Freedman DS, Khan LK, Serdula MK, Dietz WH, Srinivasan SR, Berenson GS. Relation of age at menarche to race, time period, and anthropometric dimensions: the Bogalusa Heart Study. Pediatrics. 2002;110:e43.
39. Chumlea WC et al. Age at menarche and racial comparisons in US girls. Pediatrics. 2003; 111:110–3.
40. Juul A et al. Pubertal development in Danish children: comparison of recent European and US data. Int J Androl. 2006;29:247–55.
41. Centers for Disease Control and Prevention. In: 4th national report of human exposure to environmental chemicals. 2009. Atlanta, Georgia,USA.
42. Mouritsen A et al. Hypothesis: exposure to endocrine-disrupting chemicals may interfere with timing of puberty. Int J Androl. 2010;33:346–59.
43. Casals-Casas C, Desvergne B. Endocrine disruptors: from endocrine to metabolic disruption. Annu Rev Physiol. 2011;73:135–62.
44. Patisaul HB, Adewale HB. Long-term effects of environmental endocrine disruptors on reproductive physiology and behavior. Front Behav Neurosci. 2009;3:10.
45. Bourguignon JP, Rasier G, Lebrethon MC, Gérard A, Naveau E, Parent AS. Neuroendocrine disruption of pubertal timing and interactions between homeostasis of reproduction and energy balance. Mol Cell Endocrinol. 2010;324:110–20.
46. Milton AH et al. Chronic arsenic exposure and adverse pregnancy outcomes in Bangladesh. Epidemiology. 2005;16:82–6.
47. Wirth JJ, Mijal RS. Adverse effects of low level heavy metal exposure on male reproductive function. Syst Biol Reprod Med. 2010;56:147–67.
48. US Environmental Protection Agency. National primary drinking water regulations; arsenic and clarifications to compliance and new source contaminants monitoring. Fed Regist. 2001;66(14):6976–7066.
49. Chakraborti D, Mukherjee SC, Pati S, Sengupta MK, Rahman MM, Chowdhury UK, et al. Arsenic groundwater contamination in Middle Ganga Plain, Bihar, India: a future danger? Environ Health Perspect. 2003;1111:1194–2201.
50. Dávila-Esqueda ME, et al. Effects of arsenic exposure during the pre- and postnatal development on the puberty of female offspring. Exp Toxicol Pathol. 2010; E-pub ahead of print.
51. Davey JC, Nomikos AP, Wungjiranirun M, Sherman JR, Ingram L, Batki C, et al. Arsenic as an endocrine disruptor: arsenic disrupts retinoic acid receptor– and thyroid hormone receptor–mediated gene regulation and thyroid hormone-mediated amphibian tail metamorphosis. Environ Health Perspect. 2008;116:165–72.
52. Selevan SG, Rice DC, Hogan KA, Euling SY, Pfahles-Hutchens A, Bethel J. Blood lead concentration and delayed puberty in girls. N Engl J Med. 2003;348:1527–36.
53. Naicker N, Norris SA, Mathee A, Becker P, Richter L. Lead exposure is associated with a delay in the onset of puberty in South African adolescent females: findings from the Birth to Twenty cohort. Sci Total Environ. 2010;408:4949–54.
54. Tomoum HY, Mostafa GA, Ismail NA, Ahmed SM. Lead exposure and its association with pubertal development in school-age Egyptian children: pilot study. Pediatr Int. 2010;52:89–93.
55. Williams PL et al. Blood lead levels and delayed onset of puberty in a longitudinal study of Russian boys. Pediatrics. 2010;125:e1088–96.
56. Dearth RK, Hiney JK, Srivastava V, Burdick SB, Bratton GR, Dees WL. Effects of lead (Pb) exposure during gestation and lactation on female pubertal development in the rat. Reprod Toxicol. 2002;16:343–52.
57. Ronis MJ et al. Endocrine mechanisms underlying the growth effects of developmental lead exposure in the rat. J Toxicol Environ Health. 1998;54:101–20.
58. Vandenberg LN, Chahoud I, Heindel JJ, Padmanabhan V, Paumgartten FJ, Schoenfelder G. Urinary, circulating, and tissue biomonitoring studies indicate widespread exposure to Bisphenol A. Environ Health Perspect. 2010;118:1055–70.

15 Secular Trends in Pubertal Timing...

59. Bondesson M et al. A CASCADE of effects of Bisphenol A. Reprod Toxicol. 2009;28:563–7.
60. Braun JM, Hauser R. Bisphenol A and children's health. Curr Opin Pediatr. 2011;23:233–9.
61. Beronius A, Rudén C, Håkansson H, Hanberg A. Risk to all or none? A comparative analysis of controversies in the health risk assessment of Bisphenol A. Reprod Toxicol. 2010;29:132–46.
62. Willoughby KN, Sarkar AJ, Boyadjieva NI, Sarkar DK. Neonatally administered tert-octylphenol affects onset of puberty and reproductive development in female rats. Endocrine. 2005; 26:161–8.
63. Kim HS et al. Comparative estrogenic effects of p-nonylphenol by 3-day uterotrophic assay and female pubertal onset assay. Reprod Toxicol. 2002;16:259–68.
64. Cheng G, Remer T, Prinz-Langenohl R, Blaszkewicz M, Degen GH, Buyken AE. Relation of isoflavones and fiber intake in childhood to the timing of puberty. Am J Clin Nutr. 2010;92: 556–64.
65. Wolff MS et al. Investigation of relationships between urinary biomarkers of phytoestrogens, phthalates, and phenols and pubertal stages in girls. Environ Health Perspect. 2010;118: 1039–46.
66. Bateman HL, Patisaul HB. Disrupted female reproductive physiology following neonatal exposure to phytoestrogens or estrogen specific ligands is associated with decreased GnRH activation and kisspeptin fiber density in the hypothalamus. Neurotoxicology. 2008;29:988–97.
67. Colón I, Caro D, Bourdony CJ, Rosario O. Identification of phthalate esters in the serum of young Puerto Rican girls with premature breast development. Environ Health Perspect. 2000;108:895–900.
68. Durmaz E et al. Plasma phthalate levels in pubertal gynecomastia. Pediatrics. 2010;125:e122–9.
69. Chou YY, Huang PC, Lee CC, Wu MH, Lin SJ. Phthalate exposure in girls during early puberty. J Pediatr Endocrinol Metab. 2009;22:69–77.
70. Lomenick JP et al. Phthalate exposure and precocious puberty in females. J Pediatr. 2010; 156:221–5.
71. Houde M, Martin JW, Letcher RJ, Solomon KR, Muir DC. Biological monitoring of polyfluoroalkyl substances: a review. Environ Sci Technol. 2006;40:3463–73.
72. Liu Y, McDermott S, Lawson A, Aelion CM. The relationship between mental retardation and developmental delays in children and the levels of arsenic, mercury and lead in soil samples taken near their mother's residence during pregnancy. Int J Hyg Environ Health. 2010;213:116–23.
73. Shoeib M, Harner T, Webster GM, Lee SC Indoor sources of poly- and perfluorinated compounds (PFCS) in Vancouver, Canada: implications for human exposure. Environ Sci Technol. 2011; Epub ahead of print.
74. Olsen GW et al. Decline in perfluorooctanesulfonate and other polyfluoroalkyl chemicals in American Red Cross adult blood donors, 2000–2006. Environ Sci Technol. 2008;42:4989–95.
75. Olsen GW et al. Half-life of serum elimination of perfluorooctanesulfonate, perfluorohexanesulfonate, and perfluorooctanoate in retired fluorochemical production workers. Environ Health Perspect. 2007;115:1298–305.
76. Melzer D, Rice N, Depledge MH, Henley WE, Galloway TS. Association between serum perfluorooctanoic acid (PFOA) and thyroid disease in the U.S. National Health and Nutrition Examination Survey. Environ Health Perspect. 2010;118:686–92.
77. Pinney SM et al. Perfluorooctanoic acid (PFOA) and pubertal maturation in young girls. Epidemiology. 2009;20(6):S80.
78. Christensen KY et al. Exposure to polyfluoroalkyl chemicals during pregnancy is not associated with offspring age at menarche in a contemporary British cohort. Environ Int. 2011; 37(1):129–35.
79. Lopez-Espinosa MJ, et al. Association of perfluorooctanoic acid (PFOA) and perfluorooctane sulfonate (PFOS) with age of puberty among children living near a chemical plant. Environ Sci Technol. 2011; Epub ahead of print.
80. Biegel LB, Liu RC, Hurtt ME, Cook JC. Effects of ammonium perfluorooctanoate on Leydig cell function: in vitro, in vivo, and ex vivo studies. Toxicol Appl Pharmacol. 1995;134:18–25.
81. Zhao Y, Tan YS, Haslam SZ, Yang C. Perfluorooctanoic acid effects on steroid hormone and growth factor levels mediate stimulation of peripubertal mammary gland development in C57BL/6 mice. Toxicol Sci. 2010;115:214–24.

82. Burns JS et al. Predictors of serum dioxins and PCBs among peripubertal Russian boys. Environ Health Perspect. 2009;117:1593–9.
83. Korrick SA, et al. Dioxin exposure and age of pubertal onset among Russian Boys. Environ Health Perspect. 2011; Epub ahead of print.
84. Krstevska-Konstantinova M et al. Sexual precocity after immigration from developing countries to Belgium: evidence of previous exposure to organochlorine pesticides. Hum Reprod. 2001;16:1020–6.
85. Vasiliu O, Muttineni J, Karmaus W. In utero exposure to organochlorines and age at menarche. Hum Reprod. 2004;19:1506–12.
86. Denham M et al. Relationship of lead, mercury, mirex, dichlorodiphenyldichloroethylene, hexachlorobenzene, and polychlorinated biphenyls to timing of menarche among Akwesasne Mohawk girls. Pediatrics. 2005;115:e127–34.
87. Wolff MS et al. Environmental exposures and puberty in inner-city girls. Environ Res. 2008; 107:393–400.
88. Jasik CB, Lustig RH. Adolescent obesity and puberty: the "perfect storm". Ann N Y Acad Sci. 2008;1135:265–79.
89. Newbold RR, Padilla-Banks E, Snyder RJ, Phillips TM, Jefferson WN. Developmental exposure to endocrine disruptors and the obesity epidemic. Reprod Toxicol. 2007;23:290–6.
90. Hatch EE, Nelson JW, Stahlhut RW, Webster TF. Association of endocrine disruptors and obesity: perspectives from epidemiological studies. Int J Androl. 2010;33:324–32.
91. Casals-Casas C, Feige JN, Desvergne B. Interference of pollutants with PPARs: endocrine disruption meets metabolism. Int J Obes (Lond). 2008;32 Suppl 6:S53–61.
92. Howd RA. Considering changes in exposure and sensitivity in an early life cumulative risk assessment. Int J Toxicol. 2010;29:71–7.
93. Rogers IS, Northstone K, Dunger DB, Cooper AR, Ness AR, Emmett PM. Diet throughout childhood and age at menarche in a contemporary cohort of British girls. Public Health Nutr. 2010;13:2052–63.
94. Schoeters G, Den Hond E, Dhooge W, van Larebeke N, Leijs M. Endocrine disruptors and abnormalities of pubertal development. Basic Clin Pharmacol Toxicol. 2008;102:168–75.
95. Bernal AJ, Jirtle RL. Epigenomic disruption: the effects of early developmental exposures. Birth Defects Res A Clin Mol Teratol. 2010;88:938–9.
96. Tabb MM, Blumberg B. New modes of action for endocrine-disrupting chemicals. Mol Endocrinol. 2006;20:475–82.
97. Crews D, McLachlan JA. Epigenetics, evolution, endocrine disruption, health, and disease. Endocrinology. 2006;147(6 Suppl):S4–10.

Index

A

Adipocytes
  commitment phase, 262–263
  differentiation phase, 259–262
  early-onset obesity, 257–259
  epigenetic modifications, puberty, 264–265
  glucose homeostasis, 277–278
Adiponectin, 332
Alfalfa, 58
Alpha-fetoprotein, 63, 79, 83
5-Alpha reductase, 78
Androgen receptor (AR), 78
Anogenital distance (AGD)
  bisphenol A (BPA), 84
  organochlorine pesticides, 82
  organohalogens, 79–80
  phytoestrogens, 76–77
Anteroventral periventricular nucleus
    (AVPV), 51, 62
Antiandrogens, 346
Apoptosis, 62–63
Arcuate nucleus (ARC), 83
Aroclor, 53
Aromatase, 62
Aryl hydrocarbon receptor (AhR),
    129–130

B

Birth and childhood weights
  EDC, in environment
    organochlorines, 294–305
    phthalates, 305
    polyfluoroalkyl chemicals (PFCs),
      306–307
  fetal growth disruptors

diethylstilbestrol (DES), 291–293
maternal smoking, 289–291
gestational diabetes, 287–289
methodologic challenges, 307–309
and subsequent adiposity, 285–287
Bisphenol A (BPA)
  EDCs, properties, 60–61
  male pubertal timing, 344
  mammary gland development,
    208–209
  ovarian function, 185–186
  puberty, 363–364
  thyroid hormone regulation, 157–158
Body mass index (BMI), 120, 368
Breast. *See* Mammary gland development

C

Calbindin, 74
Central nervous system (CNS)
  EDs, perinatal development, 61
  hypothalamic control, reproduction, 50
Childhood obesity. *See* Birth and childhood
    weights
Chlorpyrifos, 56
Congenital malformations. *See* Male
    urogenital tract malformations
Coumestrol, 56
Critical window of susceptibility, 4
Cryptorchidism, 226–227

D

Daidzein, 75–76
DEHP exposure
  male pubertal timing, 345

374 Index

1,1-Dichloro-2,2-bis(p-chlorophenyl)ethylene (DDE)
  female pubertal timing, EDC impact, 326–327
  male pubertal timing, EDC influence, 344
  ovarian function, 189–190
Dichlorodiphenyltrichloroethane (DDT)
  female pubertal timing, EDC impact, 326–327
  puberty, 367
  reproductive neuroendocrine targets, 60
Diethylstilbestrol (DES)
  birth and childhood weights, 291–293
  male pubertal timing, 342
  mammary gland development, 212–213
  ovarian function, 188–189
Dioxins
  GnRH cell lines, 53
  male pubertal timing, 347
  puberty, 367

E

Endocrine-disrupting chemicals (EDCs). *See* Adipocytes; Reproductive neuroendocrine targets; Thyroid hormone regulation
  GnRH cell lines, 53–56
  prenatal EDC effect, puberty
    adipogenesis, 26
    hypothalamic-pituitary-thyroid (HPT) axes, 23–24
    peripheral control, 25
    phyto-oestrogen genistein, 25
    pubic hair, 23
    scrotum enlargement, 23
  properties, 56–61
  puberty
    arsenic, 362
    lead, 363
  reproductive development
    chromophils-producing FSH and LH, 21
    DNA methylation alteration, 22–23
    foetal EDC exposure, 22
    methoxychlor, 20
    xeno-oestrogens effect, 21
  reproductive tract, development
    female, 18–20
    male, 14–18
Endocrine disruptors and puberty disorders
  experimental toxicology studies
    chlorotriazine herbicides, 126
    dicarboximide fungicides, 128

    diethylstilbestrol (DES), 128
    female puberty onset, 127
    hexavalent chromium (Cr-VI), 131
    KiSS-1 system, 132
    male puberty, 128
    PND exposure, 129
    semicarbazide (SEM), 131
    TCDD levels, 130
    tributyltin (TBT), 130
  on human studies
    dichlorodiphenyltrichloroethane, 123
    estrogen agonist/antagonist activity, 122
    high urinary enterolactone, 123
    mercury (Hg), 124–125
    phthalates metabolites, 124
    sexual maturation, 124
    soy-based milk formulas, 125
    in utero effect, 123
  regulation
    body mass index (BMI), 120
    genetic predisposition, 121
    intrauterine growth retardation, 121
    mutations and polymorphisms, 122
    precocious puberty, 120
  risk assessment, 132–134
Endosulfan effect, 350–351
Energy balance, 332–333
Environmental chemicals exposure, in utero. *See* Endocrine-disrupting chemicals (EDCs)
Environmental phenols. *See* Phenols
Epigenetic reprogramming, 264–265
Estrogen/androgen signaling pathways, 342
Estrogen receptors (ER), 53

F

Female pubertal timing, EDC impact
  EDC exposure *vs.* human female pubertal timing variation, 327–330
  as endpoint, 325–327
  female sexual maturation, 330–332
  female sexual precocity, 332–333
Fenitrothion, 82
Flutamide
  male pubertal timing, 347
Follicle-stimulating hormone (FSH), 121

G

Genistein
  mammary gland development, 207–208
  ovarian function, 186–188

Index

375

reproductive neuroendocrine targets, 76
Genital staging, 352
Glucocorticoid, 52, 259
Glucose homeostasis, ED exposure
    adipocyte number and function, 277–278
    Islet function, 274–275
    and weight gain, 275–277
Gonadotropin-releasing hormone (GnRH)
    cell lines, 53–56
    female pubertal timing, EDC impact, 326
    function, 359
    gene expression and activation, 75
    hypothalamic control, reproduction, 50
    peptide concentrations, 56
    pubertal transition, 86–87
    pulsatile secretion, 360
    sex differences, HPG Function, 51
GT1-7 cells, 53

H

Homeostasis, 332–333
Hypospadias, 17, 226–227
Hypothalamic-pituitary-gonadal (HPG)
    EDC effects, reproductive development, 21
    hypothalamic control, reproduction, 50
    neuroendocrine axes, 31
    physiology and endocrinology, 359
Hypothalamus
    control, reproduction, 50–51
    GnRH neurons, 53
    hormones and apoptosis, 62
    leptin, 215
    pesticides exposure, 100

I

Insulin, 274–275
Intrauterine growth retardation (IUGR), 333
In utero exposures, 310
Isoflavone, 58, 74, 87

K

Kisspeptin, 63, 88, 102–103, 332

L

Leptin, 326, 332
Leydig cell, 152, 155–156
Lipids and adipose-derived hormones
    leptin, 215
    n-3 polyunsaturated fatty acid, 214–215
    n-6 polyunsaturated fatty acid, 214

Luteinizing hormone (LH), 19, 21, 126, 190

M

Male pubertal timing, EDC influence
    in human
        biomarkers, 352
        chemical contaminants, 348
        endosulfan effect, 350–351
        epidemiologic studies, 349
        organochlorine exposures, 350
        PCBs impact, 351
        pubertal-timing end points, 349
    in rats and mice, 341–342
    in rodents
        antiandrogens, 346
        bisphenol A (BPA), 344
        DDE, 344
        DEHP exposure, 345
        diethylstilbestrol (DES), 342
        dioxin, 347
        estrogen/androgen signaling pathways,
            342
        flutamide, 347
        methoxychlor, 345
        4-methylbenzylidene camphor
            (4-MBC), 346
        octylphenol, 346
        prochloraz (PCZ), 347
        TCDD, 348
        vaginal opening (VO) delay, 342
Male urogenital tract malformations
    cryptorchidism, 226–227
    EDC hypothesis, 227–228
    epidemiological studies
        actual exposure data, 232–235
        biological plausibility, 235
        occupational exposure, 228–232
    hypospadias, 226–227
Mammary gland development
    alteration of, 204–207
    bisphenol A, 208–209
    brominated compounds, 218
    description, 202–204
    diethylstilbestrol (DES), 212–213
    genistein, 207–208
    lipids and adipose-derived hormones
        leptin, 215
        n-3 polyunsaturated fatty acid, 214–215
        n-6 polyunsaturated fatty acid, 214
    metals, 213–214
    organochlorines
        atrazine, 216–217
        dioxins, 217–218

Mammary gland development (*cont.*)
  perfluorinated compounds, 218–219
  sex hormones
    estradiol, 211–212
    testosterone, 211
  xenobiotics
    nonylphenol, 209–210
    phthalates, 210–211
  xenoestrogens, 207
Maternal smoking and reproductive
    development, 33–35
Menarche. *See* Female pubertal timing, EDC
    impact; Pubertal timing
Methoxychlor (MXC)
  consummatory behaviors, EDC, 244
  effects, 345
  GT1-7 cells, 53
  male pubertal timing, EDC influence, 345
  organochlorine insecticide, 20, 60, 81
  ovary, EDs, 191
  pubertal onset, 129
4-Methylbenzylidene camphor (4-MBC)
  male pubertal timing, EDC influence, 346
Multipotent stromal cells (MSCs), 256

N
Nonylphenol
  puberty, 363–364
Nuclear receptor, 261

O
Obesity. *See* Birth and childhood weights;
    Glucose homeostasis, ED exposure
Obesogen, 263
Octylphenol
  male pubertal timing, 346
  puberty, 363–364
Organochlorine pesticides
  male pubertal timing, 350
  perinatal development, 81–82
Organochlorines
  birth and childhood weights
    DDT and HCB, 303–305
    PCB, 294–303
  mammary gland development, EDC
    atrazine, 216–217
    dioxins, 217–218
  puberty, 367
Organohalogen, 58, 77–81, 88–100
Ovarian function
  antiandrogenic endocrine disruptors
    1,1-dichloro-2,2-bis(p-chlorophenyl)
    ethylene (DDE), 189–190

vinclozolin, 190
  combined endocrine disruptors effects,
    191–192
  dysregulation, 192–194
  estrogenic endocrine disruptors
    bisphenol A (BPA), 185–186
    diethylstilbestrol (DES), 188–189
    genistein, 186–188
  physiology, 179–183
  prepubertal–pubertal ovary, 183–184

P
Perfluorooctane sulfonate (PFOS), 365–367
Perfluorooctanoic acid (PFOA)
  mammary gland development, EDC,
    218–219
  puberty, 365–367
Perinatal development, EDC
  bisphenol A, 82–85
  brain sexual differentiation, 63, 74–77
  hypothalamus, sexual differentiation,
    62–63
  organochlorine pesticides, 81–82
  organohalogens, 77–81
  phthalates, 85–86
Peroxisome proliferator–activated receptor
    gamma (PPAR g), 259
Pesticide, 59–60 put some cross reference
Phenols, 363–364
Phthalates
  birth and childhood weights, 305
  EDCs, properties, 61
  puberty, 365
  thyroid hormone regulation, 158–159
Phytoestrogens
  brain sexual differentiation, 63, 73–76
  EDCs, 57–58
  endocrine disruptors, 3
  GnRH neurons, 53
  puberty, 364
  reproductive neuroendocrine targets, 87–88
Polybrominated biphenyl (PBB), 58, 123, 327
Polybrominated diphenyl ethers (PBDEs)
  brominated compounds, 218
  endocrine disruption, differentiation phase,
    262
  industrial organohalogens, 58
  mammary gland development, EDC, 218
  pubertal development, rodents, 347
  thyroid axis disruption, 147
  thyroid hormone regulation, 160
Polychlorinated biphenyls (PCBs)
  puberty, 367
  thyroid hormone regulation, 159–160

Index 377

Polyfluoroalkyl chemicals (PFCs)
  birth and childhood weights, 306–307
Precocious puberty. *See* Endocrine disruptors
    and puberty disorders
Preputial separation, 341, 345
Prochloraz (PCZ), 347
Pseudohermaphroditism, 17–18
Pubertal development, 4–5
Pubertal timing
  BMI, 360, 368
  description, 357–359
  development, human males and females,
    359
  and endocrine disruption
    arsenic, 362
    lead, 363
  environmental phenols, 363–364
  genetics and ethnicity, 361
  nutrition and age, 360–361
  obesity, 360, 368
  organochlorines, 367
  PFOA and PFOS, 365–367
  phthalates, 365
  physiology and endocrinology, 359–360
  phytoestrogens, 364
  psychosocial and neurobehavioral aspects,
    360
Pubertal transition, EDC
  bisphenol A, 101–103
  organohalogen, 88–100
  pesticides, 100–101
  phthalates, 103
  phytoestrogens, 87–88
Pyrethroid, 53, 56

R
Reproductive and nonreproductive behaviors
  appetitive behaviors, 243–244
  consummatory behaviors, 244
  organizing and activating behaviors, 242
  phytoestrogens
    and aggression, 248–249
    and anxiety-like behavior, 246–248
  sexually dimorphic behaviors, 242
  spatial learning and memory, 244–246
Reproductive neuroendocrine targets
  EDC properties, 56–61
  endocrine disruption, neuroendocrine
    systems, 52
  GnRH cell lines, 53–56
  hypothalamic control, 50–51
  perinatal development
    bisphenol A, 82–85
    brain sexual differentiation, 63, 74–77

    hypothalamus, sexual differentiation,
      62–63
    organochlorine pesticides, 81–82
    organohalogens, 77–81
    phthalates, 85–86
  pubertal transition
    bisphenol A, 101–103
    organohalogen, 88–100
    pesticides, 100–101
    phthalates, 103
    phytoestrogens, 87–88
  sex differences, HPG function, 51–52
Resveratrol, 58, 75

S
Sensitive window of susceptibility, 4
Sertoli cell, 154, 155
Sex hormones
  estradiol, 211–212
  testosterone, 211
Sexual behavior
  HPG function, 51
  organohalogens, 78
Sexually dimorphic nucleus of the preoptic
    area (SDN-POA)
  reproductive development, EDC
    effects, 21
  sex differences, HPG function, 51
Smoking. *See* Maternal smoking and
    reproductive development
Soy, 58
SRY gene, 14
Stem cells, 262

T
Tanner stages, puberty, 358
Tefluthrin, 53, 56
Testes, 153, 154
Testicular dysgenesis syndrome (TDS), 30
Testicular volume, 350–352
Testosterone, 341, 345
Thyroid hormone regulation
  bisphenol A, 157–158
  dietary effects, 148–151
  hypo-and hyperthyroidism
    females, 152–153
    males, 151–152
  male sexual development, 153–156
  and mammalian sexual maturation, 151
  metabolism, and action
    environmental disruption, 141–142
    hypothalamic–pituitary–thyroid, 140
    intracellular T3 levels, 144

378

Thyroid hormone regulation (*cont.*)
    monocarboxylic acid transporters (MCT), 143
    phosphoMAPK (pMAPK), 145
    T4 and T3, 143
    thyroid peroxidase (TPO), 140
    thyrotropin-releasing hormone, 140–141
  phthalates, 158–159
  polybrominated diphenyl ethers, 160
  polychlorinated biphenyls, 159–160
  thyroid axis and sexual development, 156–157
  thyroid axis disruption, 145–148
  triclosan, 157
Tobacco and prenatal development
  effects, 26–27
  neuroendocrine axes, 30–32
  puberty, 32–33
  reproductive tract, 27–30
Transgenerational effects, 264
Tyrosine hydroxylase, 63

V
Vaginal opening (VO) delay, 342
Vinclozolin, 190

W
Weight gain, 275–277

X
Xenobiotics
  nonylphenol, 209–210
  phthalates, 210–211
Xenoestrogens, 207